Springer Geology

For further volumes:
http://www.springer.com/series/10172

Da Wang et al.
Translated by Junfeng Geng et al.

The China Continental Scientific Drilling Project

CCSD-1 Well Drilling Engineering and Construction

Responsible editors: Peng Han, Jingfei Zhang

Da Wang
Engineering Center of Chinese Continental
 Scientific Drilling Project
Beijing
China

Wei Zhang
China Geological Survey
Beijing
China

Xiaoxi Zhang
China University of Geosciences (Wuhan)
Wu Han
China

Guolong Zhao
Consulting and Research Center
Ministry of Land and Resources
Beijing
China

Ruqiang Zuo
Consulting and Research Center
Ministry of Land and Resources
Beijing
China

Jialu Ni
Institute of Exploration Techniques
Langfang
China

Gansheng Yang
Institute of Exploration Techniques
Langfang
China

Jun Jia
Beijing Institute of Exploration Engineering
Beijing
China

Kaihua Yang
China University of Geosciences (Wuhan)
Wu Han
China

Yongyi Zhu
Institute of Exploration Techniques
Langfang
China

Wenwei Xie
Institute of Exploration Techniques
Langfang
China

Wenjian Zhu
Beijing Institute of Exploration Engineering
Beijing
China

Peifeng Zhang
Beijing Institute of Exploration Engineering
Beijing
China

Lasheng Fan
Institute of Exploration Technology
Langfang
China

Jianliang Ye
China Geological Survey
Beijing
China

Yongping Wang
Research and Design Academy of Metallurgical
 Prospecting
Beijing
China

Translated by Junfeng Geng

ISSN 2197-9545   ISSN 2197-9553 (electronic)
Springer Geology
ISBN 978-3-662-46556-1   ISBN 978-3-662-46557-8 (eBook)
DOI 10.1007/978-3-662-46557-8

Jointly published with Science Press, Beijing
ISBN: 978-7-03-043592-7 Science Press, Beijing

Library of Congress Control Number: 2015932970

Springer Heidelberg New York Dordrecht London
© Science Press, Beijing and Springer-Verlag Berlin Heidelberg 2015
This work is subject to copyright. All rights are reserved by the Publishers, whether the whole or part of the material is concerned, specifically the rights of translation, reprinting, reuse of illustrations, recitation, broadcasting, reproduction on microfilms or in any other physical way, and transmission or information storage and retrieval, electronic adaptation, computer software, or by similar or dissimilar methodology now known or hereafter developed.
The use of general descriptive names, registered names, trademarks, service marks, etc. in this publication does not imply, even in the absence of a specific statement, that such names are exempt from the relevant protective laws and regulations and therefore free for general use.
The publishers, the authors and the editors are safe to assume that the advice and information in this book are believed to be true and accurate at the date of publication. Neither the publishers nor the authors or the editors give a warranty, express or implied, with respect to the material contained herein or for any errors or omissions that may have been made.

Printed on acid-free paper

Springer-Verlag GmbH Berlin Heidelberg is part of Springer Science+Business Media (www.springer.com)

Not for sale outside the Mainland of China (*Not for sale in Hong Kong SAR, Macau SAR, and Taiwan, and all countries except the Mainland of China*).

Foreword I

Jules Verne, the originator of science fiction in the world, wrote a science fiction, Journey to the Center of the Earth, in which he tried all his imaginations to describe various scenes of the Earth's core. The Earth carries billions of human beings and life, and what does her core really look like? Freezing ice storage, or a hot lava chamber? So far, a great variety of speculations about the Earth have been bringing to scientists all kinds of interesting mysteries for a long time.

After World War II, the world experienced a period of relative stability, and geologists got opportunities to make long-term continuous research on the deep Earth. They conceived to use modern ultra-deep drilling techniques, to drill into the deep Earth, obtain the cores, and analyze the various characteristics of rocks; to place the geophysical and geochemical and many other modern scientific instruments into borehole to get the data of borehole profile; to study the evolution of the Earth's crust by using the cores and surveyed data; and to convert geology from relying mainly on inference to relying mainly on verification through survey, which could be called "the digital Earth"; and the traditional geology could be converted into "the Earth Science" by the increased understanding of the deep Earth, which would be a great leap forward.

The rise of scientific drilling should be in the middle of the twentieth century. NEDRA, an institution of ultra-deep drilling consortium, led by the minister in the Ministry of Geology in the former Soviet Union, successively implemented many scientific deep boreholes. B.N. Khakhaev, the General Manager, specifically presided at the engineering design and construction works. During the construction process, they carried forward the great Soviet tradition of "no saying, just doing" and eventually announced the success of the world's first ultra-deep drilling when drilled to the depth of 12,262 m in SG-3 Well in Kola Peninsula. The news immediately spread around the world, startled the world drilling circles, and dedicated a great amount of the latest Earth information for the Earth Science research as well, for example: 1. The upper and lower boundaries of the local upper mantle and lithosphere, and their respective thicknesses were preliminarily clarified; 2. the fluid which "shouldn't appear" (from E.A. Kazlovski, the former minister in the Ministry of Geology in the former Soviet Union) and the abiogenetic oil and gas were astonishingly found, by which the inference of a young man from the United States in the 1940s was confirmed; and 3. lots of the latest and the most authoritative geological information data were obtained.

Thus in 1988, in the former Soviet Union was held an international scientific drilling conference, at which most of the non-confidential information and achievements were released. The finding of abiogenetic oil and gas immediately attracted a delegation from Sweden, a country lacking in oil and gas. The delegation returned even before the meeting was closed, and designed and constructed two deep boreholes—Gravberg 1 and Gravberg 2. The achievements of the scientific drilling in the former Soviet Union were huge, greatly shocked the geology circle in the world, and promoted a continuous flourish and climax of scientific drilling.

A part of scientific drilling was started from the ocean, because a great quantity of solid strata have to be drilled at the beginning of land drilling, for instance: 1. In 1968, Deep Sea Drilling Project (DSDP) was executed by Glomar Challenger drilling ship; 2. in 1985, the Ocean Drilling Program (ODP) was started, with JOIDES Resolution as the academic leadership. At the 184th voyage, a borehole was drilled in the South China Sea. The program has made a series of great scientific achievements; 3. in 2003, Japan Marine Science and Technology Center (JAMSTEC) and the existing ODP members combined into the new Integrated Ocean Drilling Program (IODP).

Our country has a long history of geology. Geologists of the older generation successively put forward their suggestions and ideas to conduct scientific drilling projects a number of years ago. For instance, Li Siguang in 1950, Zhang Wenyou in 1959, Xie Jiarong in 1965, and Li Chunyu and the famous geophysical expert Gu Gongxu in 1988 successively put forward the same idea that scientific drilling was the only way to obtain geological materials from deep Earth for further development of geology in our country.

Comrade Deng Xiaoping proposed the brilliant thesis "Since the Chinese people were able to stand up, they will surely be able to stand firm forever among the nations of the world," which greatly inspired the exploration engineering workers. In September 1979, the first exploration engineering conference was held in Beidaihe; in the conference was put forward a proposal of making preparation for the first scientific borehole in the People's Republic of China, which won wide support from the exploration engineering circle. Many famous geologists, such as Jia Fuhai, Chen Mengxiong, Zhao Wenjin, Li Tingdong, Xiao Xuchang, and Xu Zhiqin gave strong support to this proposal. Hereafter, the scientific drilling project in China experienced the procedures of project argumentation and demonstration, examination, approval and site selection, and the National Laboratory of Scientific Drilling was established.

In 1992, the Ministry of Geology and Mineral Resources hosted the first seminar on China Continental Scientific Drilling. In February 1996, China officially joined ICDP. In September 1997, the China Continental Scientific Drilling Project was listed as a national major science project in the ninth Five-Year Plan period (1996–2000).

On June 25, 2001, CCSD-1 Well, the first hole of the China Continental Scientific Drilling (CCSD) Project, was opened. Taking only 1,353 days, CCSD-1 Well was drilled to 5,158-m-deep and 4,290.91-m core was recovered, with an average core recovery of 85.7 % and an average penetration rate up to 1.01 m/h. In the whole construction, the equipment and tools with totally independent intellectual property rights of China were employed, and a variety of geophysical measurements were undertaken during the drilling process. This is a tremendous contribution thanks to the close cooperation between the exploration field and many scientific research institutions of our country. The successful completion of CCSD-1 Well opened a new page for the study of Earth Science in China, and is a great and important step forward transforming from a big geoscience country into a powerful one.

The practice of CCSD-1 Well proved that only by scientific innovation, the world top-class achievement could be obtained. The constructors creatively combined the "combined drilling technology," the "flexible double-hole program," and the "feel-ahead with small diameter core drilling method" organically, and formed a complete set of unique scientific drilling technology systems with Chinese characteristics. And particularly, they successfully developed the downhole power drive (PDM + hydro-hammer) percussive rotary diamond core drilling technique system as the world origination. China Continental Scientific Drilling Project is a successful model of scientific innovation in our country.

This book, a monograph, incorporates theory, experience, and application and is a crystallization of the wisdom of three generations of drilling technicians. The publication of the book induces much thinking and enlightenment, and will surely play a great role in the continuous development and improvement of the drilling techniques in our country.

Science is endless and so is continental scientific drilling. This book will undoubtedly inspire us to continue climbing to new heights with new actions and make greater contributions to the continuous development of scientific drilling, with the success of CCSD-1 Well as a new starting point.

Beijing, October 7, 2014

Liu Guangzhi
Academician of the Chinese Academy of Engineering

Foreword II

After years of preparation, China Continental Scientific Drilling (CCSD) Project was officially selected as a major national science project in the ninth Five-Year Plan period. And after nearly four years of serious construction, the first scientific well of our country (CCSD-1 Well) was successfully completed with outstanding results, for which I feel very happy and gratified.

The successful completion of CCSD-1 Well is a memorable event in China's exploration engineering industry, which changed the previous situations of "paper drilling" or "oral drilling" (not to despise the theoretical knowledge or book knowledge, but refer to the empty talk), and this qualitative change was not easy. Now, we have really drilled a scientific drilling hole for more than five thousand meters deep, which has been implemented in Donghai, Jiangsu Province in China. This fact is very valuable. Genuine knowledge comes from practice, which is the basis of understanding objective things, and is the sole criterion for testing truth as well. Whether our various understandings toward drilling engineering are correct or not can only be judged by practice. I think we can now participate in international scientific drilling seminars with well-regulated minds and with self-confidence. This does not mean we are arrogant, but we feel sure of the genuine facts. That is the value of practice.

CCSD-1 Well is a continental scientific drilling project with Chinese characteristics. According to the goal of continental scientific drilling and in light of the requirements for full borehole coring in hard rocks in deep crust, advantages and advanced techniques of geological drilling and oil drilling were fully applied, and new improvements and innovations were made on the basis of the two technologies, so as to get excellent achievements. During the construction, there were many presentations with Chinese characteristics, which are introduced in detail in the book, and can also be found by drawing comparisons with foreign scientific drilling constructions.

CCSD-1 Well is a world-class continental scientific drilling project, which has experienced a course of having a high regard for science and boldly overcoming miscellaneous difficulties, and has left a brilliant page for the development of drilling techniques in our country. It was a tricky problem while coring (sampling) in hard rocks in deep well. To overcome this, the technique of PDM and hydro-hammer drive swivel-type double-tube impregnated diamond core drilling was originally created. This new downhole drilling tool assembly greatly improved the dynamic condition of bottomhole, stabilized the conditions of drilling tool rotation and vertical feeding, increased penetration rate and core recovery, improved straightening effect, and has become an advanced downhole drilling tool assembly for coring and straightening in deep hard rock drilling. In fact, we can fully understand the working conditions of the drilling tool at hole bottom based on the core obtained, which is not only a basis for evaluation of geological work, but also the basis for identifying downhole drilling conditions. During CCSD-1 Well drilling construction, a piece of core more than 4 m was recovered; the well-distributed scratches were the best evidence of the bottomhole drilling conditions.

The rock layers drilled in the continental scientific drilling project were hard and dense. The hole wall protection mechanism is different from that of oil drilling, in which sedimentary strata with formation pressure are to be drilled. In CCSD-1 Well, drilling fluid, the flowability and lubricity of which greatly influenced drilling process, also played the role of a medium transmitting downhole power. In view of such peculiar conditions and requirements, a special drilling fluid was successfully developed, which effectively reduced frictional resistance and wear of the drilling tool and also significantly reduced the circulation resistance in small annular clearance in deep hole, ensuring normal and efficient drilling. As a matter of fact, the drilling fluid system for the hole is a model with Chinese characteristics fully adapting to the needs of scientific drilling, and since then, the previous situation of totally contracting out the drilling fluid to oil drilling mud companies has been changed.

The implementation of scientific drilling is an advanced system engineering project, whole chains of which are mutually supported and restricted, working in synchronization. The concepts and measures worth mentioning are widespread, of which this book has given detailed explanations. I think the following points should be mentioned:

Flexible Double-Hole Program

The double-hole program, which is divided into the pilot hole and the main hole, was an initiative of the KTB project in Germany and was based on their own objective conditions. While waiting for the newly developed deep well drill rig, they used the ordinary oil drill rig to drill a pilot hole to race against time, avoiding any delay in drilling the pilot hole and the main hole. A double-hole program was also adopted in CCSD-1 Well; however, whether to move the hole position was to be decided according to the actual results of the pilot hole drilling. This decision was not only stochastically flexible, but also at a higher level in decision making, in comparison with the decision of moving hole position before drilling. Drilling practice showed that this strategic concept promoted the requirements for pilot hole construction that vertical drilling should be guaranteed for better coring. As a result, after the completion of 2,000-m-deep pilot hole, the deviation angle of the hole was only a little more than four degrees, and the main hole could be continued right in the pilot hole. Then one-hole program was realized after hole expanding. It was proved that the flexible double-hole program had saved large funds.

Selection of the Drill Rig

The new advanced ZJ70D oil drill rig was selected for drilling CCSD-1 Well. This laid a good foundation. However, the drill was mainly designed and made for oil drilling (cone bit, high weight on bit, and low rotary speed). In order to meet the requirements in scientific drilling with impregnated diamond bits in hard rocks, some necessary modifications were made on the drill rig which was completely reasonable. The original brake system was modified, and an electronic driller system was installed to fully satisfy the requirements of smooth and accurate feeding for drilling with impregnated diamond drill bit, and a surface top drive or downhole PDM drive was applied to increase the rotary speed of the drilling tool so that the drilling speed of impregnated diamond drill bit could be guaranteed. Meanwhile, in order to ensure the cleanness and the stable properties of the drilling fluids, high-performance solid control equipment was specially selected and this equipment played a key role in high-efficient safety drilling. In addition, due to full-hole coring and long construction period of CCSD-1 Well, public electric power grid was applied instead of diesel generator power supply commonly used in oil drilling, and in this way, the power supply condition was improved. Practice indicated that the above-mentioned technical measures and decisions were wise, effective, and economical.

Measures and Understanding of Borehole Bending

People are always trying to drill boreholes straight, which is almost impossible and unnecessary because of various reasons. In drilling of the 2,000-m-deep pilot hole in CCSD-1 Well, borehole deviation was only 4.1 degrees, and this was an outstanding achievement of anti-deviation, which indicated that the formation cooperated and the anti-deviation technology was efficient. In the hole section of 3,500 m below, strong deviating strata were encountered, and a variety of anti-deviation measures adopted were not effective. It was understood through calculation that to the final well depth, the deviation and displacement of the well were both in the permitted range, and then it was decided to drill along the natural deviation tendency of the rock formations, and in this way, the risks of arbitrary deviation correction could be avoided, drilling rate was increased, and the expected goal was attained, with the advantages outweighing the disadvantages. This decision was reasonable and clever. In fact, the serious point of the well deviation is the sharp elbow (also called "dogleg"). Large dogleg degree can cause rotation difficulties of drill string and drilling tool and obstacles for running casing, directly affecting drilling operations.

Special Double Drive

As downhole PDM drive was adopted in CCSD-1 Well, the drill string of thousands of meters in the upper hole section was not in rotation. Practice indicated, however, that slight rotation of the drill string at the upper hole section is of benefit. Drill bit at hole bottom not only rotates, but also needs timely feeding, and slight rotation of the drill string at the upper hole section improves the feeding state, and smooth and stable feeding can be realized.

During the construction practice of CCSD-1 Well, much experience and new knowledge were obtained. The views above are just feelings of mine, which might not be definitely right but could be references for study.

This book is a comprehensive summary of the China Continental Scientific Drilling Project and a discussion of the implementation of CCSD-1 Well in all aspects. Different from the works of "book to book" (not to despise books, but refer to the books only for publishing), this book was naturally compiled based on real drilling constructions. After my first reading, I think the book, which is a monograph on continental scientific drilling, is worth reading, learning, and thinking. "Practice makes genuine knowledge" and the key point is "to make". How to raise perceptual knowledge to rational knowledge through scientific thinking should be seriously considered. Science and technology are for real and shall not permit any impetuousness and dishonesty. We are glad of the success of CCSD-1 Well which has promoted the development of scientific drilling. This must be affirmed. However, in the course of scientific and technological development, it is only one link.

In conclusion, it should be noted that all our achievements and innovations of today are made on the basis of our predecessors, without which we could not make any achievements. For this, we have no reason to become arrogant. "Modesty helps one to go forward, whereas conceit makes one lag behind." That is what I am willing to share with you.

Beijing, January 1, 2015

Li Shizhong
Professor, China University of Geosciences (Beijing)

Preface

"Going up to the space, reaching the interior Earth, entering into the sea" are three magnificent feats of human beings to challenge the natural world to expand living space. For thousands of years, human beings have achieved great success in going up to space and entering into the sea, while still struggling hard with the exploration of the interior Earth. Scientific drilling is a great project with epoch-making significance in contemporary Earth Science research. Through the direct observation of the lithosphere by scientific drilling, the material composition and structure of the continental crust can be explained, results of geophysical telemetry of the deep Earth can be rectified, the deep Earth fluids system and geothermal structure can be studied, the distribution and incubation conditions of subsurface microbes can be explored, and then the development of deep Earth geology can be promoted, and all these are helpful to solve a series of fundamental scientific problems, such as global climate change, law of earthquakes and biological origin, etc. In conclusion, scientific drilling, which is of very important significance to solve the problems of resources, disasters, and environment during the development of human society, is also a major scientific project which can bring along the development of relevant engineering technology, and is another magnificent challenge toward the Earth after man's landing on the moon.

The China Continental Scientific Drilling (CCSD) Project is a major national science project listed in the ninth Five-Year Plan (1996–2000), as well as a project of the International Continental Scientific Drilling Program (ICDP) currently being implemented. The main task of the project is to drill a 5,000-m-deep well for continuous directional cores, rock, and fluid samples, and in situ downhole observation data in Dabie-Sulu ultrahigh-pressure metamorphic belt, a global significant convergent plate boundary, to make comprehensive geophysical surveys, identify the material composition and structure of the continental orogenic belt, and reveal the formation and exhumation mechanism of the ultrahigh-pressure metamorphic belt. The 5,000-m-deep well will be built as a long-term underground observation and experimental base.

With tremendous technical difficulties, it is the first time in our country to construct a 5,000-m-deep well in hard crystalline rocks for full-hole continuous coring, which is one of the most difficult drilling constructions in the world as well. For nearly four years, from the project feasibility study, drilling technical personnel of our country have played their wisdom and creativeness and overcome numerous difficulties during the stages of engineering design, drilling construction, research and application of the key technologies, and solved construction problems until the successful completion of the project. While absorbing the world's advanced technology on scientific drilling, they successfully created and applied a series of new technologies and equipment, formed the new scientific drilling technology system with Chinese characteristics, which withstood the severe test of hard rocks and complex formations in Sulu ultrahigh-pressure metamorphic belt, completed CCSD-1 Well with high quality and efficiency and at low cost, and made outstanding important engineering achievements. These achievements greatly promoted the progress of scientific drilling technology, as well as exploration drilling for energy and resources. The success of CCSD-1 Well not only showed that deep drilling technology in our country had obtained great progress, but also greatly enhanced China's international standing in drilling technology.

The implementation of the CCSD Project is the start of the magnificent plan of "reaching the interior earth" in China, with initiative in the history of the Earth Science research in the country. In recent years, the environmental scientific drilling and the Cretaceous scientific drilling have been started in China, and the ultra-deep scientific drilling for oil and gas resources and deep solid mineral resources will be gradually started. A new situation in Earth Science has been formed; marking China's new step that has made it from a large geoscience country to a powerful geoscience country, which is bound to make impacts on the harmonious development of the society and nature and the modernization of our country.

This book comprehensively describes the drilling technologies of CCSD-1 Well, brings together various data and information accumulated in the process of drilling, and shows the latest technologies and research achievements of scientific drilling in China. The main authors of the book all used to take the major tasks at the construction site as technical backbone, and this book is a summary of their creative thinking in drilling practice, and the crystallization of their wisdom.

Veteran drilling experts threw all their energy into the project. They laid a very strong technical foundation for the start of the project. The project gathered a large number of outstanding middle-aged and young technical experts; some of them have extensive management experience; some have solid theoretical foundation and research experiences of many years, being creative and good at solving the new problems arisen during the construction; some have worked throughout the year at drill sites for technical services and production supervision and been adept in solving complex problems happened at the drill sites; and some just graduated from schools, being quick thinking and enthusiastic, with new professional knowledge, especially modern data processing technology, and brought fresh air to the drill site.

In order to enable the project to come up to international professional standards, the majority of the construction staff received training from the International Continental Scientific Drilling Program (ICDP). It is their hard work that offered the book with a wealth of original materials.

It has to be particularly noted that Liu Guangzhi, an academician of the Chinese Academy of Engineering, has led and organized the continental scientific drilling in China for decades. He first introduced the recent progress in this field and advocated the implementation of China's Continental Scientific Drilling Project; organized the planning of scientific drilling program and the discussions of technical program in the country, compiled a Series of Exploration of the Deep Continental Crust (eight volumes), cultivated a great number of middle-aged and young scientific and technological personnel engaged in scientific drilling, and therefore laid a solid technical foundation for the success of CCSD-1 Well.

Dedicated help and support to CCSD-1 Well project were given from the Institute of Exploration Techniques of CAGS, Beijing Institute of Exploration Engineering, the Institute of Exploration Technology of CAGS, China University of Geosciences (Wuhan), China University of Geosciences (Beijing), Construction Engineering College of Jilin University, Chengdu University of Technology, Zhongyuan Petroleum Exploration Bureau of Sinopec Group, Shengli Petroleum Administration Bureau of Sinopec Group, Dezhou Oil Drilling Institute of the Academy of Oil Exploration and Development, and Drilling Research Institute of CNPC and China University of Petroleum (Beijing). The Department of Land and Resources of Jiangsu Province and Geology and Mineral Exploration Bureau of Jiangsu Province Lianyungang City, Donghai County, and other circles paid great attention to the project. The authors would like to express their heartfelt thanks to the Drilling Engineering Advisory Committee and the experts from all fields, for their suggestions and great efforts to the drilling engineering. The members of the Drilling Engineering Advisory Committee are as follows:

Wan Jinshan, Ma Jiaji, Wang Jian'an, Mao Kewei, Zuo Ruqiang, Jiang Tianshou, Tang Songran, Liu Guangzhi, Liu Xisheng, Guan Xihai, Xiang Zhenze, Song Xiangyan, Li Shizhong, Li Yanzao, Li Zhenya, Li Changmao, Su Yinao, Wu Guanglin, Chen Yuandun, Shao Jiwu, Zhou Tiefang, Hu Puyuan, Zhao Guolong, Zhao Erxin, Xi Jiazhen, Geng Ruilun, Xu Chaoyi, Huang Renshan, Xie Rongyuan, Jiang Rongqing, Han Guangde, Lei Hengren, and Yan Taining.

This book was written by Wang Da, Zhang Wei, Zhang Xiaoxi, Zhao Guolong, Zuo Ruqiang, Ni Jialu, Yang Gansheng, Jia Jun, Yang Kaihua, Zhu Yongyi, Xie Wenwei, Zhu Wenjian, Zhang Peifeng, Fan Lasheng, Ye Jianliang, and Wang Yongping. After the completion of the first draft, the book was revised and unified by Wang Da, Fan Lasheng, and Zhu Wenjian.

Beijing, January 17, 2015
Da Wang

Contents

1	**Background**		1
	1.1	Scientific Drilling—A New Field of Earth Science	1
	1.2	A Brief Introduction of China Continental Scientific Drilling Project	1
	1.3	Site Selection and Scientific and Technological Objectives	2
	1.4	Developing History of CCSD Engineering	3
		1.4.1 Early Stage of Understanding (Before 1991)	3
		1.4.2 Project Argumentation and Demonstration Stage (1991–September 1999)	4
		1.4.3 Project Preparation Stage (September 1999–June 2001)	5
		1.4.4 Project Implementation Stage (June 2001–April 2005)	6
	1.5	Technical Preparation	7
		1.5.1 Technical Training	8
		1.5.2 Pre-pilot Hole Construction	9
		1.5.3 Pre-research on Key Technologies	10
2	**Drilling Engineering Design**		15
	2.1	Assignment of Drilling	15
	2.2	Basic Situation of the Well Site	15
		2.2.1 Forecast of Lithological Profile of the Formation Encountered	16
	2.3	Lithologic Characteristic of the Rock Formations to be Encountered by Drilling	17
	2.4	Drilling Technical Program	18
		2.4.1 Combined Drilling Techniques	18
		2.4.2 Flexible Double Hole Program	19
		2.4.3 Feel Ahead Open Hole Drilling Techniques	19
	2.5	Borehole Structure and Casing Program	20
		2.5.1 Designed Borehole Structure and Casing Program for the Pilot Hole	20
		2.5.2 Designed Borehole Structure and Casing Program for the Main Hole	20
	2.6	Drilling Equipment Program	20
		2.6.1 Main Drilling Equipment	20
		2.6.2 Equipment and Instruments Should Be Added	22
	2.7	Drilling String Program	22
	2.8	Core Drilling Program	23
		2.8.1 Wireline Core Drilling	23
		2.8.2 Hydro-hammer Wireline Core Drilling Tool	25
		2.8.3 PDM Wireline Core Drilling Tool	25
		2.8.4 Turbomotor Wireline Core Drilling Tool	25
		2.8.5 Conventional Core Drilling Tool	26
		2.8.6 Hydro-hammer Core Drilling Tool	26

		2.8.7	PDM Core Drilling Tool	28
		2.8.8	Design Program of Diamond Core Drill Bit and Reaming Shell	28
	2.9	Hole Deviation Control Program		29
		2.9.1	Deviation Prevention for Cored Hole Section and Monitor Measures	31
		2.9.2	Deviation Control Measure for Cored Hole Section	31
		2.9.3	Deviation Control Measure for the Upper Section of the Main Hole Where Non-core Drilling Was Conducted	31
	2.10	Non-core Drilling and Reaming Drilling Program		31
		2.10.1	Design of Drilling Tool Assembly For Non-core Drilling	31
		2.10.2	Design of Drilling Tool Assembly for Reaming Drilling	32
		2.10.3	Selection of Non-core Drill Bit	32
		2.10.4	Design of Reaming Drill Bit	33
	2.11	Drilling Fluid Technique and Solid Control Program		33
		2.11.1	The Main Technical Problems Should Be Considered	33
		2.11.2	Design of Drilling Fluid Type	34
		2.11.3	Solid Control	34
	2.12	Well Cementation and Completion Program		34
		2.12.1	Well Cementation Program	34
		2.12.2	Principle in Design of Casing String Strength	35
		2.12.3	Well Completion Operation	36
	2.13	Design of Moving Casing		36
		2.13.1	Necessity of Adopting Moving Casing Design	36
		2.13.2	Fixing of Moving Casing	37
		2.13.3	Safety Management of Moving Casing	37
	2.14	Time and Cost Estimation		38
		2.14.1	Designed Construction Progress	38
		2.14.2	Budgetary Estimation of Cost	38
	2.15	Change and Modification of Design		38
3	**Well Site and Drilling Equipment**			47
	3.1	Well Site		47
	3.2	Drilling Equipment		49
		3.2.1	ZJ70D Drill Rig	50
		3.2.2	Drill Rig Reconstruction	53
		3.2.3	The Power System	56
		3.2.4	Corollary Equipment	57
		3.2.5	Application Evaluation on ZJ70D Drill Rig	60
4	**Construction Situation**			63
	4.1	Basic Situation of the Construction of CCSD-1 Well		63
		4.1.1	The Basic Data	63
		4.1.2	Drill Hole Trajectory	67
		4.1.3	Well Temperature Curve	68
	4.2	Simple Situation of the Construction at Different Periods		69
		4.2.1	Hole Opening and Non-core Drilling (the First Opening)	69
		4.2.2	Pilot Hole (Section CCSD-PH) Core Drilling (the Second Opening)	71
		4.2.3	The First Expanding Drilling of the Main Hole (Hole Section CCSD-MH-1K)	74
		4.2.4	The First Core Drilling of the Main Hole (Hole Section CCSD-MH, the Third Hole Opening)	77
		4.2.5	The First Sidetracking (Deviation Correction) Drilling of the Main Hole	82

	4.2.6	The Second Core Drilling of the Main Hole (Hole Section CCSD-MH-1C)	89
	4.2.7	The Second Expanding Drilling of the Main Hole (Hole Section CCSD-MH-2K)	91
	4.2.8	The Second Sidetracking (Obstacle Avoidance) Drilling and Running Casing and Well Cementation in the Main Hole	97
	4.2.9	The Third Core Drilling of the Main Hole (Section CCSD-MH-2C, the Fourth Opening)	99
	4.2.10	Testing Drilling Tools	102
	4.2.11	Well Completion	105

5 Hard Rock Deep Well Core Drilling Techniques ... 107
- 5.1 Current Status of Core Drilling Techniques ... 107
- 5.2 Experiment on Core Drilling Methods for CCSD-1 Well ... 108
  - 5.2.1 Rotary Table Drive Double Tube Core Drilling ... 108
  - 5.2.2 Rotary Table Hydro-hammer Drive Double Tube Core Drilling ... 109
  - 5.2.3 Top Drive Double Tube Core Drilling ... 109
  - 5.2.4 Top Drive Wireline Core Drilling ... 110
  - 5.2.5 Top Drive Hydro-hammer Wireline Core Drilling ... 111
  - 5.2.6 PDM Drive Single Tube Core Drilling ... 111
  - 5.2.7 PDM Drive Double Tube Core Drilling ... 113
  - 5.2.8 PDM Drive Wireline Core Drilling ... 114
  - 5.2.9 PDM Hydro-hammer Drive Double Tube Core Drilling ... 116
  - 5.2.10 PDM Hydro-hammer Drive Wireline Core Drilling ... 116
  - 5.2.11 Summary of the Tests for Core Drilling Methods ... 118
- 5.3 Down Hole Power Percussive Rotary Core Drilling System ... 119
  - 5.3.1 Constituent of the System ... 119
  - 5.3.2 Technical Data of the System ... 121
  - 5.3.3 Down Hole Rotary Drive Drilling Tool—PDM ... 127
  - 5.3.4 Down Hole Percussive Drilling Tool—Hydro-hammer ... 128
  - 5.3.5 Core Drilling Tool ... 155
  - 5.3.6 Core Drilling Technologies ... 166
  - 5.3.7 The Application Results of Hard Rock Deep Well Core Drilling Techniques ... 172

6 Diamond Core Drill Bit ... 183
- 6.1 The Physical and Mechanical Properties of the Rocks to Be Drilled ... 183
  - 6.1.1 The Properties of the Rocks to Be Drilled ... 183
  - 6.1.2 The Physical and Mechanical Properties of the Rocks ... 183
- 6.2 Selection of Diamond Core Drill Bit Types ... 185
  - 6.2.1 Core Drilling Technologies ... 185
  - 6.2.2 Types of Diamond Core Drill Bits ... 186
- 6.3 Design and Manufacture of Impregnated Diamond Core Drill Bits ... 187
  - 6.3.1 Segment Inserted Drill Bit by Twice Forming ... 187
  - 6.3.2 Sintered Diamond Drill Bit ... 189
  - 6.3.3 Electro-plated Diamond Drill Bit by Twice Forming ... 190
- 6.4 Application of Diamond Core Drill Bits ... 191
  - 6.4.1 Brief Introduction ... 191
  - 6.4.2 Application Results of Three Main Core Drill Bits ... 200
  - 6.4.3 Application Results of Other Type Core Drill Bits ... 205

7	**Reaming Drilling Techniques of Hard Crystalline Rock**		211
	7.1 Development of Pilot Reaming Bits		211
		7.1.1 KZ157/311.1 Type Reaming Bit	213
		7.1.2 KHAT 157/311.1 Reaming Bit	215
		7.1.3 Development and Improvement of KZ157/244.5 Reaming Bit	216
	7.2 Design of Drilling Tool		218
		7.2.1 Strength Check of Drilling String	218
		7.2.2 Selection of Drilling Tools	219
		7.2.3 Design of Drilling Tool Assembly	222
	7.3 Optimization of Drilling Parameters		224
		7.3.1 WOB	224
		7.3.2 Rotary Speed	224
		7.3.3 Pump Displacement	226
	7.4 Effect of Reaming Drilling		226
		7.4.1 General Drilling Conditions	226
		7.4.2 Application of Pilot Reaming Bits	228
8	**Well-Deviation Control Techniques for Strong Dipping Strata**		233
	8.1 Summary		233
		8.1.1 The Formation Conditions	233
		8.1.2 The Well Deviation Control Technology	235
		8.1.3 The Basic Conditions of Well Deviation Control in CCSD-1 Well	236
	8.2 Deviation Prevention Drilling Technology		236
		8.2.1 The Well Deviation Control in Core Drilling	237
		8.2.2 Well Deviation Control in Non-core Drilling and Reaming Drilling	242
	8.3 Drilling Techniques for Deviation Correction		244
		8.3.1 Side-Tracking Deviation-Correction Techniques	245
		8.3.2 Situation on Side-Tracking Drilling for Deviation-Correction	247
		8.3.3 Deviaton Correction at the Well Bottom of MH-1C Well Section	257
	8.4 Side-Tracking Drilling for Bypassing Obstacles		260
		8.4.1 Selection of Side-Tracking Drilling Tool	260
		8.4.2 Drilling Conditions of Side-Tracking Drilling to Bypass Obstacles	260
	8.5 Development of PDM Drive Continuous Deflector		265
		8.5.1 Working Principle of the Drilling Tool	265
		8.5.2 Practical Drilling Test at Drill Site	266
		8.5.3 Test Result Commentary	268
	8.6 The Analysis on Well Deviation Control Effect		272
9	**Drilling Fluids and Solids Control Technology**		273
	9.1 Requirements of Scientific Drilling for Drilling Fluid		273
		9.1.1 Strata Encountered and Requirements of Well Structure	273
		9.1.2 Requirements of Core Drilling	274
		9.1.3 Requirements of Non-core Drilling and Expanding Drilling	275
		9.1.4 Requirements of Borehole Log	275
		9.1.5 Requirements of Environmental Protection	275
		9.1.6 Requirements of Drilling Fluid Design	276
	9.2 Drilling Fluid System		276
		9.2.1 Selection of Drilling Fluid System	277
		9.2.2 LBM-SD Composite Drilling Fluid Material	277
		9.2.3 Drilling Fluid Mechanism and Composition of LPA Polymer	277

		9.2.4	Manufacture Technology of LBM-SD	280
		9.2.5	Evaluation Procedure of Drilling Fluid	281
		9.2.6	Performance of LBM-SD Drilling Fluid System	281
	9.3	Drilling Fluid for Core Drilling.	285
		9.3.1	Properties.	285
		9.3.2	Circulating Pressure Drop	288
		9.3.3	Lubrication Effect of Drilling Fluid.	292
	9.4	Solid Control Technique of Drilling Fluid	293
		9.4.1	Cuttings Size Analysis.	294
		9.4.2	Requirement of Solids Control Equipment to Drilling Fluid	295
		9.4.3	Analysis of Solids Control Effect	295
	9.5	Site Application of Drilling Fluid	297
		9.5.1	Application of Drilling Fluid in Non-core Drilling in the First Opening (Spudding-in)	297
		9.5.2	Application of Drilling Fluid in Pilot Hole Core Drilling	297
		9.5.3	Application of Drilling Fluid in the First Expanding Drilling in the Main Hole	298
		9.5.4	Application of Drilling Fluid in the First Core Drilling in the Main Hole	298
		9.5.5	Application of Drilling Fluid in the First Sidetrack Straightening Drilling in the Main Hole.	299
		9.5.6	Application of Drilling Fluid in the Second Core Drilling in the Main Hole	299
		9.5.7	Application of Drilling Fluid in the Second Expanding Drilling in the Main Hole	300
		9.5.8	Application of Drilling Fluid in the Second Sidetrack Drilling-Around in the Main Hole.	301
		9.5.9	Application of Drilling Fluid in the Third Core Drilling in the Main Hole	301
		9.5.10	Application Characteristics of LBM Drilling Fluid	302
10	**Casing and Well Cementation**			303
	10.1	Borehole Structure and Casing Program.	303
		10.1.1	Borehole Structure and Casing Program for the Pilot Hole.	303
		10.1.2	Borehole Structure and Casing Program for the Main Hole	303
		10.1.3	Casing Design	303
	10.2	Well Head Assembly.	306
		10.2.1	Well Head Assembly for the First Opening (Spud-in)	306
		10.2.2	Well Head Assembly for the Second Opening (Spud-in)	307
		10.2.3	Well Head Assembly for the Third and the Fourth Opening (Spud-in)	308
		10.2.4	Well Head Assembly for Well Completion.	308
	10.3	Casing Running and Well Cementing Operation	308
		10.3.1	508.0 mm Well Head Conductor	308
		10.3.2	339.7 mm Surface Casing	309
		10.3.3	273.0 mm Intermediate Casing	311
		10.3.4	193.7 mm Intermediate Casing	317
		10.3.5	127.0 mm Tail Pipe	320
	10.4	Moving Casing Techniques	322
		10.4.1	Overall Programme.	323
		10.4.2	Design of Fixing Moving Casing	323
		10.4.3	Moving Casing Strength Check	325
		10.4.4	Design of Casing Shoe and Retaining Sub.	325
		10.4.5	Design of Thread Back-off Proof for Moving Casing	327
		10.4.6	Design of Centralizer	327

		10.4.7	Operating Technology of Moving Casing.	329
		10.4.8	Application of Moving Casing Techniques.	329

11 Drilling Data Acquisition ... 333
- 11.1 General Situation ... 333
- 11.2 Analysis of Data Acquisition and Processing Requirements ... 336
  - 11.2.1 Data Acquisition System Requirements ... 336
  - 11.2.2 Data Processing System Requirements ... 336
- 11.3 Drilling Data Acquisition System ... 337
  - 11.3.1 Surface Drilling Data Acquisition System ... 338
  - 11.3.2 Down-Hole Drilling Data Acquisition System ... 341
- 11.4 Drilling Data Processing System ... 343
  - 11.4.1 Single Parameter Monitoring ... 343
  - 11.4.2 Comprehensive Monitoring ... 344
  - 11.4.3 Case History ... 345

12 Technical Economical Analysis ... 349
- 12.1 Construction Time and Cost Analysis ... 349
  - 12.1.1 Construction Time Analysis ... 349
  - 12.1.2 Construction Cost Analysis ... 349
- 12.2 Economic Evaluation of Core Drilling Techniques ... 352
  - 12.2.1 Evaluation Method ... 352
  - 12.2.2 Index System of Technical Economic Evaluation for Core Drilling Construction ... 353
  - 12.2.3 Calculation of Drilling Construction Time and Cost ... 353
  - 12.2.4 Technical Economical Indexes of Different Core Drilling Methods ... 355
  - 12.2.5 Economic Evaluation ... 355
  - 12.2.6 Technical Risk Evaluation ... 355
  - 12.2.7 Comprehensive Evaluation ... 358

References ... 361

Authors and Translators

Authors
Da Wang, China Geological Survey
Wei Zhang, China Geological Survey
Xiaoxi Zhang, China University of Geosciences (Wuhan)
Guolong Zhao, Consulting and Research Center, Ministry of Land and Resources
Ruqiang Zuo, Consulting and Research Center, Ministry of Land and Resources
Jialu Ni, Institute of Exploration Techniques
Gansheng Yang, Institute of Exploration Techniques
Jun Jia, Beijing Institute of Exploration Engineering
Kaihua Yang, China University of Geosciences (Wuhan)
Yongyi Zhu, Institute of Exploration Techniques
Wenwei Xie, Institute of Exploration Techniques
Wenjian Zhu, Beijing Institute of Exploration Engineering
Peifeng Zhang, Beijing Institute of Exploration Engineering
Lasheng Fan, Institute of Exploration Technology
Jianliang Ye, China Geological Survey
Yongping Wang, Research and Design Academy of Metallurgical Prospecting

Translators
Junfeng Geng (Chaps. 2, 5, 6 and 10)
Yongqin Zhang (Chaps. 3, 4, 7 and 8)
Longchen Duan (Chaps. 4 and 9)
Haipeng Li (Preface, Chaps. 1, 11 and 12)

Reviser of the English Version
Junfeng Geng

Background

1.1 Scientific Drilling—A New Field of Earth Science

The earth is a giant system including the atmosphere, hydrosphere, lithosphere, biosphere, and the mantle, core and planetary space. Earth science is the study on the substance composition, origin, formation, evolution and the interaction of each part of the earth system. In the late 20th century, the earth science developed towards the earth system science, focusing on the lithosphere, and gradually extended to the relationship and interaction with the other layers. Since the 1960s, the International Upper Mantle Program (IUMP), the Mohole Project, the Deep Sea Drilling Project (DSDP) and the Ocean Drilling Program (ODP) were successively implemented in the world, through these projects great progress was made in earth science, confirming the continental drift theory and creating the plate tectonics theory. And the international geoscience community gradually achieved a common view that scientific drilling is the only way to obtain real samples from the deep crust, and a long-term observation station can be established through scientific drilling, which is a telescope and endoscope penetrating into the earth, and is the most important technical method to research on the crustal structure and evolution. At the same time, it will play an important role in detecting the deep biosphere, explaining the mysteries of life evolution and understanding the mechanism of global environmental change. Scientific drilling is not only of important scientific significance, but also of great practical significance for maintaining the man-earth coordination and harmony, and promoting the economic and social development.

Scientific drilling is one of the important eye-catching frontier subjects of the modern earth science. According to the drilling areas, scientific drilling can be divided into ocean scientific drilling and continental scientific drilling. It is known that the world's earliest scientific drilling started from ocean. However, the continental crusts are older than oceanic crusts, with more hidden mysteries of the earth. Besides, continents are the places where human beings directly live, get the main mineral resources and suffer the greatest geological disaster threats. Therefore, people are eager to understand continents more and deeper through continental scientific drilling.

China is a large geological country, with many significant earth science problems attracting world-wide attention, such as the uplift mechanism of the third pole of the world—Qinghai-Tibet Plateau and its impact on the global environment, the formation mechanism of the Central Asia's largest fracture—Altun fracture, the cause and exhumation mechanism of the world's largest ultrahigh pressure metamorphic belt (Jiaonan-Dabie area), the focal mechanism of the Beijing-Tianjin-Hebei earthquake zone and the North China plate internal dynamics, etc. These problems urgently need the deep real information obtained through scientific drilling, especially the real samples of original cores, rocks, and fluids, etc., to reveal the mysteries and solve the key theoretical issues by using modern scientific research methods.

1.2 A Brief Introduction of China Continental Scientific Drilling Project

In the past forty years, China's geological, geophysical scientists and drilling engineers have been giving great attention and efforts to continental scientific drilling. They wrote articles and books, organized domestic and international academic exchange activities, propounded ideas, suggestions and proposals, and carried out early studies. They wholeheartedly called for the implementation of China's continental scientific drilling project and held seminars, workshops, demonstration meetings around the project's approval and initiation, feasibility study, site selection,

Translated by Li Haipeng.

construction scheme, etc., which promoted the smooth implementation of the project.

After long-term unremitting efforts, the CCSD Project was finally approved and listed as the national important scientific project in the ninth Five-Year Plan on June 4th, 1997 by the State Science and Technology Leading Group.

As one of the national important scientific projects, the CCSD Project is the deepest scientific drilling hole among the international continental scientific drilling projects currently being implemented. CCSD-1 Well is located in Maobei Village of Donghai County, Lianyungang City, Jiangsu Province, i.e. the south part of the Sulu UHP metamorphic belt.

CCSD-1 Well was officially opened on June 25th, 2001, and a grand commencement ceremony was held on August 4th, which opened a new page for China's geoscience research. After nearly four years of construction, in April, 2005, drilling, logging and other constructions were completed at drill site. Zeng Peiyan, Vice Premier of the State Council, attended the hole completion ceremony held in Donghai County and declared the victorious completion of CCSD-1 Well construction. In the future, monitoring instruments would be installed in the well and a world first-class long-term deep observation laboratory would be established.

It was such a special great science project that attracted national high attention. Premier Wen Jiabao pointed out, "The CCSD Project is a comprehensive project incorporating science and technology into one, and a multi-disciplinary and multi-field system integration. The implementation of the project will promote the development of China's earth science theory and earth exploration technology level, and is of very vital significance."

The successful completion of the CCSD Project marked that China had taken a new step from a big geo-country towards a powerful one, which would inevitably exert a certain influence on the well coordinated development of society and nature and modernization construction of the country.

1.3 Site Selection and Scientific and Technological Objectives

Sixteen ultrahigh pressure metamorphic belts have been found in the global collision orogenic belts. Located in the middle east of China, Dabie-Sulu orogenic belt, the convergent boundary between the North China plate (Sino-Korean plate) and the South China plate (Yangtze plate), contains rare ultra-high pressure metamorphic minerals such as coesite and diamond, etc.

In the Dabie-Sulu ultrahigh pressure metamorphic belt, Dabie and Sulu were once an integral whole, which were separated after Mesozoic era by the Tanlu fracture, the largest fracture in east China. Here is the world largest ultrahigh pressure metamorphic belt, which is the best place to study the deep dynamics in convergent boundary of continental plates generally acknowledged by the scientists at home and abroad. The collision orogeny of the modern continental convergent boundary is now in process in the world-famous Himalayas. However, the deep process can only be remotely sensed relying upon geophysical means, while in Dabie-Sulu area it was exposed the direct result of the deep process in geological history—the ultrahigh pressure metamorphic belt, making it the best location to study the crust-mantle movements in convergent boundary of continental plates in geological history.

Coesite is a messenger from the deep earth. Like a golden key to open the deep earth, it records the geological process of the rocks intervening into and exhuming back to the ground. The research on ultrahigh pressure metamorphic rocks and coesite has become a hot point in the current geoscience study. CCSD-1 Well was drilled at the root of continental collision orogenic belt, located in the deepest part among the scientific drilling boreholes completed in the world, which was to obtain the information of the deepest through the shortest distance. In addition, it is an ideal place to study the post-orogeny, which can provide a new method for researching the mineralization of ultra-high pressure metamorphic belt, as well as a physical basis for researching the earthquake mechanism and then providing information for earthquake prediction.

After a long-term study, it was decided that the 5,000 m deep main hole of the CCSD engineering was eventually located in Donghai County, south of the Sulu high pressure metamorphic belt, with the scientific goals as the following:

Through a complete survey and comprehensive study on all the continuous cores, liquid and gaseous samples and the in situ logging data, to set up a variety of multidisciplinary fine profiles of the 5,000 m deep well; to reconstruct the deep three-dimensional material composition, distribution and the three-dimensional structure of the convergent boundary between the north China plate and the Yangtze plate; to expound the deep fluids effects in plate convergent boundary, the interaction between the crust and the mantle, and the material circulation and rheology in the mantle; to look for the symptomatic minerals formed under the ultra deep mantle conditions and to reveal the ultra-high pressure metamorphic mineralization mechanism; to establish the geophysical theory model and interpretation standards of the crystalline rock area; to reveal the formation and exhumation models of the ultrahigh pressure metamorphic rocks and the deep dynamics mechanism of the plate convergent boundary; and to study the physical, chemical and biological functions of modern crust, to make comprehensive geophysical measurements, and to accurately monitor the modern crustal movements by taking the advantages of the special underground space shaped by the 5,000 m deep hole (without noise, less disturbance, high temperature and high pressure, and with less influence of atmospheric precipitation). The 5,000 m continental scientific deep hole would become the first long-term observation and experiment station in Asia (Zhiqin et al. 2005).

The engineering technical goals of CCSD-1 Well included:

By using the modern high and new deep drilling technologies to construct the first 5,000 m scientific deep hole at the east part of Dabie-Sulu ultrahigh pressure metamorphic belt (Donghai County, Jiangsu Province) in China, and through the implementation of this scientific drilling project, to research and develop a set of new drilling technology system suitable for hard rock deep hole adverse drilling conditions; to research and develop a new combined core drilling system; to perfect the hydro-hammer drilling technologies with Chinese characteristics; to promote the further development of drilling tools and materials manufacturing technologies and to enable China's drilling technologies come up to the advanced international level in the 1990s; and to establish a test base for geophysical logging instruments, new methods, and new technologies, and thus promote the development and application of logging technologies in the country.

1.4 Developing History of CCSD Engineering

The earliest documents introducing the situations of world scientific drilling and advocating to implement the scientific drilling project in China could be traced back to the 1970s. Over more than 30 years, the developing history of scientific drilling in China could be divided into five stages, the early stage of understanding, project argumentation and demonstration stage, project preparation stage, project implementation stage, and scientific drilling popularization and application stage.

1.4.1 Early Stage of Understanding (Before 1991)

After the introduction of the world scientific drilling information into China, the geological scientists and engineers in China had always been tracking its progress, paying close attention to its development, making data collection and information research, and thinking deeply about our orientation.

Early in 1965, Mr. Xie Jiarong, the late famous geologist in China and the chief engineer of the Ministry of Geology, pointed out in a conversazione that after World War II the geological research had been developing towards two directions: to the outer space (the launch of ERTS earth resource satellite) and to the interior earth (scientific deep drilling), making great progress in geology.

In 1978, in his article entitled *To Pay Attention to the Deep Geological Study in the Research on Basic Geological Theory and Mineralization Regularity*, Xiao Qinghui from the Intelligence Institute of Chinese Academy of Geological Sciences introduced the history, present situation, significance and achievements of the world deep geological research (including scientific drilling), pointed out the necessity and urgency to carry out deep geological research in our country and offered suggestions.

In 1979, in the Second Conference held by Exploration Engineering Professional Commission of the Geological Society of China in Beidaihe, Liu Guangzhi gave a lecture on the status and developing trend of drilling techniques and for the first time introduced the developing status and prospects of super deep drilling and deep ocean drilling in foreign countries. In 1983, his article *Super Deep Drilling and Deep Geology* was published in *Geological Review*.

In 1985 and 1986, *Deep Geological Study in Foreign Countries* and *Study on the Crust and Upper Mantle* were respectively published, in which were respectively introduced the formation, development, implementation and the achievements of some deep geology research projects in the US, the former Soviet Union, Germany, France, India, Japan and other countries as well as of international cooperation, the current situation, prospect, research approaches and methods of deep geological study, and the super deep drilling status and achievements in the former Soviet Union.

In May 1988, Prof. C. Marx, president of ITE of Clausthal University of Technology of Germany visited China. He introduced in detail the status of ocean drilling and the KTB project in Germany, further deepening our understanding of world scientific deep drilling activities and their contributions to earth sciences.

In 1988, Mr. Gu Gongxu, a famous geophysicist in China pointed out in his article entitled *Suggestion on Making a Long-term Program for Continental Scientific Drilling in the Near Future* (published in *China Science and Techonology Guide* Vol. 1, 1988) that the actual results of underground inference depending only upon surface geological, geophysical and geochemical observations had been unable to satisfy the requirements of rapid development of earth sciences. He proposed that China should make a long-term planning for scientific deep drilling and also made concrete suggestions on drilling goals, well location selection, funds channels and technical equipment.

From 1988 to 1993, with Mr. Liu Guangzhi as the chief editor, *A Series of Exploration of the Deep Continental Crust* was published, which played an important role in promoting scientific drilling in China.

In September 1990, Wang Da and Zhang Wei visited Germany to attend the Fifth International Symposium on

Observation of the Continental Crust through Drilling and the opening ceremony of KTB project, and made detailed understanding of the KTB main hole drilling equipment, tools and technical program and the latest development of world scientific deep drilling. After return, they edited the *Sidelights on the Fifth International Symposium on Observation of the Continental Crust through Drilling* and the *KTB Opening Ceremony*. In October 1990, the Ministry of Geology and Mineral Resources organized the deep drilling investigation group of 5 people to visit Germany. They got a detailed understanding of German continental scientific drilling purpose, planning, management, site selection and achievements, compiled a *Comprehensive Report on German Continental Scientific Drilling*, and put forward the proposals to carry out continental scientific drilling in China and to list the pre-study on scientific drilling in the eighth Five-Year Plan (1991–1995) of the Ministry.

1.4.2 Project Argumentation and Demonstration Stage (1991–September 1999)

During this period, the investigations, visits, conferences, lectures and other international exchange activities were widely carried out, with more definite purpose and notable results obtained.

From October 21st to November 3rd, 1991, four scientists and engineers from the Ministry of Geology and Mineral Resources visited Japan. They visited the Japan Science and Technology Agency, the Resources Development Department and the Earthquake Research Institute of the University of Tokyo, the Department of Earth Science of Shizuoka University, the Geological Survey of Industrial Technology Institute of MITI, the Comprehensive Research Institute of Resources and Environmental Technology and the Research Institute of Natural Calamity Prevention Techniques, met with forty experts and scholars, and understood the purpose, significance, hole site selection of Japanese scientific deep drilling activities, the organization of Japanese Scientific Drilling (JSD) and Japanese Super Deep Core Drilling Research Association (SDD) and their main research projects and the preliminary preparation for scientific deep drilling.

From April 7th to 10th, 1992, experts of the Ministry of Geology and Mineral Resources attended the 3rd International Symposium on Observation of the Continental Crust through Drilling held in Paris, France, where they introduced the preliminary preparation and study of China's continental scientific drilling.

From April 15th to 17th, 1992, the Ministry of Geology and Mineral Resources held in Beijing the first Seminar on China Continental Scientific Drilling (CCSD), which was a meeting linking the past and the future. More than sixty experts and professors from the scientific research institutes, colleges and universities, government agencies and industrial departments attended the seminar. More than twenty papers were exchanged in the seminar, with the contents on world scientific drilling progress and achievements, and the significance, necessity and feasibility, preliminary site selection, implementation of technical policies and procedures of CCSD. Participants agreed that the CCSD Project was of great significance and imperativeness, and scientific drilling was a system project with science and technology combined, which must be carried out step by step, from easy to difficult and from shallow to deep. On the meeting, the experts proposed thirty CCSD candidate locations. After the seminar, more than thirty drilling experts held a two-day seminar, at which drilling theory, technology, equipment and other aspects of scientific drilling were discussed, and the papers presented at the seminar were compiled into a book.

From June 18th to July 2nd, 1992, at the invitation of China Academy of Geological Exploration Technology, Doctor B.N. Khakakhaev, general manager of the Russia Science and Production Consortium of Ultra Deep Hole Drilling and Comprehensive Survey of the Interior Earth (NEDRA) and Doctor M.J. Vorozhibitov, director of the Ultra Deep Drilling Laboratory of the Geological Information System Institute visited China and gave lectures. They comprehensively introduced the drilling technology and the achievements of the scientific drilling of the former Soviet Union. Meanwhile, the two sides held talks on the cooperation and exchanges issues in the field of scientific drilling.

From June 7th to 24th, 1993, at the invitation of Mr. B.N. Khakhaev, a delegation of the Ministry of Geology and Mineral Resources visited Russia and took a comprehensive study on Russian scientific drilling activities. The delegation visited the NEDRA Superdeep Drilling and Comprehensive Investigation Institute, Russian Oil Drilling Technology Institute Perm Branch, Kungur Petroleum Machinery Factory, Ural scientific ultra deep drilling site, Russian National Geo-information Technology Research Institute, Cola ultra deep drilling site, NEDRA headquarters, Institute of High Pressure Physics, and mainly investigated the evolution and adjustment of the scientific drilling programs in Russia, site selection principles and procedures, ultra deep drilling construction technology, laboratory simulation technology and geo-information obtaining technology.

On September 2nd, 1993, the First International Continental Scientific Drilling Management Conference was held at KTB scientific ultra deep hole site in Windischeschenbach, Germany, and participants from fifteen countries including China attended the conference, at which was set up a preparatory group, which was authorized to draft the framework views on the International Continental Scientific Drilling Program (ICDP) and the proposal on operation and funding issues. Based upon this proposal, a memorandum of understanding (MOU) was prepared as the foundation for countries to accede to ICDP.

From February 26th to March 1st, 1996, the Eighth International Symposium on Observation of the Continental Crust through Drilling was held in Tsukuba Scientific Town, Tokyo. During the Symposium, the International Continental Scientific Drilling Program (ICDP) was formally established. China, Germany, and the United States of America, as the sponsor nations, signed a memorandum of understanding (MOU) on ICDP and attended a large seminar on the organization, management and future international cooperation of the program. In the same year, China formally submitted to the ICDP the project implementation proposal to construct China's first continental scientific well in Dabie-Sulu area. In ICDP SAG conference held in July 1996, China's proposal was ranked as the second among the sixteen proposals submitted. The ICDP Executive Committee decided to subsidize China to hold an international symposium on site selection for CCSD project.

On August 12th, 1996, at the invitation of Exploration Engineering Professional Commission of the Geological Society of China and China Academy of Geological Exploration Technology, a delegation from German Continental Scientific Deep Drilling Program, led by Prof. R. Emmermann, the director of German Research Center for Geosciences (GFZ, GeoForschungs Zentrum), held a seminar on German KTB drilling technology at the academic exchange center of China University of Geosciences (Beijing), where they introduced KTB results in earth sciences and drilling, logging and testing technologies to more than ninety experts and scholars from various departments of China.

In August 1996, the Ministry of Geology and Mineral Resources and GFZ signed a comprehensive development agreement in Beijing.

In January 1997, a research project of 5,000 m drilling engineering was started in order to lay down a scheme for further CCSD drilling technology, including drilling equipment, apparatus, techniques, construction procedures, and evaluation of the economic feasibility of the drilling construction. This scheme was of great significance to guide the construction of the first continental scientific well.

On June 4th, 1997, the State Science and Technology Leading Group examined and discussed the major scientific engineering projects recommended and reported by the State Planning Commission according to the evaluation results from experts, and agreed in principle that the project of China Continental Scientific Drilling and other three projects listed as the second batch of the national major science projects in the ninth Five-Year Plan. Therefore, the implementation of CCSD was started.

From August 18th to 20th, 1997, the International Seminar on China Continental Scientific Drilling Engineering in Dabie-Sulu UHP Metamorphic Belt was held in Qingdao City. The seminar was funded by ICDP, and co-sponsored by the Chinese Academy of Geological Sciences and GFZ, and more than seventy experts from Germany, the United States, France, Canada, Japan and China attended the seminar. Focusing on the scientific significance and goals to carry out the scientific drilling in Dabie-Sulu UHP metamorphic belt, the participants made a full exchange and discussion on the achievements of geological and geophysical comprehensive research in three drilling candidate areas in Qianshan, Anhui Province; Zhucheng, Shandong Province and Donghai, Jiangsu Province. Finally, it was decided that Donghai should be the first choice to implement the first scientific well in China.

In early April, 1998, the Science Advisory Group of ICDP examined and approved the formal proposal concerning the continental scientific drilling project in Dabie-Sulu UHP metamorphic belt put forward by Chinese geologists jointly with other nine experts from Germany, Canada, the United States and France. Formally approved by the Executive Committee and the Executive Council of ICDP, 1.5 millions USD would be aided to the project within 5 years from 1999.

In June 1999, the CCSD Engineering Center and the Operation Support Group (OSG) of ICDP made an exchange and discussion on the design scheme of drilling, logging, testing and analysis of CCSD project and the matters concerning cooperation, and reached extensive intentions and signed an agreement.

In August 1999, Mr. Jiang Chengsong, Vice Minister of the Ministry of Land and Resources led a delegation to visit GFZ and KTB drill site. The two sides had an extensive exchange of views on further cooperation in the field of scientific drilling.

1.4.3 Project Preparation Stage (September 1999–June 2001)

On September 27th, 1999, State Planning Commission approved the proposal on China continental scientific drilling project. To strengthen the organization and leadership of CCSD project, the Ministry of Land and Resources set up a leading group of China continental scientific drilling project and a project legal person—the CCSD Engineering Center, and established a science and technology advisory committee consisted of thirty experts and academicians. This marked that the CCSD project was officially started.

In order to understand in detail the geological structure, stratum lithology and occurrence, geothermal gradient and information required by comprehensive geophysical logging in the drilling location and its surrounding area as well as to make drilling technology tests and accumulate drilling experience, the Engineering Center successively constructed three pre-pilot holes and one drilling technology test hole near CCSD-1 Well site.

For well site selection, the Engineering Center organized all sides to carry out a large number of geological and geophysical researches and found that Maobei area of Donghai County accorded with the principle of well site selection. Strong reflecting bodies of high density, high electric resistance and high wave velocity exist at 3–4 km under the ground in this area, and locate in the deepest part of the UHP metamorphic belt, where 5 km deep drilling could penetrate through the four microlithons. At the same time, here is the part with the most moderate dip angle in the UHP metamorphic belt, where a short distance of drilling could penetrate through the multiple units. The temperature measurement showed that the geothermal gradient here was not high (about 2.5 °C/100 m), which was favourable to the implementation of drilling construction.

In Donghai area of the south Sulu UHP metamorphic belt, thorough surface geological and geophysical surveys were conducted, 1:5,000 and 1:10,000 geological mapping of the drilling target area were completed (1998–1999) and the geological structure frame of the selected well site area was ascertained. 160 km seismic reflection profile across the orogen was completed with the orogenic belt lithosphere structure profile established. Some shallow borehole cores had been collected during the 1970s and 1980s were rearranged, listed, analysed and studied, from which the valuable data of shallow underground geological structure in the drilling area were obtained. All these provided an important scientific basis for the final determination of the well site location. Multi-disciplinary researches were made in Donghai area, including researches on structural geology, petrology, mineralogy, geochemistry, isotope chronology, petrophysics and biology, and great achievements were obtained. On December 24th, 1997, the Engineering Center held the Symposium of CCSD Well Selection and forty domestic geological experts and scholars participated the symposium. Through discussion, the participants agreed that Maobei could be selected as the first target area because its conditions were better than the others.

In February 2000, the Ministry of Land and Resources submitted the Official Letter on the Feasibility Study Report of the China Continental Scientific Drilling Project to the State Planning Commission. From March 18th to 19th and from April 18th to 19th, 2000, the China International Engineering Consulting Company, authorized by the State Planning Commission, respectively evaluated the engineering part and the overall part of the Feasibility Study Report of the China Continental Scientific Drilling Project. On July 31st, 2000, the State Planning Commission assigned the Official Reply to the Feasibility Study Report of the China Continental Scientific Drilling Project, in which the report was officially approved.

In November 2000, the Preliminary Engineering Design of CCSD Project was completed. After the examination by the well-known experts and scholars, on January 3rd, 2001, the Ministry of Land and Resources submitted the Official Letter on the CCSD Project Engineering Design to the State Planning Commission, and on February 6th, 2001, submitted the Official Application Letter on Starting the CCSD Project to the State Planning Commission. Based upon the feasibility study report and the opinions from the Examination Commission of the Project Engineering Design, the State Planning Commission assigned the Official Reply to the Preliminary Engineering Design and Starting of the CCSD Project on August 2nd, 2001.

In November 2000, the Engineering Center issued the public bidding documents for drilling sub-project and five drilling companies including the No. 3 Drilling Company of Zhongyuan Petroleum Exploration Bureau, Chuandong Drilling Company of Sichuan Petroleum Administration Bureau, the No. 4. Drilling Company of Zhongyuan Petroleum Exploration Bureau, the No. 1. Drilling Engineering Company of Huabei Petroleum Administration Bureau, and Bohai Drilling Company of Shengli Petroleum Administration Bureau submitted a tender respectively. In December 2000, the Engineering Center organized an investigation group to comprehensively investigate the engineering equipment, personnel quality, construction experiences, qualifications and achievements, management and construction quotations of the five bidders. On February 27th, 2001, the Engineering Center issued the bid winning notice to the No. 3. Drilling Company of Zhongyuan Petroleum Exploration Bureau of SINOPEC. On March 6th, a signing ceremony of the drilling sub-project contract was held in the Ministry of Land and Resources.

In March 2001, the Engineering Center issued the bid winning notice for logging sub-project, and officially signed the construction contract with the Logging Company of Shengli Petroleum Administration Bureau, the winning bidder.

1.4.4 Project Implementation Stage (June 2001–April 2005)

On June 25th, 2001, the CCSD pilot hole was opened.

On August 4th, 2001, an opening ceremony was formally held at Maobei construction site, Donghai County of Jiangsu Province (Fig. 1.1).

On April 16th, 2002, core drilling in the pilot hole was completed at the depth of 2046.54 m.

On May 7th, 2002, the main hole construction was started, which experienced three stages of core drilling, reaming, sidetracking (deviation correction) drilling, sidetracking (avoidance of underground obstacle) drilling and the stages

1.4 Developing History of CCSD Engineering

Fig. 1.1 Opening ceremony of CCSD-1 Well

Fig. 1.2 Completion ceremony of CCSD-1 Well

of casing running and cementation, with 994 days lasted. On January 23rd, 2005, all the tasks of coring were successfully completed at the depth of 5118.2 m. After that, new drilling tools tests, liner running, cementing, drifting, logging, VSP (borehole vertical seismic profile measurement) and well completion were conducted. On April 18th, 2005, a grand completion ceremony (Fig. 1.2) was held, thus marking the completion of CCSD-1 Well (5,158 m), which totally lasted 1,395 days.

From March 30th to April 1st, 2005, the International Seminar on 10 Years Continental Scientific Drilling, Review and Prospect was held by ICDP at German Research Center for Geosciences (GFZ, Deutsches GeoForschungs Zentrum) in Potsdam, Germany and more than two hundred experts and scholars in various fields in the world attended the seminar (Fig. 1.3).

The participants widely discussed in eight fields covering climate change and global environment, meteorite impact structure, earth's biosphere and early life, volcanic system and thermal mechanism, mantle plume and rift valley, active tectonics, collision zone and convergent plate boundary and natural resources, including achievement exchange, introduction of project implementation, and plans for future research. A delegation of the CCSD project attended the seminar and showed the new development and achievements of CCSD from drilling technologies to scientific research. The successful drilling technologies and scientific significance of CCSD with a series of new achievements received high opinions from the participants.

On December 14th, 2007, the China Continental Scientific Drilling Project was checked and accepted by the State Development and Reform Commission in Beijing.

1.5 Technical Preparation

The construction of the CCSD was difficult with strict demands, requiring advanced drilling technologies and thus facing hitherto unknown challenges. To ensure the success of the construction, scientists and drilling engineers had made a large number of engineering technical preparations. To obtain foreign scientific drilling experiences in construction and management, many experts and scholars were sent abroad to study and get trained for many times, and meanwhile, some foreign experts and scholars were invited to give lectures in China. To investigate the underground geological conditions, test the drilling technology to be used in the construction and provide verification samples for

Fig. 1.3 International seminar on 10 years continental scientific drilling, review and prospect

geophysical data interpretation, a pre-pilot hole was constructed. Before and in the construction, many research projects were set, and lots of pre-studies were carried out on the construction scheme and special technologies.

1.5.1 Technical Training

In order to draw on the foreign advanced experiences and technologies in scientific drilling, in 1996, the former Ministry of Geology and Mineral Resources and the German Research Center for Geosciences (GFZ, Deutsches GeoForschungs Zentrum) signed the Memorandum of Understanding on Implementation, Administration and Operation of the International Continental Scientific Drilling Program (ICDP), in which it was specifically put forward that part of the membership dues paid by China would be used for supporting a Chinese personnel training program for a period of five years, by which the Chinese scientists and engineers would receive training at the drilling sites and accumulate experiences for the implementation of China's scientific drilling project in the future.

Since joining the international continental scientific drilling organization, China sent technical personnel to participate the continental scientific drilling training classes held by ICDP many times. In 1997, seven engineers were sent to GFZ, the headquarters of ICDP, to receive a four-month-training, mainly on drilling, logging, rock physical properties, geology, geophysics, geochemistry and information management, through which the trainees had a complete and deep understanding of scientific drilling. At the end of the course, seven trainees made a simulated design of China continental scientific drilling project, and completed a design of scientific drilling in Sulu UHP metamorphic belt.

Then, in 1998, 2000 and 2001, four to seven engineers were sent each year to GFZ to participate the ICDP training classes.

In August 1998, Ni Jialu, Yang Gansheng, Niu Yixiong and Zhang Zeming attended a continental scientific drilling engineering training held by ICDP in Hawaii, and visited the Hawaii scientific drilling site, at which they carefully investigated the purpose of scientific drilling, drilling construction methods, logging projects and methods, wellsite geologic log methods, physical parameters measurement methods and construction management, and obtained a large number of experiences concerning drilling construction management and drill site administration.

Through abovementioned technical trainings, the purpose and significance of scientific drilling were deeply understood, and the procedures and methods for conducting scientific drilling, the related technical means and technical methods, and advanced technologies and experience were obtained.

In June 1999, four experts from the Operation Support Group (OSG) of ICDP visited Beijing and carried out extensive exchanges and discussions on the design of drilling, logging, testing and analysis of the CCSD and the cooperation between the two sides.

From March to May, 2001, drilling technology training materials (two volumes) were compiled and published by the CCSD Engineering Center.

From May 23rd to 24th, 2001, the first drilling technology training class was held in the No. 3 Drilling Company of Zhongyuan Petroleum Exploration Bureau in Lankao County, Henan Province. Party A sent five experts to train the workers and technical personnel from the No. 3 Drilling Company, which was to contract for the drilling construction. More than seventy people including some leaders and drilling technicians from the Zhongyuan Petroleum Exploration Bureau, the No. 3 Drilling Company and all personnel of the 70101 Drilling Brigade attended the training class, through which the trainees got a comprehensive understanding of continental scientific drilling and the related technologies, organization and management.

From June 27th to July 3rd, 2001, the ICDP/CCSD technology training class was held. At the CCSD construction site, six experts from ICDP introduced to the engineering and technical personnel and geoscientific research personnel participating the CCSD project the field scientific planning, on-site laboratory organization, on-site geology, data management and core scanning, scientific drilling basis, KTB experience and in-the-hole test, mud and mud system, hydraulic test and fluid sampling, borehole stability in metamorphic rocks, logging basis in crystalline rocks, new development of logging techniques, and KTB/ICDP logging. Three experts from CCSD also introduced the related technologies.

1.5.2 Pre-pilot Hole Construction

To further understand the underground geological conditions of CCSD drill site and to test the drilling technologies to be used in the CCSD project, the Engineering Center entrusted Jiangsu No. 6 Geological Brigade to successively construct the Pre-pilot hole I (CCSD-PP1) in Zhimafang and the Pre-pilot hole II (CCSD-PP2) in Maobei, which mainly aimed at surveying the underground geological conditions. CCSD-PP1 was completed in November, 1997, with the hole depth of 430 m and the final hole diameter of 75 mm. Because no drilling technology test was conducted in CCSD-PP1, detailed introduction is unnecessary. Besides further survey the underground geological conditions, in CCSD-PP2 preliminary test was conducted concerning the drilling technologies to be used for the pilot hole and the main hole.

CCSD-PP2 was located in Maobei, Donghai County of Jiangsu Province, 382 m from CCSD-1 Well. The designed vertical depth was 1,000 m, with the final diameter of 75.5 mm. The engineering coordinate were: X = 3809.435 km, Y = 40378.244 km.

The drilling purposes of the CCSD-PP2 mainly included:
1. To measure the VSP (vertical seismic profile), and understand the relevant underground geological information.
2. To understand the lithology and occurrence of underground strata.
3. To make geothermal gradient and comprehensive geophysical well logging, and get strata information.
4. To test drilling technologies, methods and tools.

Strata to be drilled are mainly gneiss, interspersed with small amounts of eclogite, mixed monzonite and tectonic breccia with the following characteristics:
1. Hard to drill, with drillability grade 7–9, sometimes even to grade 10 (quartz vein).
2. Schistosity development with significant anisotropy, belonging to the strong dipping stratum; with tectonic activities such as fault.
3. With good strata integrity, and the core obtained was basically complete. Tripping was smooth, with sticking accident rarely happened.
4. Obvious fissures exist in some hole sections, resulting in complete drilling fluid loss. Slight leakage happened in most of the hole sections.

Diamond wireline core drilling method was employed for CCSD-PP2, with the main equipment and tools as follows:
1. Drill: vertical spindle hydraulic drill Type XY-6S, with 73.5 kW power
2. Derrick: Type K40 derrick, with 22 m height
3. Drill rod: S75 wireline drill rod
4. Pump: WX-200 and BW320 mud pumps were successively adopted
5. Power: diesel engine and electric generator driving methods were successively adopted
6. Drill bit: impregnated diamond core bit was mainly used for core drilling.

CCSD-PP2 hole was started on Dec. 8th, 1998, and was finished on Jun. 6th, 1999, with a total of 180 days lasted. The final hole depth reached to 1028.68 m, with the vertical depth of 1002.71 m and the final hole diameter of 75.5 mm. The total core length of the full borehole was 954.45 m, with the core recovery of 92.77 %. The hole structure and casing program had been changed during the construction, with the final hole structure shown in Fig. 1.4. In general, because the strata were relatively stable, after a small amount of casing running to isolate the overburden and the weathered layer in the upper hole section, open hole drilling was mainly conducted, and the problems of hole wall stability basically did not happen during the construction.

Two core drilling methods, conventional core drilling and wireline core drilling were employed in CCSD-PP2 hole. 75.5 mm conventional double tube diamond core drilling was adopted above 101.45 m deep, while wireline core drilling was adopted below 101.45 m. Because of the serious problems of hole deviation and drill rod broken, low drilling data were adopted when 75.5 mm diamond wireline core drilling was used, which influenced the technological and

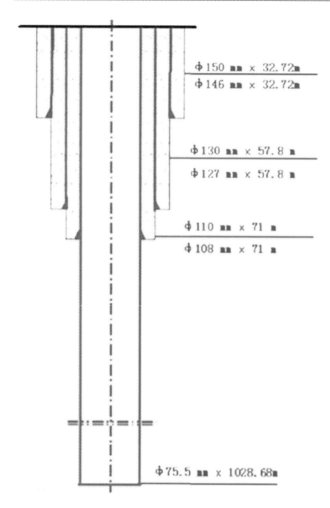

Fig. 1.4 CCSD-PP2 borehole structure

economic indexes more or less. The drilling data mainly adopted were as follows:

Bit pressure	800–1,200 kg
Rotary speed	300 r/min was mainly employed in the shallow hole section, 400 r/min occasionally employed; 200 r/min mainly employed in the deep hole section
Pumping rate	100 L/min

During drilling, the mud material LBM which was to be used in CCSD-1 Well was tried out for four times, however, without any success. The main reason was that the LBM mud could not be fully mixed due to the limited conditions and then flocculated soon once entered into circulation. Therefore, in the process of drilling, clear water added with a small amount of polyacrylamide and 126 lubricant were mainly adopted as the drilling fluid to improve its cuttings carrying capacity and lubrication effect.

Borehole deviation survey: Magnetic ball inclinometer was used to measure vertex angle and azimuth, and then hydrofluoric acid bottle inclinometer was employed to measure vertex angle after the failure of magnetic ball inclinometer.

Borehole deviation correction method: PDM/bending sub deviation correction system was utilized.

Thirty seven pieces of S75 wireline diamond core bits were used, with a total footage of 992.03 m drilled. The average bit service life was 26.81 m, while the longest was 101.4 m. The average ROP was 1.26 m/h, with the highest of 8.40 m/h. The average footage drilled per roundtrip was 1.91 m.

From June to July, 1999, hydro-hammer drilling was tested near CCSD-PP2 hole. The diameters of the hole sections penetrated with hydro-hammer were 158 mm (drilled with 158 mm non-core button impact drill bit) and 152.4 mm (drilled with 152.4 mm Type 6H637 tricone bit). The main purposes of the tests were to evaluate whether a fast drilling could be realized by using hydro-hammer in hard rocks, to understand the matching relationship between different types of drill bit and hydro-hammer, and to find out the problems of the available hydro-hammer in structure and in properties. The test results can be found in Tables 1.1 and 1.2, from which the following conclusions can be obtained.

1. Hydro-hammer drilling is an effective method for large diameter non-core drilling in hard rocks.
2. A combination of cone bit and hydro-hammer can be effectively used for large diameter non-core drilling in hard rocks.

Through drilling construction of CCSD-PP2 hole, the accurate data about the physical and mechanical properties, drillability, integrality, leakage degree, deviation degree and geothermal gradient (Fig. 1.5) of the underground strata were obtained, and temperature curves for CCSD-1 Well (Fig. 1.6) was predicted. All these provided a basis for the technical design of CCSD-1 Well.

1.5.3 Pre-research on Key Technologies

In order to make technological preparations for the implementation of the CCSD project, the Pre-research on CCSD, the research on the 5,000 m drilling engineering technical proposal and the feasibility study on the CCSD project were respectively carried out. After the official approval of the projects, the research and development of a number of key technologies were conducted in form of pre-research projects before and during the CCSD-1 Well construction.

From July 1991 to April 1994, the Chinese Academy of Geological Sciences completed the project of Pre-study on China Continental Scientific Drilling, in which a sub-project entitled the Feasibility Study on Drilling Construction

1.5 Technical Preparation

Table 1.1 Statistics of drilling rate in hydro-hammer drilling test

Hole depth (m)	Footage drilled (m)	Actual drilling time (h)	ROP (m/h)	Drilling method	Remarks
15.40–51.89	36.49	7.29	5.00	Hydro-hammer + non-core button bit	Intact and soft stratum
51.89–54.46	2.59	3.13	0.83	Hydro-hammer + non-core button bit	Hard stratum, alloy teeth seriously worn
54.46–73.59	19.47	6.53	2.98	Hydro-hammer + cone bit	Bit normal wear
73.59–75.73	1.78	1.17	1.52	Cone bit	Bit normal wear
75.73–101.14	29.05	9.97	2.91	Hydro-hammer + cone bit	Bit normal wear
Total	83.96	25.72	3.26		

Table 1.2 Comparison of drilling test results by using two types of hydro-hammer

		Footage drilled (m)		ROP (m/h)	
		KSC127	YZX127	KSC127	YZX127
Soft rocks	Button non-core bit	16.98	19.53	4.16	6.07
Hard rocks	Button non-core bit	1.16	1.41	0.95	0.74
	Cone bit	16.37	28.51	2.72	3.07
Total (based on hydro-hammer)		34.51	49.45	3.05	3.43
Total (full hole)		83.96		3.26	

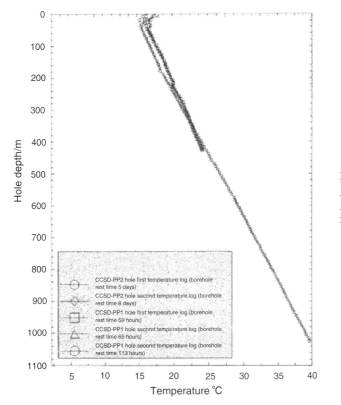

Fig. 1.5 The measured temperature curves of CCSD-PP1 and CCSD-PP2 hole

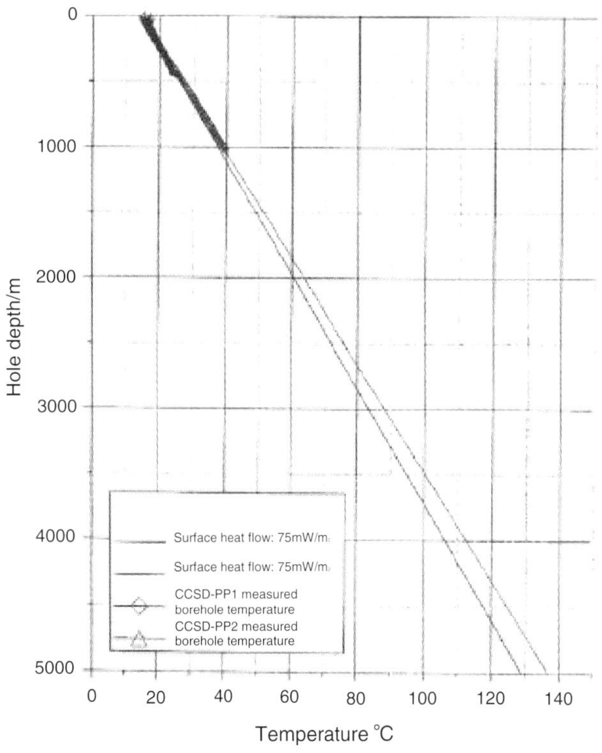

Fig. 1.6 The predicted temperature curves for CCSD-1 Well

Table 1.3 Pre-research for CCSD drilling sub-project

Category	Project	Undertaker	Checked and accepted by
Drill bit	Development of 152 mm diamond wireline core drill bit (electro-plating)	Institute of Exploration Technology	China Geological Survey
	Development of 152 mm diamond wireline core drill bit and reaming shell (heat pressing)	Beijing Institute of Exploration Engineering	China Geological Survey
	Research and development of type KZ special reaming bit for CCSD project	Institute of Exploration Techniques	CCSD Engineering Center
Hydro-hammer	Research on SYZX273 hydro-hammer	Institute of Exploration Techniques	China Geological Survey
	Development of fluidic hydro-hammer for scientific deep drilling	Jilin University	CCSD Engineering Center
Coring tools	Research on two-in-one (PDM, wireline coring) drilling tool	Institute of Exploration Techniques	CCSD Engineering Center
	Research on combined down-hole turbine motor diamond wireline core drilling tool	Institute of Exploration Techniques	CCSD Engineering Center
	Development of KS152 wireline coring hydro-percussive mechanism assembly	Institute of Exploration Techniques	CCSD Engineering Center
	Three-in-one (hydro-hammer, PDM and wireline coring) tool assembly	Institute of Exploration Techniques	China Geological Survey
	Development of KZ swivel type double tube coring tool	Institute of Exploration Techniques	CCSD Engineering Center
	Research on core orientation technology	Chengdu University of Technology	China Geological Survey
	Research on sidewall offset coring technology in CCSD project	China University of Geosciences (Beijing)	CCSD Engineering Center
Deviation prevention and correction	Research on borehole deviation prevention technology	China University of Petroleum (Beijing)	CCSD Engineering Center
	Research on PDM drive continuous deflector and its directional drilling technology	CCSD Site Headquarters, Institute of Exploration Technology	China Geological Survey
Data monitoring and processing	Development of down-hole parameters auto recording and playback device	China University of Geosciences (Wuhan)	CCSD Engineering Center
	Redevelopment of the imported drilling parameter instrument and research on site drilling database	China University of Geosciences (Wuhan)	CCSD Engineering Center
Drilling fluid	Research on the drilling fluid system for scientific drilling	Beijing Institute of Exploration Engineering	China Geological Survey
	Research on the drilling mud system for scientific drilling	China University of Petroleum (Beijing)	CCSD Engineering Center
	Research on new type formate solid-free drilling fluid	Chengdu University of Technology	CCSD Engineering Center
Comprehensive study	Research and application of new core drilling technology for deep hole in hard rocks	Beijing Institute of Exploration Engineering	CCSD Engineering Center
	Research on scientific deep hole drilling technology	Institute of Exploration Technology	CCSD Engineering Center
	Construction of experiment stand for CCSD pre-research project and test of the prototype	CCSD Site Headquarters	China Geological Survey
Others	Research on movable casing fixing and its lifting-up and running-down technology	Beijing Institute of Exploration Engineering	CCSD Engineering Center
	Research on rope arranging device for coring drawworks and its detection system	China University of Geosciences (Wuhan)	CCSD Engineering Center

Fig. 1.7 Proportion of research and development funds in pre-research of drilling sub-project

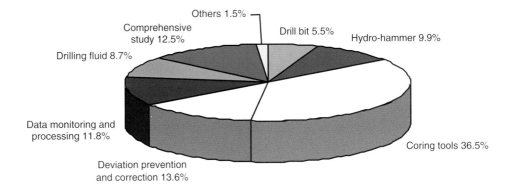

was accomplished and submitted by Zhang Wei, Geng Ruilun and Feng Qinglong of the China Academy of Geological Exploration Technology, with the main contents as follows:

1. To systematically investigate the present situation, level and developing trend of continental scientific deep drilling in the major industrial countries;
2. To put forward a variety of site selection plans for China's continental scientific deep drilling, aiming at the important basic geological problems to be resolved in the future deep geology research in our country;
3. To evaluate the present situation and level of China's deep drilling and logging techniques;
4. To expound and prove the technical (including drilling and logging) and economical feasibility of implementing continental scientific deep drilling in China, and put forward the scientific research and technological development plan;
5. To integrate the abovementioned research results, put forward the planning and related suggestions for China's continental scientific deep drilling.

During the ninth Five-Year Plan period, the research on 5,000 m drilling engineering technical proposal was officially started to extensively investigate and study the technologies needed for China's continental scientific drilling project, which included nine subjects:

1. Hole structure, drilling program and drilling technology system for 5,000 m scientific drilling construction;
2. Drilling equipment needed for 5,000 m scientific drilling;
3. Deep hole core drill string and accident handling tools, and in-the-hole core drilling system;
4. Core bit, reaming bit and reaming shell for deep hole hard rock drilling;
5. Drilling fluid and cementing slurry system for 5,000 m scientific drilling;
6. Borehole deviation prevention and correction (deflection) system for deep hole hard rock drilling;
7. Data acquisition and database system for drilling parameters in a 5,000 m scientific drilling project;
8. Core orientation method and tools for deep hole hard rock drilling;
9. Research on the information of international continental scientific drilling technology.

In 1999, the CCSD Engineering Center organized the experts in the fields of geophysical prospecting, well logging, experimental geology, drilling, drilling fluid and computer to deeply research into the feasibility of the CCSD project, and completed the Feasibility Study Report of CCSD Project.

Twenty four research and development projects concerning eight categories were carried out to solve the drilling technical problems which might be encountered during the construction of CCSD-1 Well (Table 1.3 and Fig. 1.7).

Drilling Engineering Design

As a huge systems engineering project, China Continental Scientific Drilling (CCSD) engineering project consisted of five main sub-projects and one subsidiary sub-project. The five main sub-projects included drilling sub-project, borehole geology and analysis and test sub-project, borehole logging sub-project, geophysics sub-project and information sub-project; while the subsidiary sub-project denoted civil engineering. Among these sub-projects, drilling sub-project was the key, which was the precondition of conducting other sub-projects. Only by obtaining core, rock samples, gas and fluid samples through drilling project could analysis and test sub-project be started and only after borehole completed that a passageway could be available for logging and for geophysical tests, so as to obtain the underground material information. Besides, drilling sub-project was the one which cost the largest investment and the longest time, and with extreme difficulty. Therefore the successful completion of drilling sub-project determined the success of the whole scientific drilling engineering project.

Engineering design is the standard for executing the project. The design of drilling sub-project was completed by the design company who won the bid, on the bases of collection of a vast amount of scientific data and research.

2.1 Assignment of Drilling

The overall assignment for constructing the drilling sub-project of the first borehole of China Continental Scientific Drilling engineering project (CCSD-1 Well) was to drill a borehole of 5000 m deep in the ultra high pressure metamorphic crystalline rock formations in Donghai County, Jiangsu Province. Continuous coring was to be conducted for the whole borehole, samples were to be taken and in situ logging carried out.

Translated by Geng Junfeng.

Technical requirements:
1. Hole depth: 2000 m for the pilot hole and 5000 m for the main hole
2. Final hole diameter: 156 mm
3. Hole deviation: no larger than 14° from 0 to 2000 m and no larger than 18° from 2000 to 5000 m
4. Coring: Continuous coring was to be conducted for the whole borehole, of which
   Core recovery should be no less than 80 % for the whole borehole.
   Additional core should be taken for complement in case of no core recovered in a long hole section.
   Orientational coring should be conducted for one roundtrip in every approximate 100 m.
5. Assist to conduct logging, formation fluid sampling and a variety of in-hole tests
6. Well completion: Casing cementing from 0 to 4800 m and open hole completion from 4800 to 5000 m
7. Borehole geographical coordinates:
   The pilot hole: X = 3809.530 km, Y = 40 377.874 km
   The main hole: X = 3809.530 km, Y = 40,377.980 km

2.2 Basic Situation of the Well Site

Donghai area borders on the Yellow Sea in the east, and has a semi-humid climate of North China temperature zone, with arid winter, drought spring and autumn, and liable to waterlogging in summer. The yearly average temperature is 13.7 °C, with the maximum temperature of 39.7 °C and the minimum of −18.3 °C. July is the hottest month, with the average temperature of 26.5 °C while the coldest month is January, with the average temperature of −0.6 °C. The annual precipitation is 884 mm, most probably concentrates in July, August and September, accounting for 60 % of the annual precipitation, and the annual evaporation capacity is larger than the annual precipitation. The maximum daily precipitation recorded is 204.5 mm. The annual average thunder and lightning day accounts for 20–30 days most happen from March to October. Freezing season starts from January to March, with the maximum frozen soil depth of 15 mm. In spring and summer often blows east wind and in winter often north wind, with annual average wind speed of 3.2 m/s. The days in which with over force 8 wind (on the Beaufort scale) per year amount to 24.2 days, with the maximum wind speed of 34 m/s (June 18th, 1996).

Donghai area is located in a plain, with smooth terrain. Above the bed rock is covered with loess layer of more than 3 m thick.

2.2.1 Forecast of Lithological Profile of the Formation Encountered

On the basis of massive surface geological survey, trench prospecting, shallow borehole drilling and deep geophysical prospecting, a three dimensional geologic and geophysical model of the drilling area was initially established and the forecast tectonic column of the lithologic units and rocks which would be penetrated through by a 5000 m deep borehole can be found in Fig. 2.1, in which from top to bottom the whole borehole can be divided up into five major units (layers).

Unit A: From 0 to 650 m is mainly composed of ultra high pressure eclogite, interbedded with a little thinly laminated biotite-plagioclase gneiss and schist. Many layers of garnet-peridotite may exist at the middle and the lower parts.

Unit B: From 650 to 1930 m mainly consists of different types of gneiss, including, biotite-plagioclase gneiss, epidote-biotite-plagioclase gneiss and granite-gneiss; interbedded with thinly laminated schist, amphibolite and eclogite.

Unit C: From 1930 to 3210 m is mainly composed of aegirine bearing biotite-plagioclase gneiss, with much eclogite and amphibolite or lenticular body.

Unit D: From 3210 to 4550 m mainly consists of eclogite + garnet-peridotite, being the drilling target layer of high wave velocity and high density. All the rocks were formed under the ultra high pressure metamorphic condition

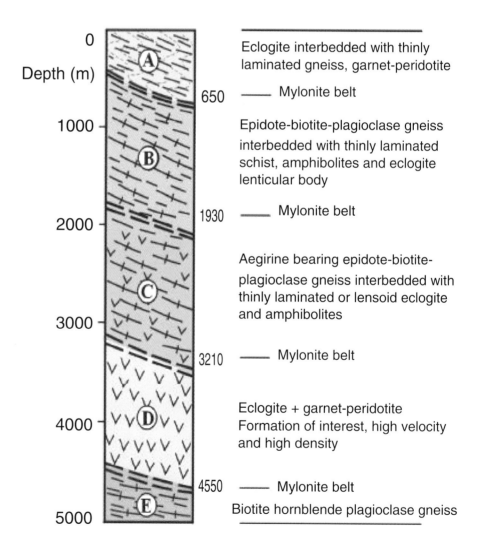

Fig. 2.1 The forecast tectonic column of the lithologic units and rocks of CCSD borehole

of earth mantle, and exhumated to basic and ultrabasic rock layers in shallow earth crust (or large lenticular body).

Unit E: From 4550 to 5000 m is mainly composed of biotite hornblende plagioclase gneiss, probably with little eclogite lenticular body. In comparison with above layer both the wave velocity and the density decrease.

Abovementioned five rock-tectonic units are all separated by tough shear zones with uneven thickness ranging from tens of meters to less than 100 m. In the tough shear zones deformation is intense and foliation develops, the rocks are harder than the upper and lower neighbouring rocks. Due to the stacking of the brittle deformation, tough shear zone may transform into brittle fault zone.

2.3 Lithologic Characteristic of the Rock Formations to be Encountered by Drilling

1. Gneiss is mainly composed of feldspar and quartz, generally with the content of more than 80 %. It may contain little biotite, fasciculate, epidote and muscovite, etc. The rock is of flake granoblastic texture, with gneissose structure which can be divided into orthogneiss and paragneiss, the initial rock of the former is granite while that of the latter is sedimentary rock. Gneiss will be the main lithology at the depth of more than 1000 m both at the main hole and the pilot hole.
2. Eclogite is mainly composed of garnet and acmite, generally with the content of more than 80 %. It may contain little secondary mineral quartz, phengite, cyanite, epidote, clinozoisite, fasciculate and cajuelite, etc. Generally the foliation is not developed, with a block structure. However, a little eclogite experienced intense plastic deformation. Sheet mineral and columnar mineral such as acmite and phengite distribute orientationally and form structural foliation and lineation. According to the content of minor minerals it can be further divided into phengite eclogite, cyanite eclogite, quartz eclogite, ordinary eclogite (very little minor mineral) and cajuelite eclogite. In Maobei area the content of cajuelite in cajuelite eclogite accounts for more than 5 %, being the mother rock of cajuelite mineral. As basically without containing light coloured minerals, this rock has dark colour and high hardness and will be the main rock type of the pilot hole at the depth of less than 1000 m.
3. Peridotite (serpentinized peridotite) is mainly composed of peridotite which contains different amount of garnet, orthopyroxene, clinoaugite and brown mica. It is mostly of grain texture and block structure. This type of rock belongs to ultrabasic rock, with high density and high hardness. Because of the alteration action at later period, peridotite, orthopyroxene and clinoaugite will be transformed into ophite or amesite while garnet will be transformed into amesite and metallic minerals, then the hardness and specific gravity of the rock reduce obviously, sometimes with obvious foliation developed. By inference, peridotite of a certain thickness would be penetrated through both in the pilot hole and the main hole, however, this peridotite would be serpentinized peridotite with different alteration degree.
4. Schist currently exposed at borehole area is mainly muscovite quartz-schist which is mainly composed of quartz and muscovite, with little garnet, cyanite and anorthose. The rock is of granolepidoblastic texture and obvious sheet structure, belongs to the rock layer which easily causes serious hole deviation. However, according to estimation this rock would be hardly seen in both the pilot hole and the main hole.
5. Amphibolite is mainly composed of fasciculate and anorthose, both of which contain approximately same content of mineral, i.e., about 70 % of the rock. The rock contains different amount of minor minerals such as garnet, quartz, muscovite, biotite and epidote. The rock is of prismatic and grain crystalloblastic texture, with foliation developed at different degrees. This rock has two occurring forms, one is occurred in single layer or interbeding with gneiss while the other is formed by retrogressive metamorphism of eclogite and in close paragenesis with eclogite in space.
6. Mylonite Mylonite and cataclastite are tectonite with two different geneses. Mylonite is a product of rock which experienced plastic deformation under relatively high temperature and high pressure. It is a rock of strongly foliated, with obvious mineral lineation developed, being the main component of tough shear zone. The mineral composition of mylonite is basically the same as that of the initial rock before deformation. Schist, gneiss and amphibolite are all the initial rock of mylonite. However, in comparison with the initial rocks, besides the much developed foliation, the mineral grain size of mylonite is finer and the hardness larger. It was estimated that lots of mylonite belts (tough shear zones) would be penetrated through both in the pilot hole and the main hole.
7. Cataclastite Cataclastite (or fault rock), the main composition of fault, is a product of rock which experienced brittle deformation under low temperature and low pressure at the shallow area of the earth's crust. Based upon the size of the broken rock after rock breaking fault rock can be further divided up into breccia, cataclastite, granulitic rock, powdery rock and fault clay (arranged in order of the size of broken rock piece from large to small). Because of the different cementing ways and bonding materials, the porosity, hardness and density of fault rock varies greatly, and some fault breccias with

poor cementation and large porosity may become lost circulation zone.

The borehole position is located in the east boundary of the Maobei eclogite body, and the structure of the borehole area is very complicated. Maobei cajuelite mine is situated at the overturned anticlinal axis, with medium occurrence. The strike of peripheral stratum is totally towards NNE, and the dip angle becomes moderate at the south of the cajuelite mine. In the area faulted structure develops and it is shown from seismic reflection information that there exist underground lots of reverse faults of NNE strike, with SE dip, disrupted by a series of orthogonal normal faults. Besides, in the area develop many series of tough shear zones of NEE strike and exists a series of faults of NW strike, with the character of heave.

According to previous borehole information, it was inferred that following problems would be encountered both in the pilot hole and the main hole of CCSD project.

1. Rock layers are mainly gneiss, eclogite, peridotite, schist, amphibolite, mylonite and cataclastite, with high hardness and poor drillability, generally with drillability grade from 8 to 9, some even from 10 to 11. Peridotite of a certain thickness would be drilled and this rock belongs to ultra-basic rock with high hardness and density. It was inferred that both in the pilot hole and the main hole would be encountered lots of mylonite belts, which is still higher in hardness and with extremely developed foliation.
2. The hardness, density and porosity of fault rock vary greatly and hard and soft rock layers alternately exist.
3. Metamorphic rock, due to uneven metamorphoses, causes a frequent alternation of hard and soft rock layers. With the addition of extremely developed foliation and sheet texture of schist, it would be the rock formation which easily causes serious hole deviation. In this area the maximum stratigraphic dip is 30–35° and the maximum foliation dip is 30–70°.
4. Rock layers with tectonization are unstable, thus precautions must be adopted to prevent hole from collapsing and sticking. Special attention must be paid to the variation of rock formations at the hole sections of 650, 1930, 3210 and 4550 m of the columnar section and appropriate drilling measures should be taken accordingly.
5. In the borehole area fault develops and lots of faults would be encountered in the borehole, and then leakage and blowout would easily happen.

2.4 Drilling Technical Program

Based upon the abovementioned rock formation situation and the technical requirements of the engineering, the basic drilling technique system and the overall construction program must be firstly determined for the drilling engineering design of CCSD-1 Well. On the basis of widely drawing on the scientific drilling experiences of other countries, a complete set of new design concept was adopted for CCSD-1 Well, and three techniques of strategic level, i.e., "combined drilling techniques", "flexible double hole program" and "advanced open hole drilling program" were put forward and organically combined into a complete set of technical program for scientific drilling in hard crystalline rock with Chinese characteristics. This technical system and construction program of strategic level determined that the concrete construction techniques of operational level were the most important technical strategies to complete the whole project efficiently, safely and economically. These achievements involved overall technical program, borehole structure, selection of drilling equipment, drilling tubing, core drilling techniques, application of hydro-hammer, reaming drilling in hard rock, vertical hole drilling techniques, deviation prevention and correction techniques, sidewall sampling techniques, diamond drill bits, borehole logging, drilling mud, leak protection and anti-plugging, cementation and data collection and treatment.

2.4.1 Combined Drilling Techniques

The combined drilling techniques denote an organic combination of geological diamond core drilling techniques as the main and large scale petroleum drilling equipment as the platform, thus become a new combined drilling technique suitable for scientific drilling, with the advantages of both geological diamond core drilling and petroleum drilling. This technical system adopted thin wall impregnated diamond core drill bit as the main cutting tool, high rotary speed, low bit weight and small pump discharge as the main drilling parameters, to overcome the difficulties in large diameter deep hole continuous core drilling in hard rock formations. It is a new combination to realize high efficient core drilling in hard rock, and a unique system of drilling techniques for scientific drilling.

Geological exploration core drilling techniques are suitable for small diameter comparatively shallow hole continuous core drilling in hard rock formations, with the main methods of impregnated diamond core drill bit, high rotary speed, low bit weight, small pump discharge and small scale equipment. Special drilling technologies such as wireline core drilling and rotary percussive drilling are widely utilized. While in oil drilling, as the equipment has large capability, is suitable for large diameter deep hole drilling, by using non-core drilling with rock bit as the main cutting tool, sometimes with PDC bit. In oil drilling the rotary speed of rotary table, the precision of bit weight control and the proportion of core drilling are low, being suitable for non-core drilling in sedimentary rock layers and unsuitable for continuous core drilling in hard crystalline rocks. It is known that scientific drilling project often need to drill deep hole or super-deep hole in hard crystalline rock formations, for instance, in CCSD-1 Well a borehole of 5000 m deep and with final hole diameter of 156 mm should be drilled in eclogite and gneiss, which can hardly completed by oil drilling techniques or geological core drilling techniques alone. In order to solve the problems of constructing a scientific borehole with large diameter and large depth, the only method was to combine geological core drilling techniques with oil drilling techniques and equipment, i.e., to use combined drilling techniques.

The way to realize this purpose was to install a set of high speed top drive system onto an oil rotary table drill rig, or to install downhole high speed motor onto the downhole drilling tool assembly, so that high rotary speed diamond core drilling could be realized.

Combined drilling technical system was adopted for CCSD-1 Well. ZJ700 electric drill with drilling capacity of 7000 m was used. This drill was produced by Baoji Petroleum Machinery Plant, with advanced level at home. To satisfy the requirement of high rotary speed for diamond wireline core drilling, top drive and wireline coring auxiliary device were to be installed. Wireline core drilling techniques, downhole power percussive rotary drilling techniques and swivel type double tube core drilling tool were used.

2.4.2 Flexible Double Hole Program

There existed lots of undefined factors in China Continental Scientific Drilling project. Through full technical and economic discussion it was decided that a "flexible double hole program" would be adopted. The "double hole program" was a new strategy for scientific deep hole construction, and in KTB in Germany had been adopted the same construction strategy, which denotes that a small size and shallower cored borehole is drilled first near the final target borehole area and then the final target borehole completed, the former is called as pilot hole whereas the latter main hole.

Besides in super deep hole, the "double hole program" can also be adopted in constructing deep hole of 4000–5000 m, where the depth of the pilot hole is only 1000–2000 m.

The double hole program for CCSD project can be found in Fig. 2.2. The designed depth of the pilot hole was 2000, 106 m from the main hole, which was designed 5000 m deep.

Different from the double hole program in Germany, the double hole program in China was a flexible one, either possibly double hole or single hole, decided by the result of pilot hole construction. Under the circumstances that the

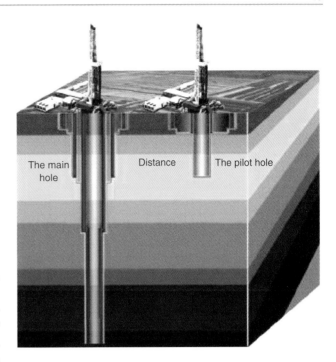

Fig. 2.2 The double hole program for CCSD-1 Well

construction quality of the pilot hole is good and borehole deviation is controlled within the allowable limits, the main hole can be directly drilled at the pilot hole position, without moving borehole site. What is necessary to do is to directly ream the pilot hole and set casing. The later construction can be conducted according to the design of the main hole, and in such a way "the two holes are combined into one" and double hole drilling is changed into one hole drilling, thus large funds and much time saved. On the contrary, if the casing program in the pilot hole is rather complicated or hole deviation is serious double holes must be drilled, that is, to drill the main hole at the location 106 m away from the pilot hole.

The "flexible double hole program" was designed in accordance with the concrete conditions of CCSD-1 Well and it was essentially a complete set of overall program of flexible application of the two construction procedures based upon different construction results.

2.4.3 Feel Ahead Open Hole Drilling Techniques

In the area of CCSD-1 Well location according to historical record the deepest borehole drilled was no deeper than 1100 m and for the geological information of 1100 m deeper the reference materials from neighboring wells were unavailable. Though surface geological work and geophysical reconnaissance were widely conducted the inferred underground condition was still untrustworthy because of the complexity of underground condition and the interpretation ambiguity of geophysical reconnaissance. Under the circumstances of unknown deep geological conditions and inadequate basis for borehole design, the adoption of "feel ahead open hole drilling method" was the optimum program for borehole construction, because its effectiveness was verified in the former Soviet Union as this construction program had been basically adopted for scientific drilling in crystalline rocks. In consideration of economy, the diameter of core drilling for CCSD-1 Well was designed to be 156 mm, instead of 215.9 mm adopted in the former Soviet Union.

2.5 Borehole Structure and Casing Program

For scientific drilling, either for single hole program or for double hole program, detailed pre-drilling data and materials are unavailable. For borehole structure design, stress must be laid on two factors, i.e. down hole safety and drilling cost, and adequate casing program must be prepared so as to isolate complicated layers. To guarantee to successfully reach to the designed borehole depth, double hole program was adopted in the initial design of the borehole program to deal with serious hole deviation and other complicated situations. In the main hole structure double tail pipe was prepared to solve a variety of difficulties which may occur in drilling process. In the main hole, except that the setting depth of surface casing, intermediate casing and completion casing was basically determined, the setting depth of 219.1 mm tail pipe and 177.8 mm tail pipe was not yet determined. Whether setting these two casings and the setting depth would be decided based upon the concrete conditions at drill site.

Metamorphic layer is of good stability. According to the scientific drilling experiences from the former Soviet Union and other countries, ultra long open hole drilling was possible in CCSD project. If so, it may be unnecessary to set 219.1 and 177.8 mm tail pipe, as well as 273.0 mm intermediate casing. Only running completion casing in 156 mm borehole was necessary. In this way casings could be saved. Furthermore, if intermediate casing was unnecessarily run until about 2000 m deep after the second opening of the pilot hole, then the double holes could be combined into one, i.e., the borehole structure of the pilot hole after the second opening could be constructed in accordance with the borehole structure of the main hole after the second opening and then the repeated construction of the upper hole section of the main hole was saved. The surface structure design (hole diameter was 444.5 mm and surface casing 339.7 mm) of the pilot hole provided possibility for this conversion.

2.5.1 Designed Borehole Structure and Casing Program for the Pilot Hole

The designed borehole structure and casing program for the pilot hole can be found in Table 2.1; Fig. 2.3.

2.5.2 Designed Borehole Structure and Casing Program for the Main Hole

The designed borehole structure and casing program for the main hole can be found in Table 2.2; Fig. 2.4.

2.6 Drilling Equipment Program

In the light of the requirement of full hole coring in CCSD project, geological drilling equipment was unable to undertake 5000 m hole drilling and in this case petroleum drilling equipment must be employed, however, the conventional coring techniques used in oil drilling industry were unable to be effectively utilized for full hole coring. Therefore, for the pilot hole and the main hole drilling a combined drilling technique (geological drilling + oil drilling) was to be used, i.e., a set of high speed (300–500 rpm) rotary top drive system and wireline coring system were installed onto a rotary table oil drill rig, so as to realize diamond wireline core drilling for large diameter deep hole. In addition, wireline coring system for core drilling needs a high precision for bit feeding, small discharge capacity of drilling fluid and wireline fishing tools. Corollary equipment should be installed and modification should be conducted on the selected oil drill in order to satisfy the needs of scientific drilling.

2.6.1 Main Drilling Equipment

Under the prerequisite to satisfy the needs of drilling CCSD-1 Well, the selected drill should be advanced and economical to a certain extent, mainly satisfying the following conditions:

1. Need of drilling depth should be satisfied: to 5000 m with 156 mm drilling tool.
2. Hook load should meet the need of lifting the heaviest drill string, and at the same time has adequate intake of tensile force to satisfy the requirement of treating complicated situations. The maximum drill string weight is 145 and 126 t after minus buoyant force; the maximum

2.6 Drilling Equipment Program

Table 2.1 Borehole structure and casing program for the pilot hole

Hole opening	Drill bit size (mm)		Drilled depth (m)	Casing (tail) size		Setting depth (m)
	mm	in.		mm	in.	
First opening	444.5	$17^1/_2$	100	339.7	$13^3/_8$	100
Second opening	215.9	$8^1/_2$	1000	177.8	7	1000
Third opening	156	$6^1/_8$	2000	127.0	5	1800

Note For the second opening, drilled depth was based on the actual situation at well site

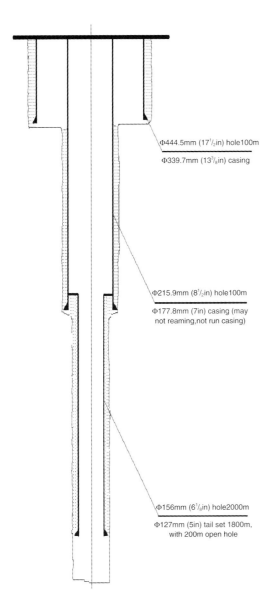

Fig. 2.3 Borehole structure and casing program for the pilot hole

casing string weight is 170 t (273 mm casing set to 2000 m deep), and 150 t after minus buoyant force.

3. Need of special drilling technologies should be satisfied: to satisfy the requirement of wireline core drilling high speed driving device should be equipped, such as Varco high speed top drive, which requires a 43 m high derrick.

Drills which can satisfy the abovementioned requirements include ZJ45, ZJ70L and ZJ70D and after technical and economic analyses it was believed that advanced and economical ZJ70D drill, with adequate drilling capacity (included the capacity to treat accidents and complicated situations), should be selected for drilling CCSD project. As CCSD project would last a long time, drill rig with low daily cost has much economic value. If ZJ45 drill could be technically modified and then meet the need of the construction, it would have much application value. In this case ZJ45 drill was selected as the alternate.

The auxiliary 3NB1600 electric driven mud pump has a maximum working pressure of 34.4 Mpa, with control of stepless change from 0 to maximum stroke realized, can work at a small discharge rate for a long time, thus the requirements of small discharge rate and high circulation pressure for wireline coring can be satisfied.

ZJ70D drill has a 5000 m bailing drum, which can be used as wireline hoist, to meet the needs of core fishing and deviation survey at fixed point.

Equipped with ZJ70D drill is a three stage solid control system consists of oscillating screen, desander (desilter) and centrifugal, among which two sets of oscillating screen are available and 200 mesh screen cloth can be used to meet the need of drilling fluid solid control for wireline core drilling.

Commonly used drilling parameter gauges and data collection system are the necessity for driller to operate the equipment. To satisfy the needs of scientific drilling, at least the ZJC-B2 eight drilling parameter gauge should be equipped with the drill. This gauge can continuously measure and record eight engineering data, including hook load,

Table 2.2 Borehole structure and casing program for the main hole

Hole opening	Drill bit size (mm)		Drilled depth (m)	Casing (tail) size		Setting depth (m)
	mm	in.		mm	in.	
First opening	444.5	$17^1/_2$	100	339.7	$13^3/_8$	100
Second opening	311.1	$12^1/_4$	2000	273.0	$10^3/_4$	2000
Third opening	244.5	$9^5/_8$	3250	219.1	$8^5/_8$	3250
Fourth opening	200	$7^7/_8$	4600	177.8	7	4600
Fifth opening	156	$6^1/_8$	5000	127.0	5	4800

Note For the second opening, drilled depth was based on the actual situation at well site

For the third opening, drilled depth and casing setting were based on the actual situation at well site

drilling footage, pump pressure, rotary speed of rotary table, torque of rotary table, pump speed, torque of tongs and outlet discharge of drilling fluid.

2.6.2 Equipment and Instruments Should Be Added

1. **High speed top drive**

In general, the rotary linear velocity of drill bit should be 1.5–3.0 m/s (equivalent to 184–367 rpm of rotary speed of rotary table) to guarantee an effective drilling of 156 mm impregnated diamond drill bit, and this requirement the conventional oil rotary table and commonly used top drive cannot satisfy. High top drive must be equipped.

2. **High precision automatic bit feeding device**

In core drilling the requirement of diamond drill bit to bit pressure control is very high, thus an automatic bit feeding device with precision of no less than 500 kg should be equipped. Three types of the device were available and it was recommended that the electronic driller device manufactured by M/D TOTCO Tool Company be used.

3. **Compound logging instrument**

In order to obtain the related data fully and accurately it was necessary to equip an oil drilling compound logging device. Based on the material information of core and chips, and in combination with drill time variation, stratigraphic profile can be timely established by the compound logging device. Gas bearing abnormal interval of strata in the borehole can be classified through monitoring total gas and methane content variation by chromatographic logging. Fluid property in the borehole can be comprehensively judged according to the aquosity of core and chips, surface gas bearing index, strata gas bearing index in combination with non-hydrocarbon gas content, drilling fluid change and fluid level show in pit (ditch). The abnormal events in drilling construction can be interpreted and forecast through real-time monitoring borehole and drilling fluid data. Therefore compound logging is the necessary auxiliary logging method for scientific drilling and such device should be equipped. In design SDL-9000 compound logging device was selected.

2.7 Drilling String Program

The main types of drilling tools used in drilling construction included:

139.7 mm non-coupling wireline drill rod and 146 mm wireline drill collar, 89 and 127 mm conventional oil drill pipe, different sizes of drill collar and casing.

The drill string may be used for non-core drilling and reaming drilling included two types: 89 and 127 mm oil drill strings. In drill string design the following problems should be mainly considered:

1. Drill rod should meet the need of tensile strength and torsional strength, in which,

   The strength of 89 mm drill string should satisfy the need of deviation correction in 5000 m deep in 156 mm borehole.

   The strength of 127 mm drill string should satisfy the need of reaming drilling in 4 500 m deep in 200 mm borehole.

2. The quantity of drill collars should satisfy the need of putting weight on bit; the size of drill collars should be suitable for deviation prevention, deviation correction and milling operations after drill pipe sticking.

3. The design of lower drilling tool assembly should satisfy the needs of deviation prevention and deviation correction.

4. For this hole drilling, the clearance between the drill string and the borehole wall and the inside diameter of the drilling tool should be considered in selection of the drill string, to decrease the resistance of drilling fluid circulation and the surge pressure created by tripping.

2.8 Core Drilling Program

A variety of core drilling techniques were adopted in design so as to satisfy the requirements of full hole coring for CCSD project.

1. Conventional core drilling was to be used when borehole was shallow.
2. Conventional wireline core drilling was to be employed when borehole reached to a certain depth, so as to decrease tripping time and increase drilling efficiency.
3. Downhole motor wireline core drilling (two combined into one) was to be employed when borehole was relatively deep, rotating torque was large and surface driving could not be used.
4. Conventional downhole motor core drilling was to be used in case that downhole motor wireline core drilling tool was not well prepared.
5. Hydro-hammer drilling, including hydro-hammer wireline core drilling (two combined into one) and conventional hydro-hammer core drilling were to be adopted in order to increase drilling rate in hard rocks.
6. Packed hole drilling tool assembly should be adopted for all the core drilling techniques so as to prevent hole deviation.

2.8.1 Wireline Core Drilling

To increase core recovery and decrease auxiliary drilling time, wireline core drilling system was widely utilized for scientific drilling projects in the countries of the world. Besides reduced tripping time and decreased cost, wireline core drilling system has the following advantages:

1. With improved core recovery and quality, the scientific research purpose of this project can be still better satisfied.
2. Logging instruments can be lowered by utilizing internal flush drill rod and drawworks.
3. Inner tube structure can be changed in accordance with the variation of rock layer.
4. Labour intensity of the operators can be reduced.

Based upon the Drilling Purpose of China Continental Scientific Drilling Project, the Designed Task of China Scientific Drilling Engineering Project and Additional Appendix, the Feasibility Study Report on China Scientific Drilling Engineering Project and the Bidding Document on Engineering Design for China Scientific Drilling Project, full hole coring was required. In consideration of techniques and economy, wireline core drilling system was determined as the optimum core drilling system in the design stage.

In comparison with the wireline core drill rod and drilling tool made in Japan, German made products had obvious

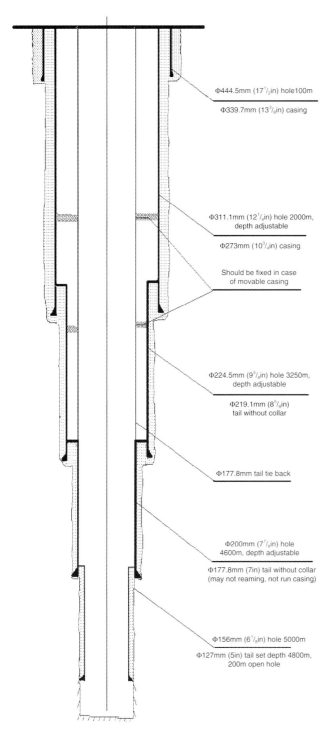

Fig. 2.4 Borehole structure and casing program for the main hole

5. Considering that scientific drilling would last a long period (3–5 years), the outer surface of the selected drill string sub and pin and box thread should be of wear resistance, with good sealing and pressure bearing capacities, to reduce the possibility of drilling tool and casing accidents.

Fig. 2.5 Wireline core drill rod and drill collar

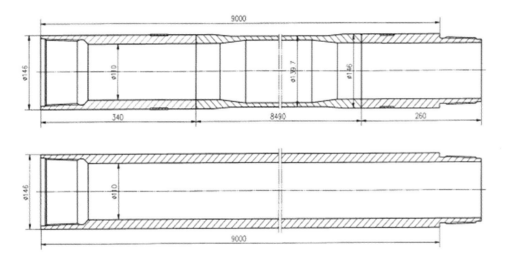

superiority both in mechanical properties and in price. For this reason it was mapped out that the wireline coring system used for CCSD project would be imported from a certain foreign company in Germany. International standards were to be adopted for the materials used for the imported wireline core drill rod and the material used for drill rod body was just the same as that used in KTB project (API5D-G105). To guarantee a long antifatigue life and high safety factor for break-out, steel grade for drill rod sub should be high and the material should be 30CrNiMo8 (equivalent to S135 in API Standard). All the pipes must be seamless. The structure of wireline core drill rod and drill collar can be found in Fig. 2.5.

The internal and external upset structure was adopted for drill rod, which was a significant improvement comparing with the external flush structure used in KTB project. The internal and external upset structure has the following characteristics:

1. The reliability of wellhead clamping can be improved
   For wireline core drilling techniques used in hard rocks, the annular area for rock crushing should be decreased as much as possible in order to increase penetration rate, for instance, the wireline diamond drill bit used in KTB project had an outside diameter of 152.4 mm and an inside diameter of 94 mm. As the drill rod had an outside diameter of 139.7 mm, the annular clearance between drill rod and unilateral hole wall was 6.35 mm only. Because the outside diameter of drill rod sub was as the same as that of drill rod (external flush drill rod), clamping of drill rod could only be realized by using frictional or similar modes, instead of the safe modes such as tongs or fork. For this frictional mode, once the teeth of slips were worn off and friction force decreased obviously, drill string very easily became out of brake and then downhole accident happened. To overcome this, in CCSD core drilling tool design, a certain foreign company was required to produce the wireline core drill rod with both ends internally and externally upset, with an outside diameter of 146 mm for upset end, that is, there was a 3 mm shoulder on each side. As the drill rod sub was also 146 mm in size, once drill string became out of brake the shoulder would move down on the slips, which increased the holding force under the action of the

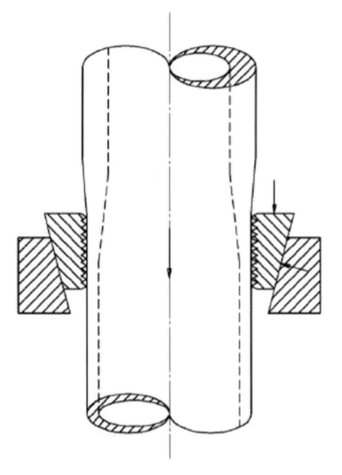

Fig. 2.6 Wireline drill rod clamped by slips

2.8 Core Drilling Program

Fig. 2.7 The imported wireline core drilling tool

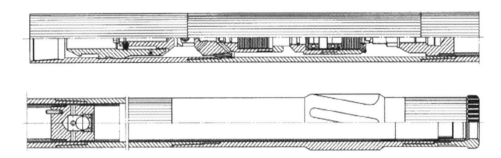

Table 2.3 The imported wireline core drilling tool

Item	Parameter	Item	Parameter
Model	SK146/94		
Drilling tool O.D. (mm)	146	Drilling tool length (mm)	8820
Drill bit O.D. (mm)	156	Core size (mm)	94
Reaming shell O.D. (mm)	156	Stabilizer O.D. (mm)	156

back chamfer, and in this way the accident such as drill string running would be avoided (Fig. 2.6).

2. The connection strength of drill rod thread can be increased The wireline drill rod used in KTB was only internal upset, with thread thickness (total thickness of pin and box thread) of $(139.7 - 110)/2 = 14.85$ mm. While for CCSD project the wireline drill rod used had a thread thickness (total thickness of pin and box thread) of $(146 - 110)/2 = 18.0$ mm, with the ultimate tensile load increased to 2200 kN from 2144 kN (from the Operation Guide of SK146 × 94 mm Wireline Drilling Tool for CCSD Project, August 2002).

This drilling tool consists of two parts, i.e. wireline outer tube and wireline inner tube assembly (see Fig. 2.7), and at the upper part of the inner tube assembly is installed a dip angle inclinometer, with positioning alarm and core blockage alarm, with specifications shown in Table 2.3.

2.8.2 Hydro-hammer Wireline Core Drilling Tool

Percussive rock fragmentation is the most effective way to increase penetration rate in hard rock formations. Though cone bit can produce percussion while in rotation, it is of low drilling rate and short service life in drilling rocks with drillability over 7–8 grade, because tungsten carbide used for its cutting elements. Diamond is brittle, but in application of diamond drill bit the drilling efficiency can be greatly improved under an appropriate percussive force which doesn't damage diamond. From this principle the method of diamond percussive rotary drilling emerged. In CCSD-1 Well project KS156 hydro-hammer wireline core drilling tool developed by the Institute of Exploration Techniques was employed (Fig. 2.8), with the specifications shown in Table 2.4.

2.8.3 PDM Wireline Core Drilling Tool

For wireline core drilling, it is required to drive the rotation of the whole drilling tool system from surface, thus consuming enormous energy to overcome the friction between the drilling tool and borehole wall. As to diamond wireline coring the high rotation still accelerates the consumption of energy and produces serious disturbance to borehole wall, easily resulting in accidents such as rock piece falling or drill pipe sticking. Downhole power is driven by drill mud, only rotating drill bit and core barrel, and the whole drill string doesn't rotate or only slowly rotates (to overcome the loss of bit weight). For this reason in the period of early study a program of combining downhole power and wireline coring was put forward. In this project LS156 PDM wireline core drilling tool assembly developed by the Institute of Exploration Techniques was to be utilized. This drilling tool assembly consists of the outer tube assembly and the wireline core drilling tool inner tube assembly combined with PDM, with the structure and principle shown in Fig. 2.9. The specifications of PDM wireline core drilling tool are shown in Table 2.5 and the main specifications of PDM used for the drilling tool can be found in Table 2.6.

2.8.4 Turbomotor Wireline Core Drilling Tool

Either PDM or turbomotor can be downhole power. In this project SV156 turbine wireline core drilling tool assembly developed by the Institute of Exploration Techniques was used. This drilling tool assembly consists of the outer tube assembly and the wireline drilling tool inner tube assembly combined with turbomotor (see Fig. 2.10), with the specifications shown in Table 2.7.

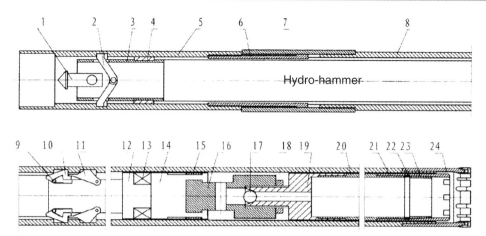

Fig. 2.8 Structure of KS156 hydro-hammer wireline core drilling tool. *1* Spear head, *2* Spring clip clamp, *3* Spring clip support, *4* Sealing sub, *5* Spring clip chamber, *6* Splined shaft, *7* Spline sleeve, *8* Outer tube, *9* Spring, *10* Power transmitting block, *11* Positioning probe, *12* Independent sub, *13* Bearing, *14* Upper separating adapter, *15* Separating ring, *16* Lower separating adapter, *17* Steel ball, *18* Nut, *19* Core barrel adapter, *20* Core barrel, *21* Core catcher seat, *22* Catching ring, *23* Core catcher, *24* Drill bit

Table 2.4 KS156 hydro-hammer wireline core drilling tool

Item	Parameter	Item	Parameter
Model	KS156	Manufacturer	IET
Drilling tool O.D. (mm)	146	Drilling tool length (m)	4 (not including core barrel)
Percussive work per single stroke (J)	100–150		
Working pump duty (L/min)	150–400	Percussion frequency (Hz)	15–30
Thread type for connecting upper and lower ends	Same as the imported Wireline tool	Working pressure drop (MPa)	1–4
		Service life (h)	120

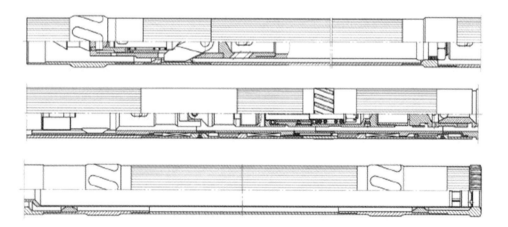

Fig. 2.9 Structure of LS156 PDM wireline core drilling tool

2.8.5 Conventional Core Drilling Tool

The conventional core drilling tool adopted in this project was selected in accordance to the Standard GB/T16950-1977, that is, S sized (139.7 mm) double tube drilling tool (see Fig. 2.11) of P type diamond core drilling double tube core barrel drilling tool (double tube drilling tool), with the main specifications shown in Table 2.8.

2.8.6 Hydro-hammer Core Drilling Tool

To increase drilling rate for conventional core drilling technique, percussive rotary drilling method was to be employed. YZX127 hydro-hammer (Fig. 2.12) manufactured by the Institute of Exploration Techniques would be adopted, the specifications of the tool can be found in Table 2.9.

2.8 Core Drilling Program

Table 2.5 Specifications of LS156 PDM wireline core drilling tool

Item	Parameter	Item	Parameter
Model	LS156	Manufacturer	IET
Borehole diameter (mm)	156	Coring length per run (m)	6
Core drilling diameter (mm)	94	Total length of drilling tool (m)	Approxiamte 14
Wireline fishing capacity (kN)	Larger than 10	Life of bearing (h)	60–100 (single set)
Thread type for connecting upper end	Same as the imported drill collar	Thread type for connecting lower end	Same as the imported wireline drill tool

Table 2.6 Specifications of LZ100 PDM

Item	Parameter	Item	Parameter
Model	LZ100 × 7.0	Manufacturer	Beijing oil machinery
Drilling tool O.D. (mm)	100	Drill bit pressure drop (MPa)	1.4–7.0
Motor flow rate (L/s)	4.7–11	Output rotary speed (rpm)	280–700
Working torque (N m)	650	Max. torque (N m)	1300
Motor working pressure drop (MPa)	5.17	Suitable temperature (°C)	120
Drilling tool length (m)	6.4	Drilling tool weight (kg)	245
Drilling tool power (kW)	19.1–47.65	Thread for connection	$2^7/_8$ REG
Drilling tool life (h)	80–100		

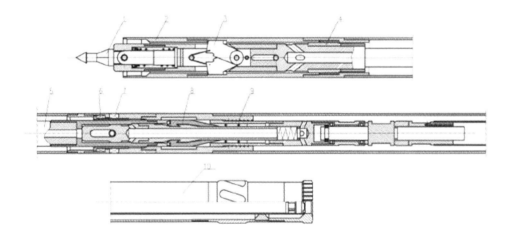

Fig. 2.10 Structure of SV156 turbo-drill wireline core drilling tool. *1* Spear head, *2* Outer tube assembly, *3* Upper spring clip package, *4* Flow plugging package, *5* Turbomotor, *6* Small bearing, *7* Large bearing, *8* Torque transmitting device, *9* Lower plugging device, *10* Drill bit and coring vessel assembly

Table 2.7 Specifications of SV156 turbomotor wireline core drilling tool

Item	Parameter	Item	Parameter
Model	SV156	Manufacturer	IET
Borehole O.D. (mm)	156	Coring length per run (m)	Less or equal to 4.5
Core drilling diameter (mm)	No less than 90	Total length of drilling tool (m)	No less than 20
Working discharge (L/s)	6–10	Life of bearing (h)	60–80 (single set)
Working pressure drop (MPa)	3.5–4	Output torque (N m)	300–500
Output rotation (rpm)	400–600	Output power (kW)	18–21
Drilling tool weight (kg)	1000	Suitable hole depth (m)	4000
Thread type for connecting upper end	Same as the imported drill collar	Thread type for connecting lower end	Same as the imported wireline core drilling tool

2.8.7 PDM Core Drilling Tool

In conventional core drilling, PDM or turbomotor can be used to drive the core drilling tool at hole bottom. As impregnated diamond drill bit is used, LZ127 × 3.5 PDM with higher rotation speed was selected, the structure of which is illustrated in Fig. 2.13 and the specifications can be found in Table 2.10.

2.8.8 Design Program of Diamond Core Drill Bit and Reaming Shell

1. **Selection of diamond core drill bit**

In accordance with the drillability, abrasiveness and crumbliness degree of the rock formations which would be encountered in borehole drilling, by reference to the standards of diamond drill bit selection recommended in related regulations and based on the experiences in the pilot hole drilling of CCSD project, priority should be given to the utilization of impregnated diamond drill bit, with the technical parameters should satisfy the following requirements.

1. Drill bit outside diameter 156 mm.
2. Drill bit inside diameter 94 mm for wireline core drill bit and 108 mm for conventional core drill bit.
3. Diamond grain size 35–40 mesh was recommended by reference to the Core Drilling Regulations (1983 version) issued by the former Ministry of Geology and Mineral Resources.
4. Diamond monocrystal strength The monocrystal strength of selected diamond should be larger than 343 N, i.e. equivalent to SMD 35 synthetic diamond or even higher (reference to the National Standards on Diamond issued by the former National Bureau of Standards on May 20th, 1986).
5. Matrix hardness of drill bit HRC 35–45 was determined according to the recommended value in the Core Drilling Regulations (1983 version) issued by the former Ministry of Geology and Mineral Resources.
6. Water opening and slot In design of water opening and slot the application of drill mud and downhole motor should be fully considered and in this way the cross section of water opening and slot should be appropriately enlarged and the quantity of water opening and slot be increased (10–16 water openings and water slots, the projected area of water openings accounts for 40–50 % of the annular rock fragmentation area) so as to reduce the flow resistance and ensure a full cooling for drill bit (Fig. 2.14).
7. Bit face profile Based on different drilling methods and drillability of the rock formations four bit face profiles were recommended from tens of bit face profiles (round face profile was mainly used for diamond core drill bit in German KTB project).

For step face profile (Fig. 2.15a), a variety of profiles such as single step, double step and triple step are available, normally used for thick wall drill bit such as wireline core drill bit, which crushes rock in larger area and with good stability, being suitable for drilling medium hard rock. In hard rock with weak abrasiveness this drill bit can still obtain a satisfactory result.

Drill bit with inner conical profile (Fig. 2.15b) has good stability and guidance at hole bottom, thus being favourable for preventing hole deviation. This profile is often adopted for wireline core drill bit.

Concentric saw teeth profile (Fig. 2.15c), also called as concentric sharp slot profile has large rock fragmentation area, and thus has a combined rock crushing action of grinding and shearing, with coarse cuttings produced, which are favourable to diamond exposure. The drill bit requires less axial weight on bit, and this is favourable to deviation prevention. Saw teeth profile drill bit is suitable for drilling in hard and compact rock formation with weak abrasiveness.

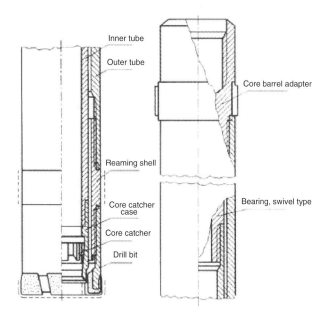

Fig. 2.11 Structure of S sized double tube core drilling tool in P type

Table 2.8 Specifications of S sized double tube core drilling tool in P type

Item	Parameter	Item	Parameter
Drill bit size (mm)	156	Outer tube O.D./I.D. (mm)	139.7/127
Inner tube O.D./I.D. (mm)	120/112	Core diameter (mm)	108
Inner tube length (mm)	6000	Total length (mm)	7500

2.8 Core Drilling Program

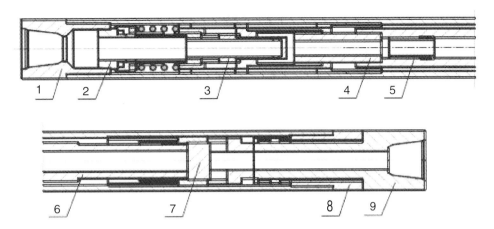

Fig. 2.12 Structure of YZX127 hydro-hammer. *1* Upper adaptor, *2* Pressure limiting valve, *3* Upper valve, *4* Upper piston, *5* Core valve, *6* Hammer, *7* Anvil, *8* Spline sleeve, *9.* Splined shaft

Table 2.9 Specifications of YZX127 hydro-hammer

Item	Parameter	Item	Parameter
Model	YZX127	Manufacturer	IET
Drilling tool O.D. (mm)	127	Drilling tool length (m)	2.5
Percussive work per single stroke (J)	150–300	Percussion frequency (Hz)	5–12
Working pump duty (L/min)	200–600	Working pressure drop (MPa)	2–5
Average service life (h)	80	Thread type for connecting upper and lower ends	3½ REG

Round profile (Fig. 2.15d) is suitable for the rock formations with high abrasiveness.

2. **Selection of diamond reaming shell**

Diamond reaming shell is used for trimming the borehole size and stabilizing the drilling tool. It was decided that impregnated diamond reaming shell would be used by reference to the selection of diamond drill bit. Spiral reaming shell with good functions of water discharge and cuttings discharge was selected. The diamond quality used for manufacture of reaming shell was equivalent to that used for diamond drill bit.

Diamond reaming shell products with unified specifications and properties were to be used for different drilling methods and different rock formations, i.e. outside diameter of the reaming shell was 156.3–156.5 mm, with 8–10 water slots, 35–40 mesh diamond grain size and approximate HRC 40 matrix hardness. The overflow area should be 45–50 % larger than the cross sectional area of the annular space between drilling tool and borehole wall.

2.9 Hole Deviation Control Program

Due to the lithological characteristics at well location, hole deviation and dogleg were the major factors which would affect the construction schedule. Thus the control standards for hole deviation and dogleg should be reasonably designed

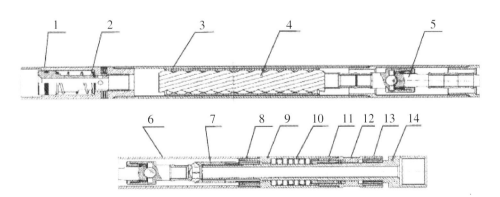

Fig. 2.13 Structure of PDM. *1* Overflow valve body, *2* Overflow valve core, *3* Stator, *4* Rotor, *5* Cardan, *6* Bending outer tube, *7* Water passing joint, *8* Upper radial bearing package, *9* Upper bearing tube, *10.* Bearing package, *11.* Step bearing, *12.* Lower bearing tube, *13* Lower radial bearing package, *14* Transmission shaft

Table 2.10 Specifications of LZ127X3.5 PDM

Item	Parameter	Item	Parameter
Model	LZ127 × 3.5	Manufacturer	Beijing Oil Machinery
Drilling tool O.D. (mm)	127	Pressure drop at drill bit water hole (MPa)	1.0–3.5
Motor flow rate (L/s)	9.5–15.8	Output rotary speed (rpm)	355–560
Motor pressure drop (MPa)	2.5	Max. bit weight (kN)	40
Working torque (Nm)	576	Max. torque (Nm)	1 152
Drilling tool power (kW)	21.4–33.78	Suitable temperature (°C)	<135
Drilling tool weight (kg)	400	Drilling tool length (m)	5.8
Drilling tool life (h)	100	Thread type of upper and lower ends	3½ REG

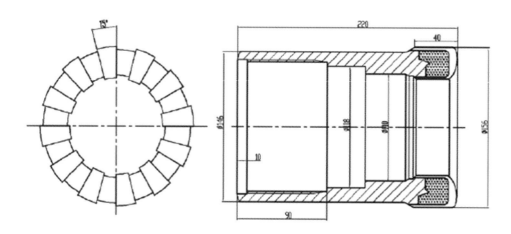

Fig. 2.14 Recommended structure for diamond drill bit

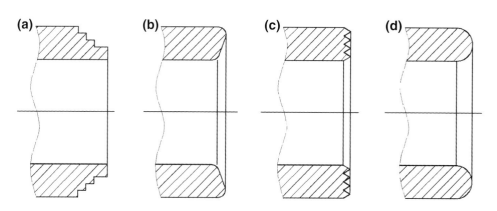

Fig. 2.15 Four recommended face profiles for diamond drill bit. **a** Step profile. **b** Inner conical. **c** Saw teeth profile. **d** Round profile

under the preconditions that the requirements of CCSD project could be satisfied and drilling cost reduced.

As the pilot hole was 2000 m deep and the cored hole section of the main hole was started from 2000 m, the deviation standard for the cored hole section could be appropriately relaxed. During core drilling operations, because of high rotary speed and small annular space, the dogleg in this hole section should be as small as possible, to prevent an excessive dogleg from increasing friction resistance for the drilling tool, which might cause an accident of drilling tool breaking. For the upper hole section of the main hole in which non-core drilling was conducted by using oil drilling technique, petroleum drilling standards could be executed for dogleg control. However, for a smooth drilling in the lower hole section of the main hole, it was required that hole deviation at the upper hole section of the main hole should be as small as possible and the hole trajectory should be controlled as smooth as possible. In accordance to this principle a standard for controlling borehole quality was laid down to reasonably control the drilling cost.

For the pilot hole:
0–1000 m, vertex angle ≤5°, maximum dogleg ≤1(°)/30 m
1000–2000 m, vertex angle ≤14°, maximum dogleg ≤ 1(°)/30 m
For the main hole:
0–2000 m, vertex angle ≤2° to 5°, maximum dogleg ≤2 (°)/30 m
2000–5000 m, vertex angle ≤18°, maximum dogleg ≤1 (°)/30 m

2.9.1 Deviation Prevention for Cored Hole Section and Monitor Measures

In core drilling the related parts of the packed hole drilling tool should be checked up at regular intervals, drilling parameters should be adjusted in time according to different situations so as to decrease deviation intensity. At the same time effective deviation monitor measure should be adopted to avoid excessive hole deviation.

1. In core drilling, make the widest use of conical profile core drill bit.
2. The outside diameter of diamond drill bit and reaming shell must be strictly inspected and any one exceeding regulation should be changed in time.
3. The straightness of the lower drill collar should be observed in lifting drill string and any one bending must be thrown off in time.
4. Weight on bit should be adjusted in time according to drilling speed change and drilling in soft and hard rock interface should be carefully treated.
5. Weight on bit should be reduced in drilling fractured layer (bit bouncing happens).
6. Hole deviation should be timely surveyed.

2.9.2 Deviation Control Measure for Cored Hole Section

Deviation correction by using wire deviation survey-while-drilling techniques is an effective and economical way to control hole deviation and it could be used according to the borehole situation in case of deviation exceeding standard.

2.9.3 Deviation Control Measure for the Upper Section of the Main Hole Where Non-core Drilling Was Conducted

In non-core drilling hole section the conventional anti-deviation techniques in oil drilling could be employed, however, hole deviation would easily happen in drilling with conventional petroleum anti-deviation techniques in crystalline rocks. In order to improve anti-deviation efficiency, in this scientific drilling project the VDS automatic vertical drilling system (Fig. 2.16) was used as the main technique and conventional petroleum anti-deviation techniques was used as subsidiary to control hole deviation.

For subsidiary anti-deviation measure, hydro-hammer technique combined with conventional petroleum anti-deviation technique was used to improve anti-deviation result. It was a supplement to VDS system.

2.10 Non-core Drilling and Reaming Drilling Program

It was known from the borehole program and construction procedure that non-core drilling would be conducted in the upper section of the main hole and in the deviation correction section of backfilled small borehole. For other hole sections core drilling would be carried on by using 156 mm drill bit and then reaming when necessary. Petroleum drilling techniques were utilized for non-core drilling and reaming drilling.

For the large diameter non-coring upper section of the main hole, staged reaming techniques would be used because direct drilling in hard rock formation which easily caused hole deviation by using large sized drill bit would result in a low drilling speed and poor anti-deviation effect. Deviation was hard to be corrected once hole deviation appeared and the cost would be rather high. Therefore, opening the hole with conventional sized drill bit and then reaming in stages to the designed hole size was a common technique used for continental scientific drilling in the world.

2.10.1 Design of Drilling Tool Assembly For Non-core Drilling

It was very easy for borehole to become deviated because stratigraphic dip is larger than 30°, anisotropy of rock layers varies greatly and the hard rock contains mica. Under these circumstances two problems were mainly considered in the design of non-core drilling tool assembly for the upper section of the main hole: (1) anti-deviation and deviation correction; (2) improving penetration rate.

Rigid packed drilling tool assembly for oil drilling and VDS (vertical drilling system) employed in KTB were to be used for anti-deviation, in which 203.2 mm thick drill collar and four stabilizers were equipped to the packed drilling tool to increase its rigidity and thus improve its holding (anti-deviation) capacity. Hydro-hammer (impactor) should be used as much as possible in order to improve efficiency of

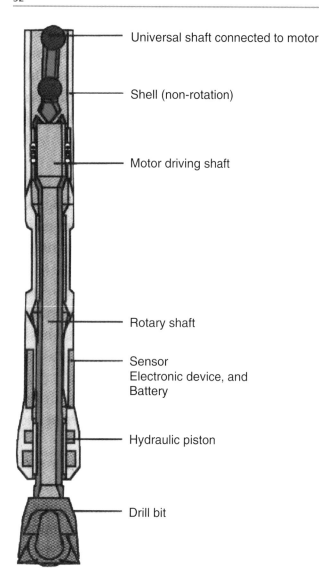

Fig. 2.16 VDS-3 vertical drilling system

rock fragmentation in hard formation and increase penetration rate. Furthermore, selection of the outside diameter of the drill collar above stabilizer should satisfy the requirement of milling operation after drill pipe sticking and the length of the drill collar was decided by the maximum weight on bit used. As with effective capacity to prevent hole deviation, VDS was utilized as the main anti-deviation and deviation correction tool in the upper section of the main hole while pendulum drilling tool and power tool + bending sub tool was used as an auxiliary measure for deviation correction.

A structure of advanced double stabilizer pendulum drilling tool +hydro-hammer was adopted for pendulum drilling tool, which utilized 203.2 mm thick drill collar to increase pendulum force and improve deviation correction result. The use of hydro-hammer was for the purposes of reducing drill bit weight, increasing drilling rate and improving deviation correction effect. Moreover, selection of the outside diameter of the drill collar above stabilizer should meet the need of milling operation after drill pipe sticking and the length of the drill collar was decided by the maximum weight on bit used.

Non-core drilling was also conducted for deviation correction in 156 mm backfilled borehole. VDS could not be used as the minimum hole size it is suitable to was 215.9 mm. In such small sized borehole the pendulum force and rigidity of the drilling tool were greatly decreased and the pendulum drilling tool could not effectively correct hole deviation. Therefore, only power tool + bending sub could be used for deviation correction in 156 mm backfilled borehole and in order to keep the dogleg not exceeding the designed requirement a "single sub" was used instead of conventional PDM and sub, and in this way the dogleg could be reduced in combination with rotary drilling. Hydro-hammer could be used to increase drilling rate. Selection of the outside diameter of the drill collar should meet the need of milling operation after drill pipe sticking and the length of the drill collar was decided by the maximum weight on bit used.

2.10.2 Design of Drilling Tool Assembly for Reaming Drilling

Because reaming drilling was to be conducted under the precondition that the drilled small sized borehole had satisfied the requirement, therefore in the design of reaming drilling tool the problems of anti-deviation and deviation correction were not taken into consideration and what considered were the problems of downhole safety and how to increase drilling speed.

Selection of the outside diameter of the drill collar should satisfy the requirement of milling operation after drill pipe sticking.

The length of the drill collar was decided by the maximum weight on bit used.

Hydro-hammer could be used in order to increase penetration rate in hard rock.

2.10.3 Selection of Non-core Drill Bit

Selection of drill bit denotes the selection of adaptability of drill bit to rock layers, with purpose that the selected drill bit adapts to rock layer, so a high drilling rate can be obtained, the service life of drill bit prolonged and drilling cost reduced.

Information of international continental scientific drilling shows roller cone bit or button bit can be used for non-core drilling in hard rock layers with high abrasiveness. As the frequent variation of the hard and soft layers of metamorphic crystalline rock which is unlike sedimentary rock that can keep relatively stable in a certain interval, wide range drill bit should be emphasized. Drill bit should be selected in accordance with the drillability and the compressive resistance of the rock, by reference to the bit type recommended by the manufacturer based upon the physical properties of the rock to be drilled.

By comprehensively considering all factors it was decided that type H617–H727 or HJ617–HJ727 roller cone bits were to be used for non-core drilling while SKHA617–SKHA717 roller cone bits were to be employed for reaming drilling.

2.10.4 Design of Reaming Drill Bit

Design of reaming drill bit included the design of reaming size series, the design of adaptability of the reaming part to the rock layer and the selection of pilot bit type, in which,

1. The design of reaming size series Rock fragmented volume, drilling speed, prevention of bit accident and influence of bit accident on bit service life should be considered. The series of reaming size also affects drilling efficiency as too much reaming size grade though increases penetration rate yet the times of reaming are also increased, leading to an unsatisfactory comprehensive drilling result, while too less reaming size grade would reduce reaming efficiency.
2. The design of adaptability of the reaming part of the drill bit to the rock layer Like the drill bit selection for non-core drilling, roller cone reaming bit (Fig. 2.17) was the best choice. Equipped with hydro-hammer with large impact power, button reaming bit could be selected for use.
3. The selection of pilot bit type As reaming was conducted under the precondition that the quality of the original borehole was up to standard thus there did not exist the problem of borehole deviation control. Because the pilot bit only served the functions of piloting, breaking the large pieces of fallen stone and clearing away the settled sand, tungsten carbide structure should be used for the pilot bit (see Fig. 2.17).

2.11 Drilling Fluid Technique and Solid Control Program

2.11.1 The Main Technical Problems Should Be Considered

1. The selected treating agent should reduce as much as possible the influence on the analysis of formation fluid The addition of any treating agent will exert influence on the analysis of formation fluid. Treating agent with less influence should be selected for use under the precondition that safe drilling can be guaranteed.
2. Scaling at inner wall of drill rod The well known scaling forms include extended scaling, scaling at vortex area, filtration scaling, eccentric scaling and spiral scaling, among which the extended scaling exerts the largest influence. Scaling at inner wall of drill rod is mainly produced by drilling tool rotation, concentration and size of solid grains in drilling fluid, and surface adsorption of solid grains. Scaling at inner wall of drill rod will seriously affect the smooth uplifting of wireline core barrel.
3. High circulating pressure consumption results from small annular clearance The annular clearance for wireline core drilling is only 5–8 mm and this annular clearance will become even small in case of filter cake existence, causing very high annular circulating pressure drop, rock avalanche at fractured zone caused by high suction pressure and formation leakage caused by high surge pressure.
4. Solid control As the most harmful solid in drilling fluid, cuttings will affect the properties of drilling fluid in the

Fig. 2.17 Reaming drill bit

whole process of drilling, increasing density, viscosity, yield point, filter loss, mud cake, abrasiveness, glutinousness and flow resistance of drilling fluid, decreasing drilling speed and increasing rotary table torque. In wireline core drilling, the increase of cuttings will accelerate scaling at pipe wall.

5. Formation leakage and borehole wall sloughing High stress metamorphic rock zone would be drilled in this borehole. Fault and fractured zone would easily cause lost circulation and borehole wall out of stable, resulting in complicated downhole situations. Spilling of formation liquid probably results in pressure kick or blowout. Thus in drilling process attention should be paid to leak protection and anti-sloughing, as well as blowout prevention.
6. Drilling tool wear Small annular clearance and high rotary speed will easily cause an increase of the friction among formation, drilling tool and casing, thus accelerating the wear of drilling tool.
7. High temperature 150 °C high temperature will exert unfavourable influence upon the flow pattern and oiliness of the most drilling fluids and upon the anti-scaling additives. The drilling fluid additive selected must stand a high temperature environment above 150 °C, with properties kept stable.

Basic requirements for drilling fluid:

1. In coring, drilling fluid should have good anti-scaling property, oiliness and rheology property.
2. In reaming drilling or non-core drilling, drilling fluid should also have good cuttings carrying capacity.
3. In deep drilling, drilling fluid should also have good property of temperature resistance.
4. Should be equipped with complete solid control equipment and appropriate measures.
5. Technical measures for preventing and treating circulation loss and borehole wall sloughing should be available.
6. Anti-brine contamination The employed drilling fluid system should maintain good stability under the condition of 10 % NaCl.

2.11.2 Design of Drilling Fluid Type

1. Surface drilling (0–100 m, 444.5 mm hole size)
   Drilling fluid system: common water base drilling fluid
   Drilling fluid make-up and mud maintaining treatment agent: NV-1 artificial bentonite, PAC-141 thickening fluid loss reducer and NaOH
2. Core drilling
   Drilling fluid system: polymer drilling fluid
   Drilling fluid make-up and mud maintaining treatment agent: LBM low viscosity extender, JT888 anti-sloughing fluid loss reducer, RH-3 lubricant, RH-4 cleaning agent, XY-27 thinner and NaOH
3. Reaming/non-core drilling
   Well depth: 0–3000 m
   1. Drilling fluid system: polymer drilling fluid
   2. Drilling fluid make-up and mud maintaining treatment agent: NV-1 artificial bentonite, PAC-141 thickening fluid loss reducer, RH-3 lubricant, SK-III thinner and NaOH
   Well depth: 3000–5000 m
   1. Drilling fluid system: polysulfonate drilling fluid
   2. Drilling fluid make-up and mud maintaining treatment agent: NV-1 artificial bentonite, PAC-141 thickening fluid loss reducer, SMP anti-high temperature fluid loss reducer, RH-3 lubricant, SK-III thinner and NaOH.

2.11.3 Solid Control

1. Solid control equipment adopted in design included oscillating screen, desander, desilter and centrifuge.
2. For surface drilling, 60–100 mesh oscillating screen was continuously utilized, and desander, desilter (with 200 mesh screen) and centrifuge were used.
3. For core drilling, 200 mesh oscillating screen was continuously utilized, and desilter (with 200 mesh screen) and centrifuge were used.
4. For reaming/non-core drilling, 100 mesh oscillating screen was continuously utilized, and desander, desilter and centrifuge could be intermittently used.
5. Analyses of solid content and constituent must be made every day and then relevant measures could be adopted based upon the analytical results.

2.12 Well Cementation and Completion Program

2.12.1 Well Cementation Program

1. According to the design of borehole structure, well cementation was to be conducted based on two programs; the first program denoted a design of five layers of casing run in a single borehole and in the second program which was a double hole program three layers of casing were to be run in the pilot hole (Table 2.11) and five layers of casing run in the main hole (Table 2.12).

Table 2.11 Design of casing program in the pilot hole

No.	Drill bit		Casing			Chock ring position (m)	Slurry return depth (m)
	Size (mm)	Drilling depth (m)	Size (mm)	Setting depth (m)	Cementing interval (m)		
1	444.5	100	339.7	100	0–100	90	0
2	215.9	1000	177.8	1000	0–1000	980	0
3	156	2000	127.0	1800	850–1800	1750	750

Table 2.12 Design of casing program in the main hole

No.	Drill bit		Casing			Chock ring position (m)	Position of tail landing funnel opening (m)	Slurry return depth (m)
	Size (mm)	Drilling depth (m)	Size (mm)	Setting depth (m)	Cementing interval (m)			
1	444.5	100	339.7	100	0–100	90		0
2	311.1	1800	273.0	2000	0–2000	1980		0
3	244.5	3000	219.1	3250	1850–3250	3210	1850	1750
4	200	4500	177.8	4600	3100–4600	4540	3100	3000
5	156	5000	127.0	4800	4450–4800	4740	4450	4350

Note (1) Whether tie back was to be conducted should be decided according to the situation after 177.8 mm tail pipe was hung. (2) Cement slurry of intermediate casing returned to surface and cementing slurry of tail pipe returned to 100 m above the funnel opening

2. In design of cement slurry, the problems of leakage and improvement of slurry displacement efficiency in narrow clearance should be put into consideration (Figs. 2.18 and 2.19).
3. Requirements for cement slurry total properties

It was required that 60 min should be added for cement slurry thickening time on the base of construction time.

It was required that cement slurry should have good rheology property because of the large friction drag in cementing and in cement slurry displacement resulted from small annular clearance in cementation of 127, 177.8 and 219.1 mm tail pipes.

For well cementation deeper than 3500 m with high formation static temperature (>110 °C) sand cement slurry system was to be adopted to prevent cement slurry from strength retrogression due to high temperature.

API filter loss of cement slurry for tail pipe cementation should be less than 100 ml.

Considering that leakage might happen in cementing in this borehole, the experiment of low density cement slurry system should be well made in advance besides the preparation of conventional density cement slurry system.

24 h compressive strength of the cement slurry should be larger than 14.0 MPa.

It was recommended that MTC cementation was to be used in 219.1 mm casing cementation in the main hole, based upon the technical requirement of small annulus cementation, in combination with the technical characteristics of drilling mud transforming into cement slurry.

2.12.2 Principle in Design of Casing String Strength

1. **Designed safety factor**
   Safety factor of tension (S_t): 1.8
   Safety factor of collapsing (S_c): 1.125
   Safety factor of internal pressure strength (S_i): 1.1
2. **Calculation of external load**
   Calculation model for strength: two-dimensional stress model
   Calculation method for buoyance: buoyance factor method
   In calculation of effective external squeezing force, the following factors should be considered: (1) inside casing 50 % space was emptied (for 219.1 mm technical tail pipe and 177.8 mm × 4500 m moving casing, 1/3 was emptied). (2) Full hole saturated salt water (density 1.15 g/cm^3) was used in calculation of fluid column pressure outside casing. (3) Mud density (1.05 g/cm^3) was used for the pressure outside casing in calculation of internal pressure.
   Calculation method of internal pressure: based on oil well kick.
3. **Other factors should be considered**
   1. Under the condition that external load was satisfied, design should be made based upon the method of minimum cost.
   2. Selection of casing thread: TM thread was used for moving casing and 177.8 mm extreme-line casing and trapezoidal thread for other casings.

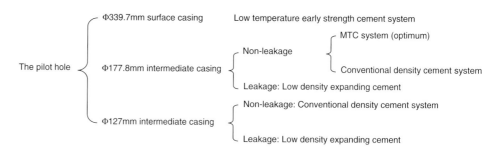

Fig. 2.18 Cement slurry design for the pilot hole

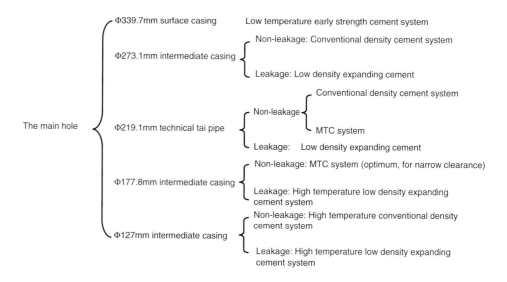

Fig. 2.19 Cement slurry design for the main hole

3. As 156 mm drill bit was to be used for final hole drilling, in selecting the wall thickness of different casings the drift diameter must satisfy the requirement of drill bit diameter for next step hole opening.

2.12.3 Well Completion Operation

After well cementation with 127 mm tail pipe, to avoid the opened hole section being filled with some cement slurry and then long-term observation instrument could not be set down, the cementing techniques of casing packer + differential pressure stage collar was employed. The packer was to be set before cement injection and then the differential pressure stage collar at the top of casing packer was opened and cement slurry was injected at the top of casing packer. After 48 h curing cement plug was drilled out by using 73 mm oil tube + 89 mm drill collar × 110 m + 89 mm PDM + 108 mm drill bit, then the borehole was completed and with protection liquid injected.

The designed well head device for completion is shown in Fig. 2.20.

2.13 Design of Moving Casing

2.13.1 Necessity of Adopting Moving Casing Design

Because lots of undefined factors exist in rock formation, adequate casing program should be prepared in design of borehole structure, so as to deal with the complicated problems may happen. In practical drilling construction, however, drilling cost must be taken into account, thus casing program and setting depth should be adjusted according to actual situation. After running casing each time, drilling with small sized drill bit is conducted first and then reaming is carried on when complicated situation is encountered and casing setting is necessary. This construction method often brings about two harmful results: (1) the annular clearance (between inside wall of casing and drilling tool) at upper hole section is much larger than that (between opened hole and drilling tool) at lower hole section, the consumption of circulating pressure is large in wireline core drilling, discharge capacity is restricted, and mud flowing

2.13 Design of Moving Casing

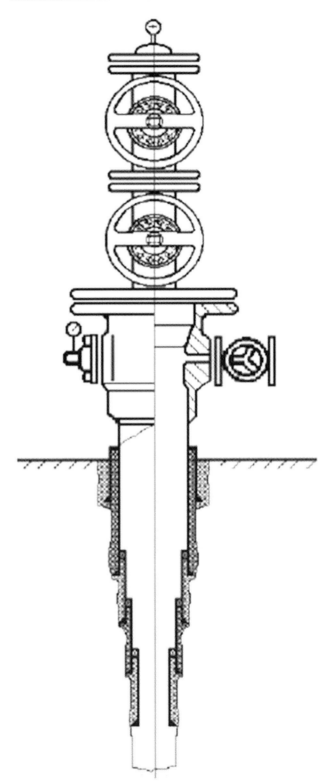

Fig. 2.20 Well head device for completion

position where hole size suddenly changes (the boundary area of opened hole and casing shoe).

To solve abovementioned problems, moving casing technique was adopted in design, i.e. after the larger sized casing is set another casing with inside diameter slightly larger than drill bit is set in the former larger casing, without cementation and can be retrieved when necessary. This is called as moving casing. In this way the cuttings carrying capacity under restricted discharge capacity at upper hole section can be improved and accident of drilling tool broken caused by collision of high rotation drilling tool against the inside wall of large sized casing can be avoided. Furthermore, as the bearing effect is produced by the movement between drilling tool and casing, accident of casing broken caused by serious wear of drilling tool to fixed casing in long time drilling process can be avoided. So the application of moving casing technique in core drilling in large diameter casing is very necessary and the experiences of scientific drilling in the former Soviet Union and in Germany indicated that this technique was necessary and feasible.

2.13.2 Fixing of Moving Casing

In this design two kinds of thread type single stage casing-head used for oil drilling, i.e. 339.7 mm ($13^{3}/_{8}$ in.) × 177.8 mm (7 in.) and 273.0 mm (or 219.1 mm) × 177.8 mm were to be utilized to solve the problems of upper fixing, suspending and retrieving the moving casing (Figs. 2.21 and 2.22). At the middle position of the moving casing was to be used a rigid centralizer to improve the stability. At the lower position of the moving casing was to be utilized a special double cone casing shoe with large contact surface and water channels to prevent the moving casing from moving downwards. The weight of the casing was separately borne by the upper suspension and the lower holding in a certain proportion, at the initial stage of fixing the upper suspension bore more weight while the weight the lower holding bore would become more along with casing elongation resulted from the increased temperature as borehole was deepened. This variation was still within the design limits.

2.13.3 Safety Management of Moving Casing

Moving casing is under the condition of long-lasting impact and wear of high rotation drilling tool, thus feasible and reasonable precaution, accident treatment and safety inspection programs must be adopted to avoid casing accidents. Furthermore, strict casing safety management measures should be taken.

velocity at upper hole section decelerates. As a result, cuttings cannot be effectively carried out; (2) the drilling tool with high rotation speed is easily broken at the borehole

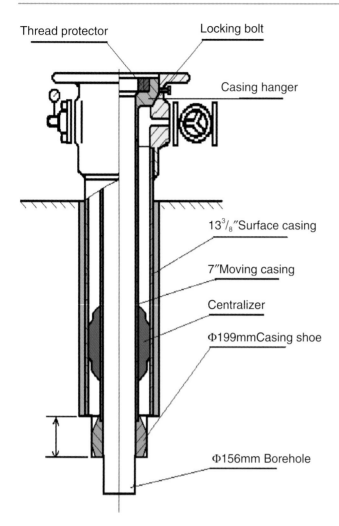

Fig. 2.21 Fixing of moving casing in the second opening drilling

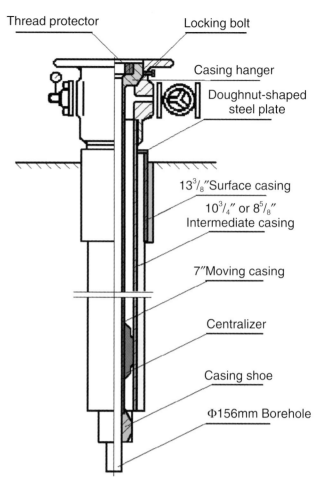

Fig. 2.22 Fixing of moving casing after the third opening drilling

In the design strict measures were worked out for anti-sticking, anti-breaking and for accident treatment.

2.14 Time and Cost Estimation

2.14.1 Designed Construction Progress

According to the initial design, drilling construction for double-hole program needed 1138 days, in which the pilot hole drilling construction needed 242 days (Table 2.13; Fig. 2.23) and the main hole drilling needed 896 days (Table 2.14; Fig. 2.24). Moreover, the construction before drilling and drill rig moving and installation needed 25 days, and completion logging and geothermal gradient logging needed 20 days. The arrangement of the total construction progress of the whole project can be found in Fig. 2.25.

2.14.2 Budgetary Estimation of Cost

Drilling engineering cost included the corollary tool cost, the construction cost for the pilot hole and the construction cost for the main hole (see Table 2.15). Drill rig daily cost was based on 35,022 RMB Yuan per day and thus the total budgetary resources reached to 96,454,000 RMB Yuan.

2.15 Change and Modification of Design

1. Change of core drilling diameter

As the both ends of 139.7 mm wireline drill rod were upset to 146 mm and the diameter of drill rod sub was also 146 mm, the wall clearance for wireline core drilling was only 5 mm (the clearance at the position of upset ends of

2.15 Change and Modification of Design

Table 2.13 Plan of the pilot hole construction progress

Sequence of spudding-in	Content	Day of operation	Accumulated days
Before drilling	Installing equipment	5	5
The first opening (spud-in) surface drilling	Actual core drilling	4	9
	Tripping for drilling	0.4	9.4
	Core fishing	1.4	10.8
	156 mm reamed to 244.5 mm	3.5	14.3
	244.5 mm reamed to 311.1 mm	4.2	18.5
	311.1 mm reamed to 444.5 mm	4.2	22.7
	Tripping for reaming	1	23.7
	Set casing and well cementation	2	25.7
	Set moving casing	0.5	26.2
The second opening (spud-in) to 1000 m deep	Actual drilling	31.3	57.5
	Core fishing	6.3	63.8
	Tripping	2.6	66.4
	Reaming drilling	25	91.4
	Tripping for reaming	1.6	93
	Set casing and well cementation	2	95
	Straightening drilling (420 m)	12	107
	Tripping for Straightening drilling	0.8	107.8
	Shifting drilling tool	2	109.8
The third opening (spud-in) to 2000 m deep	Core drilling	35	144.8
	Core fishing	21	165.8
	Tripping	9	174.8
	Straightening drilling (420 m)	12	186.8
	Tripping for straightening drilling	2.2	189
	Shifting drilling tool	3	192
	Set tail pipe and cementation	2	194
Others	Treating drilling fluid	5	199
	Equipment repair	5	204
	Hole testing and sampling	15	219
	Unpredictable	23	242

drill collar and drill rod, and drill rod sub) and 8.15 mm (the clearance of drill rod body). This clearance was too narrow. To further improve the hydraulic properties of down hole coring tool, decrease annular pressure drop and ensure safety for borehole, the diameter of drill bit was increased 1 mm, i.e. from originally designed 156 to 157 mm.

2. **Change of drilling method for the first opening (spudding-in)**

Full hole coring was required for CCSD-1 Well to provide complete geological information such as full hole core for geoscientific study. According to this guiding ideology, core drilling method was adopted in the design

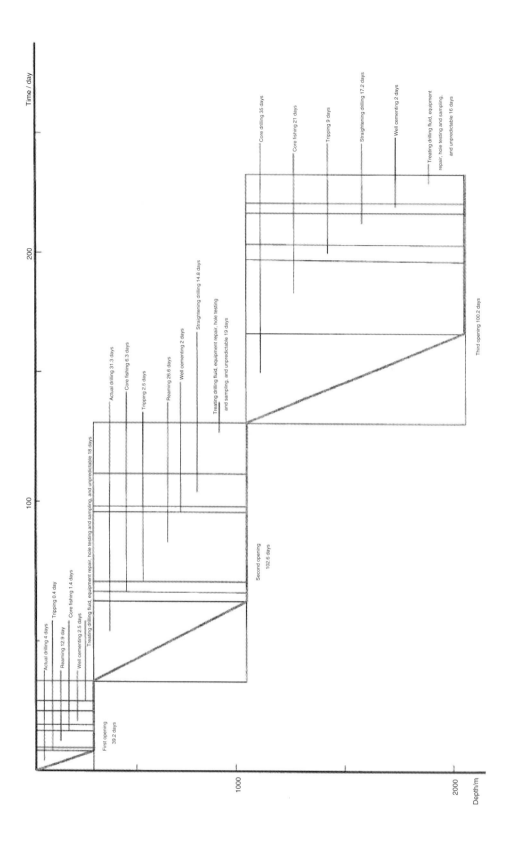

Fig. 2.23 Construction progress of the pilot hole

2.15 Change and Modification of Design

Table 2.14 Plan of the main hole construction progress

Sequence of spudding-in	Content	Day of operation	Accumulated days
Before drilling	Installing equipment	15	15
The first opening (spud-in)	244.5 mm drilling (with VDS)	4.2	19.2
	244.5 mm reamed to 311.1 mm	4.2	23.4
	311.1 mm reamed to 444.5 mm	4.2	27.6
	Tripping for reaming	1	28.6
	Set casing and well cementation	2	30.6
The second opening (spud-in)	244.5 mm drilling (with VDS)	80	110.6
	244.5 mm reamed to 311.1 mm	80	190.6
	Tripping for reaming	13.3	203.9
	Set casing and well cementation	3	206.9
	Set moving casing	1	207.9
The third opening (spud-in) (Well depth 3000 m)	Actual drilling	35	242.9
	Core fishing	21	263.9
	Tripping	15	278.9
	Straightening drilling (375 m)	13	291.9
	Tripping for straightening drilling	3.3	295.2
	Shifting drilling tool	4	299.2
	156 mm reamed to 244.5 mm	35	334.2
	Tripping for reaming	9	343.2
	Set casing and well cementation	3	346.2
	Set moving casing	1.5	347.7
The fourth opening (spud-in) (Well depth 4000 m)	Actual drilling	35	382.7
	Core fishing	49	431.7
	Tripping	21	452.7
	Straightening drilling (375 m)	13	465.7
	Tripping for straightening drilling	12	477.7
	Shifting drilling tool	5	482.7
	156 mm reamed to 200 mm	28	510.7
	Tripping for reaming	12	522.7
	Set casing and well cementation	4	526.7
	Set moving casing	1.5	528.2
The fifth opening (spud-in) (Final well depth 5000 m)	Actual drilling	35	563.2
	Core fishing	49	612.2
	Tripping	27	639.2
	Straightening drilling (375 m)	13	652.2
	Tripping for straightening drilling	16	668.2
	Shifting drilling tool	5	673.2
	Set tail pipe and cementation	2	675.2
Others	Equipment repair	40	715.2
	Equipment maintenance	20	735.2
	Drilling fluid maintenance	20	755.2
	Hole testing	40	795.2
	Unpredictable	100	895.2

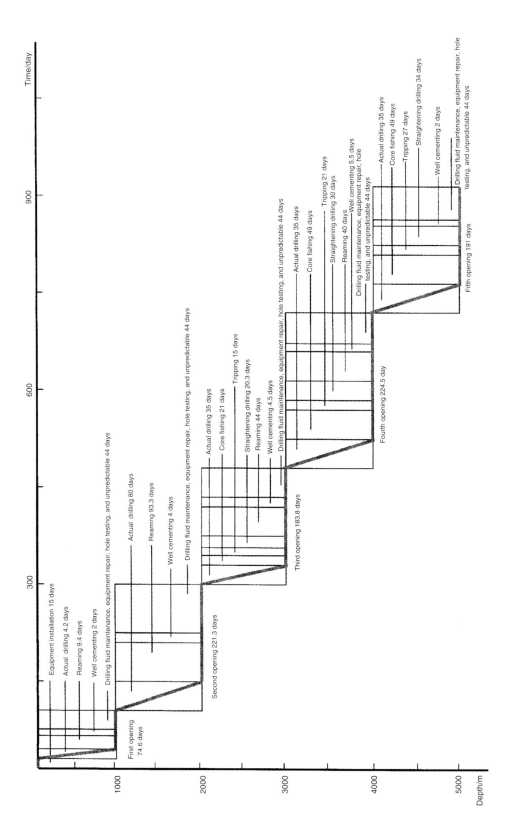

Fig. 2.24 Construction progress of the main hole

2.15 Change and Modification of Design

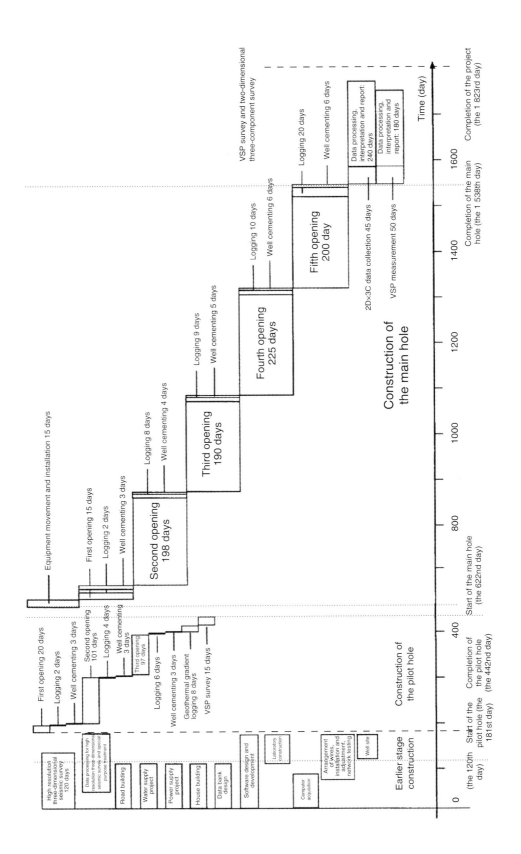

Fig. 2.25 Overall arrangement of the construction progress of the CCSD project

Table 2.15 Budgetary estimation of drilling engineering

No.	Item	Cost (in million RMB Yuan)				
1	Cost for necessary tool	30.94				
2	Construction cost for the pilot hole	14.849	Before drilling and equipment installation Material cost Cementation cost	1.185 3.245 1.204	Drill rig cost Drilling fluid cost	8.475 0.74
3	Construction cost for the main hole	50.665	Before drilling and equipment installation Material cost Cementation cost	0.185 9.514 5.736	Drill rig cost Drilling fluid cost	31.38 3.85
Total		96.454				

Note Afterwards, the total cost for drilling was readjusted to 101.59 million RMB Yuan

Table 2.16 The changed design of casing string for CCSD-1 Well structure

No.	Drill bit		Casing			Chock ring position (m)	Position of tall landing funnel opening (m)	Slurry return depth (m)
	Size (mm)	Drilling depth (m)	Size (mm)	Setting depth (m)	Cementing interval (m)			
1	444.5	100	339.7	100	0–100	90		0
2	311.1	2000	273.0	2000	0–2000	1980		0
3	244.5	4500	193.7	4500	1750–4500	4440	1850	1750
4	157	5000	127.0	4800	4250–4800	4740	4350	4250

of the first opening for the pilot hole. However, as many boreholes deeper than 100 m had been drilled in the surrounding area and CCSD-PP2 and test hole constructed near the well site of the main hole, lots of core and geological information were available for reference, cutting logging could be used instead of core, without any influence on geoscientific study. Moreover, by using non-coring method drilling program would be simplified, and drilling construction time and cost would be reduced. Also, non-core drilling in the first opening was beneficial to adopting technical measures to prevent hole deviation. Based upon this actual situation, in the 0–101 m hole section of the first opening (spudding-in), non-core drilling with 444.5 mm roller cone drill bit equipped with heavy collar was conducted and cutting sample was fished out every meter for geoscientific study, whereas the original design (core drilling first and then reaming in steps) was abandoned.

3. **Change of borehole structure and casing program**
In the process of ascertaining the main materials before starting the construction, it was found out that in the design of the fourth layer of casing string 177.8 mm (7 in) thin wall (δ = 8.065 mm) extreme-line casing was to be employed, which could only be imported from Japan because it was an unconventional type and thus unavailable in China. Although the Japanese company was capable of producing the casing

2.15 Change and Modification of Design

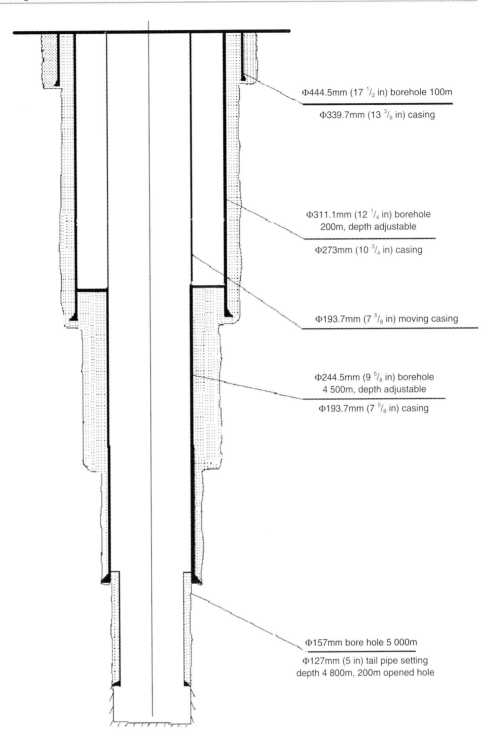

Fig. 2.26 The changed design of borehole structure and casing program

jet they were unwilling to because of our less quantity. In this connection, the borehole structure was appropriately changed by the designer at the request of China Continental Scientific Drilling Engineering Centre. The original third layer casing (244.5 mm drill bit × 219.1 mm casing) and the fourth layer casing (200 mm drill bit × 177.8 mm extreme-line casing) were combined into one layer casing, i.e. 244.5 mm ($9^5/_8$ in) drill bit × 193.7 mm ($7^5/_8$ in) casing. The changed hole structure is shown in Table 2.16 and in Fig. 2.26.

3 Well Site and Drilling Equipment

3.1 Well Site

CCSD-1 Well is located at the southwest of Donghai County, Lianyungang City, Jiangsu Province, and 18 km (Fig. 3.1) from Donghai County. There is a south-north village byway on the west side of 100 m from the well site, and a village highway (Fengquan Road) on the east side of 300 m from the well site. About 2 km along the road to the south is Maobei Village; about 3 km along the road to the north is just jointed with Xuzhou-Lianyungang Highway, from which is about 15 km to Donghai County. Shihu Village Railway Station of East Longhai Railway Line is 8 km away on the north side of the well site, and Lianyungang Port and Airport (Baitabu) are respectively 86 and 32 km away on the east side of the well site.

The well site area is hummock-and-hollow plain physiognomy with wide open terrain, which is higher in the east and lower in the west. The west is hammock mound; the east is valley depression, with the height above sea level of 24–33 m. The surface soil is brown and skeletal earth with gravels and medium-grained sands, which belongs to Sushiling sandy soil. The water system of this area is not well-developed. There are a medium-sized Ahu reservoir and a small-sized reservoir 4 km away on the west side of the well site, and an Anfengshan large-sized reservoir on the east of 2 km from the well site and a Maobei small-sized reservoir of 500 m from the well site. All the reservoirs are connected by diversion rivers in order to adjust the water flow.

Eight kilometers away on the north side of the drill site is a transformer station with 35/10 kV, which leaves a space for installing transformer to transmit the electricity power to the drill site area, and a high volt transmission line with 10 kV capacity built by the transformer station could meet the demands of drilling equipment operation and water supply.

Translated by Zhang Yongqin.

The well site construction included the foundation of drilling equipment, enclosing wall and gate, various pools, roads and drainage ditches and so on. The well site is a rectangle of 85 m × 120 m. The gate leads straightway to the headquarters of CCSD center and the working and living sections of the drilling crew.

At the drill site were built the concrete roads of 4.5 m wide, which were connected with a road had been previously built. The concrete thickness is 0.25 m, the gradient is 3 %, and the load grade is heavy standard of the Highway BZZ-100, for transporting and hoisting heavy equipment. The operation section at the drill site was paved with 20 mm thick cement in order to keep clean. At the drill site were also arranged the rooms for drilling supervision, meeting of drilling crew, drilling engineers on duty, drilling fluid examination, logging, fluid chemistry testing, and fire fighting.

There are pools for sands settling, debris, waste fluid, sewage water, disposing, storage, clean water at the well site. The sand settling pool is used for transferring the solids cleaned by the solid control equipment. the debris pool is used for heaping up the mud solids, the waste fluid pool is used for laying up the waste mud, the sewage pool is used for concentrating waste water, the clean water and storage pools are used for storing clean water for fire fighting and drilling. The contaminated materials from sand settling pool, debris pool, and waste fluid and sewage pools are cleaned up together after finishing the drilling engineering. Discharge ditches for sewage disposal are set up around the drill rig base, mud pump, diesel-engine generator and mud tank, the discharge ditches are connected with the sewage pool and thus the sewage water for cleaning drilling equipment and the mud dropped can be gathered into the sewage pool (Fig. 3.2). The arrangement for ZJ70D drill rig is shown in Fig. 3.3.

The bases for drilling equipment at well site are of cast-in situ concrete, with the basic materials of medium-grained sands, quarry stone, cement and reinforcing bar. The mixing ratio (cement: sand: stone) of the raw materials are: 1:2:4 for the base of drill derrick, 1:3:6 for the base of the drilling

Fig. 3.1 Traffic and location map of CCSD-1 Well

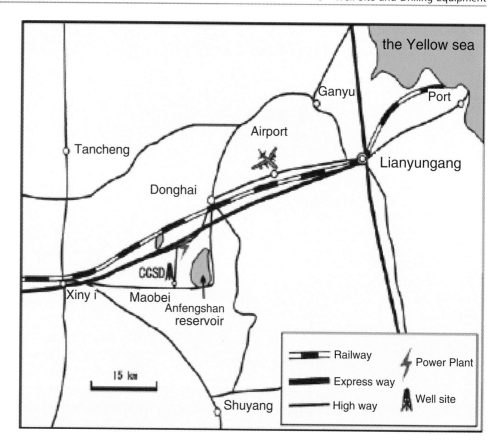

Fig. 3.2 Schematic map of road and drainage ditches at well site

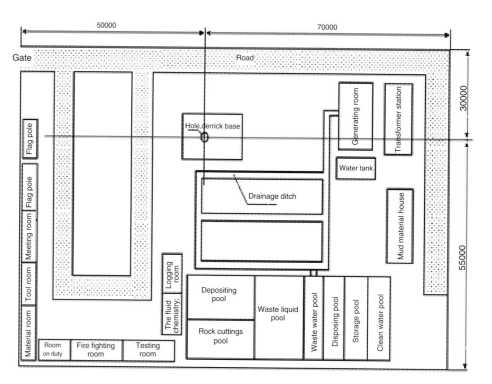

Fig. 3.3 Schematic map of ZJ70D drill arrangement

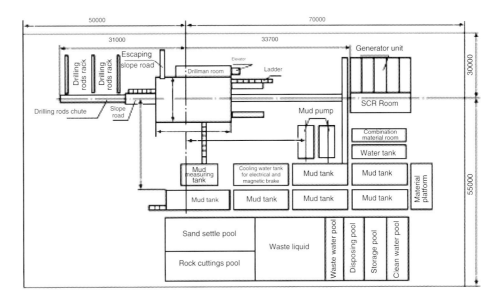

Fig. 3.4 Drill site

control chamber. The surfaces of the bases is 0.1 m thick (the ratio of cement to sand is 1:2). The loads on the bases are respectively: the load on the drill derrick base is 1500 t, the load on pump base is 300 t, the load on drilling control chamber base is 300 t, and the load on the circulation tank base is 180 t. The time for curing is 5 days and within 28 days the base cannot bear the maximum load. The photo of the well site after accomplishing the installation of the drilling equipment is shown in Fig. 3.4. The bird's eye view of the well site general arrangement after the finishing the well site construction is shown in Fig. 3.5.

3.2 Drilling Equipment

Different from conventional petroleum and geological core drilling, a set of completely new techniques were to be used for CCSD-1 Well. For drilling equipment selection,

Fig. 3.5 A bird's view of the drill site

not only the practicability, but also the advantages, economy and nationalization should be considered. In preliminary design, electric drive drill was taken as the prime choice for CCSD-1 Well and a priority was given to a 7000 m electric drive drill rig. The electric drive drill rig is divided into two types of DC and AC frequency conversion drive. Due to the limitation of market supply at that time, homemade AC frequency conversion drive drill rig was at the stage of trial-manufacture and development, and the technology for the AC drill rig was not yet well developed. No. 3 Drilling Company of Zhongyuan Petroleum Administration Bureau, the bid winner for the sub engineering of drilling for CCSD-1 Well utilized a ZJ70D drill rig shown in Fig. 3.6.

3.2.1 ZJ70D Drill Rig

A thorough and full outlook to the present developing trend of petroleum drill rigs in the world can let us see that the electric drive drill rig have become out first in petroleum drilling industry, in which the electric drive drill rig has shown a maximum superiority to the mechanical drive drill rig, especially in deep well drilling. ZJ70D drill rig was designed and manufactured by Baoji Petroleum Machinery Plant in terms of the requirement of SY/T-5690 Petroleum Machinery Models and Basic Parameters and the relevant API Specifications and a renovation and reconstruction plan of China National Petroleum and Gas Corporation. ZJ70D is the first DC electric drive petroleum drill rig put into the domestic market in order to satisfy the demands in deep petroleum and natural gas exploration and development

Fig. 3.6 ZJ70D drill rig used for CCSD-1 Well

and some new drilling technologies. The technical performance and reliability of ZJ70D already reached the world advanced level in the 1990s; it is an upgrading product of homemade petroleum drill rigs, with the main advantages as follows:

1. Simplified transmission system and high power utilization: For conventional mechanical drive drill rig, the prime engine usually transfers the power to the drill rig with help of the hydraulic torque converter, reducing gear box, transmission belt or chain, whereas the prime engine of electrical drive drill rig only needs the transmission belt or chain and universal coupling to transfer the power to drill rig without the torque converter and the reducing gear box. Therefore, electrical drive drill rig has many advantages of simplified transmission chain, smooth drive, high transmission efficiency, low mechanical failure, less mechanical wear and convenient repair.
2. Stepless speed regulation: With the help of stepless speed regulation the range of rotary speed of winch, rotary table and mud pump can be increased and thus constant torque output in a low speed can be realized. This is beneficial for the derrick and base to lift and lower drill rods and tools, and to handle down hole accidents, especially for mud pump to obtain various combinations of pump pressure and displacement and to realize a working situation of high pressure and small displacement, without changing cylinder liner to adjust the pump displacement in a large range. Therefore, the application range of the drill rig is expanded, being more suitable to the applications of multi-techniques in a single well. It is just what CCSD-1 Well needs for drilling.
3. Good mobility: Since the prime engine and the driving engine are connected by cable, and the modularization design of electrically controlled system, the arrangement of power system is flexible and not limited, being convenient for assembling and disassembling. So the weight of the whole drill rig is lighter than that of mechanical drive drill rig and the cost for transportation can be reduced.
4. Convenient operation and high automation level: Automation and intelligent operation can be easily realized, being convenient for man-machine communication. Rotary speed and torque of the drill string can be adjusted at will according to the need of drilling technology. Mud displacement can be adjusted to some extent at any moment, torque change of drill string in the well can be supervised in the process of drilling at any time, and thus drilling efficiency can be increased and in-the-hole accident effectively avoided.
5. The requirements of HSE management system can be preferably satisfied: The requirements for health, security and environment protection can be preferably satisfied. The prime engine (diesel engine) can be installed away from the drilling operation area, so the noise pollution and disturbance to the personnel at drill site can be avoided.
6. It is possible for the drill rig to get power from power supply network: Noise and air pollution from diesel engine is radically eliminated at drill site, the shutdown time from mechanical failure is cut down, and the influence to other machines and intellectualized meters from the surge of drilling load is effectively avoided. And at the same time, it is much safer at the drill site than where natural gas may exist and the possibility of burning from the use of diesel engine is avoided.
7. It is easy to fit a top drive system on an electrical drive drill rig.

The main technical parameters are listed in Table 3.1.

ZJ70D drill rig mainly consists of crown pulley, drawworks, derrick, foundation, rotary table, bailer drawworks, etc.

1. **TC$_5$-450 crown pulley**

The crown pulley denotes some fixed pulleys on the top of derrick; it consists of crown frame, main pulley, guide pulley assembly, bailer pulley, auxiliary pulley, crane frame, bumper beam, etc. Six main pulleys and ϕ 38 mm wire cable for traveling pulley make up a set of assembly pulley system, with the maximum hook load of 4500 kN, pulley outside diameter of 1524 mm. The auxiliary pulleys are used respectively for two air driven drawworks and hang tongs for drill rod. The overall size of the crown pulley is 3407 mm × 2722 mm × 2856 mm, and the weight is 9735 kg.

2. **JC70D drawworks**

The drawworks are driven by two GE752 DC motors, with the maximum output power of 1470 kW, and the maximum fast line tensile force is 480 kN. The drawworks

Table 3.1 The technical parameters of ZJ70D drill rig

Item		Parameter	Item		Parameter
Nominal drilling depth range (m)	ϕ 114 mm drill rod	5000–7000	Derrick	Model effective height/m	Model–JJ450/45-K$_5$ front opening Effective height (m)—45
	ϕ 127 mm drill rod	4000–6000			
Maximum hook load (kN)		4500	Base	Model	DZ450/9-S
Hoist input power (kW)		1470		Height (m)	9
Pulley lines		6 × 7	Motor	Model	GE752
Steel cable diameter (mm)		38		Quantity	6
Rotary table	Model	ZP-375		Total power output (kW)	800 × 6
	Openning diameter (mm)	952.5	Transmission type		AC-SCR-DC direct current drive

Fig. 3.7 The curve of hoist lifting

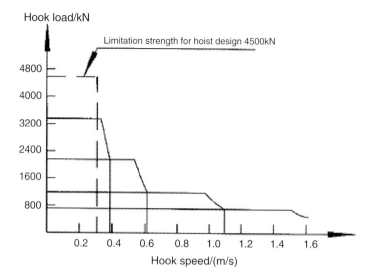

have 4 positive gears and 4 reverse gears; the lifting capacity of the 4 positive gears is shown in Fig. 3.7. The positive gear has a device against over-reeling and bumping, the diameter of drawworks drum is 770 mm, the length is 1310 mm, the diameter of brake drum is 1370 mm, the width is 270 mm, the overall size is 7670 mm × 2812 mm × 3216 mm, and overall weight is 44,000 kg.

3. **JJ450/45-K5 derrick**

The main body of JJ450/45-K5 derrick is in K structure (shown in Fig. 3.8), with front hatch. Divided into four sections and eight parts, the derrick is equipped with casing and pipe decks, the ladder to the second working platform and to the crown pulley platform, cathead pulley, deadline stabilizer and tong hang pulley. A gin-pole is equipped in JJ450-K5, with a door shaped structure consists of left and right front legs and left and right back legs, for rising, lowering and supporting the derrick. A rising device applying the gin-pole for rising is equipped, with rope and balance pulley. A hydraulic cushion device is specially set up in order to let the derrick lean steadily on the gin-pole during erecting while let the center gravity of the derrick move ahead by deadweight falling during lowering.

The maximum hook load of the second working platform is 4500 kN with 6 × 7 rope systems, without drill-pipe stand racking and wind load. The working height of JJ450/45-K5 derrick is 45 m; the height of the second working platform is optional in 24.5, 25.5 and 26.6 m. The second working platform can accommodate 6600 m of φ 127 mm drill rod (28 m length drill-pipe stand), and at the same time, it also can accommodate 4 columns of φ 254 mm drill collar and 6 columns of φ 203.2 mm drill collar. The capacity against wind force is 36 m/s without hook load and with full drill-pipe stand and the capacity against wind force is 47.8 m/s without hook load and drill-pipe stand. The wind speed should be less than 8.3 m/s when rising and lowering the derrick.

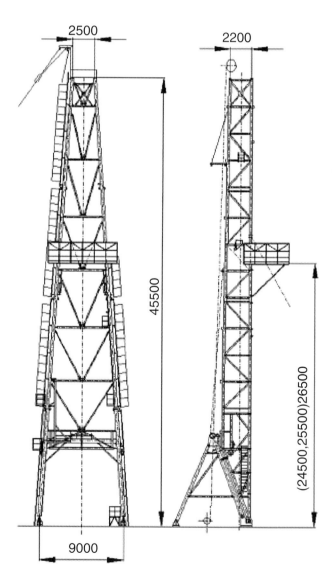

Fig. 3.8 JJ450/45-K5 derrick

3.2 Drilling Equipment

Fig. 3.9 DZ450/9-S foundation

4. **DZ450/9-S foundation**

DZ450/9-S foundation (Fig. 3.9 and Table 3.2) for ZJ79D drill rig consists of lower base, top base, drill-pipe stand setback, drawworks beam, rotary table beam, front and back columns, gin-pole, and etc. The foundation is a double rising structure and its parallelogram structure is convenient for installing equipment on the foundation at a lower position. Using hoist power and hook rope system makes the whole foundation rise to working position from the lower position. The structure of high platform surface and large space fully satisfy the demands for installing BOP in deep well drilling. DZ450/9-S foundation is designed for installing derrick, drawworks, rotary table, driller house, hydraulic power station for tongs, drill-pipe stand setback, and necessary tools and with big and small rat holes.

5. **ZP-375 rotary table**

The maximum diameter of the shaft hole of ZP-375 rotary table for ZJ70D drilling rig is 952.5 mm, with the maximum quiescent load of 5850 kN, the maximum working torque of 32,362 N m, the highest rotation speed of 300 r/min, and gear reduction ratio of 3.56. The weight of the rotary table is 8026 kg, the overall size is 2468 mm × 1810 mm × 718 mm.

6. **Bailer drawworks**

The drum diameter of the bailer drawworks is 400 mm and the length is 1460 mm, and the diameter of wire cable is 14.5 mm. During drilling CCSD-1 Well, the bailer drawworks was used for wire-line coring after necessary reconstruction.

3.2.2 Drill Rig Reconstruction

In the process of CCSD-1 Well drilling, 157 mm diameter impregnated diamond drill bits were mostly used, with wire-line core drilling technology and other advanced drilling techniques. In comparison with petroleum drilling, the annular space of drilling tool for CCSD-1 Well is smaller, the rotary speed of drill bits is higher, and drilling tool should be lowered in borehole steadily and precisely. The available petroleum drill rig is unable to meet the demands on drilling technology in CCSD-1 Well, therefore it was necessary to rebuild the drill rig and some necessary equipment.

1. **Reconstruction of brake system**

The original brake for ZJ70D drill rig was band brake and unable to meet the demand on steady and precise lowering of the drilling tool required by impregnated diamond drill bits in drilling CCSD-1 Well. Before drilling, it was required that the band brake had to be changed to the disc brake, which was recognized as a revolutionary new technology in modern petroleum drilling equipment, and only some countries have developed and used the disc brake techniques. The PSZ series disc brake devices (shown in Fig. 3.10) developed by Machinery Research Institute of Petroleum Exploration and Development Academy with Chinese independent intellectual property rights were used in CCSD-1.

The disc brake in comparison with traditional band brake has many advantages of bigger capacity of braking torque, steady and reliable braking, smooth control, more precise in lowering drilling tools, simple and flexible operation, labor saving, friction pairs with good performance of enduring high temperature and against aging, convenient repairing and adjusting, and remote control available. So the disc brake obviously improves the working conditions, increases the safety and reliability in drilling, provides a necessary

Table 3.2 Technical parameters of DZ450/9-S foundation

Item	Parameter	Item	Parameter
Platform height (m)	9	The maximum combination load	The maximum load of rotary table (kN)	4500
Platform sizes (m)	13 × 12		Stand load (kN)	2200
Rotary table beam bottom height (m)	7.62	The nominal mass (kg)		190,024
Center distance from well mouth to drawworks (m)	3.25			

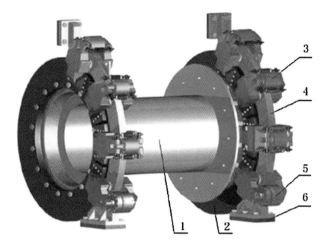

Fig. 3.10 PSZ disc brake. *1* Reel. *2* Brake disc. *3* Working tong. *4* Tong support. *5* Safety tong. *6* Transition plate

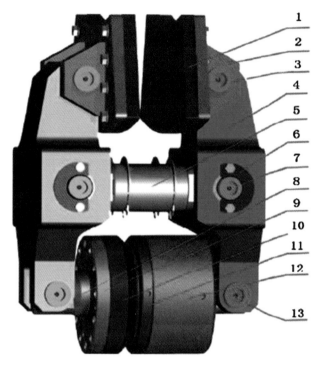

Fig. 3.12 Safety tongs. *1* Brake block. *2* Tong body. *3, 6, 13* Pin shaft. *4* Lever. *5* Fulcrum lever. *7* Detent ring. *8* Piston. *9* Cylinder lid. *10* Hydro-cylinder. *11, 12* Nut

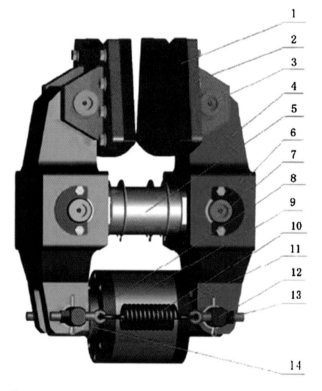

Fig. 3.11 Working tongs. *1* Brake block. *2* Tong body. *3, 7, 12* Pin shaft. *4* Lever. *5* Fulcrum lever. *6* Detent ring. *8, 11* Cylinder lid. *9* Cylinder. *10* Extension spring. *13* Nut. *14* Piston rod

condition to automatically lower drilling tools and lays a foundation for the automation of drilling.

The disc brake is a machinery-electronics integration drilling device concentrating the power transmission and executing mechanism, with five main functions as follows:

(1) Working brake: A handle of brake valve is used to control the working tongs (shown in Fig. 3.11), then a positive pressure is exerted to the brake disc and the adjustable braking torque can be provided to the drawworks, so that the demands of lowering drilling tools, adjusting weight on bit (WOB) and the velocity of lowering drilling tools to suit to different working conditions can be satisfied.

(2) Parking brake: When drawworks stop working or driller leaves the operating platform, the parking handle is started to brake safety tongs (Fig. 3.12) in order to prevent the hook from falling off.

(3) Emergency brake: To press the emergency brake button in case of emergency, then both the working tongs and the brake tongs stop working.

(4) Over-rolling-up protection: In hook rising, due to operation error or other reasons, the working brake doesn't execute when it should do, the over-rolling-up valve or prevention valve sends out a signal, and then makes the working tongs and safety tongs fully brake simultaneously in order to execute emergency braking and avoid bumping accident to crown pulley.

(5) Faulty operation protection: When releasing parking, once faulty operation happens, all the brake tongs will

Fig. 3.13 Principle graph of electronic driller

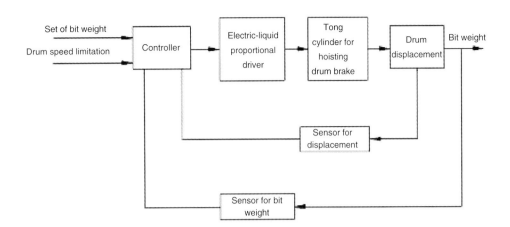

be in braking state and driller has to operate according to the regulations, i.e. to execute working brake and then release parking.

2. **Electronic Driller**

The main functions of the electronic driller include: (1) to ensure drill rig to get a constant control of WOB during drilling operation; (2) to ensure a displacement limitation for hoisting drum; (3) setting-up and displaying on WOB; (4) to alarm in case of drum speed over limitation and automatic emergency brake, etc. The electronic driller uses disc brake as its executing mechanism (Fig. 3.13), intellectually controls the hydraulic oil pressure in order to control the braking torque of the brake disc, so as to realize a real-time control on WOB. In the process of drilling, the control system monitors the real time change of WOB through WOB sensor and then feedbacks to the controller, after analysis and calculation, the control system adjusts the control output, adjusting the disc brake working tong's cylinder pressure through the electrohydraulic proportional driver to control the drum displacement and then realize the adjustment of WOB. The control system monitors the drum speed change in real-time by displacement sensor, when the drum rotation speed exceeds the speed limitation, the controller gives an emergency braking instruction to force the disc brake to brake immediately.

The 157 × 96 mm diameter impregnated diamond bits used in CCSD-1 Well require a much less WOB than in petroleum drilling. The imported M/D electronic driller claiming to be 5 kN in error provided by the drilling contractor could not reach 5 kN. In addition, the WOB used in CCSD-1 Well was less in normal conditions, only 10–35 kN. Since the small annular space between hole wall and drill rod and well deflection and other reasons, the friction drag consumed on the well wall is relatively large, and in the same roundtrip WOB would change a little as the core increased. Because of these reasons, the M/D electronic driller couldn't enhance the precision to lower drilling tools. The ZJ70D drill rig basically possessed the function of precisely lowering the drilling tools after reconstructing the disc brake. And the drillers gradually became skillful to operate the drill rig and their judging ability upon the pressure at well bottom reached a high level, fully satisfying the requirements for diamond wireline core drilling.

3. **Reconstructing ZJ70D drill rig for wireline core drilling**

The original design of core drilling in CCSD-1 Well was to use wireline core drilling technology as the main drilling method and other conventional core drilling technologies as the subsidiary. However, petroleum drill rig has much maladjustment in using wireline core drilling: firstly, the drill rig had a low rotation speed and a high speed top drive must be installed; secondly, internal flush wireline drill rod must be used in order to ensure the inner tube passing smoothly; thirdly, wireline drawworks should be installed in order to fish the inner tube assembly.

ICDP gave much help to CCSD Project and leased RB130 high speed top drive head that was used in KTB pilot hole and with rotation speed over 250 r/min to CCSD project at a reasonable price, and also assisted CCSD to contact a certain foreign company to purchase a full set of 140 mm (with 146 mm outer diameter of drill rod sub) wireline core drill rod and tools. Using the originally equipped bailing hoist, CCSD developed an automatic rope-arrangement system for ZJ70D drill rig by improvement, and also a guide rail, air-driving slips and hanger were designed and made especially for mounting RB130 hydraulic top drive head on ZJ70D drill rig. After all reconstructions accomplished, the 2000 m pilot hole was completed, and it was decided that from 2000 m deep on wireline core drilling technology would be used. However, since the welding shoulders in the inner wall of some wireline drill rod were not completely removed and then the inner tube assembly could not be fished out smoothly, all the wireline drill rods had to be repaired, and the using of wireline core drilling method had to be stopped. After the wireline core drilling rods were repaired, CCSD-1 Well nearly came to the end of drilling and the well depth exceeded the limitation capacity of RB130 hydraulic top drive head. Only after all the drilling work finished in CCSD-1 Well the wireline drill rods were used to conduct some deep drilling tests of wireline coring + PDM + hydro-hammer (three-in-one drilling tool).

3.2.3 The Power System

Usually, there are two types to supply electrical power for electrically driven drill rigs, the built in generator unit and the public electrical grid. For conventional petroleum drilling, the drilling term is usually short, the drill site is out of the way, and thus it is not economical to invest for electricity power supply equipment. So in most cases, petroleum drilling adopts the built-in generator unit to supply electricity power. The working term of CCSD-1 Well was four years, for reducing noise and air pollutions from the diesel engine and saving power, through an argumentation from all aspects, it was decided in CCSD preliminary design that the public electrical grid was to be used to supply the electricity power to CCSD-1 Well.

1. **Way of driving**

The built-in electrical transformer and distributor equipment was used at CCSD-1 drill site, and the transformer chamber could supply multi-channel electricity supply.

(1) The main channel

As the main power source, one way 600 V AC was rectified by silicon controlled rectifier (SCR) chamber and transformed into 0–750 V DC for driving six GE752 Series DC motors with 800 kW, among which four motors were used to drive the mud pumps (each pump was driven by two motors), while the other two motors were used to drive drawworks and rotary table. The series motor has a good adaptability to the impact load of the driven machines, and starting torque and overload capacity are respectively four times and four to five times of the rating torque, and at the same time, the rotation speed variation ratio is larger than other types of DC motors. These characteristics of the series motor are very suitable to the working demands of drilling equipment.

Another way 600 V AC was taken as a standby electrical power and was directly used to drive the 700 kW DC motor of the top drive system.

(2) The assistant channel

The assistant channel was from the 400 and 200 V low voltage AC channel from the transformer chamber, which was mainly used to drive the small sized electrical equipment at the well site, such as solid control device, air compressor, scientific research instrument, office and lighting electricity consumption, etc.

(3) Harmonic wave rectification

During drilling, the rectification system of SCR chamber produced a lot of harmonic waves, through an actual measurement, the harmonic wave components were respectively 5, 7, 9, 11, 13, 15, 17, 19, 21, 23, 25 times, leading to the voltage distortion rate of 60 V bus to be more than the allowable value. Due to the effect of large amount of harmonic wave current, the original reactive compensation capacitor of the power distribution system could not go into use, and the transforming-distributing power system to run in a lower power factor (0.6), and for this reason we had to pay a fine of 30,000 RMB each month for reactive loss, in the meantime, safe running and service life of the equipment and instruments at well site were all affected. For eliminating the harmonic effect, CCSD headquarters appropriated a special fund for rectifying harmonic wave in May, 2005, with the technical measures of installing filter compensator, decreasing resistance of the harmonic wave, supplying a channel for harmonic wave current and decreasing the voltage distortion rate of the bus in the transforming and distributing system. After rectifying the harmonic wave, the power factor increased to 0.97 from 0.6. We didn't need to pay the fine any longer, on the contrary, we got a bonus of 1000 RMB each month. Normal running of scientific research equipment and drilling at the well site was guaranteed, and at the same time an abnormal damage of the electrical equipment resulted from harmonic wave at well site could be avoided.

(4) The result of the power program

The selection of the power program for CCSD-1 Well was a very good engineering example, with excellent economical and social benefits obtained (shown in Table 3.3), from which it can be found that, after accomplishing CCSD-1 Well, the utilization of electricity-grid power supply actually saved 11,084,900 RMB only in power expenditure in comparison with the adoption of diesel-engine electrical-generator, according to the practical time of drilling operation and contractor's settlement standard per day

The electricity-grid power supply eliminated the air and noise pollutions to the nearby environment, improved working environment at well site, avoided a shutdown which might resulted from the malfunction of diesel engine and thus effectively increased drilling efficiency.

2. **Standby power**

For preventing an electrical power failure from public electricity grid and other accidents, a set of diesel engine generator units affiliated to ZJ70D drill rig was taken as the standby power source, among these units four 600 V AC units (Table 3.4) could independently transmit or transmit in parallel as a "main channel" to supply electricity power;

3.2 Drilling Equipment

Table 3.3 Benefit comparison on two kinds of power supply

Item		Drilling, cementing-off and accident handling	Well logging	Shutdown with workers	Shutdown without workers
	Actual operation day (day)	1266	47	43	39
Charging according to diesel engine generator	Power expenses per day (ten thousand ¥/day)	1.55	0.776	0.312	0.264
	Power expenses (ten thousand ¥)	1962.30	36.47	13.42	10.30
Total		2022.49			
Power expenses of electricity grid power supply (Actual settlement expenditure) (ten thousand ¥)		615 (electricity expenses) + 276 (expenses of electricity grid engineering) + 23 (managing expenses) = 914			
Saved expenses of electricity grid power supply (ten thousand ¥)		2022.49 − 914 = 1108.49			

Table 3.4 The technical parameters of the generator units

Generator unit	Main generator unit	Accessory generator unit	Generator unit	Main generator unit	Accessory generator unit
Unit model	CAT3512	ZYTGS	Field current (A)	70	2.2
Unit quantity	4	1	Output power (kW)	980	223
Output voltage (VAC)	600	400	Output frequency (HZ)	50	50
Output current (A)	1443	361	Power factor	0.7	0.8
Field voltage (V)	15	43	Rated revolution (r/min)	1500	1500

Table 3.5 Technical parameters of SA-350A air compressor

Item	Parameter	Item	Parameter
Axle power (kW)	35.1	Oversize (mm)	1300 × 1140 × 1600
Input voltage (VAC)	380	Mass (kg)	1070
Output pressure (MPa)	0.8	Quantity (piece)	2
Air displacement (m³/min)	5.8	Manufacturer	Beijing Fusheng

another 400 V AC unit (Table 3.4) was taken as an "accessory channel" for power supply.

3. **Air power**

At the well site was equipped with a SA-350A oil jet screw air compressor (Table 3.5) as a clean air resource for air-driven drawworks, drilling pipe power tongs and air line control system. Under the same air output pressure, the screw air compressor has higher output efficiency in comparison with the piston air compressor, without easily-worn parts, with strong reliability, higher fitting accuracy between the negative and the positive rotors and between the rotors and compressor case so that the screw compressor can reduce leakage of returning air flow and increase the volume efficiency. The process of compressing air depends on the engagement of the rotors with each other, without reciprocating motion of the air cylinder and thus vibration and noise reduced. Equipped with air-purifying system, the screw compressor guarantees smooth air flow, reliability of the parts and increases the service life.

3.2.4 Corollary Equipment

1. **Mud pump**

Wireline core drilling was considered to be used in the design of CCSD-1 Well at beginning, but positive displacement motor (PDM) hydro-hammer conventional diamond core drilling method was mainly used in drilling operations. As the diameter of core bit was 157 mm, the diameter of the outer tube was 139.7 mm, the diameter of inner tube was 110 mm, the annular spaces between the inner and outer tubes and between the outer tube and hole wall were small, the overflow area of the core bit face could not be large, as well as down-hole power (to drive PDM and hydro-hammer) used, so the pressure fall of the whole circulation system was large, F-1600 mud pump (Table 3.6) equipped with ZJ70D drilling rig could not fully meet the demands for drilling and coring. Therefore, the Baoji Petroleum Machinery Plant was entrusted to develop two pieces of high pressure F-1600HL mud pump (shown in Fig. 3.14 and

Table 3.6 Technical parameters of F-1600 mud pump

Cylinder diameter (mm)	180	170	160	150	140
Rated pressure (MPa)	22.7	25.5	28.8	32.7	34.2
Stroke number (stroke/min)	Displacement (L/s)				
120	46.54	41.51	36.77	32.32	28.15
100	38.78	34.59	30.64	26.93	23.46
80	31.03	27.67	24.51	21.55	18.77
40	15.52	13.84	12.26	10.78	9.39

Fig. 3.14 F-1600HL mud pump

Table 3.7), with increased high displacement and decreased low displacement and with rated pressure increased to 51.9 MPa from 34.3 MPa in comparison with F-1600 mud pump, fully satisfying the requirements of low pressure and large displacement for reaming drilling and high pressure and small displacement for core drilling.

2. **Solid control equipment**

Solid control system consists of oscillating screen, desander (desilter) and centrifuge, with the following characteristics:

(1) 2E48-90F-3TA oscillating screen produced by American DERRICK Company has an adjusting system while drilling, which can randomly, quickly and easily adjust the rising and lowering of the screen with hydraulic jacks within −1° to 5°. Optimum solids handling speed or flow end point can be obtained by adjusting the angle of the screen when rock formations or drilling fluid properties change. Pyramid shaped screen is used, with two layers of fine sheet and one layer of coarse sheet in the middle, in ripple shaped screen surface, with 50–125 % increased practical area, and thus the disposing capacity is increased, and screen balling and overflow can be avoided, and thus screen service life extended. Super-G vibrating motor with revolution of 1750 r/min was used to produce 7–7.3 g exciting force.

(2) The desander, desilter and oscillating screen were mounted together to form a whole set of purifying system (Fig. 3.15). One to three desanding funnels could be installed on a tray for the desander, each could dispose drilling mud at 1.9 m³/min, the maximum total disposing capacity was 5.73 m³/min. The funnels of

Table 3.7 Technical parameter of F-1600HL mud pump

Cylinder diameter (mm)	190	180	170	160	150	140	130	120
Rated pressure (MPa)	20.7	23.0	25.8	29.2	33.2	38.1	44.2	51.9
Stroke number (stroke/min)	Displacement (L/s)							
120	51.85	46.54	41.51	36.77	32.32	28.15	24.27	20.68
100	43.20	38.78	34.59	30.64	26.93	23.46	20.24	17.25
80	34.56	31.02	27.67	24.51	21.54	18.77	16.19	13.80
60	25.92	23.27	20.75	18.38	16.16	14.07	12.13	10.34
40	17.28	15.51	13.84	12.25	10.77	9.38	8.09	6.90

Note The cylinder diameters of 190, 180, 170, 160 and 150 mm are of piston structure, the cylinder diameters of 140, 130 and 120 mm are of plug structure

3.2 Drilling Equipment

Fig. 3.15 The solid control cleaning system (oscillating screen, desander and desilter) from American DERRICK Company

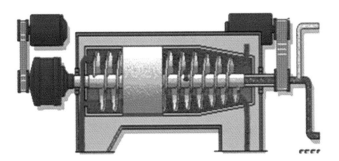

Fig. 3.16 Working principle of centrifuge

Fig. 3.17 The centrifuges (*left* is a low speed centrifuge)

desilter were arranged in circle, with equal pressure, and then the cyclone could run in optimum state.

(3) The circulating mud firstly flowed through the oscillating screen, desander and desilter cleaning system, and the solids larger than 170 mesh (91 μm) were cleaned away. However, the cuttings produced by diamond drilling were less than 180 mesh (84 μm), and these fine cuttings suspended in drilling mud could not be cleaned away by the cleaning system, only depending on centrifuge (Fig. 3.16). Therefore, two sets of LW500 × 1250-N and LWG500 × 1250-N centrifuge (Fig. 3.17 and Table 3.8) with large displacement were equipped. Both the centrifuges were manufactured by Zhejiang Jinhua Railway Machinery Plant, with the disposing capacities of more than 10 times of the conventional Model-414 centrifuge. Moreover, the separating size was decreased and the separating factor increased, thus the colloid solids were fundamentally cleaned away and a stable performance of the drilling fluid guaranteed.

3. **Well-control equipment**

Generally speaking, the high-pressure oil and gas fluids cannot be encountered in crystalline rocks. But in drilling practice of ICDP, the fluids in the earth's crust in some regions doesn't decrease as drilling hole deepens, but increases. It is more worth noticing that five large crude oil reservoirs were discovered unexpectedly below 3000–4000 m of Cambrian Period basement in the Dnepr-Donets Basin, Ukraine. If a 5000 m deep CCSD-1 Well would be drilled in the Dabie and Sulu super high pressure metamorphic belt, the possibility of meeting high pressure gas and liquids could not be ruled out, and such a deep well had never been drilled in the past in this region. For the sake of safe drilling, it was necessary to mount the well-control equipment. In terms of the characteristics of CCSD-1 Well that well diameter and annular spaces were small, FH35-35 double-ram BOP (Fig. 3.18) of 35 MPa and FKQ480-4 surface BOP control device were selected, and the same grade pressure manifold was equipped, based upon the design norm for well control.

4. **Drilling instruments**

All the operation handles for drilling parameter display instrument and for drilling parameter control were concentrated in the driller's console (Fig. 3.19) to constitute a command center for all the drilling controls. According to various drilling conditions, driller could directly control all the running parameters of mud pump, drawworks and rotary table and adjust the torque values in "real-time" in terms of the conditions at well bottom, and even under special conditions driller could execute emergency stopping in order to guarantee well bottom safety to the maximum limit. After reconstructing the braking system, the control to disc braking and to electronic driller could be realized. SK-2Z01C

Table 3.8 Technical parameters of centrifuge

Item	LWG500 × 1250-N	LW500 × 1250-N	Item	LWG500 × 1250-N	LW500 × 1250-N
Drum inner diameter (mm)	500	500	Main motor power (kW)	22	22
Drum length (mm)	1250	1250	Secondary motor power (kW)	7.5	7.5
Drum revolution (r/min)	2000–2500	1500–1800	Disposing capacity (m³/h)	40	40
Separating factor	1120–1750	630–907	Mass (kg)	2510	2510
Separating size (μm)	2–6	8–10	Oversize (mm)	3150 × 1900 × 1085	

Fig. 3.18 FH35-35 double-ram BOP

Fig. 3.19 ZJ70D driller's console

intelligence drilling-parameter instrument (Fig. 3.20) could provide drilling parameters and limit alarm. Hanging weight, WOB, torque, revolution, well depth, pump displacement and pressure and pump stroke and so on could be displayed in real-time on the sensor instrument panel, and all the drilling parameters could be connected with the computer network and thus the human-machine communication network system was set up. The remote control and the real-time recording of drilling parameters could be realized, thus the drilling automation level was effectively enhanced, and also a reliable technical guarantee was provided for the smooth operation in CCSD-1 Well.

5. **Drill pipe power tongs**

Drill pipe power tongs (hydraulic power tongs) is an indispensable tool for makeup-and-breakout of drill rods when lifting and lowering the drill rods and its technical performance and reliability will directly affect the drilling efficiency. The stepping extension and contraction of ZQ100 drill pipe power tongs (Fig. 3.21) is controlled by air cylinder, the pressure of the air source system is 0.5–1.0 MPa, the makeup-and-breakout of drill rods is realized by right or left turning of the hydraulic motor. The power of the drill pipe power tongs equipped for ZJ70D rig is supplied by a special hydraulic station installed on the drilling platform, with the rated pressure of 16.6 MPa and the rated displacement of 114 L/min. The maximum torque of the drill pipe power tongs is 100 kN m, being suitable for 89–203.2 mm drill rods (tools), the maximum stroke of the moving air cylinder is 1.5 m, with the oversize of 1700 mm × 1100 mm × 1400 mm and the mass of 2400 kg.

3.2.5 Application Evaluation on ZJ70D Drill Rig

ZJ70D drilling rig, with rebuilding and special refitment, very well satisfied the special technical demands of CCSD-1 Well. With its sufficient lifting ability, complicated down

Fig. 3.20 SK-2Z01C intelligence drlling-parameter instrument

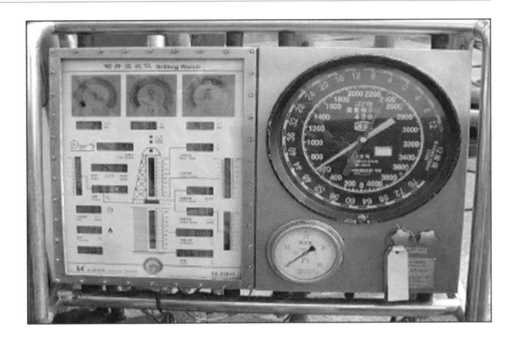

Fig. 3.21 Drill pipe power tongs

hole problems were successfully handled for many times and thus very serious accidents were avoided. The application of high pressure mud pump supplied an energy guarantee for the success of the core drilling technologies driven by a unique combination of down hole rotation and percussion in CCSD-1 Well. The reconstruction of the electricity supply brought a huge benefit for CCSD-1 Well, and the outstanding solids control system guaranteed a long service life for hydro-hammer and PDM.

During the period of drilling operation for nearly four years, the performance of the electrically-driven ZJ70D was stable and no serious breakdown occurred. It could very well meet the demands of reaming drilling with cone rock bits, PDM and hydro-hammer diamond core drilling, casing pulling up-and-in and other drilling technologies.

As a low rotation speed and big torque petroleum drill rig, ZJ70D could satisfy the demands of various drilling technologies for large diameter petroleum drilling, but it could not meet the demands for small diameter diamond core drilling technologies. Even though the designed parameter of rotary table was up to 300 r/min, but if the rotary table was used to directly drive the drill string, the actual working revolutions could not reach the revolutions required for diamond drilling because of the influence of the drilling system stability. After reconstructing the disc brake, though the stability of running-in drill rod was greatly improved, yet since small diameter diamond drilling needed small WOB, ZJ70D drilling rig had to be equipped with a high precision electronic driller system, if the problems of smooth and steady running-in drill rod were to be ideally solved.

After the pilot hole was completed, F-1600HL high pressure mud pump was used instead of F-1600 at well site. An observation of the running parameters showed that this pump retained low pressure and large displacement, as well as satisfied high pressure and small displacement, therefore the different demands of reaming drilling with cone rock bits and core drilling with diamond bits in CCSD-1 Well could be fully satisfied.

Construction Situation

The pilot hole of CCSD-1 Well was started on June 25th, 2001, reached 2,046.54 m on April 15th, 2002. The largest deviation angle was only 4.1° and the largest dogleg angle 1.25° per 30 m, the quality of the well was good enough to satisfy the scheme of combining two wells into one. Reaming of the pilot hole on May 7th, 2002 actually signified the beginning of the main hole drilling. Experienced two times of expanding drilling, one time of straightening drilling by use of sidetracking, one time of sidetracking to avoid the obstacles and three phases of core drilling, the main hole reached 5118.2 m on January 23rd, 2005, and the task of coring was successfully fulfilled. Afterwards, two types of new drilling tools were tested and on March 8th, 2005 the main hole reached the final depth of 5158 m. Then works such as running a liner hanger into the hole, cementing, drifting, completion and VSP measurement etc. were conducted. On April 18th, 2005, a grand ceremony was held at the drill site, formally declaring that all the tasks of CCSD-1 Well were fulfilled within 1395 days.

4.1 Basic Situation of the Construction of CCSD-1 Well

The whole construction progress of the drilling engineering of CCSD-1 Well can be divided up into eleven stages (Figs. 4.1, 4.2 and 4.3; Table 4.1):

1. Hole opening (CCSD-PH-0) and non-core drilling (the first opening);
2. The pilot hole (CCSD-PH) core drilling (the second opening);
3. The first expanding drilling of the main hole (CCSD-MH-1K) (including casing installation and well cementation);
4. The first phase of the main hole (CCSD-MH) core drilling (the third opening);
5. Straightening drilling of the main hole by use of sidetracking (CCSD-MH-1X);
6. The second phase of the main hole (CCSD-MH-1C) core drilling;
7. The second expanding drilling of the main hole (CCSD-MH-2K);
8. Sidetrack drilling to avoid the obstacles in the main hole (CCSD-MH-2X) (including casing installation and well cementation);
9. The third phase of the main hole (CCSD-MH-T) core drilling (the forth opening);
10. Drilling tools test (CCSD-MH-T);
11. Well completion (running liner and well cementation).

4.1.1 The Basic Data

An accumulative total drilling footage of different kinds of drilling activities in CCSD-1 Well reached 9177.71 m (not including the footage of drilling cement plug and the footage of grinding in expanding drilling), including:

Core drilling	5009.37 m (including drilling tools test);
Expanding drilling	3541.92 m (including guided reaming in straightening drilling);
Sidetracking and straightening drilling	306.32 m (including drilling tools test);
Non-core drilling	332.59 m (including drilling tools test);
Grind drilling	2.41 m.

The final well depth reached 5158 m, with the detailed data shown in Table 4.2. After finishing 5118.2 m for normal core drilling, well depth from 5118.2 m to 5158 m was completed for drilling tools test, among which 3.5 m was drilled by PDM hydro-hammer wireline core drilling, 5.3 m finished by straightening with deflector driven by PDM, and 31 m

Translated by Longchen Duan and Yongqin Zhang.

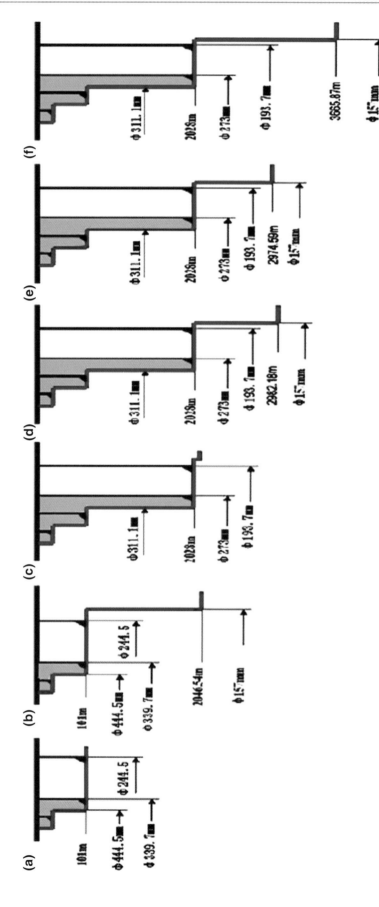

Fig. 4.1 Hole structure and casing program I in different construction section of CCSD-1 Well. **a** Spud in (CCSDPH-O) non-coringdrilling (6.87 ~ 101 m), casing running and wellcementation, and installationof moving casing. **b** The pilot hole (CCSD-PH) second opening coring drilling (101 ~ 2046.54 m). **c** The first expandingdrilling of main hole (CCSD-MH-1K) (101 ~ 2033 m), casingrunning and wellcementation, and fixingmoving casing. **d** The third openingfirst section of mainhole (CCSD-MH)coring drilling (2046.54 ~ 2982.18 m). **e** Sidetrack straighteningdrilling of main hole (CCSD-MH-1X) (2749 ~ 2974.59 m). **f** The second section of mainhole (CCSD-MH-1C) coring drilling (2974.59 ~ 3665.87 m)

4.1 Basic Situation of the Construction of CCSD-1 Well

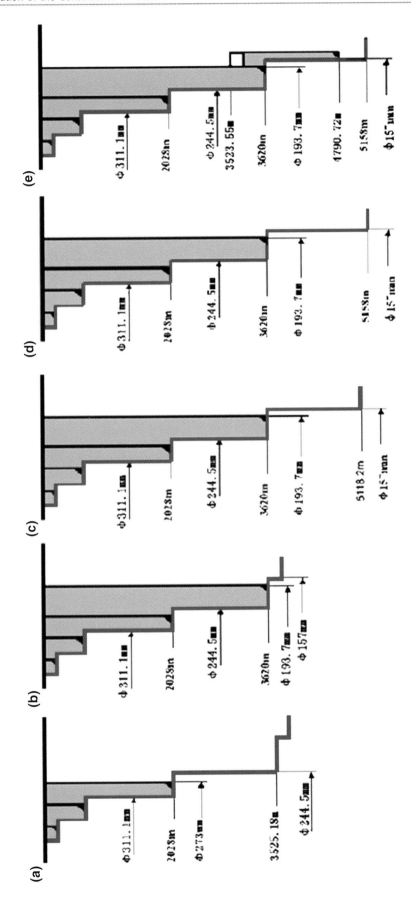

Fig. 4.2 Hole structure and casing program II in different construction section of CCSD-1 Well. **a** The second expanding drilling of main hole (CCSD-MH-2K) (2028 ~ 3525.18 m). **b** Sidetrack obstacle avoidance drilling (CCSD-MH-2X) (3400 ~ 3624.16 m) and casing installation and well cementation. **c** The forth opening coringdrilling of main hole (CCSD-MH-2C) (3624.16 ~ 5118.2 m). **d** Drilling tool test (CCSD-MH-T) (5118.2 ~ 5158 m). **e** Completion and running liner and well cementation (3523.55 ~ 4790.72 m)

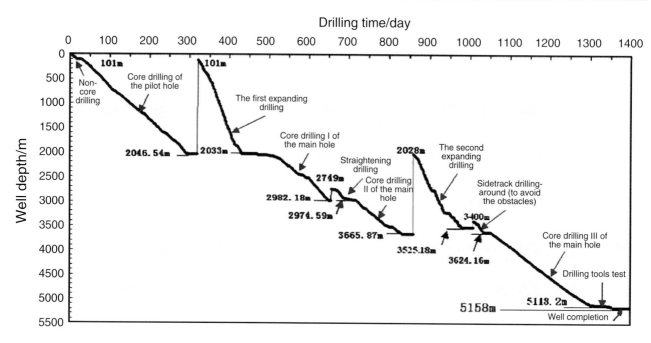

Fig. 4.3 Construction progress curve of CCSD-1 Well

Table 4.1 Drilling situation of CCSD-1 Well

Number	Well depth (m)		Diameter				Time			Remarks
	From	To	Drilling		Casing		From	To	Days	
			mm	in.	mm	in.				
I	0(6.87)	100.36	444.5	17^1/$_2$	339.7	13^3/$_8$	2001-06-25	2001-07-15	21	The first opening
	100.36	101.00	311.1	12^1/$_4$	244.5	9^5/$_8$				Non-core drilling
II	101.00	2046.54	157.0				2001-07-16	2002-05-06	295	The second opening core drilling
III	101.00	2028.00	311.1	12^1/$_4$	273.1	10^3/$_4$	2002-05-07	2002-09-20	137	The first expanding drilling of the main hole
	2028.00	2033.00	215.9	8^1/$_2$	193.7	7^5/$_8$				
IV	2046.54	2982.18	157.0				2002-09-21	2003-04-06	198	The third opening core drilling
V	2749.00	2974.59	157.0				2003-04-07	2003-06-08	63	Straighten drilling by use of sidetracking
VI	2974.59	3665.87	157.0				2003-06-09	2003-10-21	135	Core drilling
VII	2028.00	3525.18	244.5	9^5/$_8$			2003-10-22	2004-03-14	145	The second expanding drilling
VIII	3400.00	3620.00	244.5	9^5/$_8$	193.7	7^5/$_8$	2004-03-15	2004-05-07	54	Sidetrack drilling to avoid the obstacles, setting casing
	3620.00	3623.91	215.9	8^1/$_2$						
	3623.91	3624.16	165.1	6^1/$_2$						
IX	3624.16	5118.20	157.0				2004-05-08	2005-01-24	262	The forth opening core drilling
X	5118.20	5158.00	157.0				2005-01-25	2005-03-08	43	Drilling tools test
XI	5158.00	5158.00			127.0	5	2005-03-09	2005-04-01	24	Running liner and well cementation
			Total						1377	

The total time lasted should be 1395 days, including the 18 days for waiting and for completion ceremony from April 2nd, 2005 to April 19th, 2005. The first opening non-core drilling started at the hole depth of 6.87 m

completed by non-core drilling. From the hole opening on June 25th, 2001 to the final depth reached to 5118.2 m on January 24th, 2005, the construction totally lasted 1310 days, with an average footage of 6.99 m drilled per day and an average ROP of 0.95 m/h obtained, which included 1071 round trips of core drilling with an average ROP of 1.01 m/h and an average core recovery of 85.7 %; and 89 round trips of expanding drilling, with an average ROP of 1.07 m/h.

4.1 Basic Situation of the Construction of CCSD-1 Well

Table 4.2 Drilling situations at different stages of CCSD-1 Well

Stage	Core drilling				Expanding drilling		Sidetracking straighten drilling		Non-core drilling		Grinding drilling/m	Total footage drilled/m
	Footage drilled (m)	ROP (m/h)	Core recovery (%)	Footage drilled per round trip (m)	Footage drilled (m)	ROP (m/h)	Footage drilled (m)	ROP (m/h)	Footage drilled (m)	ROP (m/h)		
PH	1945.54	0.90	88.7	2.96					94.13	0.39		2039.67
The first expanding drilling					1931.89	1.05					0.11	1932
MH	934.67	1.11	77.6	6.40							0.97	935.64
The first sidetracking straighten drilling	10.75	1.04	55.6	3.58	15.31	1.18	215.20	0.69			0.04	241.30
MH-1C	634.85	1.25	83.3	8.04	19.33	1.12	40.20	0.83	15.39	0.82	0.84	710.61
The second expanding drilling					1497.18	1.07						1497.18
The second sidetracking drilling to avoid the obstacles					63.31	2.08	45.62	0.61	178.54	1.00		287.47
MH-2C	1480.06	1.03	88.0	7.96					13.53	0.52	0.45	1494.04
Sub-total	5005.87	1.01	85.7	4.67	3527.02	1.07	301.02	0.69	301.59	0.65	2.41	9137.91
Drilling tools test	3.50	0.62	68.6	1.17			5.30	0.38	31.00	0.94		39.80
Total	5009.37	1.01	85.7	4.66	3527.02	1.07	306.32	0.68	332.59	0.67	2.41	9177.71

Note The footage of drilling cement plug and the footage of grinding in the second expanding drilling are not included

4.1.2 Drill Hole Trajectory

Hole deviation survey in the drilling progress of CCSD-1 Well was conducted by use of well logging at certain depth intervals. The logging depth intervals ranged from 100 to 200 m, depending upon the drilling information, and taking the effective control of hole deviation as principle. The logging distance of data acquisition was decided by geophysical exploration and geologic conditions and one datum was collected per 0.125 m. The space trajectory of CCSD-1 Well is shown in Fig. 4.4.

The curves of deviation angle and azimuth angle of CCSD-1 Well are shown in Figs. 4.5 and 4.6, from which it can be found that both the deviation angle and azimuth angle were small above 2,000 m in the pilot hole and the bending of the hole was not serious. At hole depth of 2000 m the deviation angle was only 3.6°, and there were two relatively obvious positions at which the azimuth angle becoming mitigatory above 2000 m, i.e., at hole sections from 775 to 800 m and from 1375 to 1475 m, where the azimuth angle decreased by 1 to 1.5° because of the relatively broken rocks in the intersected zone of eclogite, gneiss and amphibolite. At this zone, drilling tool would cause a certain drooping force, as the hole diameter was relatively large. This was normal in drilling process.

Drilling did not go on smoothly since the opening of the main hole, and wireline core drilling was unsuccessful either. A low drilling efficiency was obtained when drilling with wireline core drill rods and top drive, twelve diamond bits were serially worn out which included eight new bits. Well deviation kept becoming more serious along with well depth from 8.2° at 2770 m to 16.5° at 2935 m, but the azimuth angle in this hole section did not has obvious variation. Core drilling was stopped and sidetrack straighten drilling was carried out seeing that the deviation angle was bigger than design.

Sidetrack straightening point was designed at well depth of 2497 m while straightening ended at well depth of 2974.59 m with hole deviation angle of 6.62°, and with azimuth angle a proper increase. After that, core drilling in the main hole was restarted. From Fig. 4.5 it can be found that hole deviation angle kept on increasing.

Bottom hole accident happened at well depth of 3665.87 m because of breaking of the reaming shell (including drill bit) at hole bottom. Broken rock formation at this hole section caused enlargement of hole diameter and

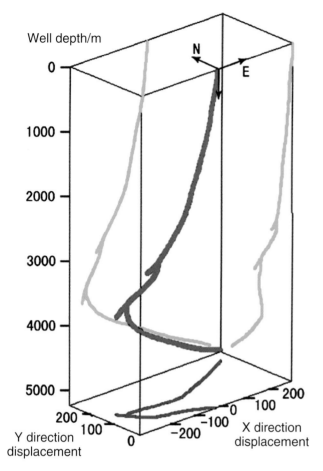

Fig. 4.4 Space trajectory of Well CCSD-1

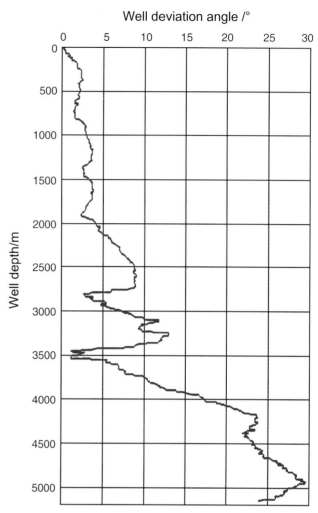

Fig. 4.5 Curve of deviation angle variation of CCSD-1 Well

then hole collapse, though a lot of methods were tried, without effective result. Finally it was decided that expanding drilling was employed after pulling up the Φ 193.7 mm (7⁵/₈ in.) moving casing. When pulling up the casing, 72 elastic plates on the casing stabilizer fell into the well, and three guiding bodies of the reaming bit were broken and left in the well in the process of expanding drilling. Under these circumstances, in-the-hole condition became very complicated. After investigation it was decided that sidetrack drilling to avoid the obstacles should be used. When sidetrack drilling reached to well depth of 3,445.62 m, deviation angle was slowed down to 1.5°. Because the deviation angle was straightened nearly to 0° and the facing angle was hard to control, the azimuth angle got a sharp variation from 350° to near 20°. Afterwards, it was gradually returned within a certain extent, which produced a hole section of space crook.

4.1.3 Well Temperature Curve

Well temperature curve is showed in Fig. 4.7, from which we can see that the well temperature varied basically based upon the temperature gradient of the stratum, remaining with normal. However, well temperature variated from 79.6 °C at the well depth of 2750 m to 72 °C within 10 m then turned back slowly to 83 °C at the well depth of 3610 m, and at 3620 m, well temperature suddenly increased to 101.24 °C. The abnormal well temperature variation in this hole section was because of several serious leakage events happened in its broken rocks. The most serious leakage exceeded 80 m³/h, so well temperature dropped down. Well temperature gradually increased once the cracks of the leaky stratum were slowly filled out by downhole sealing agent. When there was completely no leakage, well temperature would

4.1 Basic Situation of the Construction of CCSD-1 Well

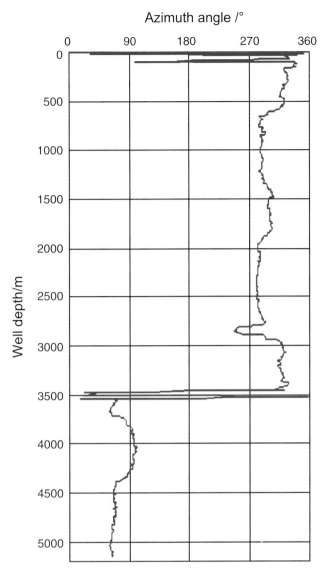

Fig. 4.6 Curve of azimuth angle variation of CCSD-1 Well

increase quickly and assume a regular variation following stratum temperature gradient along with the well extension.

4.2 Simple Situation of the Construction at Different Periods

4.2.1 Hole Opening and Non-core Drilling (the First Opening)

There was surface soil with different thickness ranging from 4 to 8 m. To ensure the reliability of the follow-up construction and the weight on bit for the first opening drilling, a 12.56 m × 12.56 m × 8.70 m foundation pit was excavated with machinery at CCSD-1 Well site before the first opening

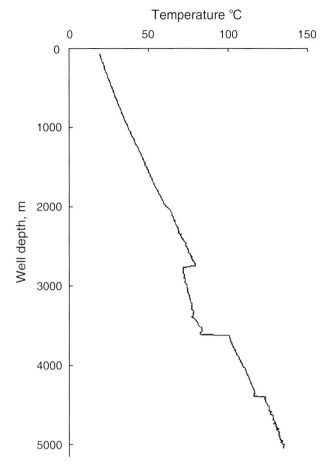

Fig. 4.7 Temperature curve of CCSD-1 Well

drilling, Φ 508.0 mm (20 in.) casing was installed from wellhead to 4 m deep and cemented, thus a good wellhead condition was prepared for pilot hole construction.

The first opening started at well depth of 6.87 m and drilling with Φ 444.5 mm rock bit reached to 100.36 m deep, Φ 339.7 mm ($13^3/_8$ in.) surface casing was installed to that depth and then executed well cementation. Then cement plug was drilled with Φ 311.1 mm rock bit to 101 m deep. In order to reduce the annular space, improve mud uplift velocity, avoid drilling cuttings detention and deposition in the casing and stabilize drilling tool at the same time, Φ 244.5 mm ($90^5/_8$ in.) moving casing was installed to 101 m. The top of the moving casing string was fixed with slips of wellhead casing head. Three rigid stabilizers and three elastic stabilizers were fixed on moving casing, the former were fixed on the first, the second and the forth casings respectively, while the latter, fixed on the seventh, the eighth and the ninth casings respectively to centre the moving casing in the well and reduce vibration of the moving casing during drilling operations. Circumstances about non-coring drilling of the first opening are as follows:

1. Spudding (6.87–28.62 m): Spudding was conducted with a drilling tool assembly consisted of rock bit + stabilizer + special heavy drill collar (Fig. 4.8a). When reaching well depth of 16.72 m, another stabilizer (Fig. 4.8b) was added and then drilling continued to 28.62 m deep. Well logging indicated that at well depth of 18 m the deviation angle was zero. At the hole section from 20 to 27 m deep, loss of circulation occurred, with 28.16 m³ loss and 1.17 m³/h loss rate. By adding one ton GD-III plugging agent, motionless plugging and circulated plugging while drilling were conducted respectively and a successful result was obtained. Strata at this hole section were comparatively broken, penetration rate reached to 2 m/h and average penetration rate to 0.92 m/h under the condition of bit pressure no more than 4 t and rotary speed of 40 r/min, except for bouncing happened some times.

2. Packed hole drilling (28.62–60.78 m): The drilling tool assembly with rock bit + stabilizer + single stem drill collar + stabilizer + double drill collar was used to drill to 60.78 m deep (Fig. 4.8c). Hole survey was conducted at the depth of 45.69 m and hole deviation angle at 40.56 m was 0.25°, and hole deviation angle at 49 m depth was found to be 0.5° after survey at the depth of 53.79 m. Stratum lithology in that hole section was homogeneous, drillability became bad, and bit bouncing and jumping were serious. Average penetration rate in this section was 0.27 m/h.

3. Drilling with single stabilizer pendulum drilling string (60.78–73.00 cm): In view of serious bit bouncing and jumping, drilling string was changed and a pendulum drilling tool assembly of rock bit + absorber + single stem drill collar + stabilizer (Fig. 4.8d) was used to drill to depth of 73.00 m and then the drilling tool assembly

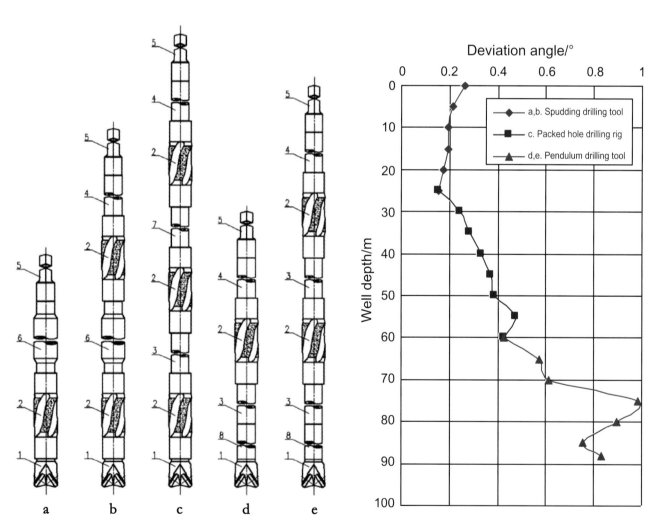

Fig. 4.8 Drilling string and well deviation angle of the first opening non-core drilling. *1* Φ 444.5 mm rock bit, *2* Φ 444.5 mm stabilizer, *3* Φ 228.6 mm single stem drill collar, *4* Φ 228.6 mm drill collar, *5* Kelly bar, *6* Φ 177.8 mm drill collar with Φ 406.4 mm lead poured heavy stabilizer, *7* Φ 228.6 mm double stem drill collar, *8* Φ 203.2 mm two-way absorber

was pulled out because of low ROP. The average penetration rate in this hole section was 0.27 m/h.

4. Drilling with double stabilizer pendulum drilling string (73.00–92.45 cm) A drilling tool assembly with rock bit + absorber + single stem drill collar + stabilizer + single stem drill collar + stabilizer (Fig. 4.8e) was used to drill to the depth of 92.45 m. Hole survey at 84.85 m showed that hole deviation angle was 1° at the depth of 79 m. Average penetration rate in this section was 0.38 m/h.

5. Drilling with single stabilizer pendulum drilling string (92.45–100.36 cm): A drilling tool assembly with rock bit + absorber + single stem drill collar + stabilizer (Fig. 4.8d) was used to drill to the depth of 100.36 m, and thus the Φ 444.5 mm first opening non-core drilling was completed. The average penetration rate in this hole section was 0.38 m/h.

The first opening started on June 25th, 2001. All the works of non-core drilling, hole logging, casing installation, well cementation, cement plug drilling and moving casing installation etc. were completed on July 15th, 2001.

4.2.2 Pilot Hole (Section CCSD-PH) Core Drilling (the Second Opening)

The second opening core drilling started on July 16th 2001, with Φ 157/Φ 97 mm diamond core drill bits, reached depth of 2046.54 m on April 15th 2002. Works such as well logging, VSP (vertical seismic profile) measurement, pulling the Φ 244.5 mm moving casing out of the hole, redox logging, susceptibility logging and three-component magnetic logging introduced from KTB in Germany, etc., were all completed till May 6th 2002. Circs about core drilling are as follows:

1. 101–112.95 m diameter change core drilling: A core drilling tool assembly (Fig. 4.9a) with drill bit + reaming shell + core barrel + reaming shell + single acting joint + PDM + three Φ 215.9 mm stabilizers + Φ 127 mm drill rod was employed to drill to 108.64 m deep (including six round trips of lifting drilling tools) by using diameter change coring drilling. Φ 215.9 mm stabilizer was left within Φ 244.5 mm moving casing. One round trip of rotary table drive double tube core drilling and one round trip of PDM single tube core drilling were conducted at this section. By using a core drilling tool assembly with drill bit + reaming shell + core barrel + reaming shell + single acting joint + PDM + two Φ 215.9 mm stabilizers + Φ 127 mm drill rod the hole size change core drilling was completed to the depth of 112.95 m. Footage drilled at this section was 11.95 m, with average penetration rate of 0.38 m/h and core recovery of 72.2 %.

2. 112.95–152.37 m core drilling with slick assembly: A core drilling tool assembly (Fig. 4.9b) with drill bit + reaming shell + core barrel + reaming shell + single acting joint + PDM + Φ 120 mm drill collar was employed to penetrate to 152.37 m deep (including 18 round trips, of which three were completed by PDM drive single tube core drilling). Footage drilled at this section was 39.42 m, with average penetration rate of 0.87 m/h and core recovery of 91.9 %.

3. 152.37–233.27 m core drilling with one Φ 156 mm stabilizer: A core drilling tool assembly (Fig. 4.9c) with drill bit + reaming shell + core barrel + reaming shell + single acting joint + PDM + Φ 156 mm stabilizer + Φ 120 mm drill collar was used to drill to 233.27 m deep (44 round trips, including one round trip conducted with PDM hydro-hammer double tube core drilling at the depth of 218.44 m). Footage drilled at this section was 80.9 m, with average penetration rate of 0.78 m/h and core recovery of 88.3 %.

4. 233.27–730.48 m core drilling with two stabilizers: A core drilling tool assembly (Fig. 4.9d) with drill bit + reaming shell + core barrel + reaming shell + single acting joint + PDM + Φ 156 mm stabilizer + Φ 120 mm single stem drill collar + Φ 156 mm stabilizer + Φ 120 mm drill collar was employed to penetrate to 730.48 m deep (227 round trips, including one round trip of rotary table drive double tube core drilling to recover the remained bit matrix at hole bottom and one round trip of PDM single tube core drilling). Footage drilled at this section was 497.21 m, with average penetration rate of 0.91 m/h and core recovery of 97.0 %, in which, 32 round trips were completed by PDM hydro-hammer double tube core drilling, with a drilling footage of 82.51 m, an average penetration rate of 1.07 m/h and core recovery of 97.5 %.

5. 730.48–742.05 m core drilling with one Φ 156 mm stabilizer: Due to the tight pull encountered during the previous round trip (resistance force of 104 kN), drilling tool assembly was changed and the drilling tool assembly with one Φ 156 mm stabilizer (Fig. 4.9c) was used to drill to 742.05 m deep within 6 round trips. Footage drilled at this section was 11.57 m, with average penetration rate of 0.79 m/h and core recovery of 102.9 %.

6. 742.05–763.53 m core drilling with two Φ 156 mm stabilizers: A drilling tool assembly with two Φ 156 mm stabilizers (Fig. 4.9d) was used to penetrate to 763.53 m deep, within 12 round trips. Footage drilled at this hole section was 21.48 m, with average penetration rate of 0.33 m/h and core recovery of 97.9 %.

7. 763.53–770.95 m core drilling with slick assembly: A slick assembly (Fig. 4.9b) was used to drill to 770.95 m deep within two round trips because of the tight pull

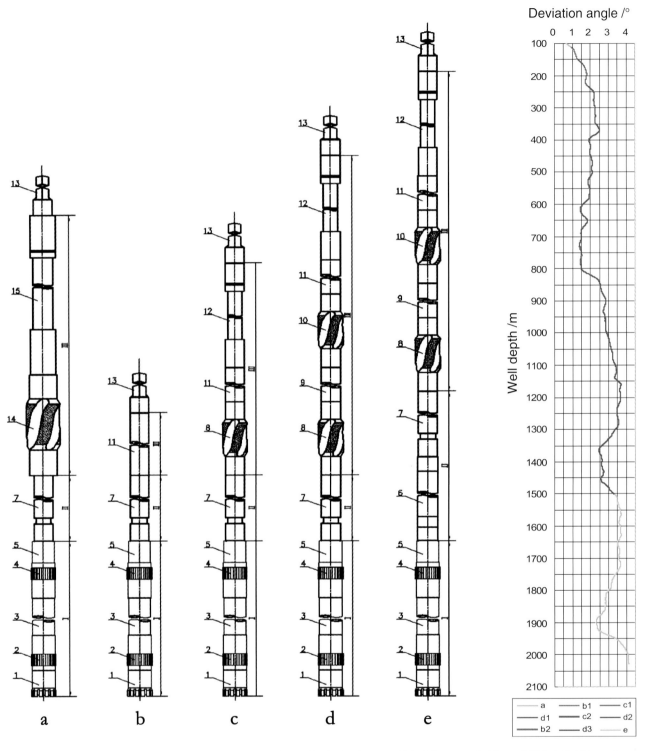

Fig. 4.9 Diamond core drilling tool with Φ 157 mm diamond bit and well deviation angle in the second opening. *1* Φ 157 mm coring bit, *2* Φ 157 mm lower reaming shell, *3* Φ 139.7 mm core barrel, *4* Φ 157 mm upper reaming shell, *5* Φ 146 mm single acting joint, *6* Φ 127 mm hydro-hammer, *7* Φ 120 mm or Φ 95 mm PDM, *8* Φ 156 mm lower stabilizer, *9* Φ 120 mm single stem drill collar, *10* Φ 156 mm upper stabilizer, *11* Φ 120 mm drill collar string, *12* Φ 89 mm drill rod string, *13* Kelly bar, *14* Φ 215.9 mm stabilizer, *15* Φ 127 mm drill rod

encountered during the previous round trip. Footage drilled at this section was 7.42 m, with average penetration rate of 1.24 m/h and core recovery of 97.7 %.

8. 770.95–2046.54 m core drilling with two Φ 156 mm stabilizers (Fig. 4.9d, e): Total footage drilled in this section was 1275.59 m (338 round trips), with an average

penetration rate of 0.95 m/h and core recovery of 85.2 %, including: (1) 159.59 m drilled within 65 round trips by PDM double tube core drilling with average penetration rate of 0.59 m/h and core recovery of 90.7 %; (2) 870.86 m drilled within 196 round trips by PDM hydro-hammer double tube core drilling with average penetration rate of 1.11 m/h and core recovery of 87.3 %; (3) 214.78 m completed within 63 round trips by water isolating double tube core drilling with average penetration rate of 0.92 m/h and core recovery of 77.2 %, in which included 135.53 m drilled within 33 round trips by PDM hydro-hammer water isolating double tube core drilling with average penetration rate of 1.11 m/h and core recovery of 82.0 %; 78.65 m drilled within 29 round trips by PDM water isolating double tube core drilling with average penetration rate of 0.72 m/h and core recovery of 69.0 %; and 0.6 m drilled in one round trip by rotary table drive water isolating double tube core drilling with average penetration rate of 0.20 m/h and core recovery of 73.3 %; (4) 13.36 m drilled within 5 round trips by jet reverse circulation double tube core drilling with average penetration rate of 0.93 m/h and core recovery of 42.7 %, in which included 6.55 m drilled within three round trips by PDM hydro-hammer jet reverse circulation double tube core drilling with average penetration rate of 0.72 m/h and core recovery of 54.3 % and 6.81 m drilled within two round trips by PDM jet reverse circulation double tube core drilling with average penetration rate of 1.33 m/h and core recovery of 30.5 %; and (5) 17.00 m drilled within nine round trips by rotary table drive double tube core drilling with average penetration rate of 0.41 m/h and core recovery of 61.0 %.

Totally 660 RIH (run in hole) roundtrips were run in the second opening core drilling, including 657 round trips of core drilling, over coring at hole depth of 957.56 m one time, running grind shoe one time at well depth of 975.94 and 1600.08 m respectively. Basically, the second opening core drilling was normal. There were three ways to drive bottom hole core drilling tools, i.e., PDM, rotary table and PDM hydro-hammer. Core drilling tools were classified into conventional swivel type double tube, single tube drilling tool, water isolating swivel type double tube and jet type reverse circulation swivel type double tube. There were 28 sizes for the outer tube length of the coring tools (Table 4.3). Before reaching the hole depth of 977.93 m (PH 393 round trip), coring tools with 2.30–5.25 m outer tube length were mainly used, while after that, coring tools with longer outer tubes began to be used. The main problem encountered in core drilling in this hole section was core blockage, not the problem of normal lifting of the drilling tool. Of all the 657 round trips of core drilling, only 144 (Table 4.3; Fig. 4.10) of which the core barrels were full and kelly bar were drilled over. In order to solve the problems of core barrel blockage and low penetration rate, hydro-hammer drilling was used from well depth of 218.44 m (PH 66 round trip) and 270 round trips were completed with this method (including five round trips by rotary table drive). By using this method, the penetration rate was improved by 52.1 % and the ratio of full core barrel increased by 131.5 % (Table 4.4) in comparison with non hydro-hammer drilling.

Core drilling in this hole section penetrated through the depths of 603 m and 707 m etc., where strata were broken and collapsed. Well logging indicated that at the hole depth of 707 m the hole diameter was increased to 345 mm (Fig. 4.11). In the long hole section from 1450 to 1950 m where the rock formations were hard, brittle and broken (Fig. 4.11), not only core recovery was low, but also drill bit wear was extremely serious, especially the inside diameter of the drill bits and this lead to drill bit earlier abandonment. During drilling, rock falling sometimes happened and this caused sticking when pulling out the drilling string or drilling tool could not reach the hole bottom when running-in drilling string. Four in-the-hole accidents happened in this

Table 4.3 Cics of core drilling with different outer tube lengths in the hole section PH

Outer tube length (m)	Round trips	Round trips with full core barrel	Footage drilled (m)	Penetration rate (m/h)	Core recovery (%)
0.38	1	0	0.80	0.19	95.0
2.30, 2.50, 2.52	149	37	282.25	0.85	96.0
3.80	114	24	294.53	0.76	97.2
4.00, 4.10, 4.18, 4.24, 4.25, 4.28	89	14	207.48	0.74	95.4
4.60, 4.65, 4.95	21	9	68.31	0.92	80.3
5.05, 5.15, 5.25	163	28	464.21	0.89	76.1
5.78	2	0	7.95	0.94	90.6
7.39, 7.49, 7.51, 7.56, 7.58, 7.61, 7.75, 7.93, 7.99, 8.08	118	32	620.01	1.13	89.4
Total	657	144	1945.54	0.90	88.7

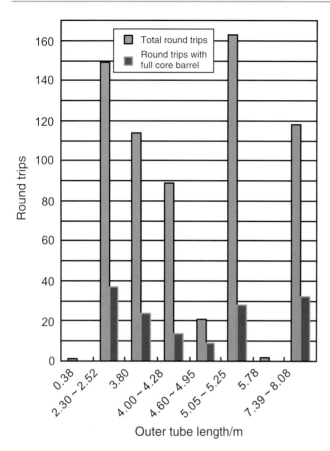

Fig. 4.10 Roundtrips and round trips with full core barrel completed by different outer tube length in section PH

4.2.3 The First Expanding Drilling of the Main Hole (Hole Section CCSD-MH-1K)

According to the Preliminary Engineering Design of CCSD Project, if the construction results of the pilot hole of CCSD-1 Well was satisfactory, and drilling condition was not complicated, the main hole could be constructed on the basis of the pilot hole by expanding drilling, without moving well location. Since the largest well deviation angle of the pilot hole was just 4.1°, core recovery was 88.7 % and with less casing grades, the requirements for combining the pilot hole and the main hole into one could be satisfied. Approved by the CCSD Leading Group, the plan of combining two wells into one (the main hole was to be drilled after expanding drilling of the pilot hole) was to be implemented. The implementation of this plan saved large funds and lots of time.

Φ 311.1 mm sized expanding drilling was started on May 7th, 2002 from well depth of 101 m, and on September 5th 2005 reached to 2033 m deep (in the hole section from 2028 to 2033 m was a well cementation pocket). Up to September 15th, 2002, attempts of running casing, circulating mud (containing consistent lubricant with small plastic balls), running Φ 273.1 mm sized casing, well cementation and drilling cement plug etc. were finished. Bottom hole grinding, mud displacement, well logging, VSP measurement and Φ 193.7 mm moving casing running etc. were accomplished from September 16th to 20th, 2002, with details showed as follows:

1. 101–417.39 m expanding drilling with double stabilizers: A drilling tool assembly of Φ 311.1 mm rock bit with guide + two Φ 203.2 mm drill collars + Φ 309 mm stabilizer + one Φ 203.2 mm drill collar + Φ 309 mm stabilizer + three Φ 203.2 mm drill collars + three Φ 177.8 mm drill collars + Φ 127 mm drill rod was utilized for expanding drilling to the hole depth of 417.39 and 316.39 m were completed (16 round trips, nine drill

hole section, two of them were reaming shells disjunction and falling into hole bottom because of quality problem, the other two were falling of centering devices of loggers into hole bottom, all the troubles were successfully treated. Hole deviation angle, diameter and azimuth angle variation in core drilling of the second opening are showed in Fig. 4.11, and the main technical indexes in Table 4.5.

Table 4.4 Cics of coring drilling with hydro-hammer in the hole section PH

Drilling method	Round trips	Round trips with full core barrel	Round trips with full core barrel/total round trips	Footage drilled (m)	Penetration rate (m/h)	Core recovery (%)
Without hydro-hammer	387	56	0.145	842.61	0.74	90.7
Hydro-hammer	270	88	0.326	1103.08	1.09	87.2
Total	657	144	0.219	1945.54	0.90	88.7

4.2 Simple Situation of the Construction at Different Periods

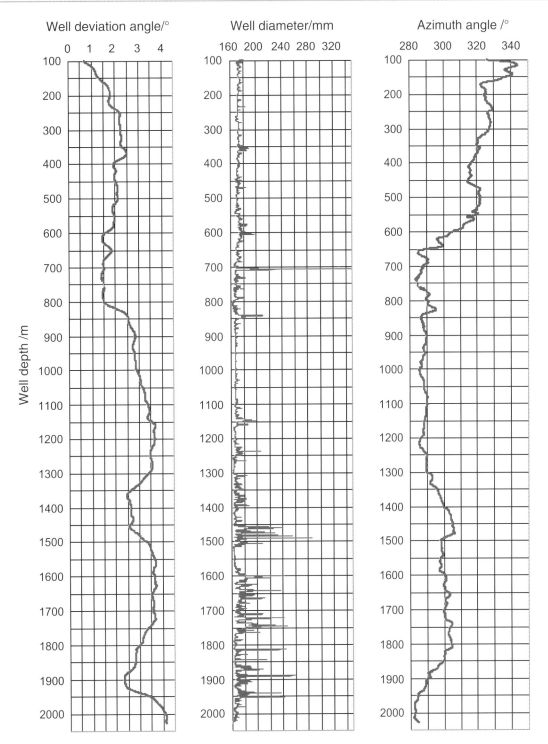

Fig. 4.11 Hole deviation angle, diameter and azimuth angle of the second opening Φ 157 mm core drilling in the hole section from 101 to 2046.54 m

bits were used). Average penetration rate was 0.72 m/h. At the hole depth of 388.79 m (Roundtrip MH-1K-15) a KSC-203 hydro-hammer was added to the drilling tool assembly and 10.07 m was drilled, with average penetration rate of 0.81 m/h.

2. 417.39–574.97 m expanding drilling with absorber and double stabilizers: A drilling tool assembly of Φ 311.1 mm rock bit with guide + Φ 203 mm absorber + one Φ 203.2 mm drill collar + Φ 305 mm stabilizer + two Φ 203.2 mm drill collars + Φ 305 mm

Table 4.5 Main core drilling technical indexes of the second opening in the hole section from 101 to 2046.54 m

Item	Index	Item	Index	Item	Index
Core drilling footage (m)	1945.54	Average core recovery (%)	88.7	Average penetration rate (m/h)	0.90
Core drilling round trips	657	Normal (full core barrel) pull out round trips	144	Average bit life (m)	17.53
Longest bit life (m)	75.23	Average footage drilled per round trip (m)	2.96	Longest footage drilled per round trip (m)	8.72
Average footage drilled per day (m)	7.23	Percentage of actual drilling in time (%)	33.5	Percentage of tripping in time (%)	48.1
Percentage of accident in time (%)	0.3	Largest hole deviation angle (°)	4.1	Average footage drilled per rig month (m/30 days)	216.97
Engineering logging (times)	24	Comprehensive well logging (times)	3	Days of core drilling (days)	269

Note The second opening core drilling started on July 16th 2001, ended on April 15th 2002, totally lasted 274 days, including 5 days of comprehensive well logging. Well logging, VSP measurement, redox logging, susceptibility logging and three-component magnetic logging were conducted from April 15th to May 6th, 2002

stabilizer + three Φ 203.2 mm drill collars + three Φ 177.8 mm drill collars + Φ 127 mm drill rod was used for expanding drilling to the hole depth of 574.97 and 157.58 m were completed (six round trips, five new drill bits and one used drill bit were used). Average penetration rate was 0.82 m/h. At the hole depth of 500.62 m a SYZX-273 hydro-hammer was added to the drilling tool assembly (Roundtrip MH-1K-19) and 8.89 m were drilled, with average penetration rate of 0.81 m/h.

3. 574.97–575.08 m hole grinding drilling: As the tungsten carbide teeth on the used rock bit were all broken, a grind shoe of 300 mm diameter was used to drill 0.11 m.

4. 575.08–779.96 m expanding drilling with absorber and double stabilizers: A drilling tool assembly of Φ 311.1 mm rock bit with guide + Φ 203 mm absorber + one Φ 203.2 mm drill collar + Φ305 mm stabilizer + one Φ 203.2 mm drill collar + Φ 305 mm stabilizer + four Φ 203.2 mm drill collars + three Φ 177.8 mm drill collars + Φ 127 mm drill rod was used for expanding drilling to the hole depth of 779.96 m (four roundtrips, three new bits and one used drill bit were used) and 204.88 m were completed. Average penetration rate was 1.12 m/h. At the hole depth of 691.4 m (Roundtrip MH-1K-25) a SYZX-273 hydro-hammer was added to the drilling tool assembly and 18.30 m were drilled, with average penetration rate of 0.85 m/h.

5. 779.96–1871.48 m expanding drilling with slick drill collar: A drilling tool assembly of Φ 311.1 mm rock bit with guide + Φ 203 mm absorber + six Φ 203.2 mm drill collars + three Φ 177.8 mm drill collars + Φ 127 mm drill rod was used for expanding drilling to the hole depth of 1871.48 m (16 roundtrips, 14 new bits and one used bit were used) and 1091.52 m were completed, with average penetration rate of 1.25 m/h. At the hole depth of 779.96 m (Roundtrip MH-1K-27) a SYZX-273 hydro-hammer was added to the drilling tool assembly and 82.77 m were drilled, with average penetration rate of 1.27 m/h. At the hole depth of 1,783.78 m (Roundtrip MH-1K-41) a KSC-203 hydro-hammer was added to the drilling tool assembly and 69.57 m were drilled, with average penetration rate of 1.07 m/h. An absorber was taken out of the drilling tool assembly at the hole depth of 931.32 m (Roundtrip MH-1K-29) and 15.41 m were drilled, with average penetration rate of 1.02 m/h. A diamond reaming shell was added to the drilling tool assembly for truing the hole wall at the hole depth of 1853.35 m (Roundtrip MH-1K-42) and 18.13 m were drilled, with average penetration rate of 1.04 m/h.

6. 1871.48–1871.99 m hydro-hammer drilling with flat bottom button bit: A drilling tool assembly of Φ 311.1 mm flat bottom button bit + SYZX-273 hydro-hammer + two Φ 203.2 mm drill collars + Φ 305 mm stabilizer + one Φ 158 mm drill collar + Φ 305 mm stabilizer + four Φ 203.2 mm drill collar + three Φ 177.8 mm drill collars + Φ 127 mm drill rod was utilized for expanding drilling to the hole depth of 1871.99 m and 0.51 m was completed, with average penetration rate of 0.31 m/h.

7. 1871.99–1921.99 m expanding drilling with limber string and tri-cone rock bit: A drilling tool assembly of Φ 311.1 mm tri-cone rock bit + Φ 203 mm absorber + two Φ 203.2 mm drill collars + Φ 305 mm stabilizer + one Φ 158.8 mm drill collar + Φ 305 mm stabilizer + four Φ 203.2 mm drill collars + three Φ 158.8 mm drill collars + Φ 127 mm drill rod was used for expanding drilling to the hole depth of 1921.99 m within two roundtrips, 50.00 m were completed and one tri-cone bit was used, with average penetration rate of 1.11 m/h.

8. 1921.99–2028.00 m hydro-hammer expanding drilling with limber string and tri-cone rock bit: A drill string of Φ 311.1 mm tri-cone rock bit + KSC-203/SYZX-273 hydro-hammer + two Φ 203.2 mm drill collars + Φ

305 mm stabilizer + one Φ 158.8 mm drill collar + Φ 305 mm stabilizer + four Φ 203.2 mm drill collars + three Φ 158.8 mm drill collars + Φ 127 drill rod was used for expanding drilling to the hole depth of 2028.00 m (two roundtrips and two tri-cone rock bit used). 106.01 m were completed, with average penetration rate of 0.98 m/h.

9. 2028.00–2033.00 m cement plug drilling: A drill string of Φ 215.9 mm rock bit + one Φ 203.2 mm drill collar + Φ 305 mm stabilizer + three Φ 178.8 mm drill collars + Φ 178 mm drilling jar + six Φ 178.8 mm drill collars + Φ 127 mm drill rod was used for expanding drilling to the hole depth of 2033.00 m. 5.00 m were drilled, with average penetration rate of 1.46 m/h.

Expanding drilling was basically normal. At early stage of the expanding drilling, bit bouncing and jumping were serious, weight on bit and rotation rate were hard to increase. Bit bouncing and jumping were obviously ameliorated by adding absorber and changing Φ 305 mm stabilizer. By taking stabilizer out of slick drill collar and retaining absorber, drilling was smooth and steady; penetration rate was high under high rotation rate, too. At later stage of the expanding drilling, limber pendulum string was employed to reduce well deviation but failed. The total RIH trips of expanding drilling were 56, included 48 trips of expanding drilling, two trips of fishing bit leg at well depth of 304.93 m by running strong magnetic fishing tool, one trip of running Φ 300 mm grind shoe at well depth of 574.97 m, one trip of cleaning-out at 1871.99 m, one trip of cleaning-out but didn't reach hole bottom at well depth of 2028.00 m, running Φ 108 mm die tap and LM70 fishing spear once respectively to fish diamond reamer and tri-cone bit at well depth of 2028.00 m, one trip of cleaning-out at well depth of 2028.00 m. Two downhole accidents happened in the expanding drilling, one concerning two broken bit legs of the guiding rock bit fell to the well bottom, the other one concerning diamond reaming shell broke at the upper soldering point in cleaning-out drilling and diamond reaming shell and tri-cone bit fell to the well bottom. Those two accidents were successfully solved. The main technical indexes of the expanding drilling are shown in Table 4.6.

4.2.4 The First Core Drilling of the Main Hole (Hole Section CCSD-MH, the Third Hole Opening)

Top drive installation, drill rod shift, top drive adjustment and well logging etc. were completed from September 21st to October 3rd, 2002. Core drilling for the third hole opening started on October 4th, 2002 and ended on April 6th, 2002 because well deviation at that time was already larger than the standard, with details as follows:

1. **Drilling in the hole section 2046.54–2071.69 m (Roundtrips MH1–MH17) with German SK wireline core drill string**

2046.54–2047.82 m (Roundtrip MH1): Hole condition and equipment were prepared as a result of wiper trip and actual drilling by top drive test with double tube core drilling tool and SK wireline drill string. Footage at this roundtrip was 1.28 m, with actual drilling time of 1.45 h and core length of 0.3 m.

2047.82–2048.74 m (Roundtrip MH2): Wireline core drilling test by top drive with SK wireline core drilling tool, including: drilling tool was run into the borehole and pumping was started when the drilling tool was only 1.5 m to the hole bottom, without back pressure when it reached to 10 MPa. Fishing of inner tube failed. Another two more fishing operations still failed even after preparing a fishing spear with weighted bar. At the third fishing operation the drilling tool assembly had to be pulled out of the hole because wire rope broke off. While the pressure test at

Table 4.6 Main technical indexes of the expanding drilling in the main hole section from 101 to 2033 m

Item	Parameter	Item	Parameter	Item	Parameter
Expanding drilling footage (m)	1931.89	Average ROP (m/h)	1.05	Roundtrips of expanding drilling	48
Average bit life (m)	55.20	Longest bit life (m)	135.53	Average footage drilled per roundtrip (m)	40.25
Longest roundtrip (m)	109.16	Average footage drilled per day (m/days)	15.84	Percentage of actual drilling in time (%)	66.1
Percentage of tripping in time (%)	14.9	Percentage of accident in time (%)	2.2	Average footage drilled per rig month (m/30 days)	475.08
Well logging (times)	7	Total RIH trips	56	Days of expanding drilling (days)	122

Note Total footage drilled was 1932 m, including hole grinding footage of 0.11 m. The first expanding drilling of the main hole was started on May 7th, 2002 and ended on September 5th, 2002, with 122 days lasted in total. Running casing, circulating mud (solid friction reducer containing little plastic balls), running Φ 273.1 mm casing, cementing well, drilling cement plug, grinding bottom hole, mud displacement, well logging, VSP measurement, and running Φ 193.7 mm moving casing etc. were accomplished between September 6th and 20th, 2002

wellhead reached 48 dashes, pump pressure suddenly increased to 18 MPa, and the inner tube was sucked flat. After changing the inner tube and testing pressure at well head, an effective fishing was obtained. Then drill string was run into the hole for test, pump pressure gradually increased at later drilling stage and penetration rate was extremely slow. Moving drill string was ineffective. Drilling was stopped and the inner tube was to be fished, however, dropped off at half way for two fishing operations, and wire rope got twisted. Drill string was stuck, released by up-lifting and undershooting. Drill bit bottom was worn to the steel body, core catcher was fallen off. Footage in this test roundtrip was 0.92 m, with actual drilling time of 2.87 h and core length of 0.78 m.

2048.74–2049.67 m (Roundtrip MH3): Top drive double tube core drilling was tested by utilizing a combination of double tube core drilling tool and SK wireline drill string. Footage drilled in this roundtrip was 0.67 m, with actual drilling time of 2.47 h, without core recovered, and bit was worn flat. Wireline coring was tested again after grind drilling for 0.26 m by running grind shoe drilling tools in the borerhole.

2049.67–2051.22 m (Roundtrip MH4): Top drive wireline core drilling was tested by employing SK wireline core drilling tool. In the later drilling stage, penetration rate slowed down rapidly, and pump pressure increased. Drilling was stopped and fishing spear was run into the borehole. Wire rope was twisted off half way of fishing inner tube. Drill string was pulled up, bit was worn flat. Footage drilled was 1.55 m, with actual drilling time of 2.88 h and core length of 0.42 m. In case that core recovery failed, grind shoe was run into the borehole to grind the remained core, and then a slim fishing cup string (Φ 65 × 300 mm) was run down.

2051.22–2055.40 m (Roundtrips MH5–MH9): A drilling tool assembly of double tube core drilling tool and SK wireline drill string was employed for top drive double tube core drilling test. No core was recovered for all of the five round trips, and of the five core drill bits used four were worn flat; one was badly worn off at the internal diameter. Coring effect and economy were extremely bad for repeated grinding and fishing drilling. Footage drilled in these five roundtrips was 3.62 m, with actual drilling time of 11.26 h and zero core recovery.

2055.40–2061.30 m (Roundtrips MH10–MH12): A drilling tool assembly of double tube core drilling tool and SK wireline drill string was employed for PDM drive core drilling. Total footage drilled in those three roundtrips was 5.90 m, with actual drilling time of 13.20 h and core recovered of 2.81 m.

2061.30–2064.57 m (Roundtrip MH13): A drilling tool assembly of double tube core drilling tool and SK wireline drill string was employed for PDM hydro-hammer drive core drilling. Footage drilled was 3.27 m, with actual drilling time of 6.44 h and core length of 1.40 m recovered.

2064.57–2066.69 m (Roundtrip MH14): A drilling tool assembly of double tube core drilling tool and SK wireline drill string was employed for PDM drive core drilling. Footage drilled was 2.12 m, with actual drilling time of 3.91 h and core length of 2.30 m recovered.

2066.69–2067.01 m (Roundtrip MH15): SK wireline drilling tool was used for top drive wireline core drilling test. Penetration rate slowed down in the later drilling stage while pump pressure increased. Drilling was stopped to run fishing spear down to the borehole, however, fishing spear fell into drill string. Drill string was pulled out of the borehole and it was found that bit matrix had been burnished. Footage drilled was 0.32 m, with actual drilling time of 0.87 h and 0.15 m core recovered.

2067.01–2069.19 m (Roundtrip MH16): A drilling tool assembly of double tube core drilling tool and SK wireline drill string was employed for PDM drive core drilling. Footage completed was 2.18 m, with actual drilling time of 6.47 h and core length of 1.3 m recovered.

2069.19–2071.69 m (Roundtrip MH17): SK wireline drill string was employed for the test of top drive coring drilling. Penetration rate slowed down in the later drilling stage, so drilling was stopped to run fishing spear down to the borehole, but wire rope was broken off. Drilling string was pulled out of the borehole and it was found that the outside diameter of the drill bit had been notched and the bit face was internally tilted worn off. Footage drilled was 2.5 m, with actual drilling time of 2.28 h and no core was recovered.

In the four ROOH (run out of hole) roundtrips (MH2, MH4, MH15 and MH17) of top drive SK wireline core drilling test, fishing for inner tube all failed, four drill bits were destroyed (matrixes of three drill bits were burnished, one bit face was internally tilted worn off). Total footage completed was 5.29 m, with average penetration rate of 0.59 m/h, core recovery of 19.9 %, and average footage drilled per ROOH roundtrip was 1.32 m. Up to this point, the test of top drive SK wireline core drilling tool in CCSD-1 Well ended up.

2. **Drilling in the hole section 2071.69–2283.86 m (Roundtrips MH18–MH58) with PDM wireline core drilling tool**

2071.69–2075.63 m (Roundtrip MH18): A drilling tool assembly of double tube core drilling tool and SK wireline drill string was employed for top drive double tube core drilling test. Drill string was lifted up because of no penetration and it was found that the thread of SK drill collar was broken. Footage drilled was 0.3 m, with actual drilling time of 1.06 h and no core recovered. In the eight ROOH roundtrips (MH1, MH3, MH5 to H9 and MH18) of top drive double tube core drilling, a total footage of 5.87 m was completed, with an average penetration rate of 0.36 m/h, core recovery of 5.1 % and an average footage of 0.73 m drilled per ROOH roundtrip. Thus, top drive double tube core drilling test in CCSC-1 Well ended up.

2071.99–2075.63 m (Roundtrip MH19): After fishing, a drilling tool assembly of double tube core drilling tool and SK wireline drill string was employed for PDM drive core drilling. Footage drilled was 3.64 m, with actual drilling time of 8.28 h and core length of 0.52 m. A core drilling tool was run down the borehole for overcoring to the hole bottom. No core was recovered, but the drill bit was still in good condition.

2075.63–2080.01 m (Roundtrip MH20): PDM wireline core drilling tool was employed for PDM drive wireline core drilling test. Four roundtrips of wireline fishing and one trip of ROOH without fishing (because of pump blocked) were carried out. Four roundtrips of wireline fishing were successful, with inner tubes recovered. Cores were obtained, except the first roundtrip. Because of pump blocked, drill string was directly lifted up in the last roundtrip, without fishing. Two pieces of bit matrix were broken off. Total footage drilled was 4.38 m, with actual drilling time of 14.14 h and core length of 3.47 m recovered.

2080.01–2130.58 m (Roundtrips MH21–MH33): Based upon well diameter measurement results, this hole section belonged to oversized section, with broken strata. Drill string under high speed rotation swayed severely in this oversized hole section, causing instability of the drill bit. Because of the broken strata, drill bit was quickly worn flat, and core could not be recovered because it was ground too slim to be taken out. The problem of drill rod breaking was also related to this. So, it was decided that PDM drive double tube core drilling was to be used to temporarily replace top drive wireline core drilling (because wireline core drilling was designed for the main hole, and hydro-hammer was unavailable at drill site at that time), to stabilize the borehole condition and to create a favorable condition for wireline core drilling. In the roundtrips from 21 to 33 (MH21–MH33), a drilling tool assembly of double tube core drilling tool and SK wireline drill string was employed for PDM core drilling. A total footage of 50.57 m was drilled, with actual drilling time of 108.47 h and core length of 25.37 m recovered.

2130.58–2132.34 m (Roundtrip MH34): PDM wireline core drilling tool was employed for PDM drive wireline core drilling test. One roundtrip of wireline fishing and another roundtrip of ROOH without fishing (as a result of pump blocked) were carried out. Inner tube and core were fished out in wireline fishing trip and in another ROOH roundtrip only drill string was lifted up, without fishing because of pump blocked, and core catcher moved up into the nipple. Total footage drilled was 1.76 m, with actual drilling time of 4.2 h and core length of 1.01 m recovered.

2132.34–2144.05 m (Roundtrips MH35 and MH36): A drilling tool assembly of double tube core drilling tool and SK wireline drill string was employed for PDM drive core drilling, for two roundtrips. Total footage of 11.71 m was completed, with actual drilling time of 15.16 h and core length of 9.10 m recovered.

2144.05–2283.28 m (Roundtrips MH37–MH57): After arrival of the fabricated hydro-hammer to the drill site, a drilling tool assembly of double tube core drilling tool and SK wireline drill string was employed for PDM hydro-hammer drive core drilling, for 21 roundtrips. Total footage drilled was 139.23 m, with actual drilling time of 114.53 h and core length of 84.83 m recovered. Jet-type reverse circulation swivel type double tube drilling tool was employed for the roundtrip MH44 at well depth of 2196.41 m because the core recovery from roundtrip MH40–MH43 was extremely low. Footage of 4.7 m was completed, with actual drilling time of 3.77 h and core length of 1.20 m recovered.

2283.28–2283.86 m (Roundtrip MH58): PDM wireline core drilling tool was employed for PDM drive wireline core drilling test. During drilling operation pump blocking occurred frequently, penetration rate was low, or pump blocking caused no penetration. So drilling was stopped to fish the inner tube, but wire rope was broken off and then drill string was pulled out. Footage drilled was 0.58 m, with actual drilling time of 2.25 h and core length of 0.35 m recovered.

During the three ROOH roundtrips (MH20, MH34 and MH58) of PDM wireline core drilling tests, eight round trips were completed and among which five were successful for fishing the inner tubes, one failed and in the other two the drill string was directly pulled out of the borehole without fishing because of pump blocked. Total footage drilled was 6.72 m, with average penetration rate of 0.33 m/h and core recovery of 71.9 %. Footage per ROOH roundtrip was 2.24 m and the footage per drilling roundtrip was 0.66 m. Up to this point, the test of PDM drive wireline core drilling tool in CCSD-1 Well was ended.

3. **Drilling in the hole section 2283.86–2460.38 m (Roundtrips MH59–MH83) with hydro-hammer wireline core drilling tool**

2283.86–2442.71 m (Roundtrips MH59–MH80): A drilling tool assembly of double tube core drilling tool and SK wireline drill string was employed for 22 round trips of PDM hydro-hammer drive core drilling. Total footage of 158.85 m was completed, with actual drilling time of 132.11 h and core length of 103.11 m recovered.

2442.71–2449.13 m (Roundtrip MH81): Hydro-hammer wireline core drilling tool was employed for the tests of top drive hydro-hammer drive wireline core drilling. Two wireline fishing roundtrips were carried out, with the inner tubes both fished out, however, with core recovered only for one roundtrip, and in the other roundtrip the core catcher moved up into the inner tube. Total footage drilled was 6.42 m, with actual drilling time of 7.15 h and core length of 6.28 m recovered.

2449.13–2458.53 m (Roundtrip MH82): A drilling tool assembly of double tube core drilling tool and SK wireline drill string was employed for PDM hydro-hammer drive core drilling, for one roundtrip. Footage drilled was 9.4 m, with actual drilling time of 7.17 h and core length of 6.82 m recovered.

2458.53–2460.38 m (Roundtrip MH83): Hydro-hammer wireline core drilling tool was employed for the test of top drive hydro-hammer drive wireline core drilling. Drilling was stopped to fish the inner tube because of decrease of penetration rate and pump blocking. Three fishing operations for inner tube failed (running overshot assembly into the borehole but meeting slack-off at well depth of 1000 m. Re-running overshot assembly into the borehole and meeting slack-off at well depth of 480 m again after 25 pieces of drill rod were pulled out of the borehole. After 16 more pieces of drill rod were pulled out, overshot assembly was successfully run down to the hole bottom, however, without the inner tube fished out at last), and then the drill string was pulled out. Footage drilled was 1.85 m, with actual drilling time of 2.1 h and core length of 1.95 m recovered. Hydro-hammer wireline core drilling tool was run again into the borehole and slack-off happened at a distance of 7.56 m to the hole bottom. Pump was blocked with 7 Mpa during reaming down, and the pressure didn't turn back by moving the drilling tool. Top drive and some pieces of drill rod were removed; slack-off happened at well depth of 724 m when running the overshot assembly into the borehole. Drill string was pulled out and it was found that the drill rod was full of mud. The drill collar (30 cm), lower stabilizer, joint and spring case were all filled with cuttings.

In the two ROOH roundtrips (MH81 and MH83) of top drive hydro-hammer drive wireline core drilling test, three roundtrips were conducted, with two successful in fishing out the inner tubes and one failed. Total footage drilled was 8.27 m, with average penetration rate of 0.89 m/h and core recovery of 99.5 %. Footage per ROOH roundtrip was 2.76 m and the footage per drilling roundtrip was 0.66 m. Up to this point, the test of hydro-hammer wireline core drilling in CCSD-1 Well was ended.

4. **Drilling in the hole section 2460.38–2462.71 m (Roundtrip MH84) with China made S157 wireline core drilling tool**

S157 wireline core drilling tool was employed for the test of top drive wireline core drilling. Drilling was stopped because of no penetration. Inner tube was fished out after running fishing spear in the borehole, but it was found that there was no core in it. Drill string was pulled out because slack-off happened when running inner tube in the borehole.

Running inner tube in the borehole and slack-off happened again, and then the drill string was pulled out again. The wear of drill bit was normal but there was no core in outer tube. Footage drilled was 2.33 m, with actual drilling time of 3.12 h and average penetration rate of 0.75 m/h. It was found by examination that the inside diameter of the male joints of some SK wireline core drill rods were smaller than the standards of Φ 110 mm because of rough fabrication, so fishing and running of inner core barrel were not able to be realized normally. Then, the test of wireline core drilling in CCSC-1 Well was ended.

5. **Drilling in the hole section 2462.71–2519.64 m (Roundtrips MH85–MH92)**

A drilling tool assembly of double tube core drilling tool and SK wireline drill string was employed for PDM hydro-hammer drive core drilling. Eight roundtrips were completed, with total footage of 56.78 m drilled. Actual drilling time of 51.1 h was used and core length of 41.44 m recovered. While lowering down the drilling tool assembly, the characteristics of top drive load were tested at the well depths of 500, 1000, 1500, 2000, 2250 and 2500 m respectively (at that time the hole depth was 2517.76 m deep) on the way of running the drilling tool assembly down to the borehole. At the well depth of 2500 m, the pressure of the top drive system was 330 bar, torque was 10,900 Nm, but rotation speed was only 156 r/min, what's more, the top drive system swayed seriously and the rotation speed was not steady either. In this case the top drive and SK wireline core drill rods were difficult to satisfy the hereafter wireline core drilling tests. And as for conventional coring, the tripping time for SK wireline core drill rods was longer than that for Φ 89 mm drill rods (at well depth of 2500 m, the pulling-out time for SK wireline core drill rods was about 10 h, and the tripping-in time was about 6 h while for Φ 89 drill rods was just 7 and 5 h). So water swivel, Φ 89 mm kelly bar and Φ 89 mm drill string were used instead of top drive and SK wireline core drill rods.

6. **Drilling in the hole section 2519.64–2982.18 m (Roundtrips MH93–MH146)**

A drilling tool assembly of double tube core drilling tool and Φ 89 mm drill string was employed for PDM hydro-hammer drive core drilling for 54 roundtrips. A total footage of 462.54 m was drilled, with actual drilling time of 320.13 h and core length of 432.21 m recovered. Core drilling in the first section (2046.54–2982.18 m) of the main hole was ended and then straighten drilling by sidetracking was to be conducted because well deviation had gone beyond the limit.

Statistics of the coring drilling effects in sequence in the first section of the main hole (CCSD-MH) was shown in Table 4.7, from which it can be learned that lots of technical work and repeated tests were conducted for the wireline core drilling for the main hole. Conventional double tube core drilling was alternately used to create a good in-the-hole

4.2 Simple Situation of the Construction at Different Periods

Table 4.7 Statistics of the coring drilling effects in sequence in the first section of the main hole (MH 2046.54–2982.18 m)

Core drilling methods	Trips	Footage (m)	Actual drilling time (h)	Core length (m)	ROP (m/h)	Core recovery (%)	Footage drilled per roundtrip (m)
Top drive double tube	1	1.28	1.45	0.30	0.88	23.4	1.28
Top drive wireline coring	1	0.92	2.87	0.48	0.32	52.2	0.92
Top drive double tube	1	0.67	2.47	0.00	0.27	0.0	0.67
Top drive wireline coring	1	1.55	2.88	0.42	0.54	27.1	1.55
Top drive double tube	5	3.62	11.26	0.00	0.32	0.0	0.72
PDM drive double tube	3	5.90	13.2	2.81	0.45	47.6	1.97
PDM hydro-hammer drive double tube	1	3.27	6.44	1.40	0.51	42.8	3.27
PDM drive double tube	1	2.12	3.91	2.30	0.54	108.5	2.12
Top drive wireline coring	1	0.32	0.87	0.15	0.37	46.9	0.32
PDM drive double tube	1	2.18	6.47	1.30	0.34	59.6	2.18
Top drive wireline coring	1	2.50	2.28	0.00	1.10	0.0	2.50
Top drive double tube	1	0.30	1.06	0.00	0.28	0.0	0.30
PDM drive double tube	1	3.64	8.28	0.52	0.44	14.3	3.64
PDM drive wireline coring	1 (5)	4.38	14.14	3.47	0.31	79.2	4.38
PDM drive double tube	13	50.57	108.47	25.37	0.47	50.2	3.89
PDM drive wireline coring	1 (2)	1.76	4.20	1.01	0.42	57.4	1.76
PDM drive double tube	2	11.71	15.16	9.10	0.77	77.7	5.86
PDM hydro-hammer drive double tube	21	139.23	114.53	84.83	1.22	60.9	6.63
PDM drive wireline coring	1	0.58	2.25	0.35	0.26	60.3	0.58
PDM hydro-hammer drive double tube	22	158.85	132.11	103.11	1.20	64.9	7.22
Top drive hydro-hammer drive wireline coring	1 (2)	6.42	7.15	6.28	0.90	97.8	6.42
PDM hydro-hammer drive double tube	1	9.40	7.17	6.82	1.31	72.6	9.40
Top drive hydro-hammer drive wireline coring	1	1.85	2.10	1.95	0.88	105.4	1.85
Top drive wireline coring (S157)	1	2.33	3.12	0.00	0.75	0.0	2.33
PDM hydro-hammer drive double tube (SK wireline drill string)	8	56.78	51.10	41.44	1.10	73.0	7.10
PDM hydro-hammer drive double tube (Φ 89 drilling string)	54	462.54	320.13	432.21	1.44	93.4	8.57
Total	146	934.67	845.07	725.62	1.11	77.6	6.40

Note 1 Roundtrips in brackets were those of wireline coring drilling without pulling out of the drill string
2 Besides core drilling, grinding drilling footage was 0.97 m in this hole section

condition for wireline core drilling. Four wireline core drilling tools including German SK wireline core drilling tools, home-made PDM wireline core drilling tools, hydro-hammer wireline core drilling tools and S157 wireline core drilling tool were employed for drilling tests, for ten ROOH roundtrips, in which German SK wireline core drilling tools were tested for four times, however, for all those four tests the fishing of inner tube failed after each round trip was finished, and all the four drill bits were destroyed. PDM wireline core drilling tool was tested for three times and eight roundtrips were drilled. Fishing of inner tube succeeded for five times, for two times the drill string was run out of the hole directly without fishing because of pump blocked, and the other failed. Hydro-hammer wireline core drilling tool was tested for two times and three roundtrips were drilled. For two trips the fishing of inner tube was successful but the other one failed. S157 wireline coring drilling tool was tested for one time and one roundtrip was drilled. Fishing of inner tube was successful, but slack-off happened when running the inner tube again in the borehole. Under these circumstances the advantage of long tripping time interval for wireline core drilling was not brought into full play, and restoration of German SK wireline core drilling rods needed time, in addition to the ability of the top drive was not enough along with increase of well depth. So, from the tests we could know that wireline core drilling could not meet the requirement of core drilling in CCSD-1 Well both from its reliability and economy. It was necessary to change other core drilling method, which could obviously improve the footage drilled per roundtrip.

Core drilling in this hole section did not go smoothly, with total 167 running-in-hole roundtrips, including 146 drilling roundtrips (containing 10 roundtrips of pulling out drill string for wireline core drilling). Running Φ 145–157 mm mill shoe + fishing cup in the hole section of 2047.82–2054.75 m for six times, in the first time there were 1 kg crushed carbides (Fig. 4.12) in the fishing cup, and diamond bit matrix and 40 mm × 35 mm nozzle body (Fig. 4.13) of rock bit etc. were fished out in the fishing cup in the third time. A Φ 65 mm overcoring drill bit (Fig. 4.14) + a fishing cup were run down the hole at the well depth of 2051.22 m for one time, a strong magnetic extractor (Fig. 4.15) was run down at the well depth of 2051.42 m for one time, a Φ 139.7 mm spear head and a Φ 168.3 mm fishing tap were run respectively at the well depth of 2052.62 m for one time, a Φ 108 mm tap was run at the well depth of 2053.82 m for one time, a Φ 139.7 mm tap and a Φ 108 mm tap were run once respectively at the well depths of 2071.99 and 2201.11 m, overcoring was conducted at well depths of 2051.22 and 2075.63 m respectively, bit matrix pieces were overcored once at the well depth of 2517.91 m, at the well depth of 2080.01 m the borehole was reamed down with the reaming tool for one time, at the well depth of 2276.33 m the drill string could not be run down to the bottom hole for one time, pump blocking happened for three times respectively at the well depths of 2449.13, 2460.38

Fig. 4.13 Tungsten carbides

and 2982.18 m when the drilling tool was run down to the bottom hole. Variations of well deviation angle, diameter, and azimuth angle in the hole section MH core drilling were shown in Fig. 4.16, and the main technical indexes shown in Table 4.8.

4.2.5 The First Sidetracking (Deviation Correction) Drilling of the Main Hole

When core drilling of the third opening reached to the well depth of 2982.18 m, well logging indicated that well deviation angle at the depth of 2935 m was 16.35° (well logging tool could not run down to the hole bottom). Furthermore, the deviation angle in the last 25 m increased linearly (deviation angle from 14.29° to 16.35° and azimuth angle from 305.79° to 307.80°). The total bending angle in this hole section was 2.13° with bending strength of 2.56 (°)/30 m. Well deviation angle at the well depth of 2982.18 m would reach 20.24°, in case of calculation based upon this developing trend in the hole section from 2910 to 2935 m. According to the Preliminary Design of CCSD Project, well deviation angle should be no larger than 18° in the hole section from 2000 to 5000 m, so it was decided that core drilling was to be stopped to conduct sidetracking straightening drilling.

Backfill sidetracking straightening drilling was taken from April 7th, 2003 to June 8th, 2003, with details shown as follows:

Fig. 4.12 Crushed carbides

4.2 Simple Situation of the Construction at Different Periods

Fig. 4.14 Overcoring bit

1. **Instauration of artificial well bottom (from April 7th to 10th, 2003)**

   A drilling tool assembly of Φ 120 mm drill collar (eight pieces) + Φ 89 mm drill rod was employed to run down to the borehole and drilling mud was circulated when the drill string reached to the well depth of 2,980 m. Grouting pipeline was connected, 3 m^3 of spacer fluid was poured, 7.5 m^3 cement slurry (JHG cement + 0.5 % expanding agent + 0.5 % dispersing agent + 0.8 % filtrate reducer + 1.0 % accelerating agent + 0.38 % water) with density of 1.97 g/cm^3 was poured, 10 m^3 of displacement slurry was poured. Drill string was pulled up to the well depth of 2,730 m and drilling mud was circulated until there was no cement slurry in the back slurry. Then the drill string was pulled out, waiting for curing.

   A drilling tool assembly of Φ 158.8 mm tri-cone bit + seven pieces of Φ 120 mm drill collar + Φ 89 mm drill rod was employed to drill cement plug from the well depth of 2718.3–2748.00 m, then drilling mud was circulated and drill string was pulled out. Drilling technical data included: bit load of 8 kN, rotation speed of 60 r/min, delivery rate of 10.51 L/s and pump pressure of 6 MPa.

   A drilling tool assembly of Φ 157 mm diamond core drill bit + Φ 157 mm reaming shell + 8.5 m long Φ 139.7 mm outer tube + upper joint + Φ 120 mm PDM + seven pieces of Φ 120 mm drill collar + Φ 89 mm drill rod was used for core drilling, with footage of 1.00 m drilled. Cement plug sample was complete with length of 0.75 m. Drilling technical data included: bit load of 10 kN, rotation speed of 135 r/min, delivery rate of 7.58 L/s and pump pressure of 5 MPa.

2. **The first stage of sidetracking drilling in the hole section 2749.00–2758.00 m (from April 11th to 19th, 2003)**

   A drilling tool assembly of Φ 157 mm natural diamond surface set drill bit (Fig. 4.17a) was employed to drill the hole section between 2749.00 and 2752.05 m. A Φ 158.8 mm tri-cone bit drilling tool assembly (Fig. 4.17b) was employed to drill the hole section between 2752.05 and 2756.34 m. A drilling tool assembly of Φ 157 mm natural diamond surface set drill bit (Fig. 4.17c) was used Well depth/m to drill the hole section between 2756.34 and 2758 m.

   For the first three roundtrips of sidetracking drilling, the particles in the returned mud at wellhead were basically cement, so no new sidetracked hole had been drilled out. A sidetracking tool assembly of Φ 158 impregnated diamond drill bit with outside diameter of 158.5 mm (Fig. 4.17d, e) was employed to run down to the borehole, but all the three times of running down the drill string slack-off was

Fig. 4.15 Strong magnetic extractor

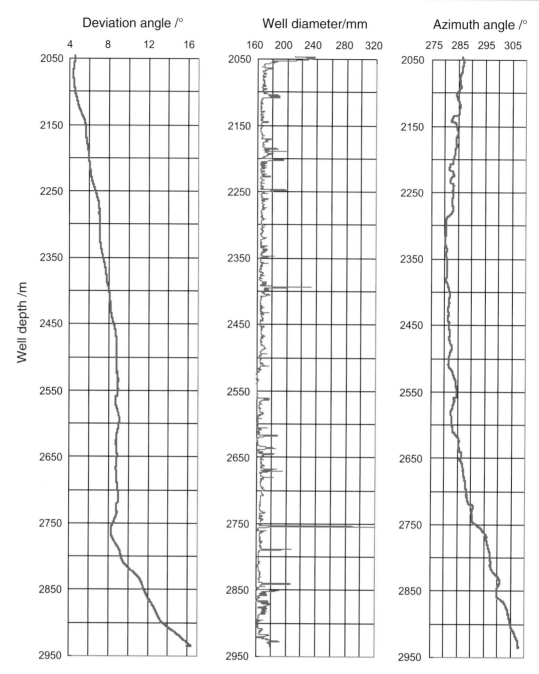

Fig. 4.16 Well deviation angle, diameter and azimuth angle of Φ 157 mm core drilling in hole section 2046.54–2982.18 m

encountered at the well depths of 2187 m (with stabilizer in the drill string), 2238 m (without stabilizer) and 2358 m (without stabilizer) respectively.

3. **The second stage of sidetracking drilling in the hole section 2758.00–2771.35 m (from April 20th to 24th, 2003)**

A drilling tool assembly of Φ 140 mm slim hole diamond impregnated drill bit (Fig. 4.17f) was used for sidetracking drilling in the hole section from 2758.00 to 2769.59 m. In the early drilling, bit load was strictly controlled for a steady feed, basically maintained at approximately 5 kN. Rotation speed was kept at 97–103 r/min, pump delivery 8.27–8.79 L/s, and pump pressure 6.7–7.1 MPa, with an average drilling rate of 0.26 m/h obtained. After a footage of 6.7 m drilled, bit load was gradually increased to 20 kN and an average drilling rate of 0.29 m/h was obtained, with rotation speed of 103–122 r/min, pump delivery of 8.79–10.34 L/s and pump pressure of 7.2–9.45 MPa. In the later drilling, particles in the returned mud at wellhead were basically cuttings. When drilling reached to the well depth of 2769.59 m, no

Table 4.8 Main technical indexes of core drilling in the main hole section 2046.54–2982.18 m

Item	Index	Item	Index	Item	Index
Core drilling footage (m)	934.67	Average core recovery (%)	77.6	Average ROP (m/h)	1.10
Core drilling roundtrips	146	Normal ROOH trips with full-barrel	78	Average bit life (m)	16.12
Longest bit life (m)	69.70	Average footage drilled per roundtrip (m)	6.40	Longest footage drilled per roundtrip (m)	9.56
Average daily footage drilled (m/days)	5.09	Percentage of actual drilling time (%)	20.1	Percentage of tripping in time (%)	56.4
Percentage of accident in time (%)	1.1	Largest well deviation angle (°)	16.39	Average footage drilled per rig month (m/30 days)	152.55
Project well logging (times)	10	Comprehensive well logging (times)	1	Days of core drilling (days)	184

Note Total footage drilled was 935.64 m, with grinding footage of 0.97 m. 185 days of core drilling including one day of comprehensive well logging was used, started on October 4th, 2002 and ended on April 6th, 2003. Top drive installation, arrangement of drill rods, top drive adjustment and well logging etc. were done from September 21st to October 3rd, 2002

penetration could be obtained even bit load was increased. Drill string was pulled up away from the hole bottom to a certain distance, and then it could not be lowered down to the original position. ROOH found that the central dead point of the drill bit was broadened, drill bit was badly worn and the diamond layer of the outside cutting edges was basically worn flat.

A drilling tool assembly of Φ 157 mm diamond bit with Φ 110 mm guiding + Φ 120 mm PDM + three pieces Φ 120 mm drill collars + Φ 89 mm drilling rods was used for drilling the hole section from 2758.00 to 2769.19 m. Redressing was conducted because slack-off had been happened during running down the drill string to the depth of 2756.53 m. Expanding drilling was carried out to 2769.19 m deep (with a distance of 0.4 m to hole bottom) and then the drill string was pulled out. Bit load for this expanding drilling was 3–10 kN, rotation speed was 161–173 r/min, pump delivery was 8.79–9.48 L/s, and pump pressure was 5.9–6.7 MPa. An average drilling rate of 1.24 m/h was obtained for this expanding drilling.

A drilling tool assembly of Φ 157 mm coring bit + Φ 157 mm reaming shell + 4.59 m long Φ 139.7 mm core barrel + Φ 157 mm reaming shell + upper joint + Φ 120 mm PDM + three pieces of Φ 120 mm drill collars + Φ 89 mm drill rod was employed for extended core drilling in the hole section from 2769.19 to 2771.35 m. RIH met slack-off at the depth of 2754.13 m, pump was turned on for redressing to the hole bottom. Drill string was pulled out at the depth of 2771.35 m because of no penetration and it was found that one piece of bit matrix had dropped out from the steel body, and core of 1.76 m long was recovered. Bit load adopted for this core drilling was 15–25 kN, with rotation speed of 173–183 r/min, pump delivery of 9.48–10 L/s and pump pressure of 7.2–7.6 MPa. An average drilling rate of 0.39 m/h was obtained.

Well logging showed that the orientation of the sidetracking was accurate, well deviation angle was reduced, and azimuth angle was basically stable, with a little increase. Well deviation angle was still stable and azimuth angle began to decrease in the extended core drilling from 2769.59 to 2771.35 m deep.

4. **Drop-off drilling with rock bit in the hole section 2771.35–2942.11 m (from April 25th to May 11th, 2003)**

A drilling tool assembly of Φ 158.8 mm tri-cone bit (Fig. 4.17b) was employed for seven roundtrip drop-off drilling at the hole sections from 2771.35 to 2797.01 m, from 2797.01 to 2812.03 m, from 2812.03 to 2850.01 m, from 2,850.01 to 2888.79 m, from 2888.79 to 2907.70 m, from 2907.70 to 2922.23 m, and from 2922.23 to 2942.11 m. In the second roundtrip a 16.5 mm thick pad was welded on the bending area of the PDM. In the fourth, fifth and sixth roundtrips an 11 mm thick pad was welded on the bending area of the PDM and in the seventh roundtrip a 17 mm thick pad was welded on the bending area of the PDM.

5. **Drop-off drilling with impregnated diamond drill bit in the hole section 2942.11–2956.91 m (from May 12th to 24th, 2003)**

A drilling tool assembly of Φ 158 mm impregnated diamond drill bit with outside diameter of 158.7 mm (Fig. 4.17e) was lowered down to the borehole and slack-off was encountered at the well depth of 2350 m. A repair drilling tool assembly of Φ 158.6 mm repairing tool + 3.92 m long Φ 139.7 mm core barrel + Φ 158.6 mm repairing tool + 3.92 m long Φ 139.7 mm core barrel + Φ 158 mm reaming shell + Φ 120 mm PDM + seven pieces of Φ 120 mm drill collar + Φ 89 mm drill rod was employed for hole repairing. Hole repairing was stopped at the well depth of 2942 m to pull out the drill string because of frequent

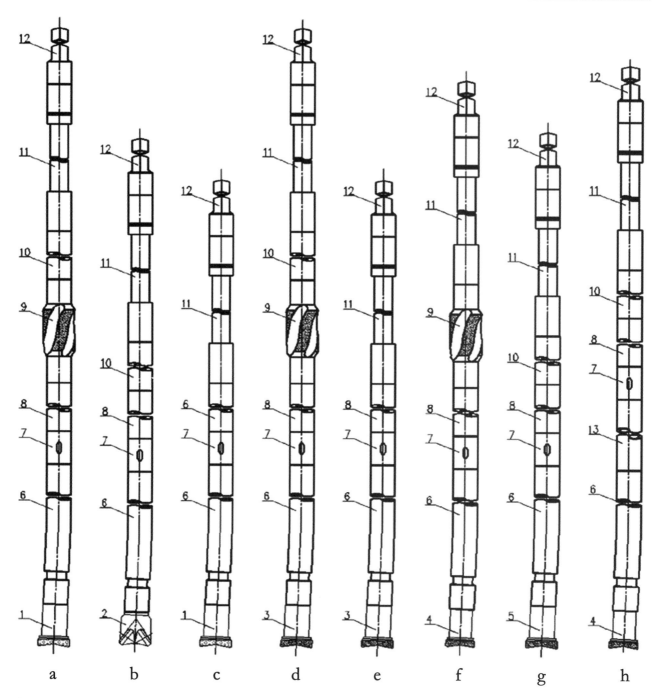

Fig. 4.17 Drilling tool assemblies for sidetracking straightening drilling. *1* Φ 157 mm natural diamond surface set drill bit, *2* Φ 158.8 mm rock bit, *3* Φ 158 mm impregnated diamond drill bit, *4* Φ 140 mm impregnated diamond drill bit, *5* Φ 157 mm impregnated diamond drill bit, *6* Φ 120 mm PDM with single bend, *7* Φ 120 mm directional sub, *8* Φ 104 mm non-magnetic drill collar, *9* Φ 156 mm stabilizer, *10* Φ 120 mm drill collar stem, *11* Φ 89 mm drill string, *12* Kelly bar, *13* Φ 120 mm drilling jar

pump blockage and bit bouncing during hole repair drilling. The matrix of repairing tools was badly worn out, and the outside diameters of lower, middle and upper reaming shell were 150, 151 and 157.3 mm respectively.

A drilling tool assembly with Φ 157 m impregnated diamond drill bit (Fig. 4.17g) was employed for drop-off drilling in the hole section from 2942.11 to 2952.96 m, with a 17 mm thick pad welded on the bending area of the PDM. No more footage was drilled after drilling reached to the depth of 2952.96 m, tight pull was encountered when moving and pulling out the drill string, but pump pressure was still normal. Pulling out the drill string repeatedly with pressure of 80–90 t (hanging load of the drill string was 60 t), downward pressure of 8–10 t, and circulating drill mud with large delivery, then tight pull was freed by pulling out the drill string with 32 t pressure and instantaneous downward pressure of 8 t.

A drilling tool assembly with Φ 157 mm impregnated diamond bit (Fig. 4.17d) could not be run down to the hole bottom. A 17 mm thick pad was welded on the bending area of the PDM. And the drill string also had a stabilizer on it.

A drilling tool assembly with Φ 157 mm impregnated diamond bit (Fig. 4.17g) was employed for drop-off drilling in the hole section from 2952.96 to 2956.91 m. A 17 mm thick pad was welded on the bending area of the PDM.

6. **Drop-off drilling with slim hole impregnated diamond drill bit in the hole section 2956.91–2974.59 m (from May 25th and June 8th, 2003)**

A drilling tool assembly with Φ 140 mm small diameter impregnated diamond drill bit (Fig. 4.17h) was used for drop-off drilling in the hole section from 2956.91 to 2961.91 m.

An expanding drill string with guiding consisted of Φ 157 mm impregnated diamond drill bit with Φ 110 mm guide + Φ 120 mm PDM + Φ 120 mm drilling jar + seven pieces of Φ 120 mm drill collar + Φ 89 mm drill rod was employed for the hole section from 2956.91 to 2961.03 m. RIH was stopped to ream because of the slack-off at the well depth of 2942 m. Pump blocking was encountered at the well depth of 2960.73 m during reaming and then the drill string was pulled out.

A repair drilling tool assembly of Φ 158.5 mm repairing tool + 3.92 m long of Φ 139.7 mm core barrel + Φ 158.6 mm repairing tool + 3.92 m long of Φ 139.7 mm core barrel + Φ 157.3 mm reaming shell + Φ 120 mm PDM + Φ 120 mm drilling jar + seven pieces Φ 120 mm drill collar + Φ 89 mm drill rod was used for repairing and redressing, during which pump blocking was frequently encountered. The drill string was pulled out after reaming to 2961.38 m deep. The outside diameters of the lower, middle repairing tools and upper reaming shell were 157.25, 156.3, and 157.3 mm respectively.

A coring tool assembly of Φ 157 mm core drill bit + Φ 157 mm reaming shell + 4.59 m long of Φ 139.7 mm core barrel + Φ 157 mm reaming shell + swivel type joint + YZX127 hydro-hammer + Φ 120 mm PDM + Φ 120 mm drilling jar + seven pieces of Φ 120 mm drill collar + Φ 89 mm drill rod was used for hole redressing to the well depth of 2961.40 m, and pump pressure increased to 12.50 MPa. Redressing repeatedly to the well depth of 2961.85 m, then was stopped. Nine pieces of drill bit matrix fell into the borehole.

A grinding and fishing tool assembly of Φ 156 mm grind shoe + fishing cup + Φ 120 mm drilling jar + seven pieces of Φ 120 mm drill collar + Φ 89 mm drill rod was employed for grinding and fishing to the well depth of 2,961.95 m. Outside diameter of the grind shoe was 155.5 mm, about 1.5 kg small chips were fished out.

A core drilling tool assembly of Φ 157 mm core drill bit + Φ 157 mm reaming shell + 4.59 m long Φ 139.7 mm core barrel + Φ 157 mm reaming shell + swivel type joint + YZX127 hydro-hammer + Φ 120 mm PDM + Φ 120 mm drilling jar + seven pieces of Φ 120 mm drill collar + Φ 89 mm drill rod was used for core drilling to the well depth of 2,966.95 m. The drill string was pulled out when the core barrel was full, with core length of 2.60 m.

A core drilling tool assembly of Φ 157 mm core drill bit + Φ 157 mm reaming shell + 4.59 m long Φ 139.7 mm core barrel + Φ 157 mm reaming shell + swivel type joint + YZX127 hydro-hammer + Φ 120 mm PDM + Φ 120 mm drilling jar + SK157 stabilizer + 15 pieces of SK146 wireline core drill collar + Φ 89 mm drill rod was employed for drifting to the well depth of 2970.54 m. Then drilling was stopped and the drill string was pulled out because of low drilling rate, high pump pressure of 10.20 MPa and wear-out of bit matrix. Core of 1.62 m long was recovered.

A drilling tool assembly with Φ 140 mm impregnated diamond drill bit (Fig. 4.17h) was lowered down the borehole and pump was started when the drilling tool assembly was 5 m away from the hole bottom. Pump pressure arose to 8.5 MPa immediately. Drill string was pulled out because of no returned pressure was found and the drill string below the directional sub was blocked with settled sands.

A drilling tool assembly with Φ 140 mm small diameter impregnated diamond bit (Fig. 4.17h) was used for drop-off drilling in the hole section from 2970.54 to 2974.59 m. The drill string was pulled out at the well depth of 2,974.59 m as a result of power failure by thunderstorm and the accident of wire MWD (measurement while drilling) deviation survey tool. Therefore the first sidetracking (deviation correction) drilling of the main hole ended up.

A total of 33 RIH roundtrips were completed in this hole section, including 15 roundtrips of sidetracking (deviation correction) drilling, three roundtrips of drilling, sweeping and fishing cement plug each for once respectively, seven roundtrips in which slack-off was encountered, two roundtrips of guide expanding drilling, two roundtrips of cleaning-out, three roundtrips of core drilling, one roundtrip of grinding and fishing. Well deviation angle, diameter and azimuth angle in this hole section were shown in Fig. 4.18, and the main technical descriptions in Table 4.9.

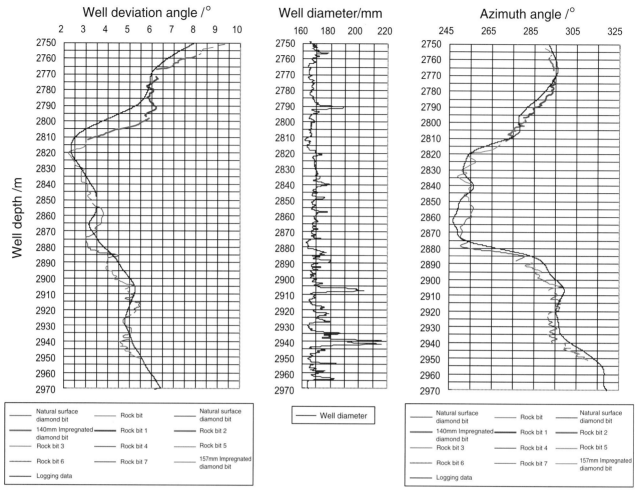

Fig. 4.18 Well deviation angle, azimuth angle and diameter of the hole section of the first sidetrack (deviation correction) drilling, obtained by wire MWD survey and logging

Table 4.9 Main technical descriptions of the first sidetrack (deviation collection) drilling in the main hole section from 2749.00 to 2974.59 m

Item	Parameter	Item	Parameter	Item	Parameter
Footage drilled by sidetrack drilling (m)	215.20	Average ROP (m/h)	0.69	Roundtrips of sidetrack drilling	15
Average footage drilled per roundtrip (m)	14.45	Longest roundtrip (m)	38.78	Footage drilled by core drilling (m)	10.75
ROP of core drilling (m/h)	1.04	Core recovery of core drilling (%)	55.6	Average footage drilled per day (m/days)	3.83
Percentage of actual drilling in time (%)	23.6	Percentage of tripping in time (%)	27.9	Percentage of accident in time (%)	0.3
Average footage drilled per rig month (m/30 days)	114.90	Well logging/times	3	Days of drilling (days)	63

Note Total footage of 241.30 m was completed, including 15.31 m by guide expanding drilling and 0.04 m by grinding and fishing. Drilling cement plug, waiting for cure, drilling out and fishing cement plug etc. were completed from April 7th to 10th, 2003. Sidetrack drilling was started on April 11th, 2003 and ended on June 8th, 2003

4.2.6 The Second Core Drilling of the Main Hole (Hole Section CCSD-MH-1C)

The second core drilling in the main hole was started on June 9th, 2003 and reached to the well depth of 3,665.87 m on October 2nd, 2003. Core drilling in this hole section ended up on November 21st, 2003 as a result of the failure of fishing the lower reaming shell and the drill bit which had been broken off to the hole bottom. In the construction of core drilling, down hole deviation correction drilling was conducted for three times and non core drilling with roller bit was completed for two times, with details as follows:

1. 2974.59–3127.54 m core drilling with slick assembly: A core drilling tool assembly (Fig. 4.19a) of drill bit + reaming shell + core barrel + reaming shell + swivel type joint + hydro-hammer + PDM + Φ 120 mm drill collar was employed for core drilling to the depth of

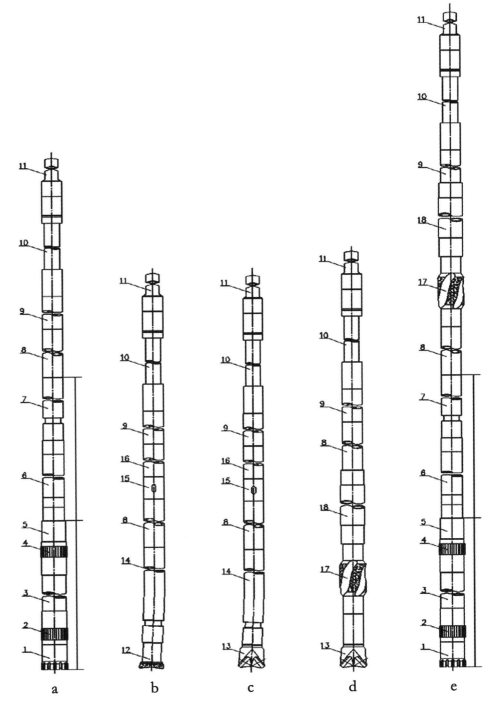

Fig. 4.19 The drilling tool assembly for the hole section 2974.59–3665.87 m. *1* Φ 157 mm coring bit, *2* Φ 157 mm lower reaming shell, *3* Φ 139.7 mm core barrel, *4* Φ 157 mm upper reaming shell, *5* Φ 146 mm swivel type join, *6* YZX127 hydro-hammer, *7* Φ 120 mm PDM, *8* Φ 120 mm jar, *9* Φ 120 mm drill collar stem, *10* Drill string (198 pieces of Φ 127 mm drill rod + Φ 89 mm drill rod in the hole section 2974.59–3400.04 m and Φ 89 mm drill rod after 3400.04 m), *11* Kelly bar, *12* Φ 140 mm impregnated deflection bit, *13* Φ 158.8 mm roller bit, *14* Φ120 mm single bend PDM, *15* Φ 120 mm directional sub, *16* Φ 104 mm non-magnetic drill collar, *17* SK157 stabilizer, *18* SK146 wireline core drill collar stem

3127.54 m within 18 roundtrips. A total footage of 152.95 m was completed, with average penetration rate of 1.33 m/h, core recovery of 73.5 % and average footage of 8.5 m drilled per roundtrip. In 16 roundtrips the drill string was pulled out with full core barrel.

2. 3127.54–3139.39 m drilling for deviation correction at hole bottom: A drilling tool assembly with Φ 140 mm impregnated diamond drill bit (Fig. 4.19b) was employed for the correcting deviation to the well depth of 3139.39 m. A guide expanding drilling tool assembly of Φ 157 mm diamond drill bit with Φ 110 mm guide + Φ 120 mm PDM + Φ 120 mm drilling jar + seven pieces of Φ 120 mm drill collar + 122 pieces of Φ 89 mm drill rod + Φ 127 mm drill rod was used for expanding drilling from 3127.54 to 3138.99 m, with average penetration rate of 1.04 m/h.

3. 3139.39–3171.28 m core drilling with slick assembly: A core drilling tool assembly (Fig. 4.19a) of drill bit + reaming shell + core barrel + reaming shell + swivel type joint + hydro-hammer + PDM + Φ 120 mm drill collar was employed for core drilling to the depth of 3171.28 m within four roundtrips. The footage drilled was 31.89 m, with average penetration rate of 1.31 m/h, core recovery of 70.2 %, average footage drilled per roundtrip of 7.97 m, and in three roundtrips with full core barrel.

4. 3171.28–3191.28 m drilling for correcting deviation at hole bottom: A drilling tool assembly (Fig. 4.19c) with Φ 158.8 mm roller bit was employed for deviation correction drilling to the depth of 3191.28 m, with penetration rate of 2.06 m/h.

5. 3191.28–3244.98 m core drilling with slick assembly: A core drilling tool assembly (Fig. 4.19a) of drill bit + reaming shell + core barrel + reaming shell + swivel type joint + hydro-hammer + PDM + Φ 120 mm drill collar was employed for core drilling to the well depth of 3244.98 m within seven roundtrips. Footage completed was 53.70 m, with average penetration rate of 1.34 m/h, core recovery of 93.9 %, average footage per roundtrip of 7.67 m, and in four roundtrips with full core barrel.

6. 3244.98–3253.33 m drilling for correcting deviation at hole bottom: A drilling tool assembly (Fig. 4.19b) with Φ 140 mm impregnated diamond drill bit was employed for deviation correction drilling to the depth of 3253.33 m. A guide expanding drilling tool assembly of Φ 157 mm diamond drill bit with Φ 110 mm guide + Φ 120 mm PDM + Φ 120 mm drilling jar + seven pieces of Φ 120 mm drill collar + 140 pieces of Φ 89 mm drill rod + Φ 127 mm drill rod was employed for expanding drilling from 3,244.98 to 3252.86 m, with average penetration rate of 1.25 m/h.

7. 3253.33–3520.91 m core drilling with slick assembly: A core drilling tool assembly (Fig. 4.19a) of drill bit + reaming shell + core barrel + reaming shell + swivel type joint + hydro-hammer + PDM + Φ 120 mm drill collar was employed for core drilling to the depth of 3,520.91 m within 38 roundtrips. Footage drilled was 266.44 m, with average penetration rate of 1.19 m/h, core recovery of 80.6 %, average footage drilled per roundtrip of 7.01 m and in 22 roundtrips with full core barrel.

Pump blocked with pump pressure of 12.50 MPa happened when core drilling reached to the well depth of 3481.01 m in the roundtrip MH-1C-R57. Tight pull was encountered when pulling out the drill string and the tight pull could not be freed by frequently moving the drill string under pulling pressure of 105 t and pushing pressure of 10 t (the hanging load of the drill string was 80 t). Tight pull still could not be solved by large delivery circulation, moving the drill string up and down, with pulling pressure of 125 t and pushing pressure of 23 t, or with pulling pressure of 130 t and downward hanging loads of 80 t. By pulling the drill string with pressure of 125 t and keeping for 10 min, pushing with pressure of 5 t and keeping for 10 min, and repeatedly moving the drill string and then tight pull was finally freed by pulling with pressure of 145 t and it was found that two pieces of bit matrix were broken off, three pieces of reaming shell matrix were broken off, too. A drilling tool assembly of Φ 156 grind shoe + fishing cup + six pieces of Φ 120 mm drill collar + Φ 120 mm drilling jar + one piece of Φ 120 mm drill collar + Φ 89 mm drill rod was run down for grinding and fishing drilling, with footage of 0.37 m drilled. One piece of matrix and a few chips were recovered.

Pump blocking happened with pump pressure of 16 MPa when core drilling reached to the well depth of 3515.90 m in the roundtrip MH-1C-R62. Tight pull was encountered when pulling out the drill string and the tight pull was freed by moving the drill string up and down, with pulling pressure of 110 t (the hanging load of the drill string was 81 t). Pump blocking happened again at the well depth of 3516.00 m in redressing drilling. Drill string was pulled out and it was found that six pieces of drill bit matrix were broken off. A drilling tool assembly of Φ 156 mm grinding shoe + fishing cup + six pieces of Φ 120 mm drill collar + Φ 120 mm drilling jar + Φ 120 mm drill collar + Φ 89 mm drill rod was run down the borehole, but slack-off was encountered at the well depth of 3486.50 m. Repeated redressing drilling was conducted to the well depth of 3492.50 m, then pulling up the drill string and lowering down to the well depth of 3486.50 m and slack-off happened again. Pulling out the drill string and then it was found that there was a scar on the edge of fishing cup. A drilling tool assembly of Φ 156 mm grinding shoe + fishing cup + SK157 wireline coring stabilizer + six pieces of SK146 wireline core drill collars + Φ 120 mm drilling jar + seven pieces of Φ 120 mm drill collar + Φ 89 mm drill rod was run down the borehole for grinding and fishing drilling. Footage of 0.47 m was drilled and a few chips were fished out.

Core drilling was stopped at the well depth of 3516.66 m (roundtrip MH-1C-R63) because of no penetration. Lower reaming shell and drill bit were broken off in the borehole. A drilling tool assembly of Φ 127 mm pin tap + Φ 120 mm drilling jar + seven pieces of Φ 120 mm drill collar + Φ 89 mm drill rod was run down the borehole and the fishes were fished out. A drilling tool assembly of Φ 157.25 mm repairing tool + 3.92 m long Φ 139.7 mm core barrel + Φ 156.3 mm repairing tool + 3.92 m long Φ 139.7 mm core barrel + Φ 157.3 mm reaming shell + Φ 120 mm PDM + Φ 120 mm drilling jar + seven pieces of Φ 120 mm drill collar + Φ 89 mm drill rod was run down the borehole for redressing to the depth of 3486.40 m and then the drill string was pulled out. The sizes of the lower, middle repairing tools and upper reaming shell were 155.5, 155.4 and 157.2 mm respectively. A drilling tool assembly of Φ 158.8 mm roller bit + Φ 120 mm drilling jar + seven pieces of Φ 120 mm drill collar + Φ 89 mm drill rod was employed for drifting to the depth of 3516.96 m.

8. 3520.91–3536.00 m non-core drilling: A drilling tool assembly of Φ 158.8 mm tri-cone bit + SK157 stabilizer + 12 pieces of SK146 wireline drill collar + Φ 120 mm drilling jar + seven pieces of Φ 120 mm drill collar + Φ 89 mm drill rod was employed for non-core drilling to the depth of 3536.00 m with average penetration rate of 0.81 m/h.

9. 3536.00–3665.87 m core drilling with SK drill collar: A core drilling tool assembly (Fig. 4.19e) of drill bit + reaming shell + core barrel + reaming shell + swivel type joint + hydro-hammer + PDM + drilling jar + SK157 stabilizer + 12 pieces of SK146 wireline drill collar + seven pieces of Φ 120 mm drill collar + Φ 89 mm drill rod was employed for core drilling to the well depth of 3665.87 m within 15 roundtrips. Footage drilled was 129.87 m, with average penetration rate of 1.28 m/h, core recovery of 99.2 %, average footage of 8.66 m drilled per round trip and in 14 roundtrips with full core barrel.

At the well depth of 3610.22 m (after roundtrip MH-1C-R71), a drill string with Φ 170 mm oil stripper + Φ 89 mm drill rod was run down the borehole to clean out the oil sludge on the wall of the Φ 193.7 mm moving casing.

To the depth of 3665.87 m (roundtrip MH-1C-R79), no footage could be further penetrated (0.59 m was drilled in this roundtrip). Pressure was gradually increased to 4 t, without back pressure. Pump delivery was changed three times and the drilling tool was moved time and again but could not reach hole bottom. Bit pressure was increased to 5 t, still without back pressure. After pulling out the drill string it was found that reaming shell and drill bit were broken off at hole bottom. Fishing treatments with all kinds of technical measures such as Φ 127 mm pin tap, Φ 127 mm pin tap with guide, Φ 60 mm × Φ 156 mm over-coring bit, Φ 127 mm pin tap with alloy guide, Φ 89 mm grind shoe and Φ 127 mm eccentric pin tap with guide etc. were used, however, without successful results (in all the 12 RIH roundtrips). What's more, serious sticking accident happened while using Φ 60 mm × Φ 156 mm over-coring bit for drifting, slack-off happened at the well depth of 3,481 m and could not be freed by pulling up the drill string with pressure of 110–190 t (the hanging load of drill string was 84 t). By pulling up the drill string for many times and keeping static for four to 10 min with pressure and then pushing downward the slack-off still could not be freed. Pulling up and pushing downward the drill string repeatedly for ten times with an increased 2 t pulling up pressure each time from 190 to 200 t then keeping static for 5 min and with downward pressure no more than 25 t the slack-off was finally freed by pulling up with pressure of 190 t. The steel body of Φ 156 mm over-coring bit was seriously damaged by pulling in two places from top to bottom (on the opposite side of water slot).

Core drilling in this hole section was basically successful though hole wall sloughing and blocks falling, and hole oversize (Fig. 4.20) were serious during drilling. Hole accidents happened in the later drilling stage, too. Total 105 RIH roundtrips were completed, including 79 roundtrips of core drilling, three roundtrips of deviation correction drilling at hole bottom, twice of expanding drilling with guiding, twice of non-core drilling, one roundtrip of redressing, three roundtrips of running grind shoe and fishing cup in the borehole, once of running pin tap for fishing, once of running oil stripper to clean out oil sludge from casing wall, once of running down the drill string and encountering slack-off, 12 roundtrips of treating the fish at hole bottom.

Variation of well deviation angle, diameter and azimuth angle in MH-1C coring drilling can be found in Fig. 4.20 and the main technical indexes in Table 4.10.

4.2.7 The Second Expanding Drilling of the Main Hole (Hole Section CCSD-MH-2K)

Moving drill rods and pulling out Φ 193.7 mm moving casing were completed from October 22nd to 25th, 2003. October 27th and 28th, 2003 were used to grind and fish the elastic steel sheet of the centralizer on the moving casing. Φ 244.5 mm expanding drilling was started at the well depth of 2,028 m on October 29th, 2003 and ended on March 14th, 2004 because of the complicated borehole situation after the well depth had reached to 3525.18 m on March 12th, 2004. Details of the second expanding drilling are shown as follows.

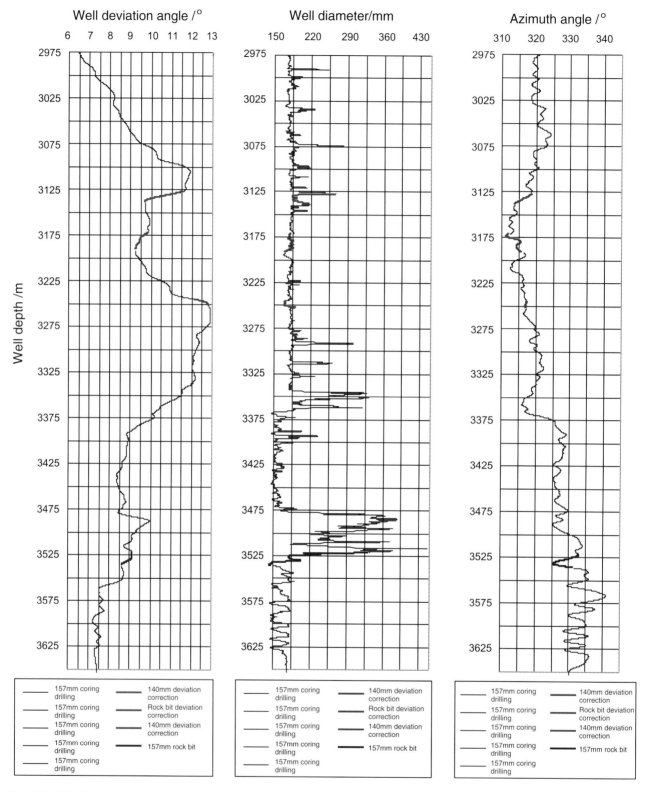

Fig. 4.20 Well deviation angle, diameter and azimuth angle of Φ 157 mm core drilling in the hole section from 2974.59 to 3665.87 m

4.2 Simple Situation of the Construction at Different Periods

Table 4.10 Main technical indexes of core drilling in the main hole section 2974.59–3665.87 m

Item	Index	Item	Index	Item	Index
Core drilling footage (m)	634.85	Average core recovery (%)	83.3	Average penetration rate (m/h)	1.25
Core drilling roundtrip	79	Normal ROOH roundtrip with full core barrel	59	Average bit life (m)	21.16
Longest bit life (m)	48.62	Average footage drilled per roundtrip (m)	8.04	Longest footage drilled per roundtrip (m)	9.51
Average footage drilled per day (m/days)	6.13	Percentage of actual drilling in time (%)	20.3	Percentage of tripping in time (%)	52.6
Percentage of accident in time (%)	11.8	Largest well deviation angle (°)	12.90	Largest hole diameter (mm)	446.2
Average footage drilled per rig month (m/30 days)	183.78	Well logging (times)	10	Days of core drilling (days)	116

Note Total footage of 710.61 m was completed, with deviation correction drilling of 40.20 m, expanding drilling with guiding of 15.31 m, non-core drilling of 15.39 m, grinding drilling of 0.84 m. Core drilling was started on June 9th, 2003 and ended on October 2nd, 2003, with 116 days in total. Fish was treated from October 3rd to 21st, 2003

1. **72 sheets of elastic steel of the centralizer on the casing fell in hole when pulling out the moving casing**

   A drilling tool assembly of Φ 249 mm grind shoe with guide + Φ 214 mm stabilizer + six pieces of Φ 177.8 mm drill collar + Φ 127 mm drill rod was run down the borehole, however, slack-off was encountered at the distance of 1.50 m away from the fillet nipple, then circulation was started. A strong magnetic overshot was run in the borehole but fishing elastic steel sheets failed. A drilling tool assembly of Φ 244.5 mm junk catcher + three pieces of Φ 203.2 mm drill collar + Φ 214 mm stabilizer + six pieces of Φ 177.8 mm drill collar + Φ 127 mm drill rod was applied for fishing but failed, too. The fishing tools were still in good condition.

2. **Expanding drilling in 2,028.00 – 2,093.65 m (Roundtrips MH-2K-1 to MH-2K-6)**

   Six different drilling tool assemblies were employed for the first six expanding drilling roundtrips, with footage of 65.65 m drilled, penetration rate of 0.76 m/h and average footage drilled per roundtrip of 10.94 m. The drilling tool assemblies used are described as follows.

   A drilling tool assembly of Φ 244.5 mm roller bit with guide + three pieces of Φ 203.2 mm drill collar + Φ 214 mm stabilizer + six pieces of Φ 177.8 mm drill collar + Φ 127 mm drill rod (expanding drilling tool assembly I) was used for grinding drilling to the depth of 2028.00 m and then expanding drilling was started. However, expanding drilling had to be stopped and then the drill string was pulled out because of low penetration rate. The outer teeth of the three cones were worn flat, and the cones were seriously worn out. Footage drilled was 6.53 m, with penetration rate of 0.93 m/h.

   A drilling tool assembly of Φ 244.5 mm roller bit with guide + Φ 203 mm absorber + three pieces of Φ 203.2 mm drill collar + Φ 214 mm stabilizer + six pieces of Φ 177.8 mm drill collar + Φ 127 mm drill rod (expanding drilling tool assembly II) was used. Because of decreased penetration rate the drill string was pulled out. The drill bit was in fairly good condition, with outer diameter of 234 mm and gauge protection outer diameter of 240 mm. Footage drilled was 19.12 m, with penetration rate of 1.03 m/h.

   A drilling tool assembly of Φ 244.5 mm tri-cone bit + Φ 203 mm absorber + three pieces of Φ 203.2 mm drill collar + Φ 214 mm stabilizer + six pieces of Φ 177.8 mm drill collar + Φ 127 mm drill rod (expanding drilling tool assembly III) was used. The outer diameter of roller bit was 242.5 mm and a sheet of elastic steel was taken out in the intermediate zone of the cones. Footage of 13.46 m was completed, with penetration rate of 0.67 m/h.

   A drilling tool assembly of Φ 244.5 mm roller bit with guide + Φ 245 mm reaming shell + Φ 203 mm absorber + three pieces of Φ 203.2 mm drill collar + Φ 214 mm stabilizer + six pieces of Φ 177.8 mm drill collar + Φ 127 mm drill rod (expanding drilling tool assembly IV) was run down in the borehole and slack-off was caused at the depth of 2,040 m. Redressing lasted for 10 h. Expanding drilling was conducted to the depth of 2072.55 m and bit bouncing happened. Moving drill string and then caused tight pull. Pulling out the drill string with pressure of 120, 130 t and finally tight pull was freed with 140 t. Redressing and expanding drilling to the depth of 2,072.64 m and then caused bit bouncing again. Pulling out the drill string with pressure of 130 t, and tight pull was freed with pressure of 150 t. Then the drill string was pulled out. The outer diameter of bit was 239.4 mm with gauge protection of 240.8 mm. Gauge protection of reaming shell was 241 mm and footage drilled was 5.53 m, with penetration rate of 1.22 m/h.

   A drilling tool assembly of Φ 244.5 mm roller bit + Φ 245 mm reaming shell + Φ 203 mm absorber + three pieces of Φ 203.2 mm drill collar + Φ 178 mm drilling jar + Φ 214 mm stabilizer + six pieces of Φ 177.8 mm drill collar + Φ 127 mm drill rod (expanding drilling tool assembly V) was used. Footage drilled was 13.38 m, with penetration rate of 0.63 m/h.

A drilling tool assembly of Φ 244.5 mm roller bit with guide + Φ 245 mm reaming shell + Φ 203 mm absorber + three pieces of Φ 203.2 mm drill collar + Φ 178 mm drilling jar + Φ 214 mm stabilizer + six pieces of Φ 177.8 mm drill collar + Φ 127 mm drill rod (expanding drilling tool assembly VI) was used. However, because of decreased penetration rate the drill string had to be pulled out. The outer diameter of the bit was 234 mm with gauge protection of 243 mm and reaming gauge protection of 240 mm. The teeth were worn out and the bit body was seriously worn. Footage drilled was 7.63 m, with penetration rate of 0.49 m/h.

3. **Expanding drilling in 2093.65–2872.81 m (Roundtrips MH-2K-7 to MH-2K-21)**

The expanding drilling tool assembly VI was applied for expanding drilling for 596.33 m within 12 roundtrips, with penetration rate of 1.13 m/h and average footage drilled per roundtrip of 49.69 m. The expanding drilling tool assembly V was used for expanding drilling for 182.83 m within three roundtrips, with penetration rate of 1.00 m/h and average footage drilled per roundtrip of 60.94 m.

In roundtrip MH-2K-9 at the depth of 2199.56 m drill string was pulled out because of bit bouncing and jumping, and it was found that a sheet of elastic steel stuck in the cones.

In roundtrip MH-2K-16 at the well depth of 2625.86 m the guide of roller bit was broken off in the borehole.

In roundtrip MH-2K-18 at the depth of 2757.99 m the guide was worn out and deformed to contact with the cones. It was found that there was a hole with diameter of 58 mm and depth of 25 mm at the front end of the guide which had been ground out by contacting with bottomhole foreign bodies. A drilling tool assembly of Φ 150 mm grind shoe + Φ 89 mm pony drill rod + Φ 178 mm drilling jar + Φ 214 mm stabilizer + six pieces of Φ 177.8 mm drill collar + Φ 127 mm drill rod was run down the hole for grinding and drifting. Lowering down the grind shoe and footage of 8.22 m was ground and then the drill string was pulled out. Wear at central face of the grind shoe was evident.

In roundtrip MH-2K-19 (2765.63 m deep), rotary table was bounced to stop because of torque increase at the depth of 2765 m. Moving the drill string for many times and then redressing and expanding drilling were conducted to the depth of 2765.63 m, bit bouncing happened and the drill string was pulled out. One of the cones was stuck by a sheet of elastic steel.

In roundtrip MH-2K-20 at the depth of 2765.75 m bit bouncing happened continuously after reaching to the hole bottom, and rotary table was stopped several times as a result of bouncing. Redressing was carried out for several times but bit bouncing still happened. Pulling out the drill string and it was found that two cones were stuck by a sheet of elastic steel. A drilling tool assembly of Φ 157 mm grind shoe + SK157 stabilizer + six pieces of SK146 wireline coring drill collar + three pieces of SK139.7 wireline drill rod + Φ 178 mm drilling jar + Φ 214 mm stabilizer + six pieces of Φ 177.8 mm drill collar + Φ 127 mm drill rod was run in the borehole for grinding and drifting for 1.32 m, and then it was found that the outside diameter of the grind shoe was worn to 153 mm and the bottom was seriously worn. After changing a Φ 158 mm grind shoe the drill string was run in the borehole again for grinding and drifting to 2849.47 m in Φ 157 mm borehole. The outside diameter of the grind shoe was worn to 152 mm and the outer area of the bottom was badly worn. A drilling tool assembly of Φ 152 mm grind shoe + Φ 120 mm drilling jar + SK157 stabilizer + five pieces of SK wireline coring drill collar + 13 pieces of SK139.7 wireline drill rod + Φ 178 mm drilling jar + Φ 214 mm stabilizer + six pieces of Φ 177.8 mm drill collar + Φ 127 mm drill rod was run in the borehole for grinding and drifting to the well depth of 2935.05 m (Φ 157 mm borehole). The outside diameter of the grind shoe was worn to 151 mm and the grind shoe itself was still in good condition.

4. **Expanding drilling by using roller bit with guide in 2872.81–3383.13 m (Roundtrips MH-2K-22 to MH-2K-30)**

A drilling tool assembly of Φ 244.5 mm roller bit with guide + Φ 245 mm reaming shell + Φ 230 mm absorber + Φ 178 mm drilling jar + Φ 214 mm stabilizer + 8 pieces of Φ 177.8 mm drill collar + Φ 127 mm drill rod was employed for expanding drilling for 510.32 m within nine roundtrips, with penetration rate of 1.06 m/h and average footage drilled per roundtrip of 56.70 m.

In roundtrip MH-2K-24 (well depth of 3052.38 m), the drill string was run into the borehole but could not reach hole bottom. 31.5 h was paid for redressing the long hole section from 2117 to 3052.38 m deep. After redressing to the hole bottom expanding drilling was conducted for 83.39 m.

In roundtrip MH-2K-25 at the well depth of 3225.48 m the guide of the roller bit was broken off in the borehole.

In roundtrip MH-2K-26 (well depth of 3230.08 m) no footage could be penetrated and it was found after pulling out the drill string that one of the cones was stuck by a sheet of elastic steel.

A drilling tool assembly of Φ 157 mm grind shoe + Φ 120 mm drilling jar + SK157 stabilizer + 9 pieces of SK146 wireline drill collar + 37 pieces of Φ 89 mm drill rod + Φ 127 mm drill rod was run to the reamed hole depth and pump was started. Then pump blocking happened. The drill string was pulled out because of non-effectiveness of moving and turning the drill string. The grind shoe and jar were blanked off by sand settling. A 10 mm × 15 mm bevel was worn out on the outer edge of the grind shoe.

A Φ 157 mm grind shoe + fishing cup assembly was run down the borehole with the former drill string for grinding and fishing from the depth of 3230.08–3230.92 m. Then the drill string was pulled out because of no penetration. A layer of 10 mm alloy coating of the grind shoe was abraded, and a Φ 75 mm × 25 mm concave was worn at bottom and a 10 mm × 8 mm bevel at the outer edge of the grind shoe. Elastic steel sheets and rock fragments were fished out.

A Φ 157 mm grind shoe was run in the borehole with the former drill string. A kelly bar was connected and pump circulation was started at the depth of 2000 m. Pump blocking happened with pump pressure of 16 MPa, the drill string was pulled out. Water channels of the grind shoe were blanked off by sand settling.

A drilling tool assembly of Φ 157 mm grind shoe with fishing cup + Φ 120 mm drilling jar + SK157 stabilizer + 9 pieces of SK146 wireline coring drill collar + Φ 214 mm stabilizer + 13 pieces of Φ 177.8 mm drill collar + Φ 127 mm drill rod was run into the borehole for grinding from the well depth of 3230.08–3321.05 m and then pulled out. The outer edge was abraded and 10 mm alloy coating of the grind shoe was worn out. Fishing cup was full of elastic steel sheets and rock fragments.

In roundtrip MH-2K-27 at the depth of 3273.04 m the guide of the roller bit was broken off in the borehole.

In roundtrip MH-2K-28 (well depth of 3315.04 m) the drill string was pulled out because of bit bouncing. The guide of the bit was seriously worn and concaves were found at the bottom face. A drilling tool assembly of Φ 157 mm grind shoe with fishing cup + Φ 120 mm drilling jar + SK157 stabilizer + 38 pieces of SK146 wireline drill collar + Φ 127 mm drill rod was employed for grinding drilling to well depth of 3529.19 m, where drilling was stopped because of bit bouncing. Then the drill string was pulled out. Water channels of the grind shoe were blanked off by sand settling, and the outer edge of the grind shoe was abraded at face bottom, 10 mm alloy coating of the grind shoe was worn out and fishing cup was full of elastic steel sheets and rock fragments.

5. **Expanding drilling with common roller bit in 3383.13 – 3525.18 m (Roundtrips MH-2K-31 to MH-2K-34)**

A drilling tool assembly of Φ 244.5 mm roller bit + Φ 245 mm reaming shell + Φ 203 mm absorber + Φ 178 mm drilling jar + Φ 214 mm stabilizer + 10 pieces of Φ 177.8 mm drill collar + Φ 127 mm drill rod was employed for expanding drilling for 142.05 m within four roundtrips with penetration rate of 1.13 m/h and average footage drilled per roundtrip of 35.51 m.

In roundtrip MH-2K-31 (well depth of 3391.33 m) drilling rate decreased and no more footage could be penetrated, then the drill string was pulled out. It was found that ten inside teeth of a cone were broken off and bit leg was badly worn. A drill string of Φ 157 mm grind shoe with fishing cup + Φ 120 mm drilling jar + SK157 stabilizer + 30 pieces of SK146 wireline drill collar + Φ 127 mm drill rod was lowered into the borehole, circulating and redressing when reaching to bottom hole, looking for small hole, no footage could be ground, the drill string was pulled out. 20 mm deep bevel was worn at the middle water channels of the grind shoe.

After roundtrip MH-2K-32 (well depth of 3487.10 m) a drill string of Φ 157 mm grind shoe with fishing cup + Φ 120 mm drilling jar + SK157 stabilizer + 19 pieces of SK146 wireline drill collar + Φ 127 mm drill rod was lowered into the borehole for grinding and fishing drilling from 3487.10 to 3531.23 m deep, slack-off was encountered and then the drill string was pulled out. The bottom of the grind shoe was seriously worn, with water channels blanked off. Fishing cup was full of elastic steel sheets and rock fragment, and the fishing cup was worn into a spiral.

In roundtrip MH-2K-33 (well depth of 3525.16 m) the drill string was lowered into the borehole and slack-off was encountered at well depth of 3447 m, then backreaming met tight pull, too. Backreaming with pressure of 150, 170 and 180 t respectively (string hanging load of 130 t), jar could not work with downward pressure of 14 t. Backreaming with pressure of 180, 190, 200 t and rotary table revolved two turns, five turns and eight turns respectively. Tight pull was finally freed by backreaming with pressure of 195 t. Expanding drilling to well depth of 3525.16 m, bit bouncing was met. Tight pull was encountered when moving the drill string in backreaming. Tight pull could not be freed by backreaming with pressure of 160 t. Backreaming with pressure of 170 t and rotary table was revolved six turns, backreaming with pressure of 180 t and drill string was repeatedly moved, then backreaming with pressure of 186 t and tight pull was solved. The drill string was lowered to hole bottom and slack-off of 10 t was met. The drill string was pulled out and it was found that the outer teeth of three cones were all broken off.

A drill string of Φ 157 mm grind shoe with fishing cup + fishing cup (outer tube) + Φ 120 mm drilling jar + SK157 stabilizer + 19 pieces of SK146 wireline drill collar + Φ 127 mm drill rod was run in the hole for grinding drilling to the depth of 3532.03 m, at which bit bouncing and pump blocking happened. Lifting the drill string and then lowering it down and slack-off happened between 3525 and 3532 m. Redressed repeatedly but no footage could be penetrated. The drill string was pulled out and it was found that the outer tube of the fishing cup was broken off into hole. The grind shoe was normally worn and the fishing cup was full of elastic steel sheets and rock fragments.

An expanding drill string was run in the borehole, bit bouncing happened at well depth of 3522.55 m, pump pressure increased, bit bouncing and sticking were encountered in redressing. Mud was circulated, the drill string moved, and the drill string was lifted (backreamed) with

150 t (string hanging load of 135 t), sticking happened. Backreaming was conducted with pressure of 150–160–200 t to move the drill string and the rotary table was rotated for 13 turns. The drill string was moved by backreaming with pressure of 200 t, pressing with pressure of 115 t repeatedly and starting the rotary table. Backreaming with pressure of 190 to 210 t and downward with pressure of 23–30–40 t to move the drill string, fixing hook, backreaming with pressure of 220 t and sticking was freed. A scar was found on bit leg. In the process of moving the drill string the downward strikes of the drilling jar were obvious, while the first five upward strikes were fairly obvious but gradually no strikes after that. The drill string was moved up and down for over ninety times.

A drill string of Φ 216 mm corncob bit + Φ 193.7 mm × 6.52 m fishing cup + Φ 203 mm absorber + Φ 178 mm drilling jar + Φ 214 mm stabilizer + 10 pieces of Φ 177.8 mm drill collar + Φ 127 mm drill rod was run in the borehole, and slack-off was encountered at well depth of 3490 m. Redressing was conducted with double pumps and fishing with single pump. Bit bouncing happened at the depth of 3523 m, torque increased, and bit sticking was freed by backreaming with pressure of 180–190 t. Before pulling up the drill string, double pump circulation with high delivery rate was employed and sand return was large. The fishing cup was full of coarse sand. The former bailing tool was run into the borehole again, and fishing cup was full of coarse sand, with few rock fragments as big as 10 mm × 65 mm.

A drill string of Φ 157 mm grind shoe with fishing cup + Φ 193.7 mm × 4.38 m fishing cup + Φ 193.7 mm × 6.52 m fishing cup + Φ 203 mm absorber + Φ 178 mm drilling jar + Φ 214 mm stabilizer + 10 pieces of Φ 177.8 mm drill collar + Φ 127 mm drill rod was run in the borehole for grinding and redressing to the depth of 3,525.16 m, at which pump was stopped for fishing twice. The lower and middle fishing cups were full, while the upper fishing cup was less, and sand samples were coarse, a few rock chips were big.

A drill string of Φ 157 mm grind shoe with fishing cup + Φ 120 mm drilling jar + 15 pieces of SK146 wireline drill collar + Φ 127 mm drill rod was run into the borehole for grinding and fishing to the depth of 3536.41 m. Full-barrel of rock chips and iron pieces were fished out, including a piece of alloy block (20 mm × 25 mm) from the Φ 157 mm reaming shell, the biggest iron piece was as big as 60 mm × 70 mm.

A drill string of Φ 150 mm grind shoe + 1 piece of Φ 89 mm drill rod + Φ 193.7 × 4.38 m fishing cup + Φ 193.7 mm × 6.52 m fishing cup + Φ 203 mm absorber + Φ 178 mm drilling jar + Φ 214 mm stabilizer + 10 pieces of Φ 177.8 mm drill collar + Φ 127 mm drill rod was run into the borehole for redressing with double pump circulation and addition of 45 m^3 high viscosity mud with viscosity of 55 s. 4.8 m^3 sands were returned and the two fishing cups were full-barrel with most of coarse sands and fewer bigger rock fragments. After changing Φ 158.8 mm tri-cone bit the former bailing tool assembly was run in the hole again for redressing to the depth of 3524 m with single and double pumping repeatedly. Pump blockage and bit bouncing happened, and tight pull was encountered during backreaming and freed by pulling with pressure of 150–185 t. The two fishing cups were full-barrel with coarse sands and a few bigger sands.

A drill string of Φ 158.8 mm tri-cone bit + 4 m long Φ 89 mm pony drill rod + Φ 193.7 mm × 4.38 m fishing cup + Φ 193.7 mm × 6.52 m fishing cup + Φ 203 mm absorber + Φ 178 mm drilling jar + Φ 214 mm stabilizer + 10 pieces of Φ 177.8 mm drill collar + Φ 127 mm drill rod was run down in the borehole and slack-off happened at the depth of 3520.80 m. Bit bouncing happened when redressing with single and double pumping alternately to the depth of 3524.70 m. Tight pull was encountered at well depth of 3521 m during repeated uplifting and lowering the drill string. Then the drill string was pulled out. Nothing was found in the upper fishing cup and few in the lower fishing cup with few bigger rock chips.

In roundtrip MH-2K-34 (at well depth of 3515.18 m), slack-off happened while drilling reached to well depth of 3512 m. Bit bouncing happened while redressing and expanding drilling reached to well depth of 3515.18 m. Sticking was encountered during backreaming and then freed by pulling with pressure of 180 t. Bit bouncing happened when redressing reached to 3522 m and freed by backreaming with pressure of 190 t. Bit bouncing happened when redressing reached to 3517.45 m and freed by backreaming with pressure of 190 t. Bit bouncing happened when redressing reached to 3,521.77 m and unfreezing failed by backreaming with pressure of 190 t. Sticking was finally freed by starting rotary table, pushing with pressure of 110 t, and backreaming with pressure of 180 t. Redressing with double pump, bit bouncing happened several times at well depth of 3521 m, then the drill string was pulled out. The second expanding drilling in the main hole ended up because of the complicated hole conditions.

Because there existed in the borehole the steel sheets from the centralizer of the moving casing, and collapsed and oversized hole section from 3475 to 3525 m, expanding drilling was not very smooth and bit bouncing was serious. The total RIH roundtrips of expanding drilling was 56, including 34 roundtrips of expanding drilling, 21 roundtrips of grinding and fishing, one roundtrip of expanding redressing to hole bottom but without any footage drilled as a result of bit bouncing. Jamming of elastic steel sheets between cones happened for five times, drill bit could not work normally and the drill string had to be pulled out. Three guides of the expanding bits were broken off into the

4.2 Simple Situation of the Construction at Different Periods

Table 4.11 The main technical descriptions of main hole expanding drilling in section 2028–3525.18 m

Item	Parameter	Item	Parameter	Item	Parameter
Footage drilled for expanding drilling (m)	1497.18	Average penetration rate (m/h)	1.07	Roundtrips of expanding drilling	34
Average bit life (m)	51.63	The longest bit life (m)	108.99	Average footage drilled per roundtrip (m)	44.03
The longest footage per roundtrip (m)	107.06	Average footage drilled per day (m/days)	10.62	Percentage of actual drilling in time (%)	44.3
Percentage of tripping in time (%)	28.7	Percentage of accident in time (%)	0.6	Average footage drilled per rig month (m/30 days)	318.55
Well logging (times)	1	Total roundtrips	56	Days used for expanding drilling (days)	141

Note From October 22nd to 25th, 2003 drill rods were transported and arranged at site; and Φ 193.7 mm moving casing was pulled out of the borehole. Elastic steel sheet was ground and fished out on October 27th. A total of 141 days between October 29th, 2003 and March 14th, 2004 were used for Φ 244.5 mm expanding drilling started from the well depth of 2028 m

borehole. The overshot was broken at the welding point and the outer tube fell into the borehole for one time. The main technical descriptions of the expanding drilling were showed in Table 4.11.

4.2.8 The Second Sidetracking (Obstacle Avoidance) Drilling and Running Casing and Well Cementation in the Main Hole

Since the normal treating methods such as milling or washover fishing became more and more risky, sidetrack (obstacle avoidance) drilling could be an economical, reliable and feasible method. From March 15th to April 7th, 2004 the backfill sidetracking (obstacle avoidance) drilling was used. From April 8th to 27th non-core drilling with roller bit was employed. And from April 28th to May 7th the jobs such as injecting sealing fluid, injecting high density mud, running casing, well cementation, and drilling cement plug etc. were conducted. Details about the sidetracking (obstacle avoidance) drilling can be found as follows:

1. **Establishment of an artificial hole bottom (March 15th to 27th, 2004)**

A drill string of 15 pieces of SK146 wireline drill collar + Φ 127 mm drill rod was run in the borehole and drilling mud circulated when the drill string was run down to the depth of 3521.50 m. Pulling up the drill string to well depth of 3511.25 m, 8 m^3 of clear water was injected as spacer fluid, 12 m^3 of cement slurry (JHG cement + 0.5 % expanding agent + 0.5 % dispersion agent + 0.8 % fluid loss agent + 1.0 % accelerating agent + 0.38 % water) with density of 1.97 g/m^3 was injected to displace 24.7 m^3 cement. Lifting the drill string up to the depth of 3,020 m and then the cement slurry in the drill string was washed out. The drill string was pulled out after no cement slurry existed in the returned fluid.

A drill string of Φ 244.5 mm tri-cone bit + Φ 203 mm absorber + Φ 178 mm drilling jar + Φ 214 mm stabilizer + 10 pieces of Φ 177.8 mm drill collar + Φ 127 mm drill rod was employed for drifting and chipping from the depth of 3264.40–3350 m and then pulling out the drill string, waiting for curing. The drill string was run in the borehole again for drilling out cement plug from well depth of 3350–3395 m. Ten hours and 55 min were used including 1 h and 27 min for pipe connection. Drilling technical parameters used: bit load 20 kN, rotation speed 40 r/min, delivery rate 27.92 L/s and pump pressure 5.5 MPa.

A drilling tool assembly of Φ 157 mm diamond coring bit + Φ 157 mm reaming shell + 7.22 m long of Φ 139.7 mm outer tube + swivel type joint + Φ 214 mm stabilizer + 10 pieces of Φ 177.8 mm drill collar + Φ 127 mm drill rod was employed for core drilling for 4.73 m. Drilling parameters used included bit load of 40–45 kN, rotation speed of 55–60 r/min, delivery rate of 24.82 L/s and pump pressure of 7.5–7.8 MPa. The sample of cement plug was fairly complete, with length of 4.73 m and core spring was broken off and carried out.

A drill string of Φ 244.5 mm tri-cone bit + Φ 203 mm absorber + Φ 178 mm drilling jar + Φ 214 mm stabilizer + 10 pieces of Φ 177.8 mm drill collar + Φ 127 mm drill rod was employed for reaming and chipping cement plug from well depth of 3395–3400 m and 1 h and 25 min were used. Drilling parameters used included bit load of 20–30 kN, rotation speed of 40 r/min, delivery rate of 25.92 L/s and pump pressure of 6.5–6.7 MPa.

2. **Sidetracking drilling in 3400.00 – 3445.62 m (March 28th to April 7th, 2004)**

A drilling tool assembly (Fig. 4.21a) with Φ 215.9 mm small diameter impregnated diamond drill bit was employed

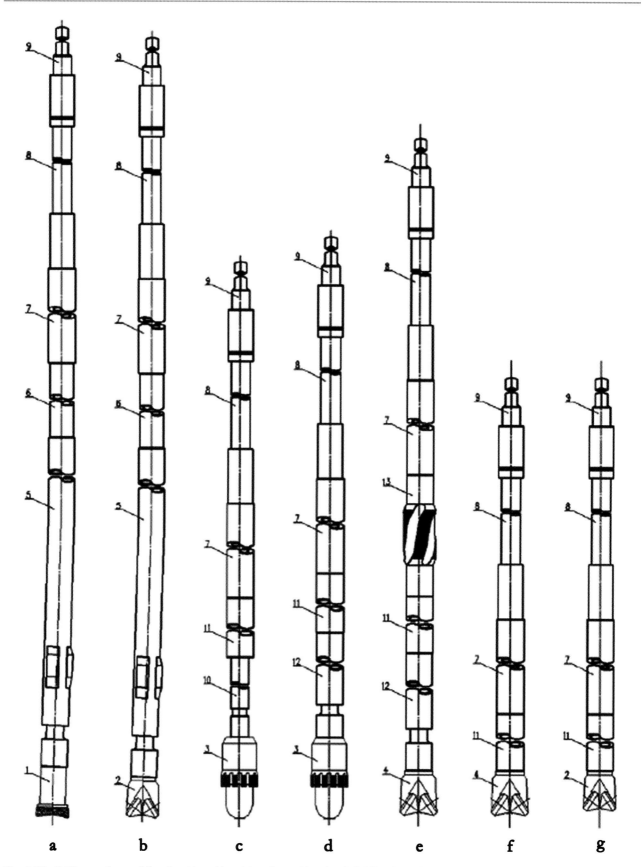

Fig. 4.21 Drilling tool assemblies for sidetracking (obstacle avoidance) drilling. *1* Φ 215.9 mm synthetic diamond impregnated drill bit, *2* Φ 215.9 mm tri-cone bit, *3* Φ 244.5 mm diamond expanding drilling bit with guide, *4* Φ 244.5 mm tri-cone bit, *5* Φ 172 mm PDM with single bend (1.75°), *6* Φ 160 mm non-magnetic drill collar, *7* Φ 177.8 mm drill collar string, *8* Φ 127 mm drill string, *9* Kelly bar, *10* Φ 120 mm PDM, *11* Φ 178 mm drilling jar, *12* Φ 172 mm PDM, *13* Φ 214 mm stabilizer

for sidetracking angle dropping drilling in the hole section from 3400.00 to 3417.37 m, during which penetration rate was strictly controlled and the tool facing direction was taken as the criterion. The readings of the tool facing direction on MWD were allowed to range within ±30° of the stipulated value. Once the tool facing direction exceeded the stipulated value, the drill string should be moved to regulate the tool facing direction. After drilling to 3417.23 m deep, penetration rate decreased gradually and pump pressure slowed down. Increasing bit load from 30 to 55 kN did not get obvious effect. In sidetracking drilling, the reading of well deviation angle from MWD at well depth of 3400 m was 9.7° and 8.4° at 3417.37 m. The drill bit run out of the borehole had outside diameter of 214 mm (the original diameter was 215.9 mm), surface set diamond grains on circumference and around central hole all dropped off, and the central hole was worn into a concentric concave with the diameter of 50 mm. There was no slim core in the drill bit. Average penetration rate was 0.32 m/h.

A drill string (Fig. 4.21b) with Φ 215.9 mm roller bit was employed for angle dropping drilling in the hole section from 3417.37 to 3445.62 m, during which the reading of well deviation angle from MWD at well depth of 3417.37 m was 8.4° and 3.3° at 3445.62 m. Average penetration rate was 1.31 m/h.

A drill string (Fig. 4.21c) with Φ 244.5 mm diamond reaming drill bit with guide was employed for the expanding drilling in the hole section from 3400.00 to 3407.42 m. Pump blocking happened when expanding drilling reached to the depth of 3407.42 m. Moving the drill string, redressing to the hole bottom and pump blocking happened again. The drill string was pulled out as a result of failures of repeated moving the drill string and redressing. Four pieces of bit matrix of the reaming drill bit (originally with 20 pieces bit matrix) were broken off. Average penetration rate was 0.97 m/h.

A drill string (Fig. 4.21d) with Φ 244.5 mm diamond reaming drill bit with guide was employed for the expanding drilling in the hole section from 3407.42 to 3443.31 m, with an average penetration rate of 2.02 m/h.

3. **Non-coring drilling in 3445.62–3623.91 m (April 8th to 27th, 2004)**

In the hole section from 3445.62 to 3576.68 m, a drill string of Φ 244.5 mm tri-cone bit + Φ 172 mm PDM + Φ 178 mm drilling jar + Φ 214 mm stabilizer + 10 pieces of Φ 177.8 mm drill collar + Φ 127 mm drill rod (Fig. 4.21e) was employed for two roundtrips of non-core drilling, with footage of 97.11 and 33.95 m completed, and penetration rate of 1.08 and 1.12 m/h respectively. Pump blocking with pressure of 14 MPa happened when second drilling trip reached to depth of 3576.68 m. Moving the drill string and redressing, pump blocking happened again at the distance of 0.48 m away from the hole bottom. There was no indication about slack-off after stopping pump to reach for the hole bottom. Because of the failures of repeated moving the drill string and redressing the drill string was pulled out of the borehole and it was found that all the teeth on the first and second rows of the cones were worn or broken off, and the bearing of the first cone was seriously worn.

A drill string (Fig. 4.21f) of Φ 244.5 mm tri-cone bit + Φ 178 mm drilling jar + 10 pieces of Φ 177.8 mm drill collar + Φ 127 mm drill rod was employed for non-core drilling in the hole section from 3576.68 to 3600.00 m, with penetration rate of 0.69 m/h.

A drill string (Fig. 4.21g) of Φ 215.9 mm tri-cone bit + fishing cup + Φ 178 mm drilling jar + 10 pieces of Φ 177.8 mm drill collar + Φ 127 mm drill rod was employed for non-core drilling in the hole section from 3600.00 to 3623.91 m, with penetration rate of 1.02 m/h.

From 3600.00 to 3620.00 m, a drill string of Φ 244.5 mm tri-cone bit + Φ 178 mm drilling jar + 10 pieces of Φ 177.8 mm drill collar + Φ 127 mm drill rod was employed for the expanding drilling to the well depth of 3620.00 m, with penetration rate of 4.02 m/h.

4. **Running casing for well cementation (April 28th to May 7th, 2004)**

A drilling tool assembly of Φ 157 mm grind shoe with fishing cup + Φ 178 mm drilling jar + 10 pieces of Φ 177.8 mm drill collar + Φ 127 mm drill rod was run into the borehole for drifting, and sealing fluid and mud with high density were injected. Then, casing was run for well cementation. A drilling tool assembly of Φ 165.1 mm tri-cone bit + 6 pieces of Φ 120 mm drill collar + Φ 89 mm drill rod was employed for drilling the cement plug and the casing accessories, to the well depth of 3624.16 m.

The sidetracking (obstacle avoidance) drilling in this time was fairly successful, with total 15 RIH roundtrips (included the roundtrips for drifting for casing cementation and drilling cement plug), including one roundtrip for sidetracking, one roundtrip for angle dropping drilling, one roundtrip for drilling cement plug, two roundtrips for chipping cement plug, one roundtrip for coring cement plug samples, three roundtrips for expanding drilling, four roundtrips for non-core drilling, one roundtrip for drifting before running casing, and one roundtrip for drilling the cement plug and the accessories. Well deviation angle, hole diameter and azimuth angle in this hole section can be found in Fig. 4.22, and the main technical indexes in Table 4.12.

4.2.9 The Third Core Drilling of the Main Hole (Section CCSD-MH-2C, the Fourth Opening)

The third coring drilling of the main hole was started on May 8th, 2004 from the depth of 3624.16 m and reached

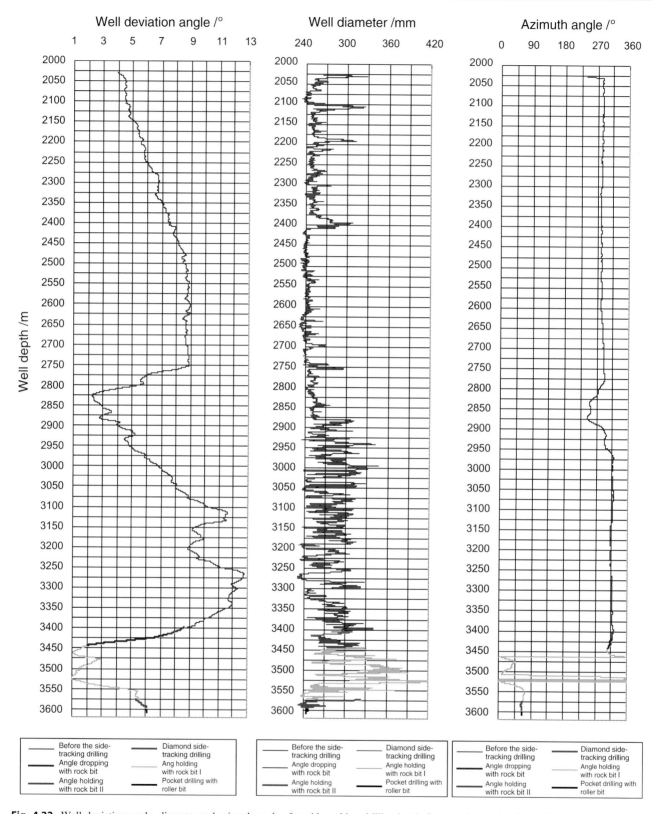

Fig. 4.22 Well deviation angle, diameter and azimuth angle after sidetracking drilling but before running casing for well cementation

to the well depth of 5,118.20 m on January 23rd, 2005, and then the construction about core drilling of CCSD-1 Well was finished. Core drilling construction ended on January 24th, 2005 after 3,500 m drill rod was left in the borehole. Details about core drilling are shown as follows:

Table 4.12 The main technical indexes of sidetracking (obstacle avoidance) drilling in the main hole section from 3,400.00 to 3,624.16 m

Item	Parameter	Item	Parameter	Item	Parameter
Footage of sidetracking (angle-drop) drilling (m)	45.62	Average ROP (m/h)	0.61	Roundrips of sidetracking straighten drilling	2
Footage of expanding drilling (m)	63.31	ROP of expanding drilling (m/h)	2.08	Footage of non-core drilling (m)	178.54
ROP of non-core drilling (m/h)	1.00	Average footage drilled per day (m/days)	6.69	Percentage of actual drilling in time (%)	34.1
Percentage of tripping in time (%)	19.3	Percentage of accident in time (%)	0	Average footage drilled per rig month (m/30 days)	200.56
Well logging (times)	2	Comprehensive well logging (times)	1	Days of drilling (days)	43

Note Total footage was 287.47 m, not including chipping, coring and drilling cement plug. Footage of guided expanding drilling was 43.31 m. From March 15th to April 7th, 2004 backfilling, sidetracking (obstacle avoidance) drilling was conducted, and from April 8th to 27th non-core drilling with roller bit was conducted, with a total of 44 days used, including one day for comprehensive well logging. From April 28th to May 7th, 2004 the works such as injecting sealing fluid, injecting mud with high density, running casing, well cementation and drilling cement plug etc. were finished

1. **Core drilling with slick drill collar in the hole section from 3624.16 to 3664.69 m (Roundtrips MH-2C-1 to MH-2C-6)**

   A core drilling tool assembly of Φ 157 mm core drill bit + Φ 157 mm reaming shell + Φ 139.7 mm core barrel + Φ 157 mm reaming shell + Φ 146 mm swivel type joint + Φ 127 mm hydro-hammer + Φ 120 mm PDM + Φ 120 mm drilling jar + six pieces of Φ 120 mm drill collar + Φ 89 mm drill rod was employed for core drilling. Footage of 40.53 m was finished in six roundtrips, with penetration rate of 0.66 m/h, core recovery of 99.2 %, average footage drilled per roundtrip of 6.76 m. In one roundtrip the drill string was pulled out with full core barrel.

2. **Core drilling with upper stabilizer in the hole section from 3664.69 to 3846.00 m**

   A core drill string of Φ 157 mm core drill bit + Φ 157 mm reaming shell + Φ 139.7 mm core barrel + Φ 157 mm reaming shell + Φ 146 mm swivel type joint + Φ 127 mm hydro-hammer + Φ 120 mm PDM + Φ 120 mm drilling jar + six pieces of Φ 120 mm drill collar + Φ 156 mm stabilizer + three pieces of Φ 120 mm drill collar + Φ 89 mm drill rod was employed.

   From 3664.69 m to 3707.84 m (roundtrips MH-2C-7 to MH-2C-11), footage of 41.35 m was completed by core drilling in five roundtrips, with penetration rate of 1.08 m/h, core recovery of 64.8 %, average footage drilled per roundtrip of 8.63 m and in two roundtrips the drill string was pulled out with full core barrel.

   As tight pull was encountered during pulling out the drill string in the roundtrip MH-2C-11, moving the drill string and jarring for unfreezing, but the core spring fell into the borehole and no core was recovered, then roller bit was employed for the drifting and chipping from 3707.84 to 3712.55 m. A drill string of Φ 158.8 mm tri-cone bit + three pieces of Φ 120 mm drill collar + Φ 120 mm drilling jar + two pieces of Φ 120 mm drill collar + Φ 156 mm stabilizer + three pieces of Φ 120 mm drill collar + Φ 89 mm drill rod was used, with penetration rate of 0.61 m/h.

   Footage of 130.75 m was completed by core drilling in 20 roundtrips from 3712.55 to 3843.30 m (roundtrips MH-2C-12 to MH-2C-31), with penetration rate of 1.04 m/h, core recovery of 83.4 %, average footage drilled per roundtrip of 6.54 m and in six roundtrips the drill string was pulled out with full core barrel. Besides, hydro-hammer was not used in the roundtrip MH-2C-29, with footage of 7.01 m drilled, penetration rate of 0.67 m/h and core recovery of 101.3 %.

   Because pump surge was serious in MH-2C-31 core drilling, moving drill string repeatedly but without any more footage penetrated, and the rubber stator of the PDM was broken down, so tri-cone bit was employed for non-core drilling from 3,843.3 to 3,846.00 m, with a drill string of Φ 158.8 mm tri-cone bit + three pieces of Φ 120 mm drill collar + Φ 156 mm stabilizer + four pieces of Φ 120 mm drill collar + Φ 120 mm drilling jar + Φ 89 mm drill rod, with penetration rate of 0.52 m/h.

3. **Core drilling with lower stabilizer in the hole section from 3846.00 to 5118.20 m (Roundtrips MH-2C-32 to MH-2C-186)**

   A core drill string of Φ 157 mm core drill bit + Φ 157 mm reaming shell + Φ 139.7 mm core barrel + Φ 157 mm reaming shell + Φ 146 mm swivel type joint + Φ 120 mm hydro-hammer + Φ 120 mm PDM + Φ 156 mm stabilizer + Φ 120 mm drilling jar + five to nine pieces of Φ 120 mm drill collar + Φ 89 mm drill rod was employed.

   Footage of 552.47 m was completed in 68 roundtrips from 3846.00 m to 4398.47 m (roundtrips MH-2C-32 to MH-2C-99), with penetration rate of 1.10 m/h, core recovery of 92.1 %, average footage drilled per roundtrip of 8.12 m. In 44 roundtrips the drill string was pulled out with full core barrel.

From 4398.47 to 4401.49 m tri-cone bit down hole drive was employed for drifting and chipping drilling, as half of the caliper plate (700 mm × 37 mm × 10 mm) of the logging device was broken into the borehole in a comprehensive well logging after roundtrip MH-2C-99. A drill string of Φ 158.8 mm tri-cone bit + Φ 120 mm PDM + Φ 156 mm stabilizer + Φ 120 mm drilling jar + eight pieces of Φ 120 mm drill collar + Φ 89 mm drill rod was used, with penetration rate of 0.96 m/h.

From 4401.49 to 5072.06 m (roundtrips MH-2C-100 to MH-2C-181), footage of 670.57 m was drilled in 82 roundtrips, with penetration rate of 1.01 m/h, core recovery of 86.8 %, and average footage drilled per roundtrip of 8.18 m. In 54 roundtrips the drill string was pulled out with full core barrel.

From 5072.06 to 5075.16 m tri-cone bit was employed for the drifting and chipping drilling because pump blockings happened frequently in the later stage of the roundtrip MH-2C-181, reaming to the hole bottom but pump blocking happened again, core spring was broken into the hole bottom with no core recovered. The drill string used consisted of Φ 158.8 mm tri-cone bit + one piece of Φ 120 mm drill collar + Φ 156 mm stabilizer + Φ 120 mm drilling jar + eight pieces of Φ 120 mm drill collar + Φ 89 mm drill rod, with penetration rate of 0.31 m/h.

From 5075.16 to 5102.18 m (roundtrips MH-2C-182 to MH-2C-184) footage of 27.02 m was completed in three roundtrips of coring drilling, with penetration rate of 0.93 m/h, core recovery of 69.3 %, and average footage drilled per roundtrip of 9.01 m. In 2 roundtrips the drill string was pulled out with full core barrel.

Half wafer (300 mm × 100 mm × 20 mm) of the logging device was broken into the borehole in a comprehensive well logging after the roundtrip MH-2C-184. A drilling tool assembly of Φ 158.8 mm tri-cone bit + one piece of Φ 120 mm drill collar + Φ 156 mm stabilizer + Φ 120 mm drilling jar + eight pieces of Φ 120 mm drill collar + Φ 89 mm drill rod was run in the borehole, slack-off was encountered at the depth of 4000 m. Moving the drill string but slack-off happened again. Moving the drill string by backreaming with pressure of 140 t (string hanging load of 95 t) and lowering pressure of 80 t, and circulating mud, the drill string was unfrozen by backreaming with pressure of 142 t and jarring at the same time. Tight pull was encountered in lifting the drill string during redressing and drifting, and was freed by backreaming with pressure of 100–180 t, lowering with pressure of 30 t but without jarring, moving the drill string, lowering with pressure of 30 t, and rotation. Single pipe was added for redressing, moving the drill string and no resistance was encountered, then the drill string was pulled out. Slight scratch was found on one or two pieces of dill collar and one serious scratch on the stabilizer.

From 5102.18 to 5102.63 m a drill string of Φ 157 mm grind shoe with fishing cup + one piece of Φ 120 mm drill collar + Φ 120 mm drilling jar + eight pieces of Φ 120 mm drill collar + Φ 89 mm drill rod was employed for grind drilling. Slack-off happened at well depths of 3830, 4040 and 4066 m respectively during running down the drill string and freed by rotation. Grind drilling was normal and wafer pieces were found in the fishing cup and in the water slots of the grind shoe.

From 5102.63 to 5118.2 m (roundtrips MH-2C-184 to MH-2C-186) footage of 15.57 m was completed by core drilling in two roundtrips, with penetration rate of 0.80 m/h, core recovery of 100.4 %, and average footage drilled per roundtrip of 7.79 m. In one roundtrip the drill string was pulled out with full core barrel.

Core drilling was successful in this hole section, with total 192 RIH roundtrips, including 186 roundtrips of core drilling, four roundtrips of non-core drilling, one roundtrip of running grind shoe and fishing cup, one roundtrip of slack-off in non-core drilling. During the drilling operations, no accident about drilling tool broken off into the borehole happened as a result of strict control of the service time and life of all kinds of drilling tools. The matrixes of diamond drill bits broke off into the borehole for five times. The rubber stator of PDM came unglued for three times and the wafer of the logging device broke off for two times. When drilling reached to well depth of 4800 m, mud loss happened, a total of 73 m^3 mud was lost when drilling to the depth of 4,900 m, during which 20 m^3 mud was lost in roundtrip MH-2C-155 and 21 m^3 in roundtrip MH-2C-157. 5.25 t of plugging agent while drilling was added in mud in three times. Hole sections from 3670 to 3750 m, from 3890 to 3930 m and from 4350 to 4385 m were hole serious enlargement sections with the largest well diameter of 333.2 mm. The main problem in the construction was that the hole deviation exceeded the limit, with the largest deviation angle of 29.67° and azimuth angle of 62.3° at the depth of 4947 m. Main technical indexes of MH-2C core drilling can be found in Table 4.13, and variations of well deviation angle, diameter and azimuth angle in Fig. 4.23.

4.2.10 Testing Drilling Tools

After accomplishment of core drilling in CCSD-1 Well, the tests of PDM hydro-hammer diamond wireline core drilling tool (three-in-one drilling tool) and PDM drive continuous deflector were conducted. From January 25th to February 28th, 2005 three-in-one drilling tool was tested (except 17 days of stop working from January 30th to February 15th). From March 1st to 4th, 2005, PDM drive continuous deflector was tested. From March 5th to 8th, 2005 tri-cone

4.2 Simple Situation of the Construction at Different Periods

Table 4.13 Main technical indexes of core drilling in the main hole section from 3,624.16 to 5,118.20 m

Item	Parameter	Item	Parameter	Item	Parameter
Footage of core drilling (m)	1,480.06	Average core recovery (%)	88.0	Average ROP (m/h)	1.03
Trips of core drilling	186	Roundtrips with full core barrel	110	Average bit life (m)	24.67
Longest bit life/m	52.06	Average footage drilled per roundtrip (m)	7.96	Largest footage drilled per roundtrip (m)	9.75
Average footage drilled per day (m/days)	5.79	Percentage of actual drilling in time (%)	24.4	Percentage of tripping in time (%)	61.8
Percentage of accident in time (%)	0.5	Largest well deviation angle (°)	29.67	Average footage drilled per rig month (m/30 days)	173.73
Well logging (times)	11	Comprehensive well logging (times)	2	Days of core drilling (days)	258

Note Total footage drilled was 1494.04 m, including 13.53 m completed by non-core drilling and 0.45 m by grind drilling. 262 days were used for core drilling, which started on May 8th, 2004 and ended on January 24th, 2005, including 4 days of comprehensive well logging

rock bit was employed for non-core drilling to the final well depth of 5158 m.

1. **PDM hydro-hammer diamond wireline core drilling tool**

Shift of drill rods, installation of wireline arrangement unit and air slips, etc. were finished from January 25th to 27th, 2005.

On January 27th, 2005, three-in-one drilling tool was assembled (Φ 157 mm core drill bit + Φ 157 mm reaming shell + Φ 146 mm outer tube + Φ 157 mm reaming shell + Φ 146 mm outer tube + 15 pieces of SK wireline drill collar + SK139.7 mm wireline drill rod) and run into the borehole to the depth of 3607 m. Mud was circulated and 23 m³ of mud was lost. From January 30th to February 15th, 2005 the work was temporarily stopped.

On February 16th, 2005 drilling mud was circulated, however, without mud returned. So rotary table was started-up for circulating mud and with little returned, and gradually no mud returned, with 55 m³ mud was lost. Running 55 pieces of drill rod out of hole and circulating mud, little mud was returned and 78 m³ mud was lost. When ROOH to well depth of 1000 m the tests of wireline fishing inner tube were conducted for three times, all with success.

On February 19th, 2005, tri-cone rock bit was used for drifting, with a drilling tool assembly of Φ 158.8 mm tri-cone bit + one piece of Φ 120 mm drill collar + Φ 120 mm drilling jar + eight pieces of Φ 120 mm drill collar + Φ 89 mm drill rod. After every 20 pieces of drilling rod were run in the borehole, mud was circulated once. Mud circulating and drilling were normal, with footage of 7.66 m drilled, penetration rate of 0.80 m/h and well depth reached to 5125.86 m.

On February 22nd, 2005, Three-in-one drilling tool + Φ 120 mm drill collar + Φ 89 mm drill rod was used for conventional core drilling test, with a drilling tool assembly of Φ 157 mm core bit + Φ 157 mm reaming shell + Φ 146 mm outer tube + Φ 157 mm reaming shell + Φ 146 mm outer tube (with inner tube assembly, Φ 98 mm hydro-hammer and Φ 95 mm PDM) + one piece of Φ 120 mm drill collar + Φ 120 mm drilling jar + eight pieces of Φ 120 mm drill collar + Φ 89 mm drill rod. Redressing was started at 10 m from hole bottom, pump blocking frequently happened and after footage of 0.54 m got no more footage could be obtained. Then the drill string was pulled out of the borehole. No core was recovered, with penetration rate of 0.55 m/h and well depth reached to 5126.40 m. The former drill string was run in the borehole after maintenance and redressing was started at 8.4 m from hole bottom. The drill string had to be pulled out because drilling was abnormal and no bit load returned. The transmission shaft of PDM was broken, with no core recovered. Footage of 0.69 m was completed, with penetration rate of 0.60 m/h and well depth reached to 5127.09 m.

The former drill string was run in the borehole on February 27th, 2005 after PDM changed. Pump blocking frequently happened at 1.8 m from hole bottom. Moving the drill string repeatedly and then drilling was basically normal after redressing to the hole bottom. Footage drilled was 2.27 m with penetration rate of 0.64 m/h, core length of 2.4 m and well depth reached to 5129.36 m.

2. **PDM drive continuous deflector drilling tool**

On March 1st, 2005 angle drop drilling test with continuous deflector was started at well depth of 5129.36 m with a drilling tool assembly of Φ 158.8 mm tri-cone bit + continuous deflector + Φ 120 mm PDM + Φ 120 mm drilling jar + Φ 120 mm directional sub + Φ 104 mm non-magnetic drill collar + nine pieces of Φ 120 mm drill collar + Φ 89 mm drill rod. The drill string had to be pulled out because of no more footage could be penetrated after 5.30 m had been drilled. The teeth of roller bit were all worn flat, with penetration rate of 0.38 m/h and well depth reached to 5134.66 m.

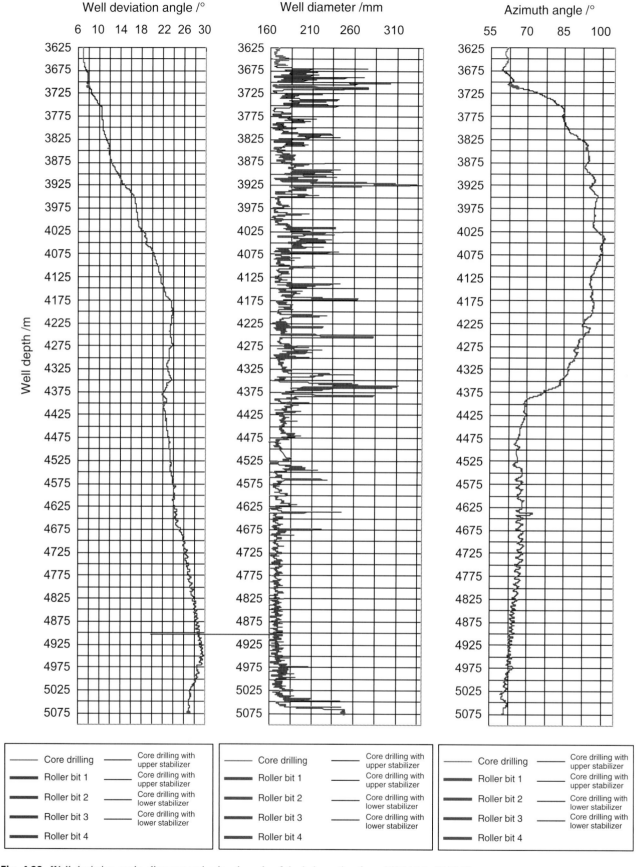

Fig. 4.23 Well deviation angle, diameter and azimuth angle of the hole section from 3624.16 to 5118.20 m

3. Non-core drilling

From March 5th to 8th, 2005 a drill string of Φ 158.8 mm tri-cone bit + Φ 120 mm PDM + one piece of Φ 120 mm drill collar + Φ 120 mm drilling jar + eight pieces of Φ 120 mm drill collar + Φ 89 mm drill rod was employed for two roundtrips of non-core drilling, with footage of 23.34 m drilled, penetration rate of 1.00 m/h and well depth reached to 5,158.00 m. Before ROOH, the borehole was flushed with large delivery rate, and then heavy mud with density as high as 1.98 g/cm^3 was injected in the hole.

The main technical indexes of drilling tool test can be found in Table 4.14.

4.2.11 Well Completion

From March 9th to 12th, the completion works such as running liner and setting hanger, and well cementation, etc. were completed. From March 13th to April 1st the works such as grinding inside step of hanger, drilling cement plug, VSP measurement, drifting, injecting completion fluid, throwing off drill string, and shifting the drilling tools at drill site were completed.

1. Running liner and well cementation

On March 9th, 2005, liner of Φ 127 mm diameter was run, connected with Φ 127.0 mm × Φ 193.7 mm hanger, which was run down by using Φ 89 mm drilling rod to the predetermined position in the borehole.

On March 10th, 2005, pump was started for circulation, injecting ball, building up pressure, and then the hanger was successfully set. Backing-off by clockwise turning, successful separation by uplifting, pressurizing and then ballseat was thrown off. Casing was run down to the depth of 4790.72 m and the top of the liner (hanger) was at well depth of 3523.55 m.

On March 11th, 2005, mud was circulated, spacer fluid was injected, cement slurry was injected, and slurry was displaced after rubber plug was pressed in, setting at last, and then well cementation was completed. Drill rod was lifted up for pump flushing, and then the drill string was pulled out.

2. Well completion

Single roller bit was employed for drifting on March 13th, 2005, with a drilling tool assembly of Φ 108.0 mm single roller bit + nine pieces of Φ 89 mm drill collar + 166 pieces of Φ 73 mm drill rod + Φ 89 mm drill rod. After the drilling tool assembly was run in the borehole to well depth of 3400 m, mud was circulated once per hour. Slack-off was caused at well depth of 3,523.55 m where hanger was fixed, drill string was moved repeatedly, however, without any result. Because that the step of Φ 73 mm drill rod was larger than the inner step of the hanger, so drifting could not be conducted any longer and in this case the drill string had to be pulled out of the borehole.

A corncob bit was employed for grinding the inner step of the hanger on March 15th, 2005. Drill string used consisted of Φ 130 mm corncob bit + three pieces of Φ 120 mm drill collar + Φ 89 mm drill rod. The drill string was pulled out after grinding the top inner step of the hanger and it was found that there were obvious wearing traces on the chamfer root of the corncob.

On March 16th, 2005, the former drill string with single roller bit was employed for drifting. When Φ 73 mm drill rod adapter (with right angle shoulder) entered into the hanger, the drill string should be moved up and down or revolved. Cement plug, ballseat, float coupling and float shoe were chipped off. Pump blocking frequently happened and mud began to leak while chipping ballseat, bit bouncing and sticking happened. Moving the drill string, backreaming with pressure of 100 to 400 t, and sticking was finally freed by backreaming with pressure of 138 t. Moving the drill string, 26 m^3 of mud was lost. The drill string was pulled out of the borehole.

On March 21st, 2005, VSP measurement was conducted.

A corncob bit was employed for redressing and drifting to well depth of 5,089.45 m on March 22nd, 2005, with a drill string of Φ 108 mm corncob bit + nine pieces of Φ 89 mm drill collar + 166 pieces of Φ 73 mm drill rod + Φ 89 mm

Table 4.14 The main technical indexes of the main hole section 5,118.20–5,158.00 m

Item	Parameter	Item	Parameter	Item	Parameter
Footage of core drilling (m)	3.50	Average core recovery (%)	68.6	ROP of core drilling (m/h)	0.62
Roundtrips of core drilling	3	Footage of straightening drilling (m)	5.30	ROP of straightening drilling (m/h)	0.38
Footage of non-core drilling (m)	31.00	ROP of non-core drilling (m/h)	0.94	Average footage drilled per day (m/days)	1.53
Total RIH roundtrips	8	Average footage drilled per rig month (m/30 days)	45.92	Days of drilling (days)	26

Note Total footage drilled was 39.80 m. Testing was started on January 25th, 2005 and ended on March 8th, with total 43 days included 17 days of stop working

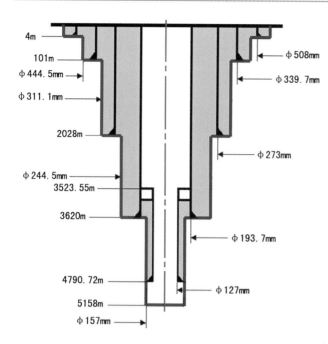

Fig. 4.24 Borehole structure of well completion and casing program of CCSD-1 Well

drill rod. 86 m³ completion fluids (clear water with 1.5 % GLUB lubricant, with PH value of 11) was injected to displace all the mud in the borehole after circulating mud with large delivery rate, then ROOH. The open hole section deeper than 4,790.72 m was used for long term observation in future. Borehole structure of well completion can be found in Fig. 4.24.

From March 25th to April 1st, 2005, VSP measurement was conducted, the drilling tools were shifted out and the whole drill site was cleaned up.

From April 1st to April 17th, 2005, the drill site was shut down.

On April 18th, 2005, a completion celebration was held at the drill site.

On April 19th, 2005, the Drill-site Headquarters of the CCSD Project Center issued an order for stop of drilling because of the well completion, formally declaring the completion of the construction of CCSD-1 Well.

Hard Rock Deep Well Core Drilling Techniques

It is well known that core is the most important carrier of deep underground geological information, either for conventional geological drilling and oil and gas drilling, or for scientific drilling which serves for earth sciences. Therefore, how to obtain high quality core under the different conditions is an eternal problem that always lies ahead of drilling engineers. Core drilling techniques are the main key techniques for CCSD-1 Well, directly affecting the success of the whole project, and its efficiency and quality.

5.1 Current Status of Core Drilling Techniques

During a long period of development in the field of geological drilling industry, a series of core drilling techniques for different geological conditions, different operation conditions and different coring requirements have been well developed and successfully applied in practice. Currently the core drilling techniques both at home and abroad can be classified into three categories, i.e., oil core drilling techniques, geological core drilling techniques and scientific core drilling techniques, which have many general characters because formations and rocks are the objects. However, these three techniques have some different characters due to their different coring purposes and different rock formations to be drilled.

Non-core drilling is mainly used for oil drilling whereas core drilling is seldom utilized, generally no more than 5 %. In oil drilling operations wireline core drilling technique is very seldom used because the proportion of coring to drilling is low, drill rod and drill pipe interchange is inconvenient and time consuming in the conversion from non-coring drilling to core drilling, and wireline drill rod cost is very expensive. In oil drilling, core recovery is realized by lifting drill pipe.

Geological exploration drilling includes solid mineral exploratory drilling, hydrological and water well exploratory drilling and engineering geological exploration drilling, the purposes of which are different, the rock formations encountered are different, and therefore the coring techniques and tools different, too.

There are a variety of core drilling techniques. According to coring method, wireline coring and coring by lifting drill pipe (conventional coring) are included. On the basis of core barrel structure, single tube, double tube and triple tube core barrels are included, whereas double tube core barrel can be classified into rigid type and swivel type. Based on rock fragmentation tools core drilling techniques can be classified into diamond core drilling, carbide core drilling, and rock bit core drilling. In accordance to rock fragmentation mode, core drilling techniques can be classified into rotary core drilling and percussive rotary core drilling. And according to circulation mode core drilling techniques can be classified into positive circulation core drilling and reverse circulation core drilling, the latter can be further divided up into non-pump, jet, water power, air and air lifting reverse circulation techniques, according to the circulation mediums employed.

The main purpose of geological exploratory drilling is to obtain core, and therefore core drilling is the main method. More and more attention has been focused on the improvement of core quality, coring efficiency and coring recovery, and on how to recover as much as core in difficult rock formations.

Rock fragmentation materials (abrasives) have experienced from carbide and calyx, natural diamond to synthetic diamond and its composite materials, and further to new type of ultra-hard materials such as β-C_3N_4, CVD (chemical vapoured diamond), CBN (cubic boron nitride), Slavutich and synthetic carbonados.

A great variety of drill bits have been developed, especially for diamond drill bits. Vast research has been conducted on diamond drill bits, in particular, the structure and manufacturing techniques of impregnated diamond drill bits. A complete set of diamond drill bit manufacturing

Translated by Geng Junfeng.

techniques which can satisfy the requirements of different rock formations has been developed.

As for coring tools, to satisfy the coring requirements in different formations and to improve coring quality, single tube coring tool has been gradually replaced by rigid type and swivel type double tube core barrels. In order to take as much as core in difficult rock formations such as very loose or broken formations, triple tube coring tool and non-pump reverse circulation core drilling and jet reverse circulation core drilling techniques have been successively developed.

To improve core drilling efficiency, research on wireline core drilling was carried out in some countries from the 1940s. In 1953 the Longyear Company obtained the first patent on wireline core drilling in the world.

Pneumatic and hydraulic down-the-hole hammers were developed, in order to increase rock cutting efficiency. Pneumatic down-the-hole hammer, as air has compressibility, can only be used for shallow hole, seldom used in geological exploratory drilling. As for hydraulic down-the-hole hammer, though a long term research has been made and very good application result obtained, its application in deep hole drilling has yet to be proved.

Since the 1970s, research activities on hydro-power reverse circulation continuous core drilling have been conducted in some countries and this technique has been widely used in the unconsolidated formation of the Quaternary Period. However, the application of this technique in medium hard and hard consolidated formation is still faced with some problems, such as low drilling rate, short drill bit service life, and core blocking during transportation.

Two coring methods are commonly used for continental scientific core drilling.

The first method is rock bit coring by lifting drill pipe, which was used in Cola super deep drilling project in the former Soviet Union, with the main advantage of fast drilling rate. Drilling in gneiss, hornblende and granite in the Cola super deep drilling project with rock bit an average drilling rate of 1.8 m/h was reached. However, by using this coring method the surface of the core was rough and uneven, and of low quality. Furthermore, this method also got a low core recovery, for example, approximate 40 % in the Cola super deep drilling project. Strictly speaking this coring method can not meet the needs of scientific drilling.

Another method is large diameter diamond wireline core drilling, which was used in the pilot hole drilling in KTB project in Germany, with the main advantages that drill pipe lifting was unnecessary and auxiliary time was dramatically reduced. Moreover, as diamond drill bit was used for drilling, the surface of the core recovered was very smooth and core recovery was high. The pilot hole of KTB was drilled 3142.6 m deep by using diamond wireline coring techniques and hole diameter 152 mm, core recovery was as high as 98 %, with the total drilling time of 449 days. However, this technique has some shortcomings, for instance, the cost of drilling tools is rather high as special wireline drill rod and drill collar are necessary and the cost of equipment is also high because high speed top drive system should be equipped.

In ocean scientific drilling projects, a variety of coring tools have been developed according to the formations drilled, including:

Wireline coring tools for deep sea continuous (soil) coring, such as rotary core barrel (RCB), hydraulic piston core sampler (HPC), advanced piston core sampler (APC), diamond core sampler (DCS), pressure core sampler (PCS), Navi core barrel (NCB), bottom hole outer core barrel assembly (BHA) and percussive rod sampler, etc.

Sampling tools for shallow seabed, such as gravity percussive seabed sampler, indentation seabed sampler, suction seabed sampler, and striking, floating ball, dragging and grab bucket seabed samplers.

Other coring tools have been developed such as rotary core barrel (RCB), small diameter rotary core barrel (SD-RCB), hydraulic piston core system (HPCS), vibration core collecting tool, and coring techniques and tools for microorganism core samples under extreme environment.

5.2 Experiment on Core Drilling Methods for CCSD-1 Well

At the upper hole section of the pilot hole and the main hole of CCSD-1 Well, totally ten core drilling methods were tested, based upon the initial design and according to the specific situations at drilling site.

5.2.1 Rotary Table Drive Double Tube Core Drilling

Rotary table drive core drilling technique is a conventional core drilling method in oil and gas drilling. Although many geological exploratory drilling experts believed that this technique could not be used for CCSD-1 Well as it could not provide high rotary speed, some experts still insisted that this technique could be used for core drilling. To guarantee a final scientific decision, a test on rotary table drive core drilling technique was conducted.

1. **Drilling tool assembly**

Rotary table drive double tube core drilling tool consisted of 157 mm diamond core drill bit, 139.7 mm double tube core drilling tool (including upper and lower 157 mm reaming shells) and upper drilling tools (stabilizer, drill collar and drill pipe). At the hole depth of 101 m when the hole size was changed for core drilling, the upper drilling tool assembly included 215.9 mm stabilizer (in the 244.5 mm moving casing) and 127 mm drill pipe. A round

Table 5.1 Result of rotary table drive double tube core drilling technique

Round trip	Footage drilled (m)	Actual drilling time (h)	Core length (m)	Penetration rate (m/h)	Core recovery (%)	Footage drilled per round trip (m)
7	11.60	29.61	6.40	0.39	55.2	1.66

trip was drilled to the depth of 102.58 m (with 0.15 m drilled, actual drilling time of 1.92 h and core length of 0.44 m). At the hole depth from 1156.19 to 1164.71 m, four round trips were drilled, with the upper drilling tool assembly of 156 mm stabilizer, 120 mm single drill collar, 156 mm stabilizer, 120 mm drill collar stem and 89 mm drill pipe (with 8.52 m drilled, actual drilling time of 18.417 h and core length of 3.63 m). At the hole depths from 1390.15 to 1391.15 m and from 2028.17 to 2030.10 m, one round trip was drilled respectively (with 2.93 m drilled, actual drilling time of 9.27 h and core length of 2.23 m).

2. **Drilling parameters**

Drilling parameters for this method: bit pressure 20–35 kN, flow rate 8–13 L/s and rotary speed 50–110 rpm.

3. **Commentary on drilling result**

Seven round trips were drilled by using rotary table drive double tube core drilling technique, with footage of 11.6 m drilled (see Table 5.1). This technique required simple drilling process and equipment. However, it could not satisfy the requirement of high linear speed (2.0–4.0 m/s) that diamond impregnated drill bit needed to effectively cut rock due to its low rotary speed (max. 0.63 m/s), thus resulting in a low penetration rate, with an average penetration rate of 0.39 m/h. A low core recovery of average 55.17 % was obtained because of the poor stability of the drill bit during drilling process resulted from the rotation of drill string. Furthermore, core blockage often occurred in the core barrel because the rock was rather hard and broken and this resulted in a short round trip, with an average of only 1.66 m.

5.2.2 Rotary Table Hydro-hammer Drive Double Tube Core Drilling

In order to improve rotary table drive core drilling technique, hydro-hammer was used for this drilling tool assembly: 157 mm diamond core drill bit, 139.7 mm double tube core drilling tool (with 157 mm upper and lower reaming shells), 127 mm hydro-hammer, 156 mm stabilizer, 120 mm single drill collar, 156 mm stabilizer, 120 mm drill collar stem and 89 mm drill pipe string.

Drilling parameters for this method: bit pressure 20–35 kN, flow rate 8–13 L/s and rotary speed 50–110 rpm.

Five round trips were drilled by using this assembly in the pilot hole (see Table 5.2).

Result indicated that by using hydro-hammer core recovery was improved. However penetration rate and round trip length were both rather unsatisfactory. The actual cause was that this technique could not provide high rotary speed for diamond impregnated drill bit to effectively cut the rock, as necessary high linear speed was the key to diamond drilling with impregnated drill bit.

At this stage, the possibility of application of rotary table drive core drilling method in CCSD-1 Well was definitely ruled out.

5.2.3 Top Drive Double Tube Core Drilling

In view of the low rotary speed resulted from rotary table drive, top drive double tube core drilling method was tested, after the RB130 top drive rented from ICDP was installed.

1. **Drilling tool assembly**

Top drive double tube core drilling consisted of 157 mm diamond core drill bit, 139.7 mm double tube core drilling tool (with 157 mm upper and lower reaming shells), SK157 stabilizer, SK146 drill collar stem and SK139.7 drill pipe string.

2. **Drilling parameters**

Bit pressure 15–45 kN, flow rate 11–13 L/s and top drive rotary speed 150–170 rpm.

Table 5.2 Result of rotary table hydro-hammer drive double tube core drilling technique

Hole section	Footage drilled (m)	Actual drilling time (h)	Core length (m)	Penetration rate (m/h)	Core recovery (%)	Footage drilled per round trip (m)
252.21–252.45	0.24	1.17	0.10	0.21	41.7	0.24
1632.79–1634.38	1.59	4.00	1.31	0.40	82.4	1.59
1641.62–1643.20	1.58	3.80	1.31	0.42	82.9	1.58
1765.65–1767.82	2.17	5.15	2.03	0.42	93.6	2.17
2030.10–2030.91	0.81	3.97	0.20	0.20	24.7	0.81
Total (average)	6.39	18.08	4.95	0.35	77.5	1.28

Table 5.3 Result of top drive double tube core drilling technique

Hole section	Footage drilled (m)	Actual drilling time (h)	Core length (m)	Penetration rate (m/h)	Core recovery (%)	Footage drilled per roundtrip (m)
2046.54–2047.82	1.28	1.45	0.3	0.88	23.4	1.28
2048.74–2049.41	0.67	2.47	0	0.27	0.0	0.67
2051.22–2051.42	0.20	1.38	0	0.15	0.0	0.20
2051.62–2052.62	1.00	3.48	0	0.29	0.0	1.00
2052.62–2053.82	1.20	3.75	0	0.32	0.0	1.20
2053.92–2054.75	0.83	1.30	0	0.64	0.0	0.83
2055.01–2055.40	0.39	1.35	0	0.29	0.0	0.39
2071.69–2071.99	0.30	1.06	0	0.28	0.0	0.30
Total (average)	5.87	16.24	0.3	0.36	5.1	0.73

3. **Commentary on drilling result**

Eight round trips were drilled with this method in the hole section from 2046.54 to 2071.99 m (MH hole section, see Table 5.3), with a footage of 5.87 m drilled, an average round trip length of 0.734 m, an average penetration rate of 0.361 m/h, and an average core recovery of 5.11 %.

The application result of this method was very poor because fracture zone was just encountered. Besides this, a diamond drill bit was consumed in each round trip and drilling tools were broken twice. According to our analysis, the reason was because the movement pattern and force bearing state of the drill pipe string under high speed rotation in the oversized hole section changed radically. Revolved both on its own axis and round the oversized hole, the drill pipe string, under the state of this complex movement, vibrated both axially and radially and thus resulted in drill bit whirl at hole bottom (see Fig. 5.1), finally leading to abnormal damage of drill bit matrix and fatigue break of drilling tool.

Moreover, the RB130 top drive could not be operated under high rotary speed because of the inadequate rated load (maximum torque 10,000 N m). To thoroughly test the potential of the top drive, it was decided to increase the protective pressure of the hydraulic system of the top drive pump station from 28 to 32 MPa, so as to observe the change of the rotary speed (January 23, 2003). Firstly, the protective pressure was adjusted to 30 MPa for 20 min stable operation, then to 32 MPa. Observation showed that after the first adjustment the rotary speed of the top drive increased from 180–190 rpm to 190–200 rpm, with the instantaneous maximum to 214 rpm. Basically the top drive worked at about 200 rpm, with rotary torque increased to 8500–9200 N m. After the second adjustment the rotary speed of the top drive did not further increase, but rotation was much stable than before and torque increased to 9500 N m. This test showed that the working potential of this top drive was rather limited.

Thus the possibility of the further application of the RB130 top drive in CCSD-1 Well was definitely ruled out.

5.2.4 Top Drive Wireline Core Drilling

Top drive diamond wireline core drilling test was conducted immediately right after the arrival of the imported wireline core drilling tools.

1. **Drilling tool assembly**

Top drive diamond wireline core drilling consisted of 157 mm diamond core drill bit, SK146 wireline core drilling tool (or China made S157 wireline core drilling tool), SK157 stabilizer, SK146 drill collar stem and SK139.7 drill pipe string.

2. **Drilling parameters**

Bit pressure 18–45 kN, flow rate 7–10 L/s and top drive rotary speed 135–145 rpm.

3. **Commentary on drilling result**

Five round trips were drilled with this method in the hole section from 2047.82 to 2462.71 m (MH hole section, see Table 5.4), with an accumulative footage of 7.62 m drilled,

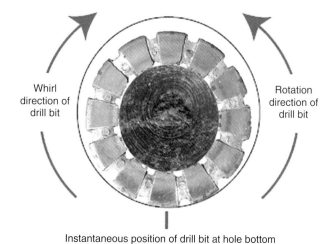

Fig. 5.1 Diamond core drill bit whirl

Table 5.4 Result of top drive wireline core drilling technique

Hole section	Footage drilled (m)	Bit number	Actual drilling time (h)	Core length (m)	Penetration rate (m/h)	Core recovery (%)	Footage drilled per round trip (m)	Remarks
2047.82–2048.74	0.92	SDD-1	2.87	0.48	0.32	52.3	0.92	SK146
2049.67–2051.22	1.55	SGL-01	2.88	0.42	0.54	27.1	1.55	
2066.69–2067.01	0.32	TGS-1	0.87	0.15	0.37	46.9	0.32	
2069.19–2071.69	2.50	G63007	2.28	0	1.10	0.0	2.50	
2460.38–2462.71	2.33	TGS-1S	3.12	0	0.75	0.0	2.33	S157
Total	7.62		12.02	1.05	0.63	13.8	1.52	

an average round trip length of 1.52 m, an average penetration rate of 0.63 m/h, and an average core recovery of 13.8 %. Successful fishing of inner tube was completed by China made S157 wireline core drilling tool only, whereas the imported SK146 tool failed.

Drilling result was not good. Besides lower penetration rate resulted from more rock cut off because of thicker bit wall, fracture zone was encountered. Wireline core drill pipe string was a full hole drilling tool assembly, mud gushed seriously during tripping in. Though it was solved by adding a floating plug, the efficiency of tripping in was affected. Meanwhile, the full hole drilling tool added with the floating plug produced tremendous pressure against the rock formation during tripping in (a violent suction was produced during tripping out), thus tripping in speed must be strictly controlled. Under this situation production efficiency was rather low and safety in the hole worsened. For the imported wireline core drill pipe, quality problem of 1 mm inner step existed at welding seam (Fig. 5.2). Lowering in and fishing inner tube assembly failed because the inner tube assembly was obstructed at the step, as the clearance was too small (Fig. 5.3). As the quality problem of the imported wireline core drill pipe could not be solved in a short time, our attempt to use diamond wireline core drilling in CCSD-1 Well was stopped.

5.2.5 Top Drive Hydro-hammer Wireline Core Drilling

Hydro-hammer wireline core drilling tool consists of conventional wireline core drilling tool and hydro-hammer. Rotary drilling is realized by bit pressure and torque transmitted through drill pipe string and outer tube core barrel from surface to core bit. Meanwhile, a percussive load, produced by hydro-hammer, is transmitted through power transmitting plate, power transmitting ring, outer splined shaft and lower outer tube to core bit, to realize rock fragmentation by percussion. Conventional wireline technique is utilized to fish core barrel (including hydro-hammer) when it is full with core. This core drilling tool is composed of outer tube assembly and inner tube assembly. The outer tube assembly consists of spring clip stop, spring clip chamber, sealing sub, stabilizer, upper outer tube, outer splined sleeve, lower outer tube, reaming shell and drill bit; while the inner tube assembly consists of spear head, recovery tube, spring clip clamp, spring clip support, sealing sleeve, hydro-hammer, power transmitting plate, swivel sub, core blockage alarming sleeve, buffer sub, buffer chamber, upper separating adapter, lower separating adapter, adjusting adapter and core catcher case (see Fig. 5.4).

1. **Drilling tool assembly**

Top drive hydro-hammer wireline core drilling consisted of 157 mm diamond core drill bit, hydro-hammer wireline core drilling tool, SK157 stabilizer, SK146 drill collar stem and SK139.7 drill pipe string.

2. **Drilling parameters**

Bit pressure 20–35 kN, flow rate 5–7 L/s and top drive rotary speed 150–180 rpm.

3. **Commentary on drilling result**

Two run-out trips and three wireline fishing trips were drilled with this method in the hole section from 2442.71 to 2460.38 m (MH hole section, see Table 5.5), with an accumulative footage of 8.27 m drilled, an average run-out trip length of 4.14 m and an average fishing trip length of 2.76 m, an average penetration rate of 0.89 m/h, and an average core recovery of 99.5 %. This test had to be stopped because inner tube assembly was blocked with rock powder.

5.2.6 PDM Drive Single Tube Core Drilling

On the basis of PDM (positive displacement motor) drive swivel type double tube core drilling technique, PDM drive single tube core drilling technique was tested in order to reduce rock cut-off area.

1. **Drilling tool assembly**

PDM drive single tube core drilling consisted of 157 mm diamond core drill bit, 139.7 mm single tube core drilling tool (including 157 mm upper and lower reaming shells),

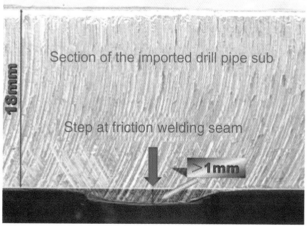

Fig. 5.2 Quality problems of the imported wireline core drill pipes

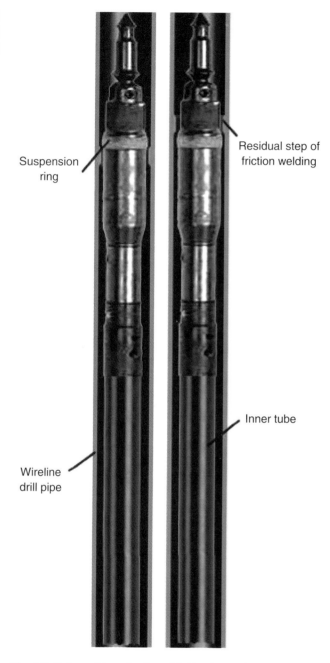

Fig. 5.3 Failure of lowering in and fishing inner tube assembly

120 mm positive displacement motor and upper drilling tools (stabilizer, drill collar and drill pipe). At the hole depth of 101 m where the hole size was changed for core drilling, the upper drilling tool assembly included 215.9 mm stabilizer (in the 244.5 mm moving casing) and 127 mm drill pipe. A round trip was drilled to the depth of 106.69 m (with 1.71 m drilled, actual drilling time of 2.67 h and core length of 0.46 m). At the hole depth from 116.75 to 120.78 m, three round trips were drilled, with the upper drilling tool assembly of 120 mm slick drill collar (Table 5.6). At the hole depth of 405.16 m, one round trip was drilled, with the upper drilling tool assembly of 156 mm stabilizer, 120 mm single drill collar, 156 mm stabilizer, 120 mm drill collar stem and 89 mm drill pipe string (Table 5.6).

2. **Drilling parameters**

Bit pressure 10–20 kN, flow rate 10–12 L/s and PDM rotary speed 320 rpm.

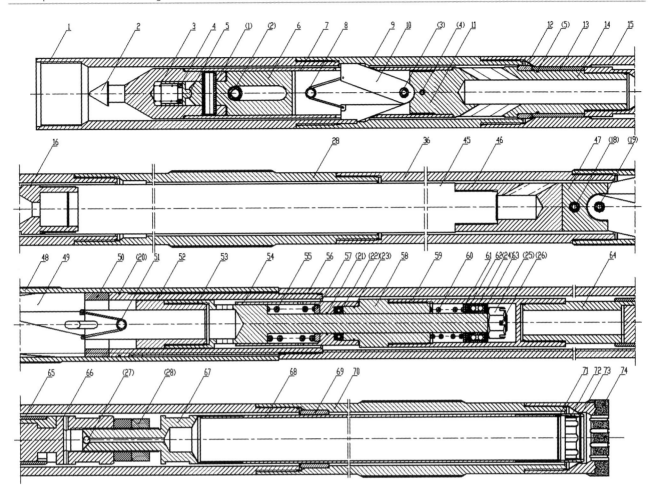

Fig. 5.4 Top drive hydro-hammer wireline core drilling tool. *1* Stop of spring clip. *2* Spear head. *3* Spear head spring. *4* Spear head spring cushion. *5* Spear head spring seat. *6* Spear head seat. *7* Recovery tube. *8* Tension spring. *9* Spring clip chamber. *10* Spring clip clamp. *11* Spring clip support. *12* Suspension ring. *13* Sealing sleeve. *14* Spring clip support. *15* Sealing sub. *16* Upper adapter. *28* Stabilizer. *36* Upper outer tube. *45* Inner spline sleeve. *46* Power transmitting plate support. *47* Power transmitting plate seat. *48* Outer spline sleeve. *49* Power transmitting plate. *50* Power transmitting cup. *51* Tension spring of power transmitting plate. *52* Outer spline shaft. *53* Swivel sub. *54* Core blockage alarming sleeve. *55* Core blockage alarming spring. *56* Lower outer tube. *57* Core blockage alarming spring cushion. *58* Buffer sub. *59* Buffer chamber. *60* Buffer spring. *61* Buffer spring cushion. *62* Left cover of 6208 bearing. *63* Right cover of 6208 bearing. *64* Upper separating adapter. *65* Stop ring. *66* Lower separating adapter. *67* Adjusting sub. *68* Core barrel. *69* Lower centralizing ring. *70* Reaming shell. *71* Stop ring of core catcher. *72* Core catcher. *73* Core catcher seat. *74* Drill bit. (*1*)–(*4*), (*18*) and (*19*) Elastic pins. (*5*), (*20*), (*21*) and (*23*) O sealing rings. (*22*) Thrust ball bearing, single direction. (*24*) Thrust ball bearing, double row radial. (*25*) Nut. (*26*) Split pin. (*27*) Steel ball. (*28*) Nut

3. **Commentary on drilling result**

Five round trips were drilled by using PDM drive single tube core drilling technique, with footage of 6.84 m drilled (Table 5.6). By using this method, both wear and power consumption of drill pipe string were rather small, damage of drill pipe string to hole wall was small, and safety in drilling operation was satisfactory. However, because of frequent core blockage, both low core recovery (32.46 %) and low penetration rate (0.57 m/h) were obtained, with round trip length of only 1.368 m. Obviously, this method could not meet the economical requirement of core drilling, and thus its possibility to be used in CCSD-1 Well was ruled out.

5.2.7 PDM Drive Double Tube Core Drilling

In the stage of initial design of CCSD-1 Well, PDM drive double tube core drilling technique was considered as an important candidate. For lack of experiences and delay of the imported wireline core drilling tool, PDM core drilling technique was taken as the main drilling method, its possibility of application would be decided by field test.

1. **Drilling tool assembly**

PDM drive double tube core drilling consisted of 157 mm diamond core drill bit, 139.7 mm double tube core drilling tool (including 157 mm upper and lower reaming shells),

Table 5.5 Result of top drive hydro-hammer wireline core drilling technique

Hole section	Footage drilled (m)	Actual drilling time (h)	Core length (m)	Penetration rate (m/h)	Core recovery (%)	Footage drilled per round trip (m)
2442.71–2447.84	5.13	4.44	5.40	1.16	105.3	
2447.84–2449.13	1.29	2.71	0.88	0.48	68.2	
Run-out trip 1	6.42	7.15	6.28	0.90	97.8	6.42
2458.53–2460.38	1.85	2.10	1.95	0.88	105.4	
Run-out trip 2	1.85	2.10	1.95	0.88	105.4	1.85
Total	8.27	9.25	8.23	0.89	99.5	4.14

Table 5.6 Result of PDM drive singe tube core drilling technique

Hole section	Footage drilled (m)	Actual drilling time (h)	Core length (m)	Penetration rate (m/h)	Core recovery (%)	Footage drilled per round trip (m)
104.98–106.69	1.71	2.66	0.46	0.64	26.9	1.71
116.75–118.35	1.60	1.75	0.87	0.91	54.4	1.60
118.35–120.28	1.93	3.92	0.44	0.49	22.8	1.93
120.28–120.78	0.50	0.67	0.15	0.75	30.0	0.50
405.16–406.26	1.1	3.00	0.30	0.37	27.3	1.10
Total	6.84	12.00	2.22	0.57	32.5	1.37

positive displacement motor (95 mm or 120 mm) and upper drilling tools (stabilizer, drill collar and drill pipe) (see Fig. 5.5). Based on the change of hole condition and formation, and the drilling method adopted for such hole condition, the upper drilling tool assembly changed accordingly. At the hole depth of 101 m where the hole size was changed for core drilling, the upper drilling tool assembly included 215.9 mm stabilizer (in the 244.5 mm moving casing) and 127 mm drill pipe (Fig. 5.5a). Eight round trips were drilled to the depth of 112.95 m (with 10.09 m drilled, actual drilling time of 26.75 h and core length of 7.73 m). Along with the increase of hole depth, the following tools were added successively to the upper drilling tool assembly: 120 mm slick drill collar (Fig. 5.5b) was added to drill seventeen round trips to the depth of 152.37 m, with 42.81 m drilled, actual drilling time of 45.05 h and core length of 42 m (two round trips were drilled at the depth from 763.53 to 770.95 m); 156 mm stabilizer (Fig. 5.5c) was added to the upper drilling tool assembly to drill 50 round trips to the depth of 233.27 m, with 98.5 m drilled, actual drilling time of 127.43 h and core length of 89.4 m (six round trips were drilled at the depth from 730.48 to 742.05 m and one round trip was drilled at the depth from 3829.54 to 3836.55 m); two 156 mm stabilizers (Fig. 5.5d) were added to the upper drilling tool assembly to drill three hundreds and one round trips, with 679.63 m drilled, actual drilling time of 915.75 h and core length of 623.68 m.

Through drilling operations and research, the optimum drilling tool assembly of PDM drive double tube core drilling (Fig. 5.5d) was formed, i.e. 157 mm diamond core drill bit, 157.3 mm reaming shell, 140 mm core barrel, 157.3 mm reaming shell, 146 mm swivel sub, PDM, 156 mm stabilizer, 120 mm single drill collar, 156 mm stabilizer, 120 mm drill collar stem and 89 mm drill pipe string.

2. **Drilling parameters**

Bit pressure 10–30 kN, flow rate 10–12 L/s and PDM rotary speed 320–340 rpm.

3. **Commentary on drilling result**

Drilling result (Table 5.7) indicated that this preparatory technique had considerable large feasibility. All the technical indexes it achieved though were not very ideal, but superior to those of oil drilling in the same rock formation. By using this method, both wear and power consumption of drill pipe string were rather small, damage of drill pipe string to hole wall was small, safety in drilling operation was satisfactory, core recovery was high (88.7 %) and core quality was good. Though penetration rate was low (0.71 m/h) and round trip length was not long (2.29 m), it was believed that by necessary technical research and development these shortcomings could be overcome.

5.2.8 PDM Drive Wireline Core Drilling

Rotation of PDM is driven by drilling fluid and the torque transmitted from PDM is transferred to outer tube through a torque transmitting apparatus, then realize rotary drilling of

Fig. 5.5 PDM drive double tube core drilling tool assembly. *1* 157 mm diamond core drill bit. *2* 157 mm lower reaming shell. *3* 139.7 mm core barrel. *4* 157 mm upper reaming shell. *5* 146 mm swivel sub. *6* 95 or 120 mm PDM. *7* 156 mm lower stabilizer. *8* 120 mm single drill collar. *9* 156 mm upper stabilizer. *10* 120 mm drill collar stem. *11* 89 mm drill pipe string. *12* Kelly bar. *13* 215.9 mm stabilizer. *14* 127 mm drill pipe. *15* SK157 stabilizer. *16* SK146 drill collar. *17* SK139.7 drill pipe. *18* Sub

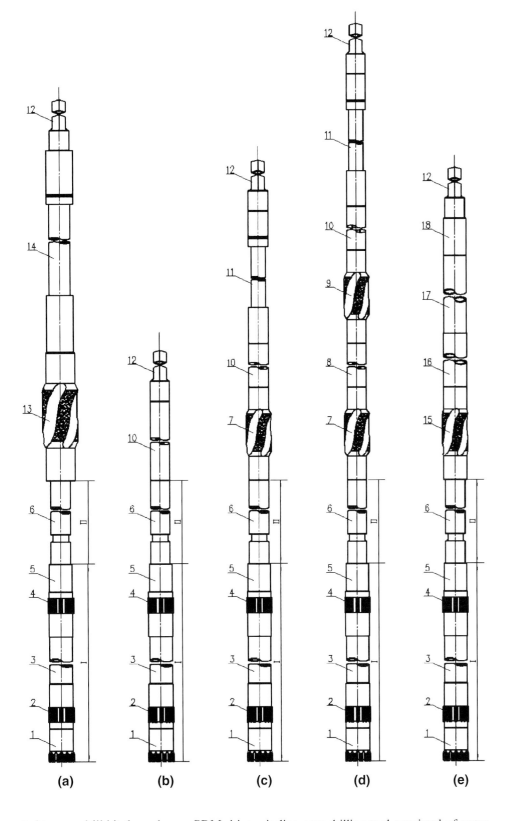

core drill bit. Pressure is transported to core drill bit through the bearing housing in the outer tube assembly. When core barrel is full of core, it (including PDM) can be fished by conventional wireline method.

PDM drive wireline core drilling tool consisted of outer tube and inner tube assemblies. The outer tube assembly was composed of counter-torsional spring clip, sealed suspension, torque transmitting apparatus, bearing housing, outer

Table 5.7 Results of PDM drive double tube core drilling technique with different upper drilling tool assemblies

Upper drilling tool assembly	Round trip	Footage drilled (m)	Actual drilling time (h)	Core length (m)	Penetration rate (m/h)	Core recovery (%)	Footage drilled per round trip (m)
Hole size change drilling	8	10.09	26.75	7.73	0.38	76.6	1.26
Slick drill collar	17	42.81	45.05	42.00	0.95	98.1	2.52
One 156 mm stabilizer	50	98.50	127.43	89.40	0.77	90.7	1.97
Two 156 mm stabilizers	301	679.63	915.75	623.68	0.74	91.8	2.26
SK wireline drill string	21	76.12	155.49	41.40	0.49	54.4	3.63
Total	397	907.15	1270.47	804.21	0.71	88.7	2.29

tube, reaming shell and core drill bit, while the inner tube assembly was composed of fishing tool, PDM, torque transmitting spring clip apparatus, universal shaft apparatus, swivel type apparatus and core barrel (Fig. 5.6).

1. **Drilling tool assembly**

   PDM drive wireline core drilling consisted of 157 mm diamond core drill bit, PDM wireline core drilling tool, SK157 stabilizer, SK146 drill collar stem and SK139.7 drill pipe string.

2. **Drilling parameters**

   Bit pressure 20–30 kN, flow rate 10–11 L/s and PDM rotary speed 300–330 rpm.

3. **Commentary on drilling result**

   At the hole depth from 2075.63 to 2283.86 m (MH hole section), three run-out trips and eight wireline fishing trips were drilled, with the total footage of 6.72 m drilled (Table 5.8). The average run-out trip length was 2.24 m, the average fishing trip length was 0.84 m, the average penetration rate was 0.33 m/h, and the average core recovery 71.9 %. The test showed that this method had the shortcomings of thick drill bit matrix, low drilling speed and short round trip length.

5.2.9 PDM Hydro-hammer Drive Double Tube Core Drilling

As PDM drive swivel type double tube core drilling technique obtained encouraging result, this method would certainly become the important technique for CCSD-1 Well drilling, provided penetration rate and round trip length be improved. All the experts and engineers believed that using hydro-hammer to this method would be the only effective way to overcome the technical difficulties, and then PDM + hydro-hammer + swivel type double tube core drilling technique was formed (see Fig. 5.7). This technique belongs to down hole power percussive rotary core drilling.

1. **Drilling tool assembly**

   Considering the strict requirement of CCSD-1 Well to hole deviation, vast tests were conducted on drilling tool assembly when this method was tested, in order to select the core drilling tool assemblies suitable for different geological formations.

   PDM hydro-hammer drive double tube core drilling consisted of 157 mm diamond core drill bit, 139.7 mm double tube drilling tool (including 157 mm upper and lower reaming shells), 127 mm hydro-hammer, PDM (95 or 120 mm) and upper drilling tools such as stabilizer, jar, drill collar and drill pipe (Fig. 5.8).

   Among these drilling tool assemblies, the most important one for feasibility test was that of Fig. 5.8a, which was used to drill 265 round trips from 218.44 to 2046.54 m in the pilot hole (Table 5.9).

2. **Drilling parameters**

   Bit pressure 10–35 kN, flow rate 9–11 L/s and rotary speed 157–189 rpm.

3. **Commentary on drilling result**

   The obtained technical indexes showed that this method was fully capable of drilling CCSD-1 Well, provided all the technical indexes were improved through technical perfection. The decisive techniques to be perfected included the property of swivel type double tube core drilling tool, core barrel length, property of hydro-hammer, and the property of diamond core drill bit. It was believed that all these properties would be inevitably improved, as long as necessary efforts were put into research. For details please see Sect. 5.3.

5.2.10 PDM Hydro-hammer Drive Wireline Core Drilling

In order to further perfect diamond wireline core drilling technique system and overcome the shortcomings of large torque of wireline coring + hydro-hammer drilling string and

5.2 Experiment on Core Drilling Methods for CCSD-1 Well

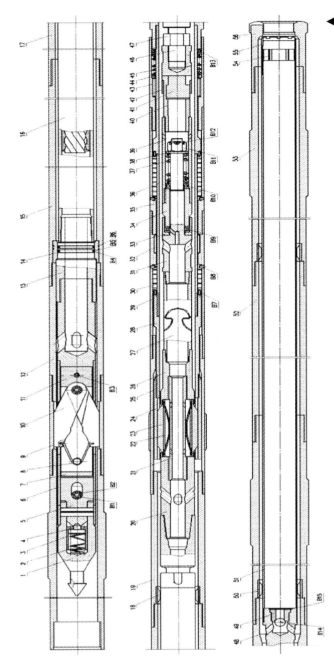

Fig. 5.6 PDM drive wireline core drilling tool. *1* Spear head. *2* Compressed spring. *3* Block sleeve. *4* Positioning block. *5* Spear head seat. *6* Recovery tube. *7* Spring clip support. *8* Tension spring. *9* Spring clip chamber. *10* Spring clip. *11* Spring clip seat. *12* Outer sub. *13* Motor adapter. *14* Suspended sealing sleeve. *15* Suspended upper outer tube. *16* C5LZ95 PDM. *17* Upper outer tube. *18* Centralizing ring for motor. *19* Lower outer tube. *20* Upper limit sub. *21* Mandrel of spring clip. *22* Spring of clip. *23* Key of spring clip. *24* Spline sleeve of spring clip. *25* Centralizing outer tube. *26* Lower limit sub. *27* Universal shaft. *28* Transmission adapter. *29* Upper bearing outer tube. *30* Compound bearing. *31* Connecting bar joint. *32* Needle bearing. *33* Bearing seat. *34* Bearing insert. *35* Lower bearing outer tube. *36* Bearing inner tube. *37* Spring. *38* Spacer. *39* Spring sleeve. *40* Upper separating adapter. *41* Lower separating adapter. *42* Stop ring. *43* Lower bearing outer sleeve. *44* Shim. *45* Buffer spring. *46* Adjusting adapter. *47* Adjusting nut. *48* Position changing sub. *49* Valve seat. *50* Centralizing ring. *51* Inner tube. *52* Outer tube. *53* Reaming shell. *54* Core catcher. *55* Core catcher case. *56* Core drill bit. *B1* Elastic pin. *B2* Elastic pin. *B3* Elastic pin. *B4* O ring. *B5* Stop ring. *B6* O ring. *B7* Pin. *B8* Screw. *B9* Bearing. *B10* Bearing. *B11* Slotted nut. *B12* Split pin. *B13* Bearing. *B14* Steel ball. *B15* Elastic stop ring

rotation driven from surface, and the shortcomings of thick bit curf and low drilling efficiency resulted from wireline coring + PDM drilling tool, a research on three-in-one drilling tool assembly of PDM + hydro-hammer + wireline coring was conducted under the basis of successful development of two types of two-in-one drilling tool assemblies. Once this three-in-one drilling tool assembly successfully developed wireline core drilling technique would have a vast range of prospects for application.

The working principle of three-in-one drilling tool (PDM + hydro-hammer + wireline coring) is that rotation of PDM is driven by drilling fluid and the torque transmitted from PDM is transferred to outer tube through a torque transmitting apparatus, then realize rotary drilling of core drill bit. Pressure is transported to core drill bit through a swivel type apparatus in the outer tube assembly. During rotary drilling, the percussive load produced by hydro-hammer is transmitted to drill bit through power transmitting plate, power transmitting cup, outer tube power transmitting joint and outer tube to realize percussive rock fragmentation. When core barrel is full of core, it (including hydro-hammer and PDM) can be fished by conventional wireline method.

PDM hydro-hammer drive wireline core drilling tool consisted of wireline core drilling tool, hydro-hammer, PDM and its connecting apparatus, drilling fluid devising apparatus, torque and reaction torque transmitting apparatus, outer tube swivel apparatus and position compensation apparatus (Fig. 5.9).

After the matching problem of flow rate between PDM and hydro-hammer had been solved and indoor adjustment of hydro-hammer finished, the three-in-one (PDM + hydro-hammer + wireline coring) drilling tool assembly was wholly adjusted and tested in the testing hole of the CCSD-1 Well in May of 2004, and a breakthrough progress was obtained. In this test a total drilling footage of 68.23 m was accomplished. The core fullness, core recovery, penetration rate and continuous working time of hydro-hammer in this testing hole all exceeded the requirements of design and contract. Under the basis of satisfactory technical results obtained in the testing hole, this drilling tool assembly was further tested in the main hole from the depth of 5125.86 to 5129.36 m in February, 2005. A total footage of 3.50 m was drilled (drilling tool was lifted due to core blockage), with an average penetration rate of 0.62 m/h. The last and the deepest core in CCSD Project was recovered. The initial

Table 5.8 Results of PDM drive wireline core drilling technique

Hole section	Footage drilled (m)	Actual drilling time (h)	Core length (m)	Penetration rate (m/h)	Core recovery (%)	Footage drilled per round trip (m)
2075.63–2076.13	0.5	1.76	0	0.28	0.0	
2076.13–2077.22	1.09	4.14	0.75	0.26	68.8	
2077.22–2078.04	0.82	2.19	0.8	0.37	97.6	
2078.04–2079.59	1.55	4.46	1.55	0.35	100.0	
2079.59–2080.01	0.42	1.59	0.37	0.26	88.1	
Run-out trip 1	4.38	14.14	3.47	0.31	79.2	4.38
2130.58–2131.53	0.95	2.29	0.33	0.42	34.7	
2131.53–2132.34	0.81	1.91	0.68	0.42	84.0	
Run-out trip 2	1.76	4.20	1.01	0.42	57.4	1.76
2283.28–2283.86	0.58	2.25	0.35	0.26	60.3	
Run-out trip 3	0.58	2.25	0.35	0.26	60.3	0.58
Total	6.72	20.59	4.83	0.33	71.9	2.24

success of the three-in-one (PDM + hydro-hammer + wireline coring) drilling tool assembly enables hard rock core drilling techniques in China to reach the world advanced level.

1. **Drilling tool assembly**

PDM hydro-hammer drive wireline core drilling consisted of 157 mm diamond core drill bit, PDM hydro-hammer wireline coring tool, SK146 drill collar stem and SK139.7 drill pipe string.

2. **Drilling parameters**

Bit pressure 20–30 kN, flow rate 6–8 L/s and PDM rotary speed 170–230 rpm.

3. **Commentary on drilling result**

PDM hydro-hammer drive wireline core drilling tool was tested in the testing hole and the main hole

in the testing hole

The test was conducted in the testing hole of the CCSD-1 Well from May 17th, 2004 to June 9th, with a drilling depth of 68.23 m, from 58.33 m deep to 130.61 m deep. This test was a successful one.

It can be seen from Fig. 5.10 that an average penetration rate of 1.2 m/h and the maximum of 2.5 m/h were achieved. From Fig. 5.11 we can get a result that the round trips which are no less than 4 m account for 54.89 m. Core recovery was as high as 98.7 %. In the test, a record of four round trips drilled per day with footage of 16.95 m was set (June 9th).

In this test 19 round trips were completed and the hydro-hammer accumulatively worked for 56.9 h, with more than 50 downhole starts. YZX98 hydro-hammer could normally start and stably worked after pump starting, except for the fourth round trip, due to blockage. It is shown that YZX98 hydro-hammer has very good working stability and starting property. In comparison with the penetration rate of 0.63 m/h in the fourth round trip during which the hydro-hammer did not work, the penetration rate of hydro-hammer was increased by 82.5 %.

in the main hole

10 run-out trips (22 wireline fishing trips) at the hole section from 5125.86 to 5129.33 m (MH-2C) were completed at the beginning of 2005, with total footage 71.73 m drilled (Table 5.10), average run-out trip length 7.17 m (average fishing trip length 3.26 m), average drilling speed 1.15 m and average core recovery 97.2 %.

This core drilling method can be used as a reserve technique for scientific drilling, for it successfully realized wireline core drilling without rotation of drill string, and thus being an ideal tool for large sized deep hole core drilling. It was indicated by test that this drilling tool was safe in connection of outer tube assembly, with good result in swivel movement, large load bearing capacity and high success rate of fishing inner tube.

5.2.11 Summary of the Tests for Core Drilling Methods

In CCSD-1 Well, ten core drilling methods were tested, with the results shown in Table 5.11, from which it can be seen that both PDM hydro-hammer drive double tube core drilling technique and PDM hydro-hammer drive wireline core drilling technique obtained satisfactory results. Since the delay of imported wireline core drilling tools, the reserve drilling method in the initial design had to be used for drilling the pilot hole of CCSD-1 Well. Through comparative tests for different core drilling methods, rotary table drive double tube core drilling (including hydro-hammer core drilling) and single tube core drilling (PDM drive, rotary table drive and top drive, etc.) were ruled out. After

5.2 Experiment on Core Drilling Methods for CCSD-1 Well

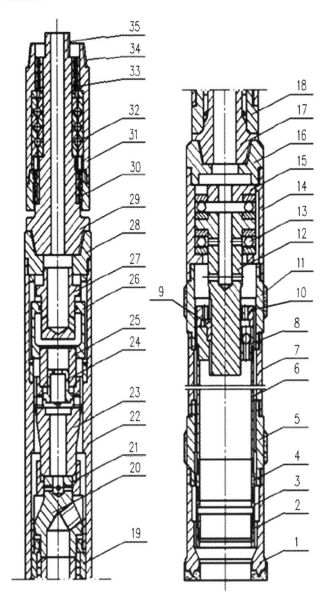

Fig. 5.7 PDM hydro-hammer drive double tube core drilling tool. *1* Core drill bit. *2* Core catcher. *3* Core catcher case. *4* Sub. *5* Lower reaming shell. *6* Core barrel. *7* Outer tube. *8* Core barrel joint. *9* Hook key. *10* Back cap. *11* Upper reaming shell. *12* Mandrel. *13* Bearing. *14* Bearing cavity. *15* Cover. *16* Upper adapter of coring tool. *17* Lower adapter with spline. *18* Spline sleeve. *19* Sealing sleeve. *20* Power transmitting case. *21* Sealing sleeve of power transmitting case. *22* Valve type percussive mechanism outer tube. *23* Hammer. *24* Valve. *25* Upper piston. *26* Cylinder sleeve. *27* Upper valve. *28* Valve type percussive mechanism upper adapter. *29* Down hole motor driving shaft. *30* Lower carbide bearing. *31* Down hole motor house. *32* Thrust ball bearing. *33* Upper carbide bearing. *34* Connecting mechanism of down hole motor house. *35* Thread connecting joint

the start of core drilling in the main hole, the imported wireline core drilling tools arrived. However, its application possibility in CCSD-1 Well was completely ruled out after a series of tests. PDM hydro-hammer drive swivel type double tube core drilling technique (down hole power percussive rotary core drilling technique) developed by the engineers at CCSD-1 Well drilling site in the light of core drilling requirement of CCSD-1 Well and on the basis of conventional core drilling techniques became the leading drilling method because of the satisfactory results achieved in the tests and gradual perfection in practice. As for PDM hydro-hammer drive wireline core drilling, initial test result showed that this brand-newly designed drilling tool and drilling technology were feasible in principle, with high technical indexes obtained, though large scale production test was not conducted due to the quality problem of the imported wireline core drill pipe.

PDM hydro-hammer down hole power percussive rotary drilling technique is a high and new technique developed independently by Chinese drilling engineers through unceasingly exploration, creation and perfection. This technique provided an important technical guarantee for the successful, fast and safe completion of CCSD-1 Well.

5.3 Down Hole Power Percussive Rotary Core Drilling System

PDM hydro-hammer drive swivel type double tube core drilling technique, a brand new core drilling method, was developed through experimental research and field application. Essentially, this technique system should be more exactly called as the down hole power percussive rotary core drilling system, in which, down hole power denotes PDM (rotary power) and hydro-hammer (percussive power) while percussive rotary represents bottom hole rock fragmentation mode, i.e., percussive fragmentation and rotary cutting (grinding). This down hole percussive rotary core drilling method was the leading technique for CCSD-1 Well core drilling.

5.3.1 Constituent of the System

Different core drilling methods are realized by corresponding drilling technique systems. Down hole power percussive rotary core drilling system consists of the following parts: PDM (rotary power), hydro-hammer (percussive power), swivel type double tube coring tool and diamond core drill bit. Only these four parts can not constitute a complete technique system because low solid mud system and drilling technology must be included. Among these six techniques, the first four parts are the key contents of down hole power percussive rotary core drilling technique and the latter two parts are the indispensable auxiliary techniques.

The constituent of down hole power percussive rotary core drilling technique system is shown in Fig. 5.12.

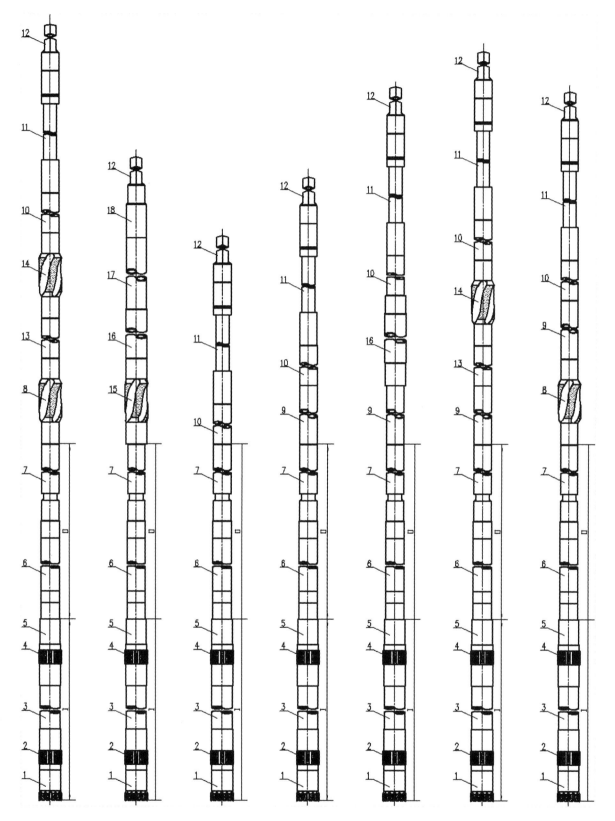

Fig. 5.8 PDM hydro-hammer drive double tube core drilling tool assembly. *1* 157 mm core drill bit. *2* 157 mm lower reaming shell. *3* 139.7 mm core barrel. *4* 157 mm upper reaming shell. *5* 146 mm swivel sub. *6* 127 mm hydro-hammer. *7* 95 or 120 mm PDM. *8* 156 mm lower stabilizer. *9* 120 mm jar. *10* 120 mm drill collar stem. *11* 89 mm drill pipe string. *12* Kelly bar. *13* 120 mm single drill collar. *14* 156 mm upper stabilizer. *15* SK157 stabilizer. *16* SK146 drill collar. *17* SK139.7 drill pipe. *18* Sub

Table 5.9 Results of PDM hydro-hammer drive double tube core drilling technique, with double stabilizers

	Round trip	Footage drilled (m)	Actual drilling time (h)	Core length (m)	Penetration rate (m/h)	Core recovery (%)	Footage drilled per round trip (m)
Valve type hydro-hammer	124	544.28	484.80	454.27	1.12	83.5	4.39
Fluidic type hydro-hammer	141	552.41	506.13	502.31	1.09	91.0	3.92
Total	265	1096.69	990.93	956.58	1.11	87.2	4.14

PDM hydro-hammer down hole power percussive rotary core drilling technique developed and perfected gradually under the basis of overcoming a series of difficulties. Drilling application in 2000 m pilot hole was a process of research and gradual perfection of this brand new technique while the application in main hole core drilling was a process of development.

According to the processes of test of core drilling method and perfection of core drilling system, the research and development of down hole power percussive rotary core drilling technique can be classified into following five steps.

Step of PDM + double tube core drilling (101.00–502.06 m): In this step the feasibility of PDM core drilling was mainly explored.

Step of initial test for PDM + hydro-hammer system (502.06–1000.05 m): Under the perfection of swivel type double tube coring tool, the feasibility of application of hydro-hammer in PDM core drilling was mainly verified.

Step of improvement of PDM + hydro-hammer system (1000.05–1503.00 m): In this step the properties of hydro-hammer was improved.

Step of perfection of PDM + hydro-hammer system (1503.00–2046.54 m): Swivel type double tube coring tool and hydro-hammer were further improved.

Step of well developed PDM + hydro-hammer system (2046.54–2982.18 m): In this step core drilling technology of PDM + hydro-hammer was perfected, drilling data were optimized and drilling mud properties improved.

5.3.2 Technical Data of the System

1. Calculation of power consumption and torque of down hole drilling tool

Power consumption and torque of down hole drilling tool during drilling operation are the basic data for choosing down hole power drilling tool. As we know, a variety of factors affect the power consumption of down hole drilling tool and the torque required. Therefore, power consumption and torque can only be estimated in theory while in practical application they should be finally determined according to down hole working conditions and tool parameters provided by manufacturer.

Bit pressure for impregnated diamond drill bit:

$$P = p \cdot A \quad (5.1)$$

in which,

P bit pressure required for impregnated diamond drill bit to effectively crush rock during pure rotary drilling, N

p recommended unit pressure (N/mm^2), the maximum value of the recommended unit pressure of impregnated diamond drill bit to effectively cut into rock in hard formation drilling is σ = 9 Mpa, p = (7–9) N/mm^2

A connecting area of bit working crown and rock, mm^2

For CCSD-1 Well drilling, outside diameter of core drill bit is D = 157 mm, inside diameter is d = 96 mm, bit crown has twelve water slots, each with 12 mm width, thus

$$A = \frac{\pi}{4}(D^2 - d^2) - 36\sqrt{D^2 - 144} + 36\sqrt{d^2 - 144}$$
$$- \frac{\pi D^2}{60}\arcsin\frac{12}{D} + \frac{\pi d^2}{60}\arcsin\frac{12}{d} = 7722 \text{ mm}^2$$

introduced into formula (5.1), then bit pressure is P = 54.1–69.5 kN.

Estimated power consumption N required to crush rock at hole bottom is

$$N = 6 \times 10^{-5} \times f P n (R + r) \quad (5.2)$$

in which,

N power consumption required for down hole drilling tool to crush rock, kW

f resistance efficient of drill bit movement along hole bottom, f = 0.25–0.3

P bit pressure, kN. P = 54.1–69.5 kN

n rotary speed of drill bit, rpm. n = 182–365 rpm

R outside radius of drill bit, mm

r inside radius of drill bit, mm

In consideration of hard rock deep hole core drilling, maximum values are adopted. The linear velocity 2 m/s of drill bit is adopted thus n is approximately 243 rpm.

introduced into formula (5.2), then N = 18.7–57.8 kW and torque is: $T = \frac{9736 N}{n}$ = 999–1541 Nm

Considering that centrifugal action produced at high speed rotation of long drilling tool will result in a large

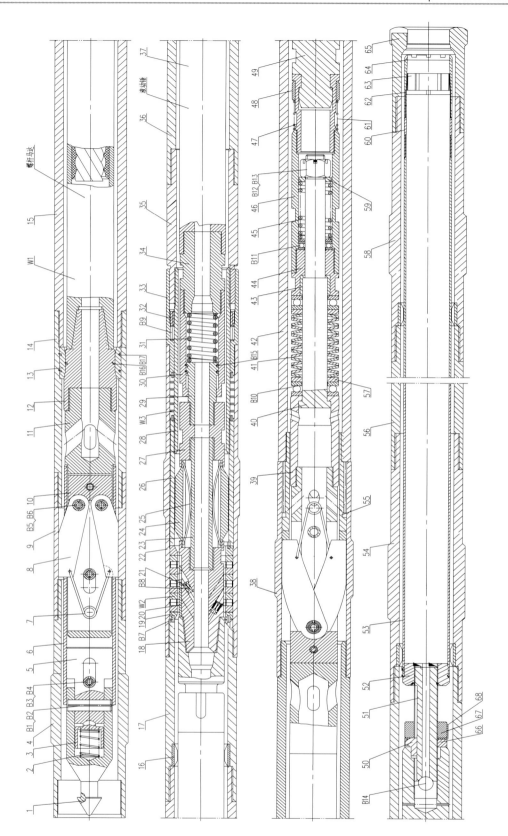

◀ **Fig. 5.9** PDM hydro-hammer drive wireline core drilling tool. *1* Spear head. *2* Spring. *3* Positioning block. *4* Stop of spring clip. *5* Spear head seat. *6* Recovery tube. *7* Spring. *8* Spring clip. *9* Spring clip chamber. *10* Spring clip seat. *11* Spring clip support. *12* Suspension adapter. *13* Suspension ring. *14* Outer tube joint. *15* Motor outer tube. *16* Centralizing ring. *17* Stabilizer adapter. *18* Diverting joint. *19* Guide ring of bearing. *20* Nozzle spacing sleeve. *21* Alloy nozzle. *22* Bearing outer sleeve. *23* Torsion transmitting bar. *24* Flexible key. *25* Flat spring. *26* Torsion transmitting sleeve. *27* Flexible key joint. *28* Central tube. *29* Compensation bar. *30* Compensation sleeve. *31* Compensation spring. *32* Bearing cover. *33* Swivel sub. *34* Hydro-hammer conversion joint. *35* Torsion transmitting joint. *36* Outer tube of hydro-hammer. *37* Hydro-hammer. *38* Outer spline sleeve. *39* Alarming joint. *40* Valve spindle. *41* Butterfly spring. *42* Short outer tube. *43* Copper sleeve. *44* Sliding joint. *45* Spring. *46* Spring sleeve. *47* Lifting joint (box). *48* Protection ring. *49* Lifting joint (pin). *50* Key. *51* Screw bar. *52* Inner tube joint. *53* Inner tube. *54* Upper reaming shell. *55* Male spline sleeve. *56* Outer core barrel. *57* Washer. *58* Lower reaming shell. *59* Stop ring. *60* Inner tube sub. *61* Spring. *62* Stop ring of core catcher. *63* Core catcher. *64* Core catcher case. *65* Core drill bit. *66* Locking washer. *67* Adjusting nut. *68* Spacer. *B1* Elastic pin. *B2* Elastic pin. *B3* Elastic pin. *B4* Elastic pin. *B5* Elastic pin. *B6* Elastic pin. *B7* Pin. *B8* O sealing ring. *B9* Needle bearing. *B10* Bearing. *B11* Bearing. *B12* Locking nut. *B13* Pin. *B14* Steel ball. *B15* O ring. *B16* O ring. *B17* O ring. *W1* PDM. *W2* Bearing. *W3* Bearing

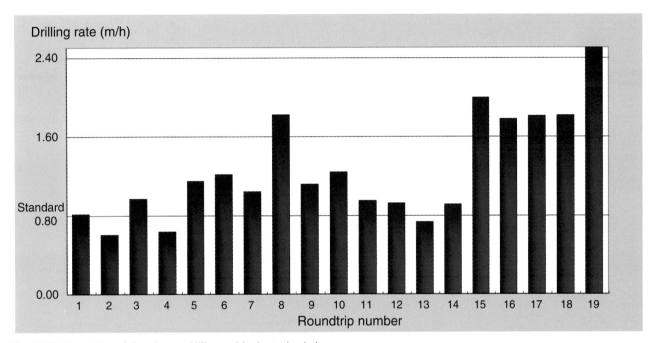

Fig. 5.10 Round trips of three-in-one drilling tool in the testing hole

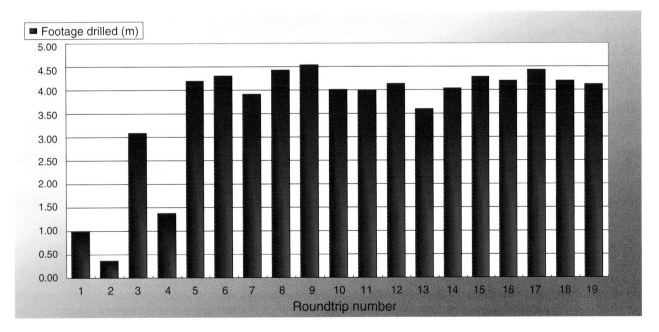

Fig. 5.11 Round trip length of three-in-one drilling tool in the testing hole

Table 5.10 Results of PDM hydro-hammer drive wireline core drilling technique

Hole section	Footage drilled (m)	Actual drilling time (h)	Core length (m)	Penetration rate (m/h)	Core recovery (%)	Footage drilled per round trip (m)
58.33–59.33	1.00	1.23	0.20	0.81	20.0	
Run-out trip 1	1.00	1.23	0.20	0.81	20.0	1.00
59.33–59.69	0.36	0.60	0.45	0.60	125.0	
59.69–62.77	3.08	3.18	2.37	0.97	77.0	
Run-out trip 2	3.44	3.78	2.82	0.91	82.0	3.44
62.77–64.15	1.38	2.18	2.03	0.63	147.0	
64.15–68.35	4.20	3.67	4.41	1.15	105.0	
68.35–72.65	4.30	3.55	4.09	1.21	95.0	
Run-out trip 3	9.88	9.40	10.52	1.05	107.0	9.88
72.65–76.57	3.92	3.77	3.96	1.04	101.0	
Run-out trip 4	3.92	3.77	3.96	1.04	101.0	3.92
76.57–81.00	4.43	2.43	4.21	1.82	95.0	
Run-out trip 5	4.43	2.43	4.21	1.82	95.0	4.43
85.06–89.60	4.54	4.08	4.27	1.11	94.0	
89.60–93.61	4.01	3.23	4.33	1.24	108.0	
93.61–97.61	4.00	4.23	3.88	0.94	97.0	
97.61–101.74	4.13	4.47	4.13	0.92	100.0	
101.74–105.34	3.60	4.92	3.60	0.73	100.0	
Run-out trip 6	20.28	20.93	20.21	0.97	100.0	20.28
105.34–109.39	4.05	4.43	4.17	0.91	103.0	
109.39–113.67	4.28	2.15	4.28	1.99	100.0	
113.67–117.87	4.20	2.37	4.20	1.77	100.0	
117.87–122.30	4.43	2.45	4.43	1.81	100.0	
122.30–126.50	4.20	2.32	4.20	1.47	100.0	
126.50–130.62	4.12	1.65	4.12	2.50	100.0	
Run-out trip 7	25.28	15.37	25.40	1.64	100.0	25.28
Subtotal	68.23	56.91	67.32	1.20	98.7	9.75
5125.86–5126.40	0.54	0.99	0.00	0.55	0.0	
Run-out trip 8	0.54	0.99	0.00	0.55	0.0	0.54
5126.40–5127.09	0.69	1.15	0.00	0.60	0.0	
Run-out trip 9	0.69	1.15	0.00	0.60	0.0	0.69
5127.09–5129.36	2.27	3.52	2.40	0.64	105.7	
Run-out trip 10	2.27	3.52	2.40	0.64	105.7	2.27
Subtotal	3.50	5.66	2.40	0.62	68.6	1.17
Total	71.73	62.57	69.72	1.15	97.2	7.17

friction between reaming shell and hole wall and a frictional drag from the cuttings, drilling torque and power should have a certain reserve. A safety factor of 1.4 is adopted, then $N = 26$–81 kW and $T = 1399$–2157 N m.

2. **Calculation of working flow rate of down hole drilling tool**

Determination of flow rate of drilling fluid mainly depends on following factors: the minimum flow rate must be enough to carry the cuttings from hole bottom and cool drill bit; adequate rotary speed and output power must be supplied for PDM; and flow rate should not be too high, otherwise erosion to hole wall and bit crown will be serious and whole circulation pressure drop will be too much.

For impregnated diamond drill bit, estimation of flow rate (Liu 1991) is as follows:

$$Q = 6 \cdot V \cdot A \qquad (5.3)$$

Table 5.11 Data statistics of different experimental drilling techniques

Hole section	Footage drilled (m)	Actual drilling time (h)	Core length (m)	Penetration rate (m/h)	Core recovery (%)	Footage drilled per round trip (m)
Rotary table drive double tube core drilling	11.60	29.61	6.40	0.39	55.2	1.66
Rotary table hydro-hammer drive double tube core drilling	6.39	18.08	4.95	0.35	77.5	1.28
Top drive double tube core drilling	5.87	16.24	0.30	0.36	5.1	0.73
Top drive wireline core drilling	7.62	12.02	1.05	0.63	13.8	1.52
PDM drive wireline core drilling	6.72	20.59	4.83	0.33	71.9	2.24
PDM drive double tube core drilling	907.15	1270.47	804.21	0.71	88.7	2.29
PDM drive single tube core drilling	6.84	12.00	2.22	0.57	32.5	1.37
Top drive hydro-hammer wireline core drilling	8.27	9.25	8.23	0.89	99.5	2.76
PDM hydro-hammer drive double tube core drilling	1096.69	990.93	956.58	1.11	87.2	4.14
PDM hydro-hammer drive wireline core drilling	71.73	62.57	69.72	1.15	97.2	7.17 (run-out)

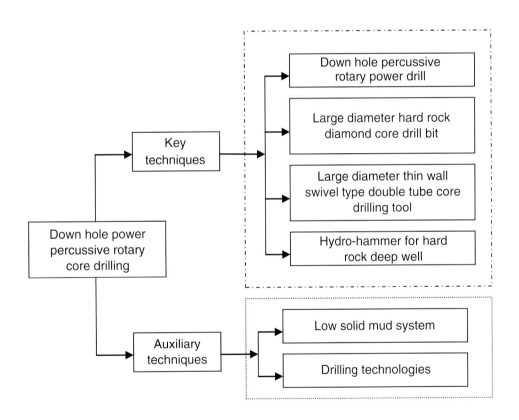

Fig. 5.12 Down hole power percussive rotary core drilling technique system

where

- Q flow rate of drilling fluid, L/min
- V uplifting velocity in drill bit annulus, m/s. For impregnated diamond drill bit, $V = 0.4$–1.0 m/s
- A annular cross section of drill bit, cm^2

$$Q = 6 \cdot V \cdot \pi \cdot \frac{D^2 - d^2}{400} \quad (5.4)$$

$Q_{\min} = 291$ L/min $= 4.85$ L/s; $Q_{\max} = 727$ L/min $= 12.12$ L/s

The calculated values of flow rate basically coincide with the input flow rate of 4LZ120 × 7Y PDM. It was indicated that the mud flow rate of PDM core drilling at CCSD-1 Well should be maintained at 7–12 L/s.

3. **Parameters matching for down hole percussive rotary power drilling tool**

Parameters matching for down hole power drive percussive rotary core drilling tool is mainly the matching between hydro-hammer and PDM, both of which belong to down hole motor, but each has its own working parameters when they work separately. Once the two motors work connectively their working parameters are both interrelated and restricted each other. Moreover, the design of swivel type double tube core drilling tool and diamond drill bit should be improved accordingly. So in the research on down hole power drive percussive rotary drilling tool assembly it should be comprehensively considered.

1. Property parameters

The type of PDM was selected and its property parameters determined according to torque and rotary speed required for diamond core drilling. Percussive work is the most important factor to determine rock crushing result. In hard rock deep hole core drilling, in order to prolong the round trip length, long core drilling tool was adopted. Thus the property parameters for hydro-hammer were as follows:

Percussive work: 60–200 J; percussive frequency: 10–25 Hz

2. Hydraulic power parameters

Both hydro-hammer and PDM work under a certain pump rate and pump pressure drop. When the same mud pump was used to provide power, distribution of hydraulic power parameters directly influence the working property of this compound drilling tool.

PDM is a volumetric motor. The much the flow rate through the motor, the higher the rotary speed will be. The output rotary speed n from PDM is Su (2001):

$$n = \frac{Q}{e^2 T} n_0 \quad (5.5)$$

in which,

Q flow rate, L/s
e eccentric distance of rotor, cm
T lead of stator,
n_0 unit rotary speed of motor shaft

n_0 can be calculated by following equation, under the condition that lost circulation is ignored.

$$n_0 = \frac{1}{[2\pi(Z_2 - 1) + \frac{8}{C_e}] Z_2} \quad (5.6)$$

in which,

Z_2 corrugated teeth number of rotor
C_e dimensionless parameter of the ratio between eccentric distance and radius of spiral tooth

The output torque of PDM is in proportion to the interior working pressure drop of motor, with the following expression.

$$M = M_o \cdot \Delta P \cdot D_p \cdot l \cdot e \quad (5.7)$$

where

ΔP working pressure drop of motor, Pa
D_p tool designed diameter, cm
l lead of the spiral surface of rotor
M_o mechanical unit moment of rotor, M_o can be derived from the following equation:

$$M_o = \frac{Z_2 - 1}{2} + \frac{2C_e}{\pi} \quad (5.8)$$

Rotary speed of PDM increases along with the increase of flow rate, its torque increases with the increase of resisting moment of drilling tool, and at the same time the pressure drop of motor increases accordingly. In case of overload resisting moment PDM will stop and then sealing will be broken.

Percussive work and percussive frequency of hydro-hammer increase with the increase of pumping rate, and working pressure drop increases accordingly. Single percussive work A of hydro-hammer can be expressed as

$$A = K \cdot P_o \cdot F_l \cdot S_o \quad (5.9)$$

In which,

K hydraulic effective coefficient 0.6–0.8
P_o input pressure of water pump, Pa
F_l sectional area of piston, cm^2
S_o percussion stroke, cm

Percussive frequency n (Hz) is expressed as

$$n = \frac{K}{1+R} \sqrt{\frac{p_o F_l}{m S_o}} \quad (5.10)$$

where

R time ratio between piston back stroke and percussion stroke
m mass of piston hammer, kg

Due to the large working flow rate of PDM, its percussive work and percussive frequency would be too high in case that conventional design was adopted. Therefore, drilling mud diversion should be considered during the design of hydro-hammer. The input pumping rate was 420–1140 L/min for 4LZ120 × 7Y PDM while the pumping rate was about 300–400 L/min for hydro-hammer with 127 mm outside diameter.

3. Water power

Water power consumption was the principal question to be considered during the design of PDM-hydro-hammer down hole power drive percussive rotary drilling tool.

An input pumping rate of 420–1140 L/min was needed for 4LZ120 × 7Y PDM, and its working pressure drop was less than 4 MPa. A pumping rate of 300–400 L/min was needed for hydro-hammer with 127 mm outside diameter, and its working pressure drop was less than 3.5 MPa. The total working pressure drop would be less than 7.5 MPa in case of adding up those of PDM and hydro-hammer.

The circulation pressure drop of inner hole of drill stem and annular space, and the pressure drop of 9.5 m long swivel type double tube core drilling tool (including drill bit) are related to drilling mud properties and flow rate (Liu 1991).

$$P_{fa} = \frac{0.2 f \rho L V^2}{D_i} \quad (5.11)$$

$$P_{fp} = \frac{0.2 f \rho L V^2}{D_w - D_e} \quad (5.12)$$

in which,

- P_{fa} flowing friction pressure drop in tube, MPa
- P_{fp} flowing friction pressure drop in annular space, MPa
- f Fanning friction factor, related to Reynolds number
- V average flow rate, m/s
- ρ fluid density, g/cm^3
- L pipeline length, m
- D_i tube inner diameter, cm
- D_w annular space outside diameter, cm
- D_e annular space inner diameter, cm

From formulas (5.11) and (5.12) we can see that the flowing friction pressure drop of inner hole of drill stem and annular space increases linearly with the length of pipeline, and increases along with the square of flow rate. By adopting low solid, low density, low viscosity and good lubricant drilling mud and moderate mud pumping rate, the total pressure drop of the circulation pressure drop of inner hole and annular space of 5000 m drill stem and the pressure drop of 9.5 m long swivel type double tube core drilling tool is no larger than 8 MPa.

In CCSD-1 Well drilling operation, down hole power drive percussive rotary core drilling was used to the depth of 5000 m, mud pumping rate was about 10 L/s. During normal drilling pressure drop in the tube was about 5.0 MPa, pressure drop in annular space was approximately 0.5 MPa, pressure drop of double tube drilling tool was about 2.0 MPa, pressure drop of PDM was nearly 3.5 MPa and that of hydro-hammer was around 2.5 MPa. The total pressure drop was about 13.5 MPa, not exceeding 15 MPa.

5.3.3 Down Hole Rotary Drive Drilling Tool—PDM

1. **Selection of down hole rotary drive drilling tool**

Down hole drive drilling tool includes electric drilling tool, turbine drilling tool, PDM and vane drilling tool. Currently the most popularized down hole rotary drives in drilling industry are mainly turbine motor and PDM, also called as turbine drilling tool and PDM drilling tool.

Turbine drilling tool, a down hole power drilling tool with drilling fluid as the power, is a power motor. Drilling fluid pumped out from mud pump rotates the transmission shaft through the vanes of turbine stator and rotor, and transmits rotary speed and torque to drill bit, thus to crush rock.

With the characteristics of high working temperature, high rotary speed and low torque output per unit length (each pair of rotor and stator), conventional turbine drilling tool is of low efficiency in general. In order to obtain a high torque output and a large power to meet the requirement of drilling, turbine rotor and stator must be increased, i.e., the length of turbine must be increased. However, the length increase of turbine drilling tool also increases the pressure drop of the turbine, being unfavourable to deep hole drilling. In case that a given flow rate of drilling fluid is pumped into the turbine, the torque on drill bit is in inverse proportion to the rotary speed of turbine shaft, the more load on drill bit, the lower the rotary speed will be, and vice versa. Not only the torque changes along with the variation of rotary speed, other working indexes such as efficiency η and power N will change. Under the condition of braking, rotary speed is zero, and its efficiency is also zero. Different from torque, efficiency and power, when rotary speed changes, the pressure drop on turbine drilling tool basically does not change, normally within little variation both under the conditions of braking and idling.

PDM, a down hole power engine with drilling fluid as the power, is a volumetric motor. Drilling fluid comes out from mud pump enters into PDM through by-pass valve, produces a certain pressure difference at inlet and outlet of the motor, pushes motor rotor rotate, and transmits rotary speed and torque to drill bit through universal shaft and transmission shaft, to crush rock.

Different from turbine motor, PDM is characterized by its large torque and low rotary speed. Its rotary speed is only related to flow rate and structure, and has nothing to do with working conditions (such as bit pressure and torque). The rotary speed of PDM is in direct proportion to the input drilling fluid flow rate. Its output torque is in direct proportion to the pressure drop of motor, and has nothing to do with rotary speed. The more the pressure drop is, the larger the torque will be. However, during the variation of both pressure drop and torque, rotary speed changes very less. Therefore, during actual application output torque and rotary speed of motor can be basically controlled provided the flow rate and the pump pressure of mud pump are controlled. Obviously, in PDM drilling operations pump pressure gauge can be used to monitor down hole working situations, to judge and display the down hole working situations by observing pressure variation. Being made of rubber, the stator of PDM can not resist high temperature. Generally they are made of butadiene-acrylonitrile rubber, with high temperature resistance of approximate 150 °C. In case of higher temperature resistance, special rubber should be used, with rather high cost.

As the target rock formation of CCSD-1 Well is a variety of ultra-high pressure metamorphic rocks, diamond drilling techniques must be utilized, for which, especially for impregnated diamond drill bit, rotary speed needed to guarantee rock cutting is a necessity for a satisfactory drilling efficiency. For 157 mm diameter impregnated diamond drill bit, its reasonable rotation speed should be in a range from 282 to 365 rpm, whereas the rotary speed of turbine drilling tool is much higher than this range. Therefore a deceleration mechanism should be used to meet the requirement of diamond drilling. But due to the space limitation it is very difficult and unreliable to install a deceleration mechanism in small diameter turbine drilling tool. Meanwhile, in deep hole drilling the working load on motor can not be precisely controlled and thus resulting in a great variation of rotary speed, which is unfavourable to the requirement of constant rotary speed that diamond drilling technique requires. Furthermore, due to the requirement for motor working load, turbine motor stage must be increased, and in case of adding with deceleration mechanism, the turbine drilling tool will be too long to be used in drilling field applications. For PDM, its motor rotary speed can be changed by changing the number of motor stator and rotor and its working load can be changed by increasing or decreasing the motor stage. Once the basic parameters determined, the working characteristics of the motor are favourable to smooth working of diamond core drilling system. And the down hole working situation can be judged by observing the variation of pump pressure gauge at surface, being favourable to optimizing drilling parameters.

To sum up, PDM was chosen as the down hole drive for diamond core drilling system for CCSD-1 Well.

2. **Optimization of PDM**

According to the formulas (5.1), (5.2) and (5.4), the flow rate PDM needs is in the range from 5.5 to 12.12 L/s, rated working power is approximately 26–81 kW and torque from 1399 to 2157 N m

PDM model was determined according to the flow rate, working power and torque calculated and by reference to the properties of available PDMs with diameter of 157 mm and smaller in Chinese market. There were some manufacturers including Zhongcheng Machinery Manufacturing Co, which is affiliated to Dagang Oil Field Co. Group, Beijing Petroleum Machinery Works, Gaofeng Petroleum Machinery Co. in Guizhou Province, and Tianjin Lilin Petroleum Machinery Co. to produce PDM. The 4LZ120 × 7Y PDM produced by Zhongcheng Co. and the 4LZ120 × 7-VI PDM produced by Lilin Co. are suitable to down hole power percussive rotary core drilling for CCSD-1 Well. In field drilling applications, the C5LZ95 × 7 PDM from Beijing Petroleum Machinery Works and the 4LZ120 × 7Y PDM of Zhongcheng Co. were mainly utilized in the pilot hole of CCSD-1 Well, while the 4LZ120 × 7Y PDM produced by Zhongcheng and the 4LZ120 × 7-VI PDM produced by Lilin Co.were used in the main hole of CCSD-1 Well (Table 5.12).

The application results of PDMs are shown in Tables 5.13, 5.14 and in Fig. 5.13.

5.3.4 Down Hole Percussive Drilling Tool—Hydro-hammer

Though the application of PDM solved the problem of inadequate rotary speed of rotary table which could not meet the requirement of impregnated diamond drill bit to cut rock, it was difficult to solve another problem that large rock fragmentation power needed to cut hard rock. To solve this difficulty, it was decided that percussive rotary drilling techniques successfully used in geological exploratory drilling should be utilized.

Percussive rotary drilling, under the basis of rotary drilling, exerts percussive energy of a certain frequency onto rock crushing tool. The rotating drill bit not only exerts static pressure and torque, but also adds a continuous percussive dynamic load on the rock. Because the high speed and strong percussive action is exerted in very short time, high stress and high strain concentration on the local area of the rock bears this percussion can not be transmitted to the surrounding rock in time and under these circumstances plastic deformation of the rock can not be formed. This is favourable for large volumetric fragmentation of the rock, resulted

Table 5.12 PDM property parameters

Manufacturer	Beijing works		Zhongcheng		Lilin
Motor model	C5LZ95 × 7	LZ120 × 7	4LZ120 × 7Y	4LZ95 × 7Y	4LZ120 × 7-VI
Input flow rate (L/s)	5–13.33	6.33–15	7–19	4–11	8.25–16.5
Output rotary speed (rpm)	140–380	245–600	128–348	116–320	128–256
Motor pressure drop (Mpa)	6.5	4.0	4.0	3.2	6.78
Recommended bit pressure (kN)	40	35	50	30	49
Max bit pressure (kN)	80	70	80	55	100
Workins torque (N m)	1490	790	2086	1075	2137
Max torque (N m)	2235	1580	3340	1720	3019
Output power (kW)	21.8–59.3	20–50	76	36	81

from the expansion of cracks in the rock. From this, hard rock fragmentation by milling, or fine grain volumetric fragmentation becomes into large volumetric fragmentation. At the same time, as hydro-hammer exerts percussive load continuously on the rock, fatigue breaking and strength failure of the rock will emerge because of rock molecular oscillation. This is favourable for rock crushing and shearing. Therefore, adoption of percussive rotary drilling techniques in the rock with hardness higher than grade VI–VII will greatly improve penetration rate, which is the main advantage of percussive rotary drilling techniques. It is known that the rock with higher brittleness has lower percussive resistance, and thus percussive rotary drilling technique makes full use of the characteristics of higher brittleness and lower shearing strength of the hard rock (Wang et al. 1988).

The tentative idea of utilizing percussive rotary drilling techniques came from Europe. During 1867–1887, some down-hole type hydro-percussive mechanisms came out successively. In 1887, a patent of new drilling technique invented by W. Bushman, a German, was granted in U.K. The key technique of this patent was the percussive rotary drilling resulted from continuous percussion on rotating drill bit, by a hydro-percussive mechanism, which was driven by a pump.

From the 1950s, some practical hydro-percussive mechanisms were developed in the United States, Canada and the former Soviet Union, with the main purposes of oil drilling and solving drilling tool sticking. Thus the diameter was large and the hammer was as weight as 300 kg, with low percussive frequency. In geological exploration drilling industry, the most effective result was achieved in the former Soviet Union, where the research on hydro-percussive rotary drilling was started during 1900–1905. However, its application in drilling activities was not realized until 1970. In the 1960s, five types of double acting hydro-percussive mechanisms with the diameters ranging from 48 to 160 mm were developed in Hungary. The hydro-percussive mechanism was mounted on the special trailer, equipped with related pumps, desander, coring tools, drill bits and accident treatment tools. This trailer was very active and could be used to serve several drilling rigs at drill site, not only for geological exploration drilling, but also for water well and engineering drilling activities. In Japan the research on hydro-percussive drilling was started in the 1970s. The most successful example was WH-120 N double acting hydro-percussive mechanism developed by Tone Company, with air liquid mixed actuating medium as its most obvious characteristic.

In China, the research on hydro-percussive rotary drilling techniques was conducted in 1958. Till 1965 seven hydro-percussive mechanisms of different structures were developed and then comparative tests and rock sample drilling tests were carried out in laboratory. In 1971 SC-89 and JSC-75 fluidic type hydro-percussive mechanisms were developed by the No. 9 Geological Brigade of Liaoning Provincial Bureau of Geology and Mineral Resources and the former Changchun College of Geology. These were the first ones that were widely used in the country. After 1975, besides geological industry, hydro-percussive rotary drilling technique was also developed in other industries, such as oil and gas drilling, with more and more successful results. Hydro-percussive rotary drilling technique has experienced its steady development and its key technique is becoming a main branch of down hole power drilling.

From the 1970s to the 1980s, the research on hydro-hammer entered its prosperity period. Varieties of hydro-hammers were developed in geological and metallurgical industries for small diameter core drilling, with an accumulative footage over one million meters drilled. These hydro-hammers included positive acting, double acting, negative acting and compound types. According to a rough estimation, with hydro-hammer drilling, drilling efficiency could be increased by 30–50 %, borehole quality and core recovery obviously improved, footage drilled per round trip prolonged and material consumption decreased.

Table 5.13 PDM application in core drilling at different hole depths[a]

	Manufacturer	PDM model	Hole section (m)	Roundtrip	Footage drilled (m)	Actual drilling time (h)	ROP (m/h)	Core recovery (%)
1	Beijing	LZ120	101.00–470.70	92	172.19	219.21	0.79	89.4
2	Beijing	LZ95	128.27–302.10	55	113.21	115.83	0.98	97.9
3	Beijing	LZ95-2	302.10–386.01	45	83.91	105.09	0.80	97.0
4	Beijing	LZ95-3	470.70–610.81	58	140.11	127.66	1.10	95.2
5	Beijing	LZ95-4X	610.81–685.13	30	74.32	60.72	1.22	98.3
6	Beijing	LZ95-5	685.13–726.22	15	41.09	42.61	0.96	100.7
7	Beijing	LZ95-6	726.22–805.01	25	62.16	79.97	0.78	95.2
8	Beijing	LZ95-7X	746.90–832.16	12	21.93	68.73	0.32	100.5
9	Beijing	LZ95-8X	805.01–826.86	10	21.85	59.26	0.37	94.8
10	Beijing	LZ120-1	832.16–884.35	12	37.84	40.93	0.92	97.0
11	Beijing	LZ95-9X	848.78–863.13	5	14.35	22.10	0.65	100.7
12	Beijing	LZ120-2	884.35–886.14	1	1.79	6.96	0.26	81.0
13	Beijing	LZ95-10X	886.14–966.34	25	79.28	85.69	0.93	99.3
14	Beijing	LZ95-2X	957.82–958.74	2	0.92	2.60	0.35	117.4
15	Beijing	LZ95-11X	966.34–1019.50	14	53.16	39.11	1.36	96.4
16	Beijing	3LZ120-1X	1019.50–1765.65	5	7.57	13.12	0.58	98.3
17	Beijing	3LZ120-3X	1021.99–1763.31	3	7.10	5.62	1.26	106.9
18	Beijing	5LZ95-12X	1022.94–1068.78	17	45.84	53.49	0.86	91.3
19	Beijing	5LZ95-13X	1068.78–1119.22	10	27.92	34.00	0.82	105.4
20	Beijing	3LZ120-2X	1078.20–1246.88	14	62.91	62.68	1.00	80.0
21	Beijing	5LZ95-14X	1119.22–1135.49	7	16.27	28.20	0.58	89.3
22	Beijing	5LZ95-15X	1135.49–1156.19	8	20.70	34.91	0.59	95.2
23	Beijing	5LZ95-16X	1164.71–1300.30	21	91.37	73.40	1.24	97.1
24	Beijing	5LZ95-17	1300.30–1548.16	30	108.80	102.83	1.06	59.0
25	Zhongcheng	4LZ120-2	1367.02–1526.96	33	138.06	122.00	1.13	66.2
26	Beijing	5LZ95-18	1548.16–1599.23	11	51.07	45.55	1.12	55.9
27	Zhongcheng	4LZ120-1	1599.23–1621.70	10	22.47	39.69	0.57	86.1
28	Zhongcheng	4LZ120-2X	1621.70–1632.79	3	11.09	13.95	0.79	103.3
29	Zhongcheng	4LZ95-1	1634.38–1641.62	2	7.24	10.66	0.68	96.3
30	Zhongcheng	4LZ95-2	1643.20–1675.36	5	32.16	25.80	1.25	79.2
31	Zhongcheng	4LZ120-3	1675.36–1717.35	10	49.02	44.26	1.11	86.8
32	Beijing	5LZ95-3X	1724.38–1775.86	7	29.56	28.59	1.03	77.2
33	Beijing	5LZ95-17X	1745.90–1755.28	3	9.38	9.92	0.95	94.2
34	Zhongcheng	4LZ120-1X	1755.28–1758.25	2	2.97	4.15	0.72	80.8
35	Zhongcheng	4LZ120-4	1775.86–1895.97	23	120.11	117.38	1.02	87.9
36	Zhongcheng	4LZ120-5	1895.97–2046.54	20	147.83	107.95	1.37	98.7
37	Beijing	C5LZ95-1	2055.40–2064.57	4	9.17	19.64	0.47	45.9
38	Beijing	C5LZ95-2	2064.57–2084.59	5	12.52	28.58	0.44	48.3
39	Beijing	C5LZ95-3	2084.59–2091.25	2	6.66	10.51	0.63	58.9
40	Zhongcheng	4LZ120-2	2091.25–2234.55	19	100.16	116.97	0.86	53.3
41	Zhongcheng	4LZ120-1	2108.88–2260.59	10	67.42	85.68	0.79	65.7

(continued)

Table 5.13 (continued)

	Manufacturer	PDM model	Hole section (m)	Roundtrip	Footage drilled (m)	Actual drilling time (h)	ROP (m/h)	Core recovery (%)
42	Beijing	C5LZ95-1A	2260.59–2276.33	2	15.74	9.23	1.71	70.5
43	Beijing	C5LZ95-2A	2276.33–2366.64	12	89.73	67.54	1.33	67.2
44	Beijing	C5LZ95-3A	2366.64–2421.34	8	54.70	46.26	1.18	64.8
45	Zhongcheng	4LZ120*7Y453	2421.34–2672.39	7	46.92	44.83	1.05	75.5
46	Zhongcheng	4LZ120*7Y454	2468.21–2511.56	5	43.35	35.98	1.20	67.7
47	Zhongcheng	4LZ120*7Y455	2511.56–2661.74	19	149.51	93.46	1.60	92.5
48	Zhongcheng	4LZ120*7Y481	2672.39–3094.45	30[b]	252.35	165.46	1.53	89.0
49	Zhongcheng	4LZ120*7Y491	2796.29–2945.99	17	149.70	119.09	1.26	94.4
50	Zhongcheng	4LZ120*7Y489	2945.99–2982.18	5[c]	37.95	40.69	0.93	89.2
51	Zhongcheng	4LZ120*7Y492	3094.45–3235.74	14[d]	109.44	88.11	1.24	57.7
52	Zhongcheng	4LZ120*7Y490	3235.74–3345.74	11[e]	101.65	90.48	1.12	89.7
53	Zhongcheng	4LZ120*7Y549	3345.74–3464.51	15	118.77	86.74	1.37	83.7
54	Zhongcheng	4LZ120*7Y542	3464.51–3516.47	9	51.31	45.41	1.13	66.0
55	Zhongcheng	4LZ120*7Y540	3516.96–3647.09	13	115.04	90.40	1.27	96.5
56	Zhongcheng	4LZ120*7Y609	3647.09–3665.87	3	18.78	18.37	1.02	94.8
57	Zhongcheng	4LZ120*7Y630	3624.16–3658.32	5	34.16	42.34	0.81	99.3
58	Lilin	4LZ120*7-16	3658.32–3725.44	8	62.41	72.84	0.86	67.3
59	Lilin	4LZ120*7-17	3725.44–3815.59	13	90.15	78.47	1.15	92.9
60	Zhongcheng	4LZ120*7Y685	3815.59–3843.30	5	27.71	33.67	0.82	63.5
61	Zhongcheng	4LZ120*7Y684	3846.00–3858.96	2	12.96	17.34	0.75	101.2
62	Zhongcheng	4LZ120*7Y627	3858.96–3960.00	12	101.04	82.73	1.22	91.2
63	Lilin	4LZ120*04-5-2	3960.00–4065.90	13	105.90	100.61	1.05	91.4
64	Zhongcheng	4LZ120*7Y688	4065.90–4131.19	9	65.29	72.51	0.90	98.8
65	Lilin	4LZ120*04-5-1	4131.19–4206.77	9	75.58	66.07	1.14	99.8
66	Zhongcheng	4LZ120*7Y689	4206.77–4207.42	1	0.65	2.11	0.31	53.9
67	Zhongcheng	4LZ120*7Y510	4207.42–4213.46	1	6.04	7.97	0.76	100.0
68	Lilin	4LZ120*04-5-3	4213.46–4316.72	12	103.26	93.53	1.10	92.1
69	Lilin	4LZ120*G04-8-1	4316.72–5118.20	8	69.48	67.93	1.02	87.7
70	Lilin	4LZ120*G04-8-2	4343.61–4437.82	10[f]	91.19	64.96	1.40	81.9
71	Lilin	4LZ120*G04-4-3	4437.82–4478.46	5	40.64	30.46	1.33	58.0
72	Lilin	4LZ120*G04-8-3	4478.46–4604.87	15	126.41	103.85	1.22	98.9
73	Lilin	4LZ120*G04-8-4	4604.87–4673.36	9	68.49	66.87	1.02	93.3
74	Lilin	4LZ120*G04-8-7	4673.36–4948.49	8	70.23	75.80	0.93	90.5
75	Lilin	4LZ120*G04-8-13	4710.60–4804.27	11	93.67	109.01	0.86	96.9
76	Lilin	4LZ120*G04-8-5	4804.27–5072.06	12	95.30	102.83	0.93	48.6
77	Lilin	4LZ120*04-5-8	4831.32–4915.50	11	84.18	90.13	0.93	98.0
78	Lilin	4LZ120*04-5-7	4948.49–5003.81	7	55.32	57.66	0.96	90.5
Total				1041	4958.48	4797.74	1.03	85.8

Note [a] The time needed for PDM to redressing was not included
[b] For the hole section of the first sidetracking to correct deviation, drilling was conducted twice, and for MH-1C hole section, 14 round trips were drilled
[c] For the hole section of the first sidetracking to correct deviation, 1 round trip was drilled
[d] Additional round trip of pilot reaming was drilled, with 11.45 m/11 h
[e] Additional round trip of pilot reaming was drilled, with 7.88 m/6.3 h
[f] Additional round trip of tri-cone rock bit drilling, with 3.02 m/3.13 h

Table 5.14 PDM application in CCSD-1 Well

	Hole section	Manufacturer	PDM type	Quantity	Roundtrip	Footage drilled (m)	Actual drilling time (h)	ROP (m/h)	Core recovery (%)	Average life Time (h)	Average life Footage drilled (m)
Based on stages	PH	Beijing	LZ95	12	296	706.29	809.37	0.87	97.3	67.45	58.86
			5LZ95	9	114	400.91	410.89	0.98	79.5	45.65	44.55
			LZ120	3	105	211.82	267.10	0.79	90.7	89.03	70.61
			3LZ120	3	22	77.58	81.42	0.95	84.2	27.14	25.86
			Sub total	27	537	1396.60	1568.78	0.89	90.5	58.1	51.73
		Zhongcheng	4LZ95	2	7	39.40	36.46	1.08	82.4	18.23	19.7
			4LZ120	7	101	491.55	449.38	1.09	85.2	64.2	70.22
			Sub total	9	108	530.95	485.84	1.09	85.0	53.98	58.99
		Total		36	645	1927.55	2054.62	0.94	89.0	57.07	53.54
	MH	Beijing	5LZ95	6	33	188.52	181.76	1.04	64.2	30.29	31.42
		Zhongcheng	4LZ120	7.5	98[a]	727.50	615.51	1.18	82.0	82.07	97.00
		Total		13.5	131	916.02	797.27	1.15	78.3	59.06	67.85
	MH-1C	Zhongcheng	4LZ120	6.5	81[b]	654.18	523.46	1.25	82.0	80.53	100.64
	MH-2C	Zhongcheng	4LZ120	7	35	247.85	258.67	0.96	91.9	36.95	35.41
		Liling	4LZ120	7	73	576.80	559.31	1.03	91.1	79.9	82.40
			4LZ120G	8	79[c]	658.43	624.84	1.05	83.7	78.11	82.30
			Sub total	15	152	1235.23	1184.15	1.04	87.2	78.94	82.35
		Total		22	187	1483.08	1442.82	1.03	88.0	65.58	67.41
	Grand total			78	1044	4980.83	4818.17	1.03	85.8	61.77	63.86
Based on size	Φ 95 mm			29	450	1335.12	1438.48	0.93	86.9	49.60	46.04
	Φ 120 mm			49	594	3645.71	3379.69	1.08	85.4	68.97	74.40
Based on manufacturer	Beijing			33	570	1585.12	1750.54	0.91	87.3	53.05	48.03
	Zhongcheng			30	322	2160.48	1883.48	1.15	83.9	62.78	72.02
	Lilin			15	152	1235.23	1184.15	1.04	87.2	78.94	82.35

Note [a] Including three round trips of core drilling for sidetracking to correct deviation
[b] Including two round trips of pilot reaming
[c] Including one round trip of tri-cone rock bit drilling

In the 1990s, the Institute of Exploration Techniques mainly focused its attention on large diameter hydro-hammers for water well, oil and engineering drilling, with a variety of large diameter hydro-hammers such as ZC-800, YQ-150, YQ-178, YS-219 and SYC-178 developed. Meanwhile, the institute carried on a deep research on improving energy utilization ratio and single percussive power, for instance, the development of YZX series hydro-hammers realized an overall drilling rate of 3–6 m/h in granite with drillability grade over VII, equipped with YZX 127 hydro-hammer and button drill bit. The former Changchun College of Geology carried on a deep research on fluidic hydro-hammer and the single percussive power was then increased by adding the stroke of the hammer. The research on large diameter hydro-hammer non-core drilling techniques achieved a satisfactory progress.

In the field of scientific drilling, attention was once paid to hydro percussive rotary drilling techniques. For instance, the institute led by Professor Marx in Klaustal University planed to use hydro-hammer technique in German KTB project and made researches on this field. They visited China, imported fluidic type hydro-hammers from the former Changchun College of Geology and made laboratory tests, carried on a series of academic and personnel exchanges, established professional testing facilities, made deep researches on design, application and rock fragmentation mechanism of hydro-hammer, and then designed a prototype positive acting hydro-hammer by reference to Chinese research results. They made some progress in material selection of hydro-hammer parts, and in application of percussive rotary drilling techniques in deep hole drilling. However, for various reasons, this research result was not used in KTB project.

International Ocean Drilling Project (IODP) paid close attention to China's hydro percussive rotary drilling techniques. The scientists from IODP visited the Institute of

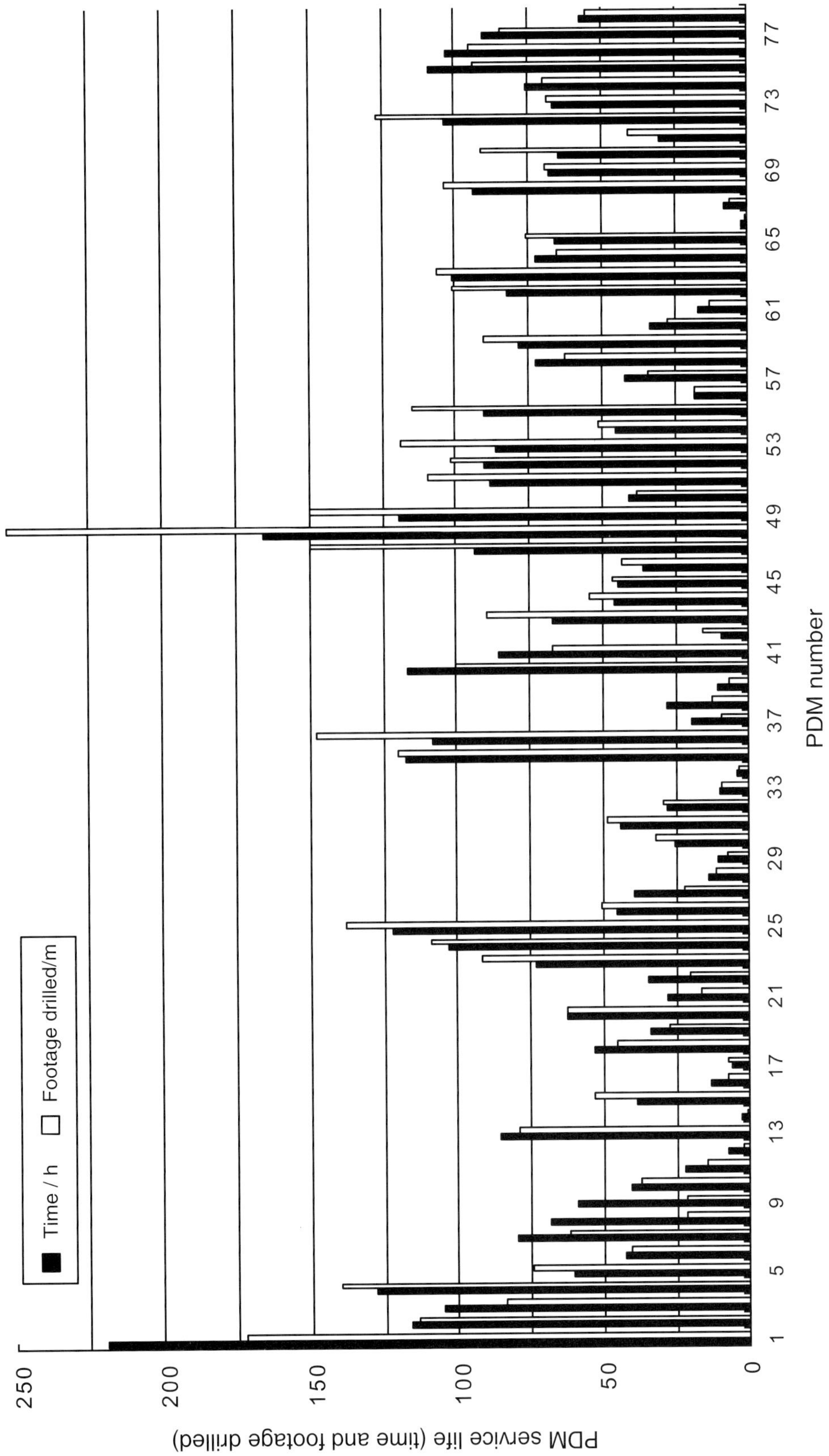

5.3 Down Hole Power Percussive Rotary Core Drilling System

Fig. 5.13 PDM application results

Exploration Techniques to investigate hydro-hammer techniques. In recent years, companies in Germany, Australia and the United States have felt interest in hydro-hammer techniques and successively imported the techniques for trial.

As CCSD-1 Well is a deep well of 5000 m drilled in hard crystalline rocks, hydro percussive rotary core drilling techniques was to be adopted during the period of initial design stage, in order to improve drilling efficiency and increase footage drilled per round trip. Diamond core drilling with hydro-hammer techniques has the following advantages:

(1) Improving drilling efficiency in hard rock formation. Rock fragmentation by dynamic load can improve drilling efficiency and be beneficial to diamond exposure, thus diamond polishing on drill bit crown can be avoided.
(2) Reducing core blockage and increasing footage drilled per round trip. Vibration produced by hydro-hammer during working can free core blockage, being beneficial to entrance of core into core barrel, reducing core blockage, increasing footage drilled per round trip, decreasing auxiliary working time, and improving efficiency per rig month.
(3) Being beneficial to reducing borehole deviation. Borehole deviation due to over large bit pressure can be avoided as hydro-hammer drilling can greatly improve drilling efficiency without increasing bit pressure, and at the same time development of hole deviation can be decreased.
(4) Reducing drilling cost.

1. **Hydro-hammer types**

The mechanisms which produce percussive action during drilling operations can be classified, based on the driving methods, into pneumatic, hydraulic, oil pressure, electric and mechanic types. Percussive power can either be produced from surface equipment or from down hole equipment. Because the percussive energy will be obviously consumed during the process of transmission and on the other hand serious destroy will be produced on the percussion bearing parts, thus it is hoped that this equipment (drilling tool) can be put into borehole together with drilling tool in deep hole drilling operations, so the percussive force comes from the equipment directly acts on drill bit or core barrel, reducing energy consumption, improving energy utilization ratio and decreasing down hole drilling tool accidents. For deep hole drilling, hydro-hammer, a down hole percussive power mechanism, is a suitable option. Down the hole hammer includes hydraulic and pneumatic types. Pneumatic down the hole hammer is driven by the compressed air from air compressor at surface and the driving medium is clean, with the advantages of large output single percussive power, high drilling rate and long service life, and with the shortcomings of high borehole wall stability requirement due to lack of mud protection of borehole wall, being suitable for shallow borehole, and large power consumption. Pneumatic down the hole hammer is widely used in mining, blasting and in engineering constructions. For deep hole hard rock drilling, only hydraulic down the hole hammer—hydro-hammer can be used. The basic working principle of hydro-hammer is that the liquid energy supplied by mud pump directly drives the hammer up and down, forming a continuous percussion against drill bit. This belongs to percussive rotary drilling techniques, with the advantage of being suitable to deep hole drilling. However, due to the impurity in driving medium (drilling mud) and its complex composition, and the adverse working circumstances, the output percussive energy is normally not very large, the continuous working life of the drilling tool is relatively low, and the stability in operations is inferior to that of pneumatic down the hole hammer.

In China, a considerable variety of hydro-hammers has been developed in the field of geological exploration industry, among those two types have obtained good economic and technical benefits, i.e. non-valve type (fluidic type and fluidic-suction type hydro-hammers) and valve type (positive acting, reacting and double acting hydro-hammers, see Fig. 5.14). For CCSD-1 Well drilling, two types of downhole percussive rotary power devices, fluidic type hydro-hammer and valve type double acting hydro-hammer were mainly developed and trial used.

(1) Structure and working principle of fluidic hydro-hammer

Structure of fluidic hydro-hammer

Fluidic hydro-hammer mainly consists of fluidic elements, inner cylinder, piston, hammer, anvil and outer tube (Fig. 5.15).

Working principle of fluidic hydro-hammer

High pressure mud which enters from upper adapter produces high speed fluid through the nozzle of the fluidic element. Wall adhesion effect of the fluid forces the fluid to adhere to one side of the wall, entering into the inner cylinder through the outlet of the element, and pushing piston and hammer moving. Drilling mud on the other side of the inner cylinder is discharged out of the fluidic element, from the discharge hole of the element, through the discharge way of the non-adhesion side, and then enters into the swivel type double tube drilling tool through the channel between the element and the outside of inner cylinder, anvil and the central hole of lower adapter. The fluidic element is of double stable symmetry type, thus the fluid can adhere alternately to either side of the wall. The change-over of the fluid is realized in such way: the water flushing pressure of the fluid in the cylinder greatly increases instantaneously when the piston hammer moves to the upper and lower dead

5.3 Down Hole Power Percussive Rotary Core Drilling System

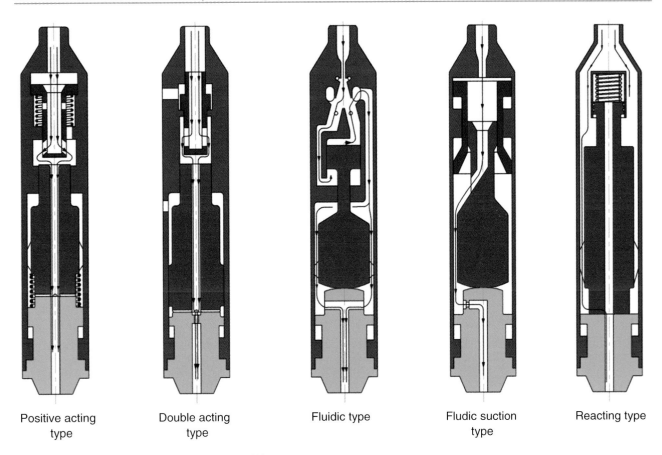

Fig. 5.14 Main hydro-hammer types developed in China

Positive acting type — Double acting type — Fluidic type — Fludic suction type — Reacting type

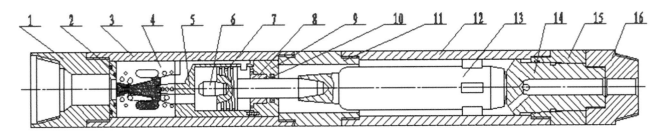

Fig. 5.15 Structure of fluidic hydro-hammer. *1* Upper adapter. *2* Adjusting spacer. *3* Outer cylinder. *4* Fluidic element. *5* Inner cylinder. *6* Adjusting conical shaft. *7* Piston. *8* Copper sleeve. *9* Cylinder cover. *10* Sealing ring. *11* Middle adapter. *12* Outer tube. *13* Hammer. *14* Anvil. *15* Octagonal sleeve. *16* Lower adapter

ends, producing pressure signal, which pushes the fluid adhering to the other side through the signal holes on the element lower plate and top plate, and the controlling channel. Thus the change-over of the fluid and the direction change of the hammer can be realized (see Table 5.15).

Characteristics of fluidic hydro-hammer

The wall adhering and change-over principle of fluid decides that the hammer is not affected by in the hole surrounding pressure and hole depth. This is favourable to deep hole drilling.

Table 5.15 KSC127 fluidic type hydro-hammer

Item	Parameter	Item	Parameter
O.D. (mm)	127	Pressure drop (MPa)	3–4
Effective length (mm)	2200	Percussion power (J)	60–80
Hammer mass (kg)	57	Percussion frequency (Hz)	16–25
Pump rate (L/min)	500–600	Adjusting travel (mm)	8–25

Only piston hammer is a moving part in the whole set of hydro-hammer. And this is beneficial to improving the service life and reliability of the hydro-hammer.

The percussion travel of piston hammer has a relatively large adjusting range, which changes the percussion frequency and per single percussion power.

All the flowing channels inside the hydro-hammer are regular opened, i.e. drilling mud can flow through even if the hydro-hammer does not work. When hammer strikes against the anvil with the highest percussion speed, a differential return circuit for the fluid in the cylinder can be realized, i.e. the drilling mud at discharge side of the cylinder can enter into the other side of the cylinder, to replenish the inadequate instantaneous mud flow during piston high speed movement.

(2) Structure and working principle of valve type double acting hydro-hammer

Structure

For valve type double acting hydro-hammer, the control of fluid is realized by the coordination of the movement and the position between valve and hammer, so that a stable reciprocating movement of the hammer can be produced (Fig. 5.16).

Working principle

Starting state: The valve inside hydro-hammer and the hammer are in the lower limit. When drilling mud flows through the hydro-hammer, a pressure difference between mud channel above throttle hole (the area shown by red arrow in Fig. 5.16) and annulus is built up, because of the throttle hole in the centre of the anvil.

Valve upward: The upward power for valve results from the acting force of drilling mud to the difference between upper and lower sealing areas. An accelerated motion is exerted and the valve begins to accelerate upwards.

Valve to upper limit: The upper valve, in the process of accelerated motion, is the first to reach the upper limit and then stops moving.

Hammer upward: At this time, the hammer begins to accelerate upwards, under the acting force resulted from the difference between upper and lower sealing areas. The speed of hammer acceleration is lower than that of valve acceleration. In case that both the flow rate and pressure are low, upward acceleration of the hammer will be later than that of the valve. When pressure is high, the upward acceleration of hammer and valve may be started at the same time. However, because the acceleration of the hammer is low and thus its speed is lower.

Fluid channel sealed by closure of hammer and valve: The hammer moves upwards at a relatively low speed and reaches to the upper limit later than the upper valve. Its upper limit position contacts with valve, while the lower end of the valve closes the central hole of the hammer, thus to stop the fluid channel.

Hammer and valve downward: When the upper piston end of the hammer approaches the lower end of the upper valve, a flow dam is formed between the lower end of the upper valve and the inner hole of the upper piston and then the upper valve starts to accelerate downwards, form its original static state. At the same time, the hammer decelerates. When the hammer contacts with the upper valve, both overcome mud pressure difference and decelerate upwards, until zero speed. Then the hammer and the upper valve enter into a downward percussion stroke. The fluid from the lower chamber discharges through the throttle hole in the centre of the anvil.

Valve reaches lower limit and hammer continuously moves downwards: Under the action of pressure difference, hammer and upper valve accelerate downwards. As the valve reaches the lower limit first and stops moving, the hammer continuously moves downwards. Then a fluid channel is opened between the upper piston of the hammer and the upper valve, and a pressure in lower chamber of the hammer is built up. At this time, under the uplift action created by area difference between upper and lower pistons, the hammer continuously moves downwards and then enters into a decelerated percussion stroke.

Hammer strikes against anvil: The hammer continuously moves downwards until it strikes against the anvil, and transmits its kinetic energy to the anvil. Then it returns to the original state, i.e. the valve and the hammer enter into a return acceleration stage for next percussion.

The advantages of valve type double acting hydro-hammer:

Volumetric principle is adopted, with high energy utilization ratio, thus with more output percussion work;

With simple structure, easy for starting and operation;

Easy for maintenance with low cost, as the easily-worn parts are low in cost;

With wide data adjusting range, being suitable for different drilling technologies, such as non-coring button bit drilling and diamond core drilling.

2. **Early research and test of hydro-hammer**

A research project of KS-152 wireline coring hydraulic percussion mechanism was set up during the starting period of CCSD-1 project. During the period from June to July, 1999, in order to make a necessary preparation for hydro-hammer technique for CCSD-1 Well, a large diameter hydro-hammer drilling test was conducted at the later drilling stage of the Pre-pilot Hole CCSD-PP2. During the test both fluidic type and valve type double acting hydro-hammers were comparatively tested.

The first test on hydro-hammer was carried on at Maobei Village, Donghai County, Jiangsu Province. The test hole was about 20 m from the CCSD-PP2 Hole, with the design depth of 100 m. The upper hole section, with the diameter of

5.3 Down Hole Power Percussive Rotary Core Drilling System

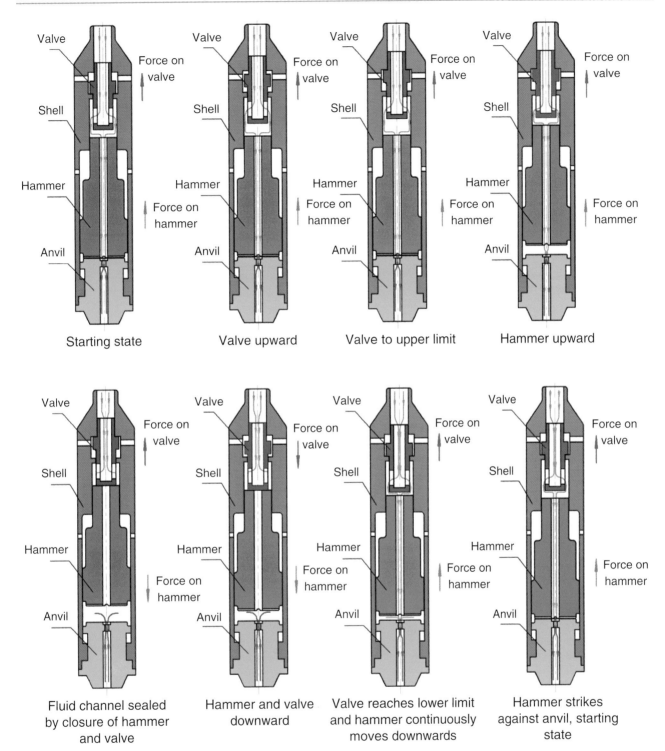

Fig. 5.16 Structure and working principle of valve type double acting hydro-hammer

158 mm, was to be drilled with button percussion drill bit, while the lower section, with the diameter of 152.4 mm, was to be drilled with cone bit.

Following conclusions were obtained (Zhang and Xie 2005):

(1) Hydro-hammer is an effective way to drill large diameter hole in hard rock

Test result indicated that the drilling rate of hydro-hammer reached to 5 m/h in soft gneiss while that in hard gneiss was approximately 3 m/h, both were much higher than those

Table 5.16 Drilling rate at upper 100 m section of CCSD-PP2 Hole

Hole depth (m)	Drilling time (h)	Footage drilled (m)	Drilling rate (m/h)	Drilling method
21.33–44.29	25.23	22.96	0.91	Alloy core drilling
44.29–85.22	62.97	40.93	0.65	75 mm diamond core drilling
88.44–101.45	14.14	13.01	0.92	75 mm diamond core drilling
101.45–117.18	14.56	15.73	1.08	75 mm diamond wireline core drilling

obtained in the corresponding hole section in CCSD-PP2 Hole, which was only 20 m in distance (see Table 5.16). The increase of drilling rate mainly resulted from the percussive dynamic load of hydro-hammer, which improved the percussive dynamic load of cone bit teeth onto hole bottom rock.

(2) Cone bit can be used as the auxiliary bit for large diameter hydro-hammer non-core drilling in hard rock

During the second drilling phase, because of the hard rock suddenly encountered, the percussive energy supplied by hydro-hammer was not enough to produce volumetric fragmentation for rock, in case of button percussive drill bit used. At that time, rock fragmentation was a fatigued one, resulting in a low drilling rate. Drill bit teeth wore very quickly and then drilling rate became even lower, entering into a vicious circle.

During the third drilling phase, cone bit was used and drilling rate was obviously increased from 0.83 to 2.94 m/h, with slight teeth wear. Effective drilling of cone bit is due to follows:

(1) At any drilling instantaneousness, the specific power for rock fragmentation of cone bit on rock formation under the same percussive energy from hydro-hammer is higher, because only less teeth contact rock. (2) Cone bit has a percussive fragmentation action, easily producing volumetric fragmentation on hard rock, added with percussion action of hydro-hammer. (3) Because of volumetric fragmentation of rock and alternate wear of teeth, cone bit wears slightly, thus keeping a long effective service.

(3) Both fluidic type and valve type double acting hydro-hammers are suitable for large diameter non-coring drilling in hard rock

In this test, two types of hydro-hammers, i.e. KSC127 fluidic type double acting hydro-hammer and YZX127 valve type double acting hydro-hammer were adopted. Both could be used for high effective large diameter non-coring drilling.

A success was obtained in the test. During this test the feasibility of large diameter hydro-hammer non-coring drilling in hard rock, the adaptability of auxiliary drill bits, and drilling parameters were understood. Application results of different hydro-hammers were examined, for further improvement.

In this test, valuable experience was accumulated for application of hydro-hammer techniques in CCSD-1 Well.

Core drilling was not conducted in the test. A comparison shows that there was no obvious difference between the two types of hydro-hammers in non-coring drilling (see Table 5.17). Generally, valve type hydro-hammer has better stability and higher drilling rate. However, there were some problems in its parts, still need improvement.

3. **Application and improvement of hydro-hammer in the pilot hole**

(1) Basic situation

The pilot hole of CCSD-1 Well was opened in June, 2001. After the first opening (changing hole size and setting casing), core drilling was started. At the initial stage, since the core drilling tool was not perfected, hydro-hammer technique was not yet adopted. YZX127 valve type double acting hydro-hammer and KSC127 fluidic type hydro-hammer were adopted for trial in August, 2001 after PDM core drilling (core recovered by lifting drilling tool) became normal. The application of hydro-hammer in the pilot hole can be classified into three stages. During the first stage hydro-hammers were tried out in normal drilling operations and some problems of unsuitability of both the hydro-hammers to the drill site conditions were discovered. In the second stage hydro-hammers were improved and their properties were increased. However, the hydro-hammers could not satisfy the need of core drilling. A breakthrough was made in YZX127 valve type double acting hydro-hammer at the third stage and its properties was much improved, basically satisfying the need of core drilling. From then on, core drilling with hydro-hammer was continuously utilized, with obvious economic benefit.

The first stage (218.44–789.60 m)

YZX127 valve type double acting hydro-hammer was first used in PH66 round trip (hole depth 218.44 m). The hammer worked for 20 min before it stopped working. Then drilling tool was lifted for checking and it was found that the upper valve inside the hammer was blocked with silt. Drilling in this round trip showed that with YZX127 hydro-hammer drilling rate could be greatly improved, as its penetration rate reached 2.97 m/h, much higher than that of conventional rotary drilling. Afterwards, YZX127 hydro-hammer was intermittently used and stable performance obviously increased drilling rate and footage drilled per round trip. In PH142 round trip, drill bit slippery occurred,

Table 5.17 Test data and performance comparison of large diameter hydro-hammers in hard rock drilling

	Item	YZX127 hydro-hammer	KSC127 hydro-hammer	Ratio
Parameter	Hydro-hammer length (includes stabilizer and button bit), (m)	1.78	2.39	
	Rotary speed (button bit/cone bit), (rpm)	40–48/70–80	40–48/70–80	
	Working pump rate, (L/min)	350–400	450–500	
	Working pressure, (MPa)	3.0–3.8	3.0–3.8	
	Average power consumption ratio	1	1.27	0.789
	Button (quantity × size)/(piece × mm)	8 × φ16 + 11 × φ14	8 × φ14 + 11 × φ12	
Result	Total footage drilled, (m)	49.45	34.51	1.433
	Total working time, (h)	14.41	11.31	1.274
	Average drilling rate, (m/h)	3.43	3.05	1.125
	Footage drilled with button bit in soft rock, (m)	19.53	16.96	
	Drilling rate of button bit in soft rock, (m/h)	6.07	4.17	1.456
	Footage drilled with button bit in hard rock, (m)	1.41	1.18	
	Drilling rate of button bit in hard rock, (m/h)	0.74	0.97	0.763
	Footage drilled with cone bit in hard rock, (m)	28.51	16.37	
	Working time of cone bit in hard rock, (h)	9.28	6.02	
	Drilling rate of cone bit in hard rock, (m/h)	3.07	2.72	1.130
	Ratio to cone bit rotary drilling rate, (%)	202	179	
Other properties	Round trip	12	17	
	Footage drilled per round trip, (m)	4.12	2.03	2.030
	Normal working round trip	12	11	
	Ratio of normal working round trip, (%)	100	64.71	1.545
	Troubles	4	6	
	Percentage of trouble-free round trip, (%)	66.67	64.71	1.030
	Pump pressure in hole flushing at 500 L/min	1.6–2.0	3.0–3.8	
	Working stability	Stable	Stable	
	Effectiveness of anti-idle striking mechanism	Effective	Intermittently work	

resulting in a very low drilling rate (0.03 m/h). Then in PH143 round trip YZX127 hydro-hammer was added and then drilling rate reached as high as 1.21 m/h. The remarkable influence of hydro-hammer on drilling rate was fully shown. Nevertheless, in many cases, the valve of the hammer was blocked with the silt in drilling mud and the hammer could not be started, or could not continuously drill a round trip. Sometimes even the pump was interrupted. Under these circumstances the trial of this hydro-hammer was stopped and then laboratory improvement was made under the basis of analysis of existing problems. During the first stage 14 round trips were drilled with YZX127 hydro-hammer, with total footage of 22.88 m drilled and an average penetration rate of 0.74 m/h.

KSC127 fluidic type hydro-hammer was used for trial from PH238 round trip (hole depth 578.90 m), with a result of obvious improvement of penetration rate and total footage drilled per round trip. In this stage KSC127 fluidic type hydro-hammer was more stable than YZX127 hydro-hammer. However, its stability deteriorated after the use for a period of time. Increase of pumping pressure was resulted in even KSC127 fluidic type hydro-hammer did not work. Moreover, KSC127 hydro-hammer still had some problems such as poor erosion resistance of the elements, sealing problem, difficult adjustment of fitting clearance and unsatisfactory stability. Though improvement was made, the result was unsatisfactory. In this stage KSC127 hydro-hammer was used for twenty round trips, with a total footage of 60.85 m drilled and an average penetration rate of 1.26 m/h.

In this stage hydro-hammers were used to drill thirty four round trips (Table 5.18). From Table 5.8 it can be seen that in this stage the desired results were not achieved with the application of hydro-hammers.

The second stage (789.60–1209.61 m)

In the second stage more round trips were drilled with hydro-hammers than with conventional rotary drilling techniques (Table 5.19). But the round trip length was not improved obviously, only 35 % increase, due to the

Table 5.18 Comparison of hydro-hammer applications from PH66 round trip to PH325 round trip

	Round trips	Round trip length (m)	Footage drilled (m)	Core recovery (%)	Penetration rate (m/h)
Hydro-hammer	34	2.46	83.23	97.4	1.06
Non hydro-hammer	226	2.16	487.43	97.3	0.82
Ratio	0.15	1.14	–	1.00	1.29

Table 5.19 Comparison of hydro-hammer applications from PH326 round trip to PH470 round trip

	Round trips	Round trip length (m)	Footage drilled (m)	Core recovery (%)	Penetration rate (m/h)
Hydro-hammer	78	3.29	256.46	97.6	1.02
Non hydro-hammer	67	2.44	163.55	90.3	0.54
Ratio	1.16	1.35	–	1.08	1.89

unsatisfactory working stability and reliability, as well as the short core barrel used. Meanwhile, normal working of the hydro-hammer greatly increased penetration rate, as high as 89 % increase.

In general, drilling results of YZX127 hydro-hammer and KSC127 hydro-hammer were acceptable. Some improvement was made to the hammers and then the hammers were intermittently used in the process of PDM core drilling. During this stage, the performance of YZX127 hydro-hammer was not much improved, while KSC127 hydro-hammer showed a satisfactory result. Thus in this stage KSC127 hydro-hammer was mainly used, with more footage drilled. From PH326 (hole depth 789.60 m) to PH470 (hole depth 1208.01 m), YZX127 hydro-hammer was used to drill six round trips, with a total footage of 13.40 m and an average penetration rate of 0.64 m/h; while KSC127 hydrohammer was used to drill seventy two round trips, with a total footage of 243.06 m, an average penetration rate of 1.06 m/h and an average round trip length of 3.38 m.

The third stage (1209.61–2046.54 m)

Based upon the problems of YZX127 hydro-hammer in the first and second stages, a thorough improvement was made and then tried in December, 2001, after indoor test and improvement. The working stability of YZX127 hydro-hammer was improved by leaps and bounds, with the working stability probability raised from 30 to 90 %. The problem of pump blockage caused by blockage of circulation channel when hydro-hammer did not work was solved. The service life of hydro-hammer was greatly increased, with the maximum of five round trip continuous drilling without examination and maintenance.

During the third stage, both YZX127 hydro-hammer and KSC127 hydro-hammer became practical and were extensively used in core drilling operations. In order to select a better hydro-hammer to increase its success and guarantee its application result, at drill site a comprehensive comparison was made between the two hydro-hammers. Before lowering into the well, either of the hydro-hammers was tested at well head and it could only be lowered into the well after a success in the test. Both the hydro-hammers were intermittently used for drilling each round trip. In this stage YZX127 hydro-hammer showed a good result, accounting for 60 % of the total footage drilled. From the 471 round trip (hole depth 1209.61 m) to the final 657 round trip (hole depth 2046.54 m) KSC127 hydro-hammer accumulatively drilled 51 round trips, with a total footage of 250.89 m drilled, an average drilling rate of 1.07 m/h and an average round trip length of 4.92 m; whereas YZX127 hydro-hammer accumulatively drilled 107 round trips, with a total footage of 512.00 m drilled, an average drilling rate of 1.16 m/h and an average round trip length of 4.79 m.

While improving hydro-hammers, the working environment of the hydro-hammers was improved accordingly, so as to improve the working stability and service life of the hydro-hammers and PDM. 1 % GLUB lubricant was added into the drilling fluid to improve its lubricating effect. And mud solid control was enhanced. All these measures obviously increased the working stability of the hydro-hammers, whereas decreased the wear of core drilling tools and resistance of core entering into the inner tube. Consequently both the drilling rate and the footage drilled per round trip were further improved.

In the third stage most of the round trips were completed by hydro-hammer drilling (see Table 5.20). Due to the improvement of the working stability and reliability of the hydro-hammers, the footage drilled per round trip was obviously increased, by as high as 89 %. Core recovery was increased from 55 to 82.5 %. All these indicated that hydro-hammer had a considerable effect upon increasing core recovery.

In the stage of pilot hole core drilling, a lot of problems were encountered in the application of hydro-hammers. Because of the different structures and different working principles of the two types of the hydro-hammers, the

5.3 Down Hole Power Percussive Rotary Core Drilling System

Table 5.20 Comparison of hydro-hammer applications from PH471 round trip to PH657 round trip

	Round trips	Round trip length (m)	Footage drilled (m)	Core recovery (%)	Penetration rate (m/h)
Hydro-hammer	158	4.83	762.89	82.5	1.13
Non hydro-hammer	29	2.55	74.04	55.0	0.88
Ratio	5.45	1.89	–	1.50	1.28

problems encountered and the methods to solve were different, too.

(2) Application and improvement of KSC127 fluidic type hydro-hammer

Problems and improvement

i. Sealing leakage at side passageway of inner cylinder

Leakage easily happened at the O ring around the rectangular fluid passageway at the side of the inner cylinder of fluidic type hydro-hammer, because of large mud flow rate and high pressure. Once the leakage happened, mud erosion would produce ditches in very short time, and then link up the high pressure and low pressure chambers. In this case the working condition for hydro-hammer was destroyed and then hydro-hammer stopped working. To solve this problem, two improvement measures were adopted for the sealing structure of the inner cylinder.

The original side passageway of the inner cylinder adopted a single O ring structure (Fig. 5.17a), which was improved into a two O ring structure (Fig. 5.17b). The improved structure showed a good sealing result, without any leakage during the test. Leakage of the inner cylinder was effectively solved.

The original side passageway of the inner cylinder adopted an outside passageway structure, i.e. the fluid passageway was formed by a match sealing between the inner and the outer cylinders. Then it was decided that an inner passageway structure was to be adopted, i.e. arc shaped discharge holes were machined at the inner cylinder to form side passageway for fluid. In this way the problem of leakage was completely solved, see Fig. 5.18. Two new inner cylinders were produced and the problem of leakage was solved in the test. However, the structure was rather complicated, with short service life.

Fluidic elements, as the easily worn parts in fluidic hydro-hammer, have an average service life of 60–100 h in core drilling operations. However, the service life of the fluidic elements in the trial in CCSD-1 Well was rather short, generally only more than 10 h, because of the serious erosion to the working chamber. Analyses of the reasons at drill site showed that the excessive pump rate resulted in an ultra-high fluid, quickening the wear of the working chamber of the fluidic elements. Although mud solid was well controlled, the micro grains which were smaller than solid control standards had a large density. High speed mud fluid produced a serious erosion and wear to the elements, resulting in a hydro-hammer cease.

Two improvement measures were adopted at the drill site:

Tungsten carbide lining was added to the working chamber of the fluidic element, so the main fluidic erosion was limited to the tungsten carbide lining and the service life of the fluidic element was increased (Fig. 5.19).

To increase the cross section of the nozzle, so as to decrease the jet speed of the fluid. The nozzle width was increased from original 5 to 8 mm and its cross section

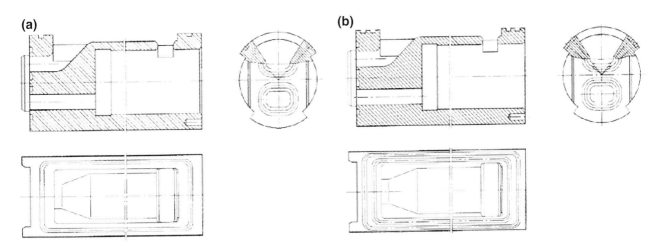

Fig. 5.17 Improvement of the side passageway sealing structure of the inner cylinder. **a** Single sealing. **b** Double sealing

Fig. 5.18 Structure of inner passageway

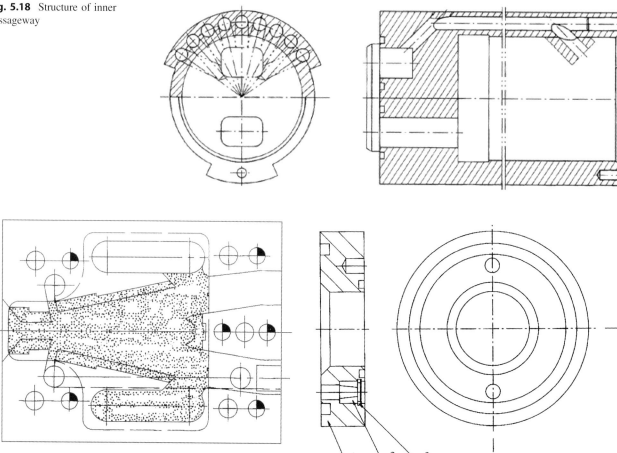

Fig. 5.19 Fluidic element after improvement

Fig. 5.20 Adjusting spacer with flow diversion holes. *1* Spacer *2* Alloy nozzle *3* Spring clip

increased by 60 %, whereas the fluidic speed decreased from original 92.3 to 53.6 m/s. In this way the erosion and wear to the element was decreased while the service life increased.

ii. The designed rated flow rate did not match the flow rate at drill site

Based upon the designed borehole diameter and mud uplift velocity, the designed rated flow rate of hydro-hammer was 5–6 L/s. PDM was adopted at drill site, with the pump rate of 9–12 L/s. Excessive pump rate caused a flow rate increase at passageways of the hydro-hammer, especially an excessive flow rate at the fluidic element nozzle, resulting in a wear acceleration and service life decrease of the hydro-hammer. Moreover, excessive pump rate produced an increase of pumping pressure, leading to leakage at the sealing areas of the hydro-hammer.

A method of flow diversion was used to solve the problem of excessive pumping rate at drill site. At the adjusting spacer two flow diversion holes were opened, see Fig. 5.20. Hole diameter was decided based upon calculation and the diverted flow rate was one third to one fourth of the total flow rate. The total flow rate which entered into the hydro-hammer was distributed to the fluidic elements after diversion and the mud which entered into the hammer cylinder for working only accounted for two thirds to three fourths of the total pumping rate. Then the hydro-hammer worked in a rated limit, both the service life and reliability were improved. Field application indicated that the result was satisfactory. But due to time limitation the flow diversion holes were not inserted with tungsten carbide nozzles and for this reason the flow diversion holes were quickly eroded, resulting in a too short service life. After inserting tungsten carbide nozzles, this method became a feasible one.

Application of KSC127 fluidic type hydro-hammer

i. Application result of KSC127 fluidic type hydro-hammer in the first trial

From September 20th to 22nd, 2001 KSC127 fluidic type hydro-hammer was tested for the first time, alternately with conventional drilling. Eight round trips were continuously counted from round trip 237 to round trip 244 (conventional drilling for 4 round trips and hydro-hammer drilling for other 4 round trips), with the comparison shown in Tables 5.21 and 5.22.

The preliminary tests showed that with hydro-hammer drilling rate was increased by 154 %, round trip length

5.3 Down Hole Power Percussive Rotary Core Drilling System

Table 5.21 Comparison of KSC127 fluidic type hydro-hammer drilling with conventional rotary drilling

Date	Round trip	Drilling technique	Hole section (m)	Footage drilled (m)	Actual drilling time (h)	Core length (m)	Core recovery (%)	Penetration rate (m/h)	Remarks
Sep. 20	237	Conventional drilling	577.51–578.90	1.39	2.00	1.35	97.1	0.70	Lifting as core blockage
Sep. 20	238	Hydro-hammer drilling	578.90–581.73	2.83	1.67	2.78	98.2	1.69	No feeding travel left
Sep. 20–Sep. 21	239	Hydro-hammer drilling	581.73–583.94	2.21	4.08	2.23	100.9	0.54	No feeding travel left
Sep. 21	240	Conventional drilling	583.94–585.24	1.30	2.17	1.16	100	0.60	Lifting as core blockage
Sep. 21	241	Conventional drilling	585.24–586.60	1.36	2.00	1.16	85.3	0.68	Lifting as core blockage
Sep. 21	242	Conventional drilling	586.60–587.84	1.24	1.17	1.34	108.1	1.06	Lifting as core blockage
Sep. 21	243	Hydro-hammer drilling	587.84–591.06	3.22	2.00	3.19	99.1	1.61	No feeding travel left
Sep. 22	244	Hydro-hammer drilling	591.06–595.42	4.36	3.67	4.44	101.8	1.19	Full core barrel

Table 5.22 Comparison of drilling rate in eight round trips shown in Table 5.21 between hydro-hammer drilling and conventional rotary drilling

Drilling method	Footage drilled (m)	Actual drilling time (h)	Core length (m)	Core recovery (%)	Penetration rate (m/h)	Average footage drilled per round trip (m)	Ratio of average footage drilled per round trip
Hydro-hammer dilling	12.62	11.42	12.64	100.2	1.11	3.16	239
Conventional drilling	5.29	7.34	5.01	94.7	0.72	1.32	100

Table 5.23 Comparison of drilling results between rotary drilling and hydro-hammer drilling

	Round trip	Footage drilled (m)	Actual drilling time (h)	Core length (m)	Penetration rate (m/h)	Average footage drilled per round trip (m)	Core recovery (%)
Hydro-hammer drilling	72	243.06	230.17	237.64	1.06	3.38	97.77
Rotary drilling	58	141.16	251.23	128.62	0.56	2.43	91.12
Ratio	–	–	–	–	1.89	1.39	1.07

increased by 239 %, and core recovery by 100 %. During the four round trips tested, drilling tool was lifted only because core barrel was full, or no feeding travel was left for further drilling. Core blockage never happened. The rationality of hydro-hammer drilling was verified in the tests.

ii. Analysis of the application result of percussive rotary drilling

130 round trips from PH340 round trip (hole depth 823.79 m) to PH469 round trip (hole depth 1208.01 m) were counted, in which 72 round trips were drilled by KSC127 fluidic type hydro-hammer and 58 round trips were completed by conventional rotary drilling technique, with the result comparison shown in Table 5.23. The application result such as drilling rate, footage drilled per round trip and core recovery can be found in Fig. 5.21.

In pilot hole drilling with fluidic type hydro-hammer, some problems had existed such as short service life and large percentage of abnormal round trips, because some structures of the hydro-hammer could not be adaptable to the large mud flow rate. The application of hydro-hammer in pilot hole drilling was still a test and those problems were all solved in the stage of pilot hole drilling.

(3) Application and improvement of YZX127 valve type double acting hydro-hammer

Existing problems and solution

In drilling the pilot hole of CCSD-1 Well with YZX127 hydro-hammer, difficult problems were encountered. To solve those problems, the hammer was improved and adjusted twice in the percussive rotary drilling test stand and then was continuously used at the hole section below 1200 m. Improvement mainly included the following (Xie et al. 2005):

i. Matching of pumping rate

For pilot hole core drilling, PDM was used as the rotary power. Because PDM and hydro-hammer were both down hole motors powered by circulating drilling mud, thus the pumping rate for hydro-hammer and that for PDM should be matched (7–12 L/s), otherwise a normal drilling could not be realized because of the interaction.

In design of YZX127, wide pump rate and percussion work limits were considered and large adjustable margin for feeding travel and pumping displacement were left. Based upon the pumping displacement required by PDM at drill site, the pumping rate for hydro-hammer was increased from 3–6.5 L/s to 7–12 L/s, by adjusting the parameters of the hydro-hammer parts, thus a matching of the pumping rate for hydro-hammer and that for PDM was realized.

ii. Matching of percussion work

For pilot hole core drilling, diamond double tube swivel type core barrel drilling tool was to be adopted. Hydro-hammer was to be used to increase penetration rate, reduce core blockage, and improve footage drilled per round trip and core recovery. As impregnated diamond drill bit should be used, the percussion work of hydro-hammer could not be too high, or drill bit service life would be decreased and in the hole accidents occurred.

As the upper valve stroke and the free stroke of YZX127 hydro-hammer had a wide adjusting scope, the upper valve

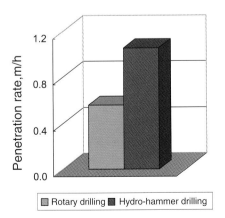 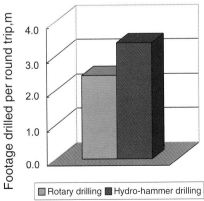 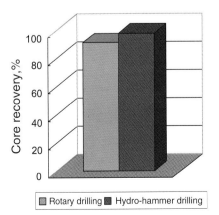

Fig. 5.21 A comprehensive comparison between KSC-127 hydro-hammer and rotary drilling

stroke was adjusted to below 20 mm and at the same time the free stroke was adjusted to over 10 mm, so as to reduce the effective working stroke of hydro-hammer and increase the resistance stroke. In this way small percussion work output could be realized and thus hydro-hammer could be suitable for the working condition of large pumping displacement.

iii. Service life of piston

In the laboratory test for the service life of YZX127 hydro-hammer, the diameter maximum fitting clearance only increased 0.23 mm after a 70 h continuous work of the parts in the medium of clear water. However, at drill site the power medium was drilling mud, though with low sand content, the accumulation of fine and hard solid content would greatly reduce the service life of the piston in hydro-hammer. For the piston of common alloy steel, the diameter fitting clearance would increase to 0.6 mm, after 6 h work. And for the piston treated with nitriding process, the nitriding coating of 0.2–0.4 mm would be worn out after a 6–8 h work. So the wear of parts by solid content in drilling mud could not be underestimated.

To solve this problem, wear resistant material was heat-sprayed onto the auxiliary matching surface and thus greatly increased the service life of the piston. During drilling operations one set of spray treated hydro-hammer continuously worked for five round trips (PH532–PH536), without any examination and repair, with accumulative 17.58 h worked, total footage of 26.30 m drilled, average footage 5.26 m drilled per round trip (full core barrel) and average penetration rate of 1.5 m/h. Field test showed that this hydro-hammer could be further used (see Fig. 5.22).

Through theoretical analyses and calculation it was found that parameter matching of the piston should be properly considered in hydro-hammer design. When the piston wears to a certain degree, the relationship of parameter matching is not yet destroyed. In this case the hydro-hammer can still work normally though its energy utilization efficiency deteriorates. In the later stage of the pilot hole drilling the hydro-hammer was improved according to this thinking and the piston made by surface quenching of common alloy steel came up to the piston treated by alloy powder heat spray in service life.

iv. Fatigue strength

Before the application of YZX127 hydro-hammer in pilot hole drilling, the problem of percussive fatigue breakage of the parts occurred. To solve this problem, material selection and heat treatment and part structure design were started with.

In material selection and heat treatment, the reasons were analyzed first and then the material strength and percussive fatigue resistance were stressed. A study on available high strength alloys was made and some steels were selected. Then a comparison of heat treatment and machining properties of the steels was made and availability of supply was

Fig. 5.22 The hydro-hammer worked for five continuous round trips

considered. On this base the materials for key parts and the heat treatment technologies were determined.

In part structure design, sharp angles and transition shoulders of the parts on which stress concentration easily arises should be rounded so as to simplify the structure of the parts and avoid stress concentration. To solve the problem of upper valve breakages in the test of large diameter hard rock drilling, the upper valve with new structure was designed to greatly increase the strength.

To verify this new design, this hydro-hammer was tested for a 70 h continuous working in laboratory test stand before field application. Test result showed that the strength of the parts satisfied designed requirements. During the period of pilot hole drilling, only the first batch upper valves broke after 6–8 h working and then such problem never happened again.

v. Adaptability to drilling mud

Adaptability of YZX127 hydro-hammer to drilling mud was the most difficult problem which consumed much time, manpower and material resources. Understanding and solution of this problem experienced following three phases.

The first phase

The initially improved prototype hydro-hammer was used, mainly improved were the decreased percussive work

5.3 Down Hole Power Percussive Rotary Core Drilling System

Fig. 5.23 Mud cake in hydro-hammer

Fig. 5.24 Sealing surface of upper valve was eroded

of the hammer and the increased pumping rate limits. During this phase the hammer could not be started in many cases. Field examination found that the upper valve or the mandrel valve was blocked because of the solid dirt existed in the matching clearance of the upper valve or the mandrel valve. This is due to the piston matching clearance, which is similar to the small cracks in rock. Based upon the plugging properties of drilling mud, once the solid grains which are similar to the clearance in size enter into the piston clearance, bridging will be formed. Then the mud components lose water, under the action of mud pressure difference, plugging the clearance as drilling mud plugs rock cracks, and leading to a blockage of the upper valve or the mandrel valve, and a failure of hydro-hammer. After that, the stable fluid flow resulted from hydro-hammer failure produces centrifugal effects under the action of high speed rotation of PDM, producing thick mud cake on the inside wall of outer tube (Fig. 5.23). In addition, when the parts were blocked clearance sealing was eroded over a long time because of lacking displacement and early scrapped (Fig. 5.24).

Blockage of parts had not been seen both in laboratory adjustment and at field clear water drilling test, mainly due to the low driving power from drilling mud. As the initial sticking force was not too high, so the problem of sticking by solids in drilling mud could be solved by increasing the driving power of the upper valve and mandrel valve.

Field application results showed that in this phase were completed 14 round trips, among which normal drilling was realized only in 7 round trips, with a total footage of 22.88 m drilled, an average footage of 1.63 m drilled per round trip, and an average penetration rate of 0.74 m/h. During this phase hydro-hammer only showed an inkling of increasing drilling rate.

The second phase

To counter the problems existed in the first phase, hydro-hammer was improved for the second round. The upper and lower area difference of upper valve was increased, the quality of mandrel valve was improved, the central hole diameter of mandrel valve was decreased, and the upper piston sealing structure of upper valve was improved. Because in lowering drilling tool assembly to the hole bottom in each round trip the drill bit needed a break-in procedure or be lifted off the hole bottom for circulation, hydro-hammer could not start working and piston was in statics. In this case the solids in drilling mud still more easily accumulated and caused a blockage. Then a starting valve package was designed so that a sudden starting effect could be realized when the idle strike proof mechanism of hydro-hammer was closed. In this way the success of starting hydro-hammer at hole bottom could be increased.

The improved hydro-hammer was adopted in the second phase; however its starting property was not obviously improved. The problem of blocking by drilling mud solids was not effectively solved by using this thinking. During this phase only a total footage of 13.44 m was drilled, with an average footage of 2.23 m drilled per round trip and an average penetration rate of 0.63 m/h.

The third phase

Based on the experiences in the first and the second phases, spiral groove size of sealing was enlarged and sealing surface length was lengthened, the matching clearance of piston was properly increased so as to reduce the probability of solids in drilling mud stay at the clearance of piston, to increase the moving power for hammer, upper valve and mandrel valve, and to realize a reasonable parameter matching. In laboratory testing stand the improved hydro-hammer was adjusted and tested for the third time.

The hydro-hammer after the third improvement was used in CCSD-1 Well and a satisfactory result was obtained. The working stability of the hydro-hammer was greatly improved, with a starting success ratio over 95 %. Except the failures caused by improper assembly, upper valve breaking or excessive wear, hydro-hammers which were properly assembled all could work normally. During this phase hydro-hammers were used for 107 round trips, with a total footage of

Fig. 5.25 Application of YZX127 hydro-hammer in the pilot hole

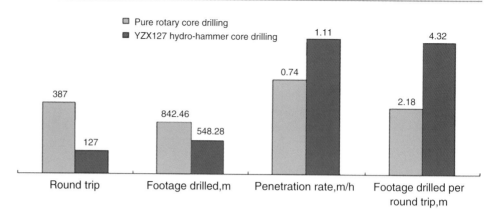

Table 5.24 Comparison of application of YZX127 hydro-hammer in the pilot hole

Phase	Hole section (m)	Round trip	Footage drilled (m)	Penetration rate (m/h)	Footage drilled per round trip (m)
The first phase	218.44–415.13	14	22.88	0.74	1.63
The second phase	789.60–809.64	6	13.40	0.64	2.23
The third phase	1209.61–2025.17	107	512.00	1.16	4.79
Later period of the third phase	1832.80–2025.17	18	129.34	1.32	7.19

Note As the abnormal round trips drilled by hydro-hammer are not deleted, the average increased results are lower than those in reality

512.00 m drilled, an average penetration rate of 1.16 m and an average footage of 4.79 m drilled per round trip. In comparison with the first phase, drilling rate was increased by 57 % and footage drilled per round trip was increased by 194 %.

What should be mentioned is that in the later stage of this phase the structure of the hydro-hammer was further perfected. At the same time properties of drilling mud were improved by adding lubricant to decrease the frictional resistance for the parts of hydro-hammer, and thus the working property and wear were greatly changed. The hole section from 1832.8 to 2025.17 m was the best period in which hydro-hammer drilling technique was used in the pilot hole, with the drilling rate even surpassing that in the shallow hole section of 500 m. In this phase hydro-hammers were used for 18 round trips, with a total footage of 129.34 m drilled, an average penetration rate of 1.32 m and an average footage of 7.19 m drilled per round trip (full core barrel). In comparison with the first phase, drilling rate was increased by 78 % and footage drilled per round trip was increased by 340 %. Through continuous improvement in pilot hole drilling, YZX127 hydro-hammer completely satisfied the requirements for drilling operations.

Application of YZX127 valve type double acting hydro-hammer

From August of 2001 to April of 2002, YZX127 hydro-hammer was used to drill 127 round trips, with a total footage of 548.28 m drilled. Figure 5.25 shows an overall statistic result: average penetration rate was 1.11 m/h (with the maximum 2.97 m/h) and average footage drilled per round trip was 4.32 m (with the maximum 8.72 m). In comparison with pure rotary driving technique, penetration rate was increased by 52.3 % while footage drilled per round trip by 100.1 %.

Through continuous improvement in above three phases, the application result of YZX127 hydro-hammer was gradually improved (Table 5.24). The application results of YZX127 valve type double acting hydro-hammer in different phases can be found in Fig. 5.26, from which it can be seen that YZX127 hydro-hammer was not stably perfected with a satisfactory result until the third phase.

Because of the property improvement, YZX127 hydro-hammer was more and more used. From hole depth 1209.61 m deep to 2025.17 m deep, where was the final depth of the pilot hole, 59.2 % of the total round trips were completed by YZX127 hydro-hammer (see Table 5.25; Fig. 5.27).

In drilling the pilot hole, YZX127 hydro-hammer manifested increased drilling rate, more stable working state, low repair cost, wide adjusting limit of structure parameters and wide adaptability at drill site.

(4) Summary of hydro-hammer application in the pilot hole

To sum up the application of hydro-hammer in the pilot hole, following understandings can be obtained.

  i. Valve type hydro-hammer can be used for deep hole drilling. In drilling the pilot hole from shallow to deep,

Fig. 5.26 Comparison of application of YZX127 hydro-hammer in the pilot hole

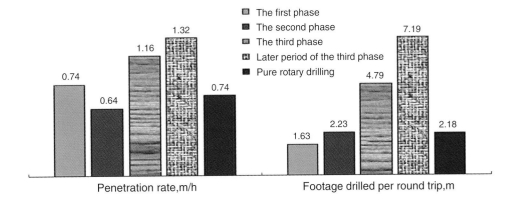

Table 5.25 Comparison of application of YZX127 hydro-hammer in the hole section from 1209.61 to 2025.17 m deep

	Round trip	Footage drilled (m)	Drilling rate (m/h)	Footage drilled per round trip (m)
All core drilling	181	815.56	1.10	4.48
Non-hydro-hammer drilling	27	69.11	0.94	2.56
YZX127 hydro-hammer drilling	107	512.00	1.16	4.79

it was not found that the working properties and efficiency was obviously affected. As the working stability and adaptability of hydro-hammer were improved, both the drilling rate and the footage drilled per round trip at deeper hole section were higher than those at shallower hole section. This shows that the valve type hydro-hammer is not sensitive to back pressure and thus the statement that the valve type hydro-hammer can not be used in deep hole drilling as it is sensitive to back pressure was negated.

ii. Improving drilling mud properties. The application results of hydro-hammer could be obviously improved by strictly controlling the solid content and increasing the lubricating property of drilling mud.

iii. Further improving the properties of valve type hydro-hammer. The service life of the parts remained to be further improved. The problem of fatigue strength of the parts had been basically solved but the service life of the matching surface of the piston sealing still needed to be further studied to prolong the continuous working life of the hydro-hammer after maintenance.

iv. The structure of the hydro-hammer should be further simplified. A relatively complicated structure was designed in order to reduce the cost of changing the easily worn parts. On the bases of improved service life, a further simplified structure would lead to problem decrease in application and increase of working stability.

Results achieved

The application of hydro-hammer in pilot hole drilling was a process of gradual improvement, which could be classified into three phases, according to the application results. In the first phase hydro-hammer was initially tested. Two hydro-hammers were not suitable to drill site conditions while fluidic type hydro-hammer worked well. During the second phase hydro-hammer was explored and improved. The properties of the hydro-hammer still could not meet the requirements for drilling, though the properties were improved. The technique was used in most of the round trips for core drilling. During the third phase a breakthrough was made in YZX127 valve type double acting hydro-hammer, with the properties obviously improved, being basically suitable to the application at drill site. The addition of lubricant into drilling mud improved the working environment for hydro-hammer and thus further improved the working properties of the hydro-hammer, with obvious economic benefits achieved. Till the later stage of the pilot hole core drilling, the properties of hydro-hammer became more and more stable, with drilling rate, footage drilled per round trip

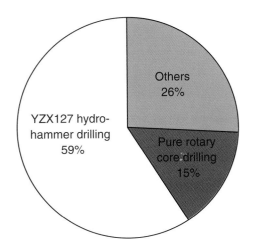

Fig. 5.27 Ratio of round trips drilled by YZX127 hydro-hammer in the hole section from 1209.61 to 2025.17 m deep

Table 5.26 Comparison of pilot hole core drilling data

Classification	Round trips	Footage drilled (m)	Penetration rate (m/h)	Footage drilled per round trip (m)
Pure rotary drilling	387	842.46	0.74	2.18
YZX127 hydro-hammer	127	548.28	1.11	4.32
KSC127 hydro-hammer	143	554.80	1.08	3.88
Pilot hole core drilling	657	1945.54	0.90	2.96

Table 5.27 Comprehensive statistics of drilling index in the second opening of CCSD-1 well

Date	Hole depth (m)	Footage drilled per month (m)	Footage drilled per rig month (m/30d)	Actual drilling time (%)	Tripping time (%)	Average drilling rate per month (m/h)
2001.07	184.87	83.87	157.26	31.84	36.49	0.69
2001.08	410.31	225.45	218.18	37.91	39.87	0.80
2001.09	685.13	274.81	274.81	37.56	45.17	1.02
2001.10	874.37	189.24	183.14	41.83	42.51	0.61
2001.11	1082.46	208.09	208.09	30.01	48.88	0.96
2001.12	1276.98	194.52	188.25	29.90	44.66	0.87
2002.01	1518.97	241.99	234.18	29.82	56.51	1.09
2002.02	1704.05	185.08	198.30	30.52	51.09	0.90
2002.03	1942.78	238.73	232.03	29.26	56.41	1.10
2002.04	2028.17	85.39	284.63	33.06	55.03	1.23
Total		1927.17	200.05	32.23	47.57	0.93

and core recovery obviously improved (Table 5.26). Because hydro-hammer drilling technique was adopted, excessive bit pressure was avoided during core drilling and that was very important for deviation control for the pilot hole.

Besides the advantage of improving drilling rate, hydro-hammer had the properties of reducing core blockage and increasing footage drilled per round trip, which was increased from 2.18 m by pure rotary drilling to 4.09 m, and thus tripping could be reduced by 60 % and much tripping time saved. All these played a decisive role in completing the pilot hole ahead of time and greatly reducing drilling cost.

From Table 5.27 it can be seen that along with the deepening of the borehole both the footage drilled per rig month and the average penetration rate per month were steady increased and the maximum of 284.63 m/30 d and 1.23 m/h were achieved before the borehole was completed. Along with the deepening of the borehole, tripping time was not increased, the utilization ratio of actual drilling time was stably kept and the average penetration rate per month was improved. All these reflected the great effects of the well developed hydro-hammer on increasing drilling rate and lowering drilling cost.

Problems existed

Unprecedented difficulties were encountered in drilling the pilot hole with two types of hydro-hammers. Through our efforts these problems were overcome and the technical level of the hydro-hammers was further improved, to suit to the complicated conditions at drill site. All these difficulties were seldom encountered in conventional geological exploration drilling activities.

For KSC127 fluidic type hydro-hammer, though it worked well at the initial period, because the fluidic element, as its key part, was rather sensitive to erosion and required a precise machining and assembling, the working results was not obviously improved at the later period, even through successive improvement. As for YZX127 valve type double acting hydro-hammer, though it did not suit well to drilling conditions at the beginning, its adaptability was solved very well after bold innovations, with working stability greatly improved at the later period of pilot hole drilling operations.

Both the hydro-hammers had the problem of insufficient sealing and inner parts were frequently eroded off. This led to a short service life of the whole hydro-hammer, requiring frequent repairs.

4. **Application and improvement of hydro-hammer in the main hole**

The research on hydro-hammer in main hole core drilling was mainly carried out in the following three aspects:

i. Application and improvement of YZX127 valve type hydro-hammer

As the wireline core drill rod supplied from Germany could not reach the standard in inside diameter and the RB130 top drive could not meet the requirement in torque, wireline core drilling drive at surface could not be realized in main hole core drilling. By study it was decided that the down hole power percussive rotary core drilling techniques would be used. Hydro-hammer + PDM core drilling technique was to be used for main hole drilling. As YZX127 valve type double acting hydro-hammer was well developed, it was mainly used in main hole core drilling. During the period of main hole drilling, the problems existed in some parts of the hydro-hammer were solved and the adaptability of the hydro-hammer to deep hole high temperature environment was studied so that the working properties of the hydro-hammer became more stable. For whole core drilling in the main hole, an accumulation of 2946.56 m was completed by hydro-hammer, with an average penetration rate of 1.15 m/h and an average footage drilled per round trip of 7.90 m achieved. The economic benefits were very obvious.

  ii. Research on the application of KS-152 wireline coring hydro-hammer

According to the initial engineering design, wireline core drilling technology was to be adopted for the initial stage of the main hole core drilling. KS-152 wireline coring percussion mechanism assembly was tested alternatively during the test of wireline core drilling, with five round trips drilled, among which two round trips failed because of mud circulation failure resulted from down hole drilling tool blockage by mud sediments. For the other three round trips, successful results were obtained.

  iii. Test research on PDM + hydro-hammer + wireline core drilling tool

On the bases of success of PDM + wireline core drilling tool and hydro-hammer + wireline drilling tool, a "three in one" combined drilling tool of PDM + hydro-hammer + wireline core drilling tool was under research so as to develop a high technique of down hole power percussive rotary wireline core drilling, which could solve the problems of inadequate top drive and initiated down hole accidents, increase penetration rate, footage drilled per round trip and core recovery. The test of "three in one" drilling tool was conducted in the test hole and the main hole separately. From May 17th, 2004 to June 9th, 2004 the drilling tool was tested in the test hole at the depth from 58.33 to 130.61 m. A total footage of 68.23 m was completed and the task was satisfactorily accomplished. In the main hole the drilling tool was comprehensively tested at the depth of 5125.68 m, with a footage of 3.5 m drilled and a penetration rate of 0.62 m/h obtained. Moreover, the last piece of core in CCSD-1 Well was successfully recovered. Though the down hole power percussive rotary wireline core drilling technique was not yet well developed, its outstanding virtues were verified in the production drilling in CCSD-1 Well. This technique can be a reserve for the future continental scientific drilling projects in China.

(1) Mature application of YZX127 hydro-hammer
  Problems existed and the solutions

At the stage of core drilling in the main hole, the main technical problems should be solved mainly included the reliability of the whole hydro-hammer and its property of high temperature resistance. At the initial period as hydro-hammer often failed and its parts easily damaged by erosion due to the complicated sealing structure, then it was considered that the sealing property of hydro-hammer should be improved along with the deepening of hole and the increase of down hole temperature.

By decreasing the part quantity of hydro-hammer and the quantity of radial sealing structure, the failure of the sealing surface resulted from vibrating wear during working process could be avoided. For axial sealing, a metal to metal plane sealing was adopted. A pulse pressure produced by vibration and percussion of the parts during working could compensate automatically once leakage happened, or crush and discharge the foreign objects which entered into the sealing surface. Along with the hole deepening, working temperature for hydro-hammer increases accordingly. Silica gel sealing materials which can bear 180 °C should be used for hydro-hammer sealing so as to satisfy the needs of high temperature sealing. Examination and repair of hydro-hammer at the later period showed that above-mentioned techniques and methods greatly reduced the failure while increased the service life of the parts. The changing frequency and the repair of the parts dramatically decreased and the stability greatly improved.

  Application results

In the main hole, a total footage of 2946.56 m was drilled by hydro-hammer, with an average penetration rate of 1.15 m/h and an average footage of 7.90 m drilled per round trip. In comparison with simple PDM core drilling by lifting drilling tool for core recovery, both penetration rate and footage drilled per round trip were increased greatly (see Tables 5.28 and 5.29).

(2) Application research on KS157 wireline coring hydro-hammer

In order to develop a drilling tool with still better effect under the basis of wireline core drilling technique, a pre-research project—Research on KS157 Wireline Coring Percussion Mechanism Assembly which combined wireline coring with hydro-hammer technique was set up in 2000. Wireline coring hydro-hammer combines the advantages of wireline coring and hydro-hammer into one, which can not only increase penetration rate, but also effectively avoid core blockage, increase footage drilled per round trip and improve core recovery. Through a comprehensive consideration of the overall size of the wireline coring hydro-hammer it was decided that the inside diameter of the diamond drill bit should be 85 mm. A percussion tool with large percussive work was required as the rock fragmentation area by diamond drill bit is large. Thus a hydro-hammer with large percussive work and necessary size needed to be developed.

Table 5.28 Comparison of application results of hydro-hammer in the main hole

	Round trips	Footage drilled (m)	Penetration rate (m/h)	Footage drilled per round trip (m)
PDM core drilling, by lifting drilling tool for core recovery	23	85.29	0.50	3.71
PDM + hydro-hammer core drilling, by lifting drilling tool for core recovery	373	2946.56	1.15	7.90
Comparison result			230 %	213 %

Table 5.29 Drilling results of hydro-hammer at different stages of the main hole

Main hole stages	Round trip	Footage drilled (m)	Penetration rate (m/h)	Footage drilled per round trip (m)
MH core drilling (including MH-IX section of deviation correction)	109	838.66	1.32	7.69
MH-1C core drilling	79	634.85	1.25	8.04
MH-2C core drilling	185	1473.05	1.03	7.96
Total in the main hole	373	2946.56	1.15	7.90

Table 5.30 Percussion test of KS157 wireline coring hydro-hammer at wellhead

Piston stroke per min.	Flow rate (L/s)	Pressure (MPa)	Remarks
24	4.14	0.77	Starting
26	4.48	0.95–1.13	Normal working at wellhead
30	5.17	1.35–1.75	Normal working at wellhead
32	5.52	2.96–3.08	Normal working at wellhead
35	6.03	3.51–3.7	Normal working at wellhead
36	6.21	3.88	Normal working at wellhead
40	6.90	4–4.48	Normal working at wellhead

Hydro-hammer parameters
  Outside diameter: 102 mm
  Hammer weight: 25 kg
  Valve stroke: 30–50 mm
  Hammer stroke: 40–60 mm
  Percussive work: 100–200 J
  Percussion frequency: 15–30 Hz
  Working flow rate: 200–400 L/min
  Working pressure: 1–4 MPa

After the success of YZX127 hydro-hammer, the development and improvement of KS157 hydro-hammer was immediately started, by drawing the successful experiences from YZX127 hydro-hammer, and indoor adjustment and inspection were completed.

Summary of field application

In the test at the wellhead, KS157 hydro-hammer was easily started and worked reliably, with the detailed data seen in Tables 5.30 and 5.31. Figure 5.28 shows KS157 wireline coring hydro-percussive mechanism assembly which was to be lowered into the well.

It can be found from the comparison of the drilling data between wireline coring hydro-hammer and conventional wireline core drilling tool shown in Table 5.32 that the penetration rate of wireline coring hydro-hammer was increased by 41 % in comparison with conventional wireline core drilling tool, though its bit kirf area was 8.9 % larger than conventional wireline coring bit.

Field production application showed that wireline core fishing tool was reasonably designed, safe and reliable in operations and 100 % success ratio in both lowering and fishing. Hydro-hammer showed very good working performances.

5. **Technical summary of hydro-hammer**

(1) Summary

In summary of the application of hydro-hammer in CCSD-1 Well, hydro-hammer played a very important role in shortening the time limit of the project and reducing drilling cost. Hydro-hammer promotes drilling technical progress in our country and embodies the drilling techniques with Chinese characteristics (Table 5.33).

i. Shortening the time limit of the project and reducing drilling cost

PDM hydro-hammer was used in 638 round trips in CCSD-1 Well. According to the footage drilled per round trip by conventional PDM core drilling, 1773 round trips needed to be drilled. In comparison with PDM hydro-hammer drilling, 1135 extra round trips needed to be drilled. Supposing each round trip needed 12 h, a total work time of

Table 5.31 Data for KS157 wireline coring hydro-hammer in actual operation

Test no.	Drilling time (h)	Footage drilled per round (m)	Core size (mm)	Penetration rate (m/h)	Starting depth (m)	End depth (m)	Remarks
1	4.44	5.13	85	1.16	2442.71	2447.84	1
2	2.71	1.29	85	0.48	2447.84	2449.13	2
3							3
4	2.10	1.85	85	0.88	2458.53	2460.38	4
5							5
Total	9.25	8.27		0.89			

Fig. 5.28 KS157 wireline coring hydro-percussive mechanism assembly

568 days would be increased. Based on a calculation that the cost per day was 40,000 RMB Yuan, a total of 22.72 million RMB Yuan were saved. Penetration rate was increased by 159 %, and according to the calculation based upon Table 5.33, the increased penetration rate shortened 71 actual drilling days and saved 2.84 million RMB Yuan. To sum up above two items, 639 days and 25.56 million RMB Yuan were saved. The input and output ratio exceeded 1:10.

ii. Promoting drilling technical progress in our country and embodying the drilling techniques with Chinese characteristics

Remarks

1 and 2. Hydro-hammer worked normally and inner tube assembly was lifted out successfully after lowering fishing tool.

After core recovered, inner tube assembly was lowered into hole at wellhead. Hydro-hammer worked for 2.71 h and then stopped working because of decreased free stroke of hydro-hammer resulted from thread loose of internal spline shaft. When inner tube assembly was lifted out it was found that core catcher had moved upwards and core had dropped into outer tube assembly. Drilling tool was lifted out.

3. Drilling tool was stopped 7.56 m apart from hole bottom. Cutting chips at hole bottom entered into the clearance between inner and outer tubes at core catcher, producing an excessive high pump pressure. Drilling tool was lifted out after releasing pump pressure.

4. Hydro-hammer worked for 1.5 h and then stopped working because of core blockage resulted from thread loose of core catcher. Pump pressure was very high. When fishing inner tube assembly, overshot was stopped by the core which had dropped into the drill rod during the second test round trip. Drill tool was lifted out.

Table 5.32 Comparison of drilling efficiency between KS157 wireline coring hydro-hammer and conventional wireline core drilling tools

	Drilling time (h)	Footage drilled per round trip (m)	Core size (mm)	Penetration rate (m/h)	Starting depth (m)	End depth (m)	Comparison
KS157 wireline coring hydro-hammer	9.25	8.27	85	0.89	2442.71	2460.38	1.41
Conventional wireline core drilling tool	12.02	7.62	93	0.63	2047.82	2462.71	1.00

Table 5.33 Comparison of drilling efficiency of hydro-hammer in CCSD-1 Well

	Round trips	Footage drilled (m)	Penetration rate (m/h)	Footage drilled per round trip (m)
PDM conventional core drilling (by lifting drilling tool for core recovery)	398	909.31	0.71	2.28
PDM + hydro-hammer core drilling by lifting drilling tool for core recovery	638	4043.25	1.13	6.34
Comparison			159 %	278 %

5. Drilling tool was stopped 3 m apart from hole bottom. Cutting chips at hole bottom entered into the annular clearance between inner and outer tubes and then stopped pump working. After releasing pump pressure, inner tube assembly was lifted out and drilling tool was lifted out after wireline broken. Examination at surface found that the whole length of 11.5 m of the annular clearance between inner and outer tubes was completely blocked up with cutting chips and the inner tube assembly could not taken out from outer tube assembly. Test had to be stopped.

The successful application of hydro-hammer in CCSD-1 Well set a record in the utilization of hydro-hammer in a well deeper than 5000 m with complicated drilling mud environment, verified the application properties of hydro-hammer percussive rotary drilling techniques in deep hole, promoted hydro-hammer techniques to a new level, and became a milestone for the development of hydro percussive rotary drilling techniques.

(2) Problems existed

Following problems existed in the application of hydro-hammer in CCSD-1 Well:

i. Impurities harm hydro-hammer

Metal chips produced by part wear in mud circulation system often lead to piston blockage, or even piston failure, once the chips enter into the sealing of hydro-hammer. In addition, the large sized inert lost circulation materials in drilling mud might block the piston clearance of hydro-hammer as they might block the formation fractures. In this case the piston of hydro-hammer would be blocked.

During the period of initial core drilling in MH-1C, small plastic balls of different sizes were used as solid lubricant in setting casings. As these plastic balls were similar to drilling mud in density, and could not be completely screened out by vibrating screen. When changing drilling mud system, the plastic balls in mud tank and in pipeline could not be completely removed and thus the remaining balls would enter into the piston of hydro-hammer and block the piston. In the down hole with high temperature, the softened balls had a certain toughness, resulting in a still more serious harm (Fig. 5.29). The rubber chips fall off from PDM stator would enter into hydro-hammer with circulating mud and block the mud channel, resulting in a hydro-hammer failure.

Fig. 5.29 Plastic balls in the clearance of piston

Fig. 5.30 Rubber chips entered into hydro-hammer after PDM degummed

Figure 5.30 shows the rubber chips cleared out of hydro-hammer in examination and repair.

ii. Damage of outer tube

The outer tube of hydro-hammer is the key part that affects hydro-hammer strength. The outer tube should not be

Fig. 5.31 Outer tube damaged by hydraulic pneumatic wrench

too thick as the thick tube wall will affect the sizes of the parts and further affecting the output of hydro-hammer. However, in field operations repeated makeup and breakout will damage the thread of outer tube, deteriorate its strength and lead to an outer tube deformation, which results in blockage of the parts and even outer tube failure (Fig. 5.31). Therefore, in future design, strength of outer tube should be comprehensively considered, based upon field application conditions.

iii. Drilling efficiency decreased while hole depth increased

It can be found from the hydro-hammer application parameters in three coring stages in the main hole that along with the increase of hole depth hydro-hammer was gradually perfected while its drilling efficiency decreased, with the reasons as follows:

(i) Rock strength increases along with the increase of rock buried depth. Elastic modulus of rock increases with the increase of pressure (Wuhan Geological College et al. 1980) and this causes a decrease of drilling efficiency. According to the research by Klaustal University, Germany, compressive strength and plasticity of rock increase to varying degrees along with the gradual increase of environmental surrounding pressure. In this case, the percussion work needed will be dramatically increased if the same rock fragmented volume is to be produced. Considering the increase of percussion work of hydro-hammer may lead to an inadequate strength of down hole coring tool, and the high cost to treat deep hole accidents, conservative drilling parameters were adopted in the later stage and this to a certain extent affected the hydro-hammer drilling efficiency.

(ii) Rock formation change during drilling operations. As the hole depth deepened, rock formation changed. The deeper the hole was, the less the proportion of eclogite and amphibolite would be. Gneiss with high strength was the main rock formation to be drilled and in the hole deeper than 4000 m gneiss was almost the only rock. Drilling eclogite with PDM hydro-hammer the penetration rate often reached to approximate 2 m/h, while in amphibolite the penetration rate only reached to approximate 1.2–1.5 m/h. Drilling efficiency would be further affected by core blockage if rock formation was relatively broken.

(iii) Though it was not found that the percussion work of hydro-hammer obviously decreased with the increase of back pressure, this phenomenon still exists in theory and in laboratory experiment. Although its tendency was not obvious, yet the back pressure exerted a slight inference on rock fragmentation results of valve type double acting hydro-hammer, along with the increase of hole depth.

The application of hydro-hammer drilling techniques in CCSD-1 Well shortened the time limit of the project and economized large funds. This technique became one of the main drilling techniques in the project. The application of hydro-hammer in CCSD-1 Well played an important role in greatly promoting the development of hydro-percussive rotary drilling techniques, widening its application, and establishing China's international leading position in hydro-hammer techniques.

5.3.5 Core Drilling Tool

Core drilling tool is another important component for down hole power percussive rotary core drilling system. As we know, there are a variety of core drilling tools in oil and gas drilling industry and in geological exploration industry. However, the new type core drilling system developed for CCSD-1 Well could not indiscriminately copy these tools, as the working condition of CCSD-1 Well was quite different from those in other drilling systems.

5.3.5.1 Basic Condition

1. **Working condition for drilling tool**

CCSD-1 Well was to be drilled by down hole power hydro-percussive rotary drilling technique, with rotary driving by PDM and high frequency percussion by hydro-hammer, in hard rock formations. In comparison with conventional core drilling, the borehole diameter of CCSD-1 Well was larger and the core was larger in size, and thus larger core lifting force was needed. And in comparison with oil core drilling, CCSD-1 Well had the characteristics of higher rock drillability grade, higher drill bit rotary speed and larger vibration on drilling tool.

Drilling tool worked under the following conditions:

Table 5.34 Technical indexes of hard rock drilling tools at home and abroad

Manufacturer	Model	Outer tube OD (mm)	Inner tube ID (mm)	Drill bit OD (mm)	Core size (mm)	Rock fragmentation coverage of drill bit (%)
Christensen		146.05	92.25	max165	88.90	67.94
Hycalog		146.05	92.08	max165	88.90	67.94
Baker Haghes		139.70		max165	88.90	67.94
Sichuan Oil Prospecting Bureau	Chuan 6 series	133.00	76.00	max165	70.00	80.12
Shengli oil field	Y-6-70	133.00	76.00	max165	70.00	80.12

Rock fragmentation coverage of drill bit was calculated based on 157 mm O.D. used for CCSD-1 Well

Eclogite and gneiss were the main rock formation to be drilled, with drillability from 7 to 9 grade. Some rock layers with high quartz content had drillability of over 10 grade. Most of the rock formations were relatively complete.

Rotary speed was 200–300 rpm and rotary linear velocity of bearing was generally over 2 m/s.

Axial pressure was 20–50 kN.

Drilling mud density was 1.03–1.06 g/cm^3, with sand content ≤0.1 %. Pump discharge was 9–15 L/s.

As PDM down hole drive, outer tube bore a high frequency radial vibration. Worked with hydro-hammer, outer tube bore a high frequency axial percussive load.

When core blockage happened, inner tube bore pressure and a complex radial and axial vibration load from PDM and hydro-hammer.

During breaking the core, drilling tool bore a downward pulling force, with the peak value of more than 100 kN.

2. **Technical requirements for drilling tool**

Because appearance scanning was to be conducted on all the core from CCSD-1 Well, thus core recovery must meet the standards, and core should be regular and smooth in appearance, as cylindricity and apparent smoothness of core have great influence on the quality of scanning imagery. In order to drill the hole safely and in high quality, the drilling tool must have the following technical properties:

Reliable move of inner tube is the basic requirement for coring tool, as it can avoid a relative rotation between core (as well as broken end of core) and inner tube, and keep high core recovery and regular core appearance.

Free water way of drilling tool can guarantee that an excessively high hydro pressure drop will not be produced when drilling mud of 9–12 L/s or even with larger flow rate passes through.

Good and effective lubrication will guarantee a reliable move of inner tube and a long service life of bearing in adverse working environment.

Simple structured drilling tool can reduce production and repair costs, and being easy for assembling and disassembling at drill site.

In order to catch core firmly, core catcher must have enough catching force, as rock formation is very hard and core is long and heavy, sometimes the peak value of core lifting force can reach to over 100 kN.

Thread should be strong enough in strength and easy for make-up and break-out.

3. **Necessity for developing special drilling tool for CCSD-1 Well**

Among the geological drilling tubes available in China, the only suitable one was that with 146 mm O.D. However, this tube was only 4.75 mm in wall thickness and high strength thread could not be machined.

Oil core drilling tool, with very high safety factor, was designed for soft sedimentary formations and for oil drilling mud system, based on surface rotary driving. In order to prevent drilling tool sticking by rock formation expansion and reduce the pressure loss by large flowrate circulation of high viscosity mud, the clearance between the outer tube of oil drilling tool and well wall and the clearance between inner tube and outer tube both are rather large, thus rock fragmentation coverage of drill bit is very high (Rock fragmentation coverage of drill bit = (rock fragmentation area/borehole area) × 100), and the domestic drilling tool is still higher, as shown in Table 5.34 (Li et al. 1993). In soft rock drilling, rock fragmentation coverage of drill bit does not exert an obvious influence on drilling rate, while in hard rock drilling this influence is great, as well as the increase of drill bit production cost and application cost. It can be found from Table 5.34 that in China there was no available drilling tool suitable for CCSD-1 Well and though there were available drilling tools in international market yet the cost was rather high. The charge standards of Baker Hughes Co. in U.S.A. for hard rock coring tool of 140 mm O.D. and 89 mm I.D. was 200 US Dollars/h, and 1000 US Dollars per person per day for field service. For sale, the whole drilling tool cost 61,500 US Dollars and spare parts cost 15,200 US Dollars per set (two sets were needed, each was recommended for twelve round trips). It was not verified whether the swivel type structure could work reliably for a long time under the compound high frequency vibration of hydro-hammer and PDM. Besides this, for oil drilling rock formation is soft, coring quantity is not large, rotary speed of drilling tool is slow, drilling time per round trip is short, and the dynamic load that the bearing bears is relatively small, thus the service life of the bearing will have no obvious problems. The convincing statistics of oil drilling tool service life in hard metamorphic rock have not yet been found. As continuous coring was to be required for 5000 m deep in hard rock formations, the service life of the bearing would be a problem which must be seriously considered. Therefore, the available drilling tools in oil drilling industry could not be the first choice for CCSD-1 Well.

In summary, for CCSD Project, coring tools (by lifting drilling tool for core recovery) which can bear high rotation and compound high frequency vibration under hard rock condition must be designed by ourselves, based upon the practical conditions of CCSD-1 project.

4. **Drilling tool design and application**

Based upon abovementioned requirements, the Scientific Drilling Project Centre entrusted the Institute of Exploration Techniques with the design of KZ swivel type double tube and single tube drilling tools. Because the rock formation was not suitable for single tube drilling (coring failed in all the three round trips), actually all the coring in the hole was completed by swivel type double tube drilling tool. The research and development of double tube drilling tool experienced two runs, the first run of drilling tool failed after drilling 105 m because of designed defects, high cost, difficult machining and assembling, and early failure of swivel structure. During the second run, the drilling tool was completely transformed, with swivel structure and bearing service life dramatically improved and core blockage reduced. Especially after adding lubricant into drilling mud, a record was set that the whole drilling tool worked for 150 h of actual drilling time, without any maintenance. Completed a footage of 4742.41 m, KZ swivel type drilling tool was appraised by the Scientific Drilling Project Centre.

At the drill site, TG-1 drilling tool developed by Beijing Institute of Exploration Engineering drilled 196.05 m and the SL-1 drilling tool developed by the Institute of Drilling Tools, Shengli Oil Drilling Academy completed 18.73 m. In rock formation where core was difficultly taken the jet reverse circulation swivel type double tube drilling tool developed by the Institute of Exploration Techniques was used for drilling not too many meters. As the TG-1 and the SL-1 drilling tools are similar to the KZ drilling tool, therefore only the KZ drilling tool is emphatically introduced in this chapter.

5.3.5.2 The First Run of Design and Trial of the KZ Drilling Tool

1. **Drilling tool structure**

The first run of drilling tool structure is shown in Fig. 5.32.

 i. In the early stage of drilling CCSD-1 Well, oriented coring at some hole sections was required so as to reveal the occurrence of the formation. Core orientator chamber, as an assembled part, should have an independent sealed swivel structure, to be connected with the outer assembly when it is going to be lowered into the hole.
 ii. Swivel structure of centripetal bearing for positioning and thrust bearing for bearing force was adopted, with

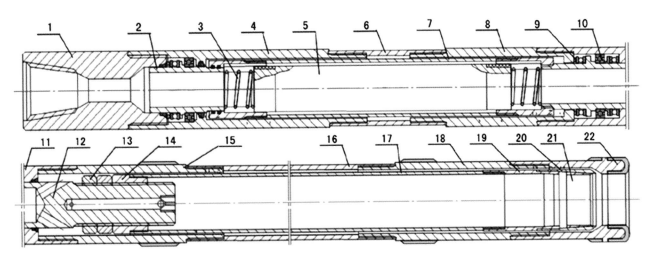

Fig. 5.32 First run of KZ-1 drilling tool (prototype). *1* Upper adapter. *2* Upper sealing ring of orientator chamber. *3* Shock absorbing spring. *4* Upper adapter of orientator chamber. *5* Core orientator chamber. *6* Outer tube of orientator chamber. *7* Inner tube of orientator chamber. *8* Lower adapter of orientator chamber. *9* O ring. *10* Bearing. *11* Upper adapter of core barrel. *12* Mandrel. *13* Back cap. *14* Upper adapter of inner tube. *15* Upper reaming shell. *16* Outer tube. *17* Inner tube. *18* Lower reaming shell. *19* Sub of inner tube. *20* Core spring case. *21* Core spring. *22* Drill bit

sealed lubrication, and bearing chamber on outer assembly.

iii. Transitional matching was used for bearing positioning, so as to prevent strong vibration.

iv. Drilling mud enters into mandrel through hole, and flows into the clearance between inner and outer tubes through water bifurcation hole.

v. The upper and the lower reaming shells were used for stabilizing the drilling tool, repairing borehole and maintaining borehole diameter.

vi. The outer tube thread teeth for geological drilling purposes was to be adopted for inner tube thread because its connecting strength would be strong enough to directly bear the breaking and pulling force during core lifting. In this case, the buffering mechanism for core pulling in inner tube assembly could be saved.

vii. Trapezoidal thread of 5° with thread pitch of 8 mm and thread height of 2 mm was adopted for outer tube thread.

2. **Application result**

In the hole section from 101.00 to 212.42 m deep (PH1–PH59), PDM downhole drive pure rotary drilling was utilized for this run of prototype of drilling tool, except for one round trip drilled by rotary table drive.

Because of improper selection of centripetal bearing, the service life of the bearing was rather short and usually damaged after two round trips (at most three round trips) in borehole. As sealed lubrication was adopted and assembly precision was stressed, assembling and disassembling of drilling tool were rather difficult. Frequent change of the bearings made the maintenance at drill site extremely heavy. The application result of the drilling tool was unsatisfactory resulted from the poor swivel property, leading to poor quality core, low core recovery, low penetration rate, short footage drilled per round trip, mud scaling inside inner tube, and easy damage of core spring and core spring case, as shown in Fig. 5.33.

The drilling tool was used for 55 round trips, with a total footage of 105.68 m drilled, accumulative core of 93.64 m recovered, an accumulative drilling time of 149.5 h consumed, an average core recovery of 88.61 % obtained, an average footage of 1.92 m drilled per round trip and an average penetration rate of 0.71 m/h achieved.

3. **Technical analyses and the direction for improvement**

i. Sealed lubrication was adopted for the bearing and thus mud was discharged from central hole of mandrel. The outside diameter of the mandrel designed in this run was as large as 90 mm, whereas the inside diameter of the bearing chamber was only 120 mm because of the limitation of hole size. In this case, it was difficult to choose heavy duty bearing to overcome the downhole compound vibration load. The failure of swivel property mainly resulted from the damage of bearing, which was also the cause of a series of adverse results.

ii. Sealing failed quickly at hole bottom because of the limited space, adverse down hole working circumstances and difficult precise machining. Failed sealing not only speeded up the wear of the bearing, but also let the solid grains in drilling mud fill the sealing clearance and wear the mandrel.

iii. The side water hole on the adapter of inner tube was originally designed for pressure release for inner tube. However, a local reverse circulation was produced in the inner tube by suction power to the inner tube chamber when drilling fluid passes through the annulus between inner and outer tubes at a high velocity. In case of failure of drilling tool swivel motion, the inner tube and the outer tube rotated simultaneously and thus mud scaling was resulted from mud solids because of centrifugal force.

iv. For thread connection between mandrel and inner tube adapter, only a nut was designed for anti-loosing, insufficiently resisting the strong shock produced by downhole power drilling tool. The relative rotation between mandrel and inner tube adapter increased the

Fig. 5.33 Adverse effects caused by failure of swivel property. **a** Rounded core. **b** Mud scaling inside the inner tube. **c** Wear and tear of core spring

length of adjusted inner tube assembly and thus drilling tool swivel motion failed as a result of core spring case pushing against drill bit steel body.

v. Frequent change of bearings and complicated assembling and disassembling brought a lot of difficulties for drilling tool maintenance at drill site.

vi. The mating of core spring and drill bit was adopted according to core drilling standards that core spring inner diameter was 0.3 mm smaller than drill bit inner diameter. When the inner surface of core spring wore, catching force was not enough. Quenching quality of core spring surface was difficultly controlled and then core spring broken frequently happened.

The guiding ideology of this round of design was to maintain the service life of the bearings with sealed lubrication, to improve the impact resistance of the bearings with accurate position, and to decrease the hydraulic loss with large mandrel through hole. In fact, the drilling mud used in CCSD-1 Well was very low both in solid content and in viscosity, and with good lubrication. Moreover, the rated pump pressure of the pump at drill site was really very high. All these conditions enable the drilling tool to decrease the flow passage area and mandrel, so as to increase the bearing chamber for selecting heavy duty bearings. Open lubrication is allowed for application. Therefore it was decided in the second round of design that the mandrel size should be decreased to increase the bearing chamber. The bearings are compulsorily lubricated by all flow drilling mud. Centripetal bearings would be abandoned and fitting precision reduced. Heavy duty thrust bearings with good impact resistance should be selected to remedy the positioning defect.

5.3.5.3 The Second Run of Design and Application of the KZ Drilling Tool

1. **Drilling tool structure**

The second run of drilling tool structure is shown in Fig. 5.34, with the following characteristics.

i. Full pumping rate open lubricated bearings simplified drilling tool structure and made drilling tool assembly and disassembly much convenient. In fluid distribution, most of the medium flows through the bearing chamber passage to realize a compulsory lubrication to the bearings while a small amount of medium flows horizontally through the side water hole on the mandrel connected to the bearing chamber to remove the dirt in the bearing body and lubricate the balls (as long as there is a pressure difference between the side water hole on the mandrel and the bearing chamber passage, fluid will horizontally flows through).

ii. As shown in Fig. 5.35, the bearings were installed in the outer assembly, instead of installing in the inner assembly based upon the conventional way. As open lubrication reduced the diameter of mandrel, enough space was provided for installing heavy duty thrust bearings, then the pressure bearing and impact resistance abilities of the drilling tool under high rotary speed were both improved, and the service life of the drilling tool assembly was increased. The mandrel acts as suspending inner tube, realizing swivel moving and adjusting the distance from the inner tube to drill bit. For the convenience of making up and breaking out the inner tube, the thread of mandrel was designed into M60 × 3. Because of the poor self-locking ability of

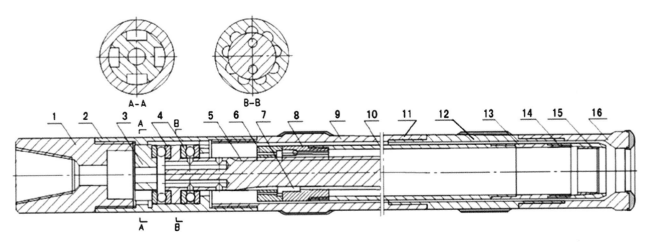

Fig. 5.34 Second run of KZ-1 drilling tool. *1* Upper adapter. *2* Bearing seat. *3* Cover. *4* Bearings. *5* Mandrel. *6* Back cap. *7* Antiloose key. *8* Adapter of inner tube. *9* Upper reaming shell. *10* Inner tube. *11* Lower reaming shell. *12* Outer tube. *13* Sub. *14* Core spring case. *15* Core spring. *16* Drill bit

Fig. 5.35 Sectional drawing of swivel structure

the conventional thread, to prevent the inner tube from axial displacement under impacted rotation during drilling process, the back cap locking design in the first round was abandoned, and the replaced design was to cut a half-opened slot axially on mandrel and inner tube adapter. Once the position of the inner tube on the mandrel was well adjusted, a stock key was inserted into the half-opened slot, so as to effectively restrain the displacement of the inner tube. The back cap exerts a pre-tight force onto the inner tube, to avoid thread damage caused by the inner tube loose and to avoid the key escaping from the slot and falling into the annulus.

iii. The pressure release hole on the inner tube adapter was changed into a direct opened design, to prevent drill mud in the inner tube from local reverse circulation and to avoid mud cake growth in the inner tube in case of swivel failure.

iv. 5° trapezoidal straight thread with thread pitch of 4 mm and thread height of 0.75 mm was adopted for inner assembly connection. Tests showed that the tensile strength of this thread was higher than 150 kN, which was much higher than the needed core breaking force. Therefore core pulling buffer in the inner assembly was eliminated and in the process of coring the thread in the inner assembly directly bore the force of core breaking. In this way the drilling tool structure was further simplified, the space in the inner assembly was saved, and longer footage per round trip could be completed with the same length drilling tool.

v. 5° trapezoidal straight thread with thread pitch of 8 mm and thread height of 2 mm was adopted for outer assembly connection. The thread profile was easy for machining and less tube wall thickness was occupied. Both the tensile strength and the torsional strength could satisfy the need for drilling (under 10 kN m upper thread torque, tensile strength was no less than 400 kN). However, poor sealing compensation property was the weakness of this thread, as under high pumping pressure leakage and thread sticking easily happened. To avoid this, a copper washer was placed at the thread which was often made up and broken down to guarantee the reliability of sealing.

vi. The inside diameter matching between core spring and drill bit was changed into—(0.5–0.8) mm and core spring body was not quenched after tempered.

vii. The inner tube assembly was 9.5 m in total length for long round trip drilling. To increase the tensile strength and service life of the inner tube, brazing method was used for the inner tube adapter connection. Male thread was adopted for the inner tube lower thread, and the outer diameter of the thread was thickened to connect the sub, which protects the inner tube lower thread and prevents the core spring from up moving.

viii. As an assembled part, the orientator cabin would be connected between the mandrel and the inner tube adapter when entering into the borehole and then a swivel movement was realized depending upon the drilling tool bearings. In comparison with the first round design, this round design had the strong points of simple structure and reliable swivel movement, however, with the shortcomings that the core containing space was occupied in the inner tube assembly and then the footage drilled per round trip for core drilling was shortened.

ix. Rock fragmentation area of the drill bit was only 62.61 %, being the lowest among all the available drilling tools. As for thread connections, drilling tool stabilization, hole repair and hole size protection, and core catching, all had been proven effective in the first round trial, were still used.

Fig. 5.36 A comparison of pure rotary drilling indexes between the two round drilling tools

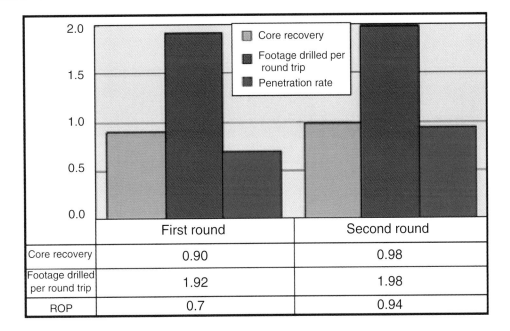

	First round	Second round
Core recovery	0.90	0.98
Footage drilled per round trip	1.92	1.98
ROP	0.7	0.94

x. In the later stage of drilling operations, the inner tube drooped down due to the increase of hole deviation and then core sometimes pushed the core spring into the inner tube. To overcome this, a stabilizing ring made of nylon 6 rod material was put between the drill bit and the reaming shell, to guarantee the inner tube always being in the centre. In this way the inner diameter of the sub lower end was reduced, effectively stopping the core spring from up moving.

2. **Drilling tool technical indexes**

   Nominal diameter of the drill hole: 157 mm
   Nominal diameter of the core: 96 mm
   Diamond drill bit size: 157.2–157.5 mm
   Reaming shell diameter: 157.5–157.8 mm
   Rock fragmentation coverage area of the drill bit: 62.61 %
   Outer tube specifications: 139.7 mm/118.7 mm
   Inner tube specifications: 110 mm/100 mm
   The longest core contained: 9.5 m
   The maximum working bit pressure borne: 40 kN
   Average service life of bearings: no less than 80 h

3. **Application results**

(1) In comparison with the first round drilling tool

The initial application of the drilling tool showed superiority in comparison with the first round drilling tool. The appearance of the core was much better and cracking face was no longer rounded. The service life of both core spring and core spring case increased, especially the service life of the bearings obviously increased, while mud scaling inside the inner tube disappeared. Drilling rate and core recovery were obviously increased, though the footage drilled per round trip was no distinctly improved during pure rotary drive. A comparison between this round drilling tool and the first round drilling tool based upon the statistics of 61 round trips of pure rotary drive drilling at the same rock formation was shown in Fig. 5.36. The increase of new drilling tool properties was due to the improvement of swivel property, which greatly reduced core blockage, as the friction between cores and between core and the inner tube was obviously decreased.

(2) Developing process of core barrel length

Along with the perfection of hydro-hammer technique in CCSD-1 Well, core blockage during drilling obviously decreased. Under these circumstances, to improve the footage drilled per round trip so as to further increase the utilization ratio of actual drilling time became an important task laid ahead of us. Through study, we decided to increase the length of core barrel from original 3 and 4.5 m to 6 m and finally to 9 m, in this way the total tripping time in deep drilling was effectively reduced and the technical and economic indexes of CCSD-1 Well were optimized. The development of core barrel length in pilot hole drilling is found in Table 5.35.

(3) Total drilling technical indexes

The total technical indexes obtained by the drilling tool are shown in Table 5.36.

4. **Technical analysis**

   i. In general, rock formation in CCSD-1 Well is relatively intact, though local broken and fissure occupied layers exist (Fig. 5.37). In this case, the overall technical requirement that the core recovery should be higher than 80 % could be satisfied by using this drilling tool. Worked with hydro-hammer, this drilling tool improved both core recovery and core quality. At drill site operations, several pieces of complete core longer than 3 m were recovered, among them the longest reached 4.25 m (Fig. 5.38).

Table 5.35 Developing process of core barrel length

Date	Well depth (m)	Core barrel length (m)	Round trip	Footage drilled (m)	Footage drilled per round trip (m)	Note
~2001.10.28	101.00–852.68	3–4.5	352	751.6	2.14	Pure rotary drilling and initial stage of hydro-hammer
2001.10.29–2001.11.26	852.68–1054.45	4.5–6	66	201.77	3.06	Improvement stage of hydro-hammer
2001.11.26–2002.03.16	1054.45–1807.6	6–9	211	753.20	3.57	
2002.03.17–2002.04.15	1807.65–2046.54	9	38	238.89	6.29	Perfection stage of hydro-hammer

Table 5.36 Application results of KZ-1 drilling tool

Hole	Round trip	Footage drilled (m)	Core length (m)	Actual drilling time (h)	Footage drilled per round trip (m)	Penetration rate (m/h)	Core recovery
Pilot hole	529	1604.88	1458.55	1745.27	3.03	0.92	90.88
Main hole	396	3031.85	2548.24	2743.12	7.66	1.11	84.05
Total	925	4636.73	4006.79	4488.39	5.01	1.03	86.41

Note Well depth from 212.42 to 5118.2 m was completed by KZ-1 drilling toal, whereas well section above 212.42 m was drilled by the first round drilling tool

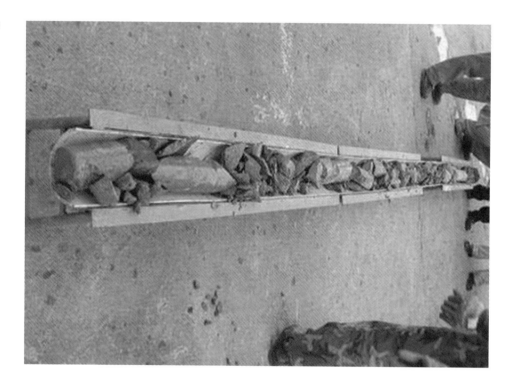

Fig. 5.37 Broken core taken by KZ-1 drilling tool

ii. At the initial usage of this drilling tool, an observation and study was conducted on bearing service life for more than 100 round trips. Without using hydro-hammer, the maximum bearing service life for continuous pure drilling reached to 150 h whereas working with hydro-hammer with an average of more than 80 h. Long service life of the bearings resulted from heavy duty bearings adopted, full pump rate forced lubrication and high quality mud used.

iii. During main hole drilling, an observation was conducted on the service life of tubing thread and it was found that the service life of tubing thread reached over 100 h in case of utilization rate of hydro-hammer in each round trip reached to 100 %. After drilling deeper than 3800 m, to absolutely guarantee down hole safety, it was stipulated that the thread of outer tube assembly must be scrapped either after 80 h drilling or after break-out for 10 times.

Fig. 5.38 A 4.25 m complete piece of core recovered by KZ-1 drilling tool

iv. Under the condition that the requirement for use is satisfied, the structure should be simplified and machining accuracy should be reasonably loosened, so that the drilling tool machining, assembling and maintenance would be highly simplified. In material selection and thread strength design, safety factors were fully considered, therefore, in-the-hole accidents caused by drilling tool failure never happened during the process of 4600 m drilling. Thick wall tube of moderate steel grade was selected for outer tube and in drilling practice the safety of thread connection was guaranteed; tubing deformation was avoided and the percussive work from hydro-hammer was effectively transmitted, though the 8.5 m long tubing was under an axial load and a complex percussive vibration. Practice showed that the material selection and the thread strength design were basically reasonable.

v. The application and perfection of hydro-hammer techniques greatly brought the ability of the drilling tool into full play. Along with the field practices and our deeper understanding, the core containing capability of the drilling tool was increased from originally designed 6 to 9.5 m after 1800 m deep and thus the drilling indexes were dramatically improved by using of the drilling tool working with the hydro-hammer. It can be found from Table 5.36 that both the footage drilled per round trip and the penetration rate in the main hole were higher than those in the pilot hole, with reasons of the usage of long core barrel and nearly 100 % success of the hydro-hammer, which exerted a decisive influence on great reduction of drilling cost in CCSD-1 Well. As for the lower core recovery in the main hole in comparison with the pilot hole, the reason was not due to the drilling tool itself, but because that the intactness of the rock formation in the main hole was inferior to that in the pilot hole.

vi. This drilling tool is low in cost. In comparison with the imported drilling tool for hard rock coring, the direct cost of the imported drilling tool was more than ten times higher than that of this drilling tool for 1000 m drilling in the same rock formation, even if the higher drill bitcost and lower drilling rate caused by the smaller core size (89 mm) of the imported drilling tool were not considered. Furthermore, this drilling tool could be processed near the drill site and so the transportation and storage costs could be reduced to minimum, because the machining accuracy was not very high and the thread machining technology was relatively simple.

vii. To increase the footage drilled per round trip is the most effective way to improve the technical and economic benefits for deep hole coring by lifting drilling tool for core recovery. A remarkable benefit was achieved by increasing the core containing capability of the drilling tool from originally designed 6 to 9.5 m based upon drilling practices. As for whether there is any possibility for further improvement, our answer is in the affirmative. We believe that there will be much work to do on theoretical analysis and material selection in the future.

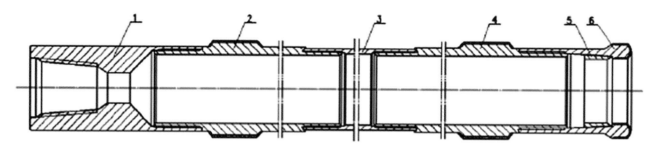

Fig. 5.39 Single tube coring tool. *1* Upper adapter. *2* Upper reaming shell. *3* Outer tube. *4* Lower reaming shell. *5* Core spring. *6* Single tube core drill bit

viii. This drilling tool was designed only for relatively intact hard rock formations. Though it can be used in broken rock formations, core recovery can not be guaranteed in structure loose, tectonic complex and water sensitive formations.

5.3.5.4 Other Drilling Tool Structure and Their Application Results

1. **Single tube coring tool**

The single tube coring tool (shown in Fig. 5.39) was developed by the Institute of Exploration Techniques. This coring tool consists of upper adaptor, upper reaming shell, outer tube, lower reaming shell, core spring and core drill bit. Core is stored in the outer tube. In lifting drilling tool, core is cut and caught by the inner conical fit between core spring and drill bit. This coring tool was used for five round trips (Table 5.37), with low core recovery.

2. **TG-1 core drilling tool**

TG-1 core drilling tool (shown in Fig. 5.40) was developed by the Beijing Institute of Exploration Engineering, with the aim of improving core recovery in broken rock formation by using the local reverse circulation in the annulus between inner and outer tubes. In the design of swivel system, bearing chamber outer assembly and opened lubrication were also used. In comparison with KZ-1 drilling tool, three points of improvement were completed.

i. The upper bearing one grade larger was selected. During drilling operations, the inner tube is often at the upper dead point because of the friction of core and therefore the force the upper bearing bears is larger than that the lower bearing bears. Consequently, the upper bearing wears faster than the lower bearing. The upper bearing one grade larger is beneficial to improving the service life of the total drilling tool.

Table 5.37 Application results of the single tube drilling tool

Round trip	Coring tool	Initial depth (m)	Footage drilled (m)	Actual drilling time (h)	Penetration rate (m/h)	Core recovery (%)	Drive mode
PH5	2.5 m single tube	104.98	1.71	2.67	0.64	26.9	LZ120
PH14	2.5 m single tube	116.75	1.60	1.75	0.91	54.4	LZ120
PH15	2.5 m single tube	118.35	1.93	3.92	0.49	22.8	LZ120
PH16	2.5 m single tube	120.28	0.50	0.67	0.75	30.0	LZ120
PH160	4 m single tube	405.16	1.10	3.00	0.37	27.3	LZ120
Total			6.84	12.00	0.57	32.5	

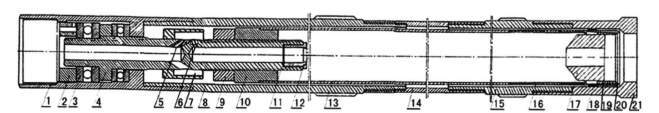

Fig. 5.40 Structure of TG-1 drilling tool. *1* Upper adapter. *2* Cover. *3* Bearings. *4* Mandrel. *5* Spray nozzle. *6* Back flow hole. *7* Fluidic retainer ring. *8* Upper reaming shell. *9* Back cap. *10* Adapter of inner tube. *11* Anti-loose key. *12* Cuttings screen. *13* Inner tube. *14* Outer tube. *15* Lower reaming shell. *16* Sub. *17* Core mark. *18* Stabilizing ring. *19* Core spring. *20* Core spring case. *21* Drill bit

Table 5.38 Application results of TG-1 drilling tool (with water isolated double tube)

Coring tool	Footage drilled (m)	Core length (m)	Actual drilling time (h)	Penetration rate (m/h)	Core recovery (%)	Drive mode	Remarks
4 m	21.04	21.30	12.35	1.70	101.2	5LZ95 PDM	6 round trips
5.25 m	65.77	33.98	64.43	1.02	51.7	5LZ95 or 4LZ120 PDM	22 round trips
8.08 m	109.24	93.06	121.90	0.90	85.2	5LZ95, 3LZ120 or 4LZ120	27 round trips
Total	196.05	148.34	198.68	0.99	75.7		

Table 5.39 Overall application results of TG-1 and KZ-1 drilling tools

Drilling tool	Round trips	Footage drilled (m)	Core length (m)	Drilling time (h)	Drilling rate (m/h)	Footage drilled per round trip (m)	Core recovery (%)	Bearing service life (h)	Remarks
TG-1	50	160.83	125.04	174.63	0.921	3.22	77.75	80.42	20 round trips drilled by hydro-hammer
KZ-1	137	350.39	335.56	464.27	0.755	2.56	95.71	77.38	7 round trips drilled by hydro-hammer

ii. A spray nozzle and a spray bearing chamber were designed in the mandrel, hoping to improve core recovery in broken rock formations by using the local reverse circulation at hole bottom.

iii. A graphite core mark was used in the inner tube. When core is completely taken out the core mark will slip down onto the core spring case and in this way drillers can judge whether core is fully taken out from the core barrel.

For TG-1 drilling tool, three core barrels with different lengths were developed, most of which were used for PDM drive drilling, with the application results shown in Table 5.38.

Overall application results of TG-1 and KZ-1 drilling tools can be found in Table 5.39, which is based on the statistics of the bearing service life between TG-1 and KZ-1 for a total of 187 round trips drilled above 1500 m hole section.

Through analyses based upon the two tables in coordination with the geological conditions, the effective length of the inner tube, application of hydro-hammers and their effective working time, a conclusion was obtained that the two drilling tools made no distinction between the superior and the inferior, in drilling indexes such as drilling rate, footage drilled per round trip and core recovery, etc. As to the low core recovery of TG-1 shown in the table, that was due to the broken and loose rock formations drilled by TG-1 for over ten round trips at the later stage, during which less core was obtained and therefore the average core recovery was largely affected, as less total footage was completed by TG-1 drilling tool. Also, the function of reverse circulation added to TG-1 did not show any superiority and the reason that its footage drilled per round trip and drilling rate higher than KZ mainly resulted from its higher ratio of hydro-hammer application. However, as higher grade bearings were used for TG-1, the service life of the bearings was greatly increased. An analysis on bearing failure showed that for KZ-1 the bearing failure was mainly the upper bearing exceeding wear while for TG-1 was the lower bearing normal wear.

3. **Jet reverse circulation drilling tool**

Broken zone was encountered at the hole depth from 1562 to 1605 m and in this hole section both KZ-1 and TG-1 drilling tools obtained rather low core recovery. During the 558th round trip in the pilot hole, the jet reverse circulation double tube swivel type drilling tool developed by the Institute of Exploration Techniques in cooperation with CCSD Drill Site Headquarters was used to recover 0.70 m core, from which it was found that silt filling materials existed in this formation (as shown in Fig. 5.41). Coring in this rock formation could not be solved by jet reverse circulation drilling tool. In fact, core recovery in the later round trips with this drilling tool was still unsatisfactory.

The structure of this drilling tool is shown in Fig. 5.42, with the characteristics as follows:

i. For swivel structure, the outboard bearing was still adopted so that the increased bearing grade could be used.
ii. In the inner tube assembly were placed the reed type core spring and two core catching tools.
iii. A circle of slant holes were drilled on the sub which is at the lower end of the inner tube assembly and on the core spring case, so that some water positively circulates through the annulus between the inner and the outer tubes to the holes, and then into the inner tube and suspend the core.

Fig. 5.41 Filling materials recovered by jet reverse circulation drilling tool

iv. Bottom discharge type was adopted for drill bit design, and at the bottom of the water hole was set a protection ring, which isolates the core from the fluid.

For jet reverse circulation drilling tool, two core barrels were manufactured and used in CCSD-1 Well for five round trips, with the results shown in Table 5.40.

4. **SL-1 core drilling tool**

A conventional sealing and lubrication mode of bearings in the inner tube assembly was adopted for the structure of SL-1 prototype during the first round drilling operation. However, the bearings were broken only after drilling for two round trips (round trip number PH368 and PH369) with the actual drilling time of 8.43 h. Later on, an open lubrication mode with bearings in the outer tube assembly was utilized based upon the experiences of KZ drilling tool design and then the situation became much better.

For the second round of SL-1 prototype, two grades of API steel were increased for outer tube material in comparison with KZ outer tube. However, female thread was broken only after drilling for 3.05 h in the round trip PH456 and a piece of metal of approximately 2 cm^2 dropped into the well and an accident of drilling tool sticking nearly happened during lifting the drilling tool. The drilling tool was freed by starting pump and rotation, with repeated up and down movement, but drill bit matrix fell off.

Drilling results in eight round trips can be found in Table 5.41.

5.3.6 Core Drilling Technologies

Down hole power percussive rotary core drilling, differs not only from oil drilling (mainly in sedimentary rock formations) with the characteristics of large diameter, deep well, low rotary speed and high drill bit pressure, but also from conventional core drilling (mainly in hard rock formations) with the characteristics of small diameter, shallow hole depth, low drill bit pressure and high rotary speed (as shown in Table 5.42). The obvious differences between down hole power percussive rotary core drilling and oil drilling as well as conventional core

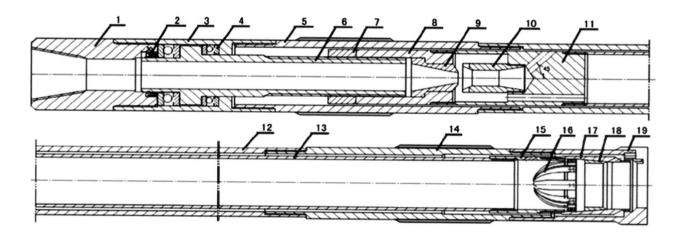

Fig. 5.42 Structure of jet reverse circulation drilling tool. *1* Upper adapter. *2* Cover. *3* Bearing chamber. *4* Bearings. *5* Upper reaming shell. *6* Mandrel. *7* Back cap. *8* Nozzle seat. *9* Nozzle. *10* Fluidic chamber. *11* Inner tube adapter. *12* Outer tube. *13* Inner tube. *14* Lower reaming shell. *15* Sub. *16* Reed core spring. *17* Core spring case. *18* Inner slot core spring. *19* Drill bit

5.3 Down Hole Power Percussive Rotary Core Drilling System

Table 5.40 Application results of jet reverse circulation drilling tool

Round trip	Depth started (m)	Footage drilled (m)	Core length (m)	Actual drilling time (h)	ROP (m/h)	Core recovery (%)	Driving mode	Coring tool
PH558	1578.29	4.33	0.70	1.70	2.55	16.2	5LZ95-18	5.25 m jet reverse circulation double tube
PH559	1582.62	2.22	1.30	3.23	0.69	58.6	5LZ95-18	5.25 m jet reverse circulation double tube
PH563	1598.52	0.49	0.00	1.25	0.39	0.0	5LZ95-18	5.25 m jet reverse circulation double tube
PH622	1816.44	4.08	1.55	4.97	0.82	38.0	4LZ120-4	4.18 m jet reverse circulation double tube
PH625	1830.56	2.24	2.15	3.28	0.68	96.0	4LZ120-4	4.18 m jet reverse circulation double tube
Total (average)		13.36	5.70	14.43	0.93	42.7		

Table 5.41 Application results of SL-1 drilling tool

Round trip	Depth started (m)	Footage drilled (m)	Core length (m)	Actual drilling time (h)	ROP (m/h)	Core recovery (%)	Driving mode	Coring tool
PH368	898.76	2.25	2.03	3.47	0.65	90.2	LZ95-10X	5.78 m water isolated double tube
PH369	901.01	5.70	5.17	4.97	1.15	90.7	LZ95-10X	5.78 m water isolated double tube
Sum		7.95	7.20	8.43	0.94	90.6		5.78 m water isolated double tube
PH373	912.30	3.21	2.09	3.93	0.82	65.1	LZ95-10X	4.28 m water isolated double tube
PH374	915.51	0.97	1.85	2.83	0.34	190.7	LZ95-10X	4.28 m water isolated double tube
PH378	928.96	3.10	2.96	7.72	0.40	95.5	LZ95-10X	4.28 m water isolated double tube
PH440	1123.59	2.17	2.20	5.38	0.40	101.4	5LZ95-14X	4.28 m water isolated double tube
PH453	1155.46	0.73	0.67	3.97	0.18	91.8	5LZ95-15X	4.28 m water isolated double tube
PH456	1163.07	0.60	0.44	3.05	0.20	73.3	Rotary table 108 rpm	4.28 m water isolated double tube
Sum		10.78	10.21	26.88	0.40	94.7		4.28 m water isolated double tube
Total		18.73	17.41	35.32	0.53	93.0		

drilling must be deeply analyzed and studied. For CCSD-1 Well drilling, a drilling technology of down hole power percussive rotary core drilling with PDM + hydro-hammer + diamond double tube core drilling tool was adopted. As it was a brand-new drilling technical system, the research on drilling technology was of especially important.

The most important characteristic of the down hole power percussive rotary core drilling system lies in organically combining two different types of down hole engines into a core drilling tool and to enable it to be the PDM hydro-hammer down hole power percussive rotary core drilling techniques, which is completely different from available conventional core drilling techniques.

5.3.6.1 Design of the Drilling Tool Assembly

Before this time, different core drilling techniques adopted each had its own applications and limitations (Table 5.42). But all of these techniques could not satisfy the requirements of core drilling for CCSD-1 Well for a variety of reasons. The emergence of PDM hydro-hammer down hole power percussive rotary core drilling techniques was a creative result under specified conditions. As for the down hole power percussive rotary core drilling tool structure which was composed of PDM + hydro-hammer + diamond core drilling tool and used for CCSD-1 Well, the following two problems were mainly considered in designing the drilling tool assembly: (1) deviation prevention must be mainly considered in design of drilling tool assembly, because the rock formations encountered by hard rock deep well diamond core drilling generally all have capability to cause hole deviation. (2) tool sticking treatment must be considered in design of drilling tool assembly, as tool sticking accidents caused by rock piece falling often happen in hard rock deep well.

1. **Design of deviation prevention**

Because the rock formations in CCSD-1 Well has a strong capability to cause hole deviation, deviation prevention must be carefully considered in tool assembly design. Besides the favourable condition such as low bit pressure used for PDM hydro-hammer core drilling which is beneficial to preventing hole deviation, the design of hole deviation prevention was mainly considered in selecting drilling tool assembly.

Table 5.42 Characteristics of different core drilling techniques

Technique	Technical characteristics	Advantages	Shortcomings
Geological drilling	Diamond wireline core drilling; Spindle or top head drive drill with high rotary speed	Equipment and tools are light in weight; High drilling rate; Low drilling cost	Lower drilling capacity, only for small hole size (less than 100 mm) and shallow hole depth (less than 2000 m)
Oil drilling	Deep well and large diameter; Rotary table drill rig; Tri-cone rock bit non-core drilling, with core drilling less than 5 %; Core is lifted by tripping	Equipment has strong drilling capacity; Drilling cost is low because of less coring	Unsuitable for overall continuous core drilling, as coring is realized by tripping
Combined drilling techniques	Top head drive system is added on oil rotary table drill, for diamond wireline core drilling; Overall continuous core drilling for whole well depth and core is recovered by wireline techniques; Thin kerf impregnated diamond drill bit is used	Tripping can be greatly reduced and actual drilling time greatly increased; High drilling rate; Higher core quality and core recovery; Top drive system is favourable for treating complicated down hole problems	Techniques are difficult because the requirements for design and materials for wireline core drill rod are rather high; The requirement for drill rig is rather high because a high rotary top drive system must be added on oil rotary table drill rig
Scientific deep drilling techniques of the former USSR	Overall continuous core drilling for whole well depth; Core was lifted by tripping; Tri-cone rock bit was used for core drilling; Rotary table oil drill rig was used and drill bit was driven by PDM; Light aluminium alloy drill pipe was used	High core drilling rate; PDM driving is favourable for reducing drill string wear and power consumption, and increasing hole wall stability; Techniques could be easily realized in practice, with less risk; Light aluminium alloy drill pipe greatly reduced the requirement for drill rig capacity and greatly increased the safety of the drill string	Drilling speed was low because of overall continuous core drilling for whole well depth by tripping; Low core recovery; For core drilling, tri-cone rock bit is only suitable for larger hole size (no less than 8 ½ in.), otherwise core is too small. Higher drilling cost is resulted, as the hole size is relatively larger

It is well known that the conventional drilling tool assemblies used for deviation prevention and correction mainly consist of the follows:

i. Full hole drilling tool This tool assembly can produce small drill bit dip angle (in comparison with pendulum drilling tool). Moreover, the straightness of the three points (three stabilizers) can guarantee the straightness of hole and restrict drill bit horizontal moving. This drilling tool assembly is suitable to the rock formations which easily cause hole deviation.

ii. Pendulum drilling tool This drilling tool can gradually reduce hole deviation by using the horizontal component force (pendulum force) created by drill collar weight under the point of tangency of the whole drilling string in a deviated hole. This drilling tool assembly is suitable to common rock formations and to reduce hole deviation.

iii. Flexible drilling tool assembly (a) Drill rod flexible drilling tool assembly, a piece of drill rod is added above the first stabilizer of the lower drill string and between the first stabilizer and drill bit is connected with a piece of drill collar with a certain length, is also called as flexible pendulum drilling tool. (b) Flexible joint drilling tool, a technical joint (flexible joint) is added between the first stabilizer and drill bit, can be used for the rock formations which easily cause hole deviation.

iv. Eccentric drilling tool An eccentric sub is placed in a suitable position of a pendulum drilling tool. The design principle of this drilling tool assembly is to improve the larger drill bit outboard dip angle of the pendulum drilling tool, suitable for deviation correction in the rock formations which cause moderate hole deviation.

During core drilling in CCSD-1 Well, pendulum drilling tool, flexible drilling tool assembly and eccentric drilling tool could not be effectively utilized because of space limitations. Under the circumstances, full hole drilling tool was mainly considered for use, with the following assembly: diamond core drill bit + lower reaming shell + core barrel + upper reaming shell + hydro-hammer + PDM + jar + stabilizer + drill collar + stabilizer + drill collar + drill string.

In the design of the core drilling tool assembly for CCSD-1 Well, in order to increase the footage drilled per round trip, the length of the core barrel was designed as long as possible, generally reached to 9 m. Two reaming shells, one upper and one lower reaming shell were placed on the drill string, approximately 8.5 m apart, being a typical arrangement of full hole drilling tool assembly.

5.3 Down Hole Power Percussive Rotary Core Drilling System

Table 5.43 Drilling tools weight

Drilling tool	Length (m)	Mass in air (kg)	Mass in drill mud (kg)
120 drill collar (unit mass 80.8 kg/m)	55	4444	3933
89 drill rod (unit mass 19.8 kg/m)	4945	97,911	86,651
Total	5000	102,355	90,548

Note Other drilling tools are calculated based on the approximate mass values

In consideration of preventing hole deviation, it was decided that the full hole drilling tool assembly was the main choice. Full hole drilling tool assembly is a multi stabilizer (reaming shell + stabilizer) tool assembly. The size of the stabilizers is close to that of the borehole, with the characteristics of full hole and rigidity. Down hole drilling tool should have the characteristics of rigidity, straightness, full hole and heavy, which can guarantee the straightness of the drilling tool and its coaxiality with borehole axis. The down hole power percussive rotary core drilling tool assembly adopted for CCSD-1 Well fully satisfied these requirements.

Rigidity: The core barrel was 139.7 mm × 10.54 mm and N80 (API Standards) oil casing. For the outer tube of the down hole power drilling tool, top quality tubing with the thickness of no less than 10 mm was adopted.

Straightness: Core barrels were meticulously selected, to guarantee the coaxiality and concentricity of the connecting thread of the large sized drilling tool.

Full hole: The outside diameter of the reaming shells was always larger than that of drill bits, to keep a full hole state. At the same time more than one stabilizers were placed on the drill collar, to prevent the large sized drilling tool from deviating at the hole bottom.

Heavy 120 mm sized drill collars of approximate 50 m were placed on the drill string, so as to keep the drill rods in a tensile state during drilling operations.

2. **Design of anti-sticking**

In the design of core drilling tool assembly for CCSD-1 Well, the capability of treating tool sticking accidents was considered from three aspects, i.e., the down hole sticking-free tool was to be placed on the drilling tool; to improve the tensile safety factor of the whole drilling tool assembly; and to improve the lubrication of drilling mud.

An integral jar (jar-while-drilling) with hydraulic upward and downward jarring was selected as the down hole sticking free tool for drilling tool assembly. To help realize downward jarring, the jar should be placed under the drill collar. In addition, to help realize an up-lifting jarring and realize sticking free, the jar should be placed on the upmost large sized tool (stabilizer), as sticking accident generally happens at the place where the large sized drilling tools stay.

To improve the tensile safety factor of the whole drilling tool assembly depended on improving the tensile strength of the drill rods and therefore high strength drill rods were adopted in the design. The main drilling tools used for core drilling in CCSD-1 Well includes drill bits of 157 mm size, drill collars of 120 mm diameter, drill rod of 89 mm diameter, 4LZ120 × 7Y PDM with 3015 N m maximum output (Mmax). The drill rod of 89 mm diameter has the following technical specifications: type I, steel grade S135, wall thickness 9.35 mm, tensile strength (Te) 2175 kN, twisting strength (Me) 4530 N m, and unit mass 19.8 kg/m. Solid free drilling mud was used, with density of 1.03 kg/cm^3. The structural mass of 5000 m drilling tool can be found in Table 5.43. When the hole depth reached to 5000 m, the load of the first piece drill rod at the surface was 905.48 kN, and then

i. Suppose that drill rod does not bear rotation torque, then the tensile safety factor

$$S1 = Te/Wd = 2175/905.48 \approx 2.4$$

ii. Suppose that drill rod bears a complex load of tension and torsion, and bears a torque of the maximum output torque of PDM, then the tensile safety factor

$$T < Te(1 - (Mmax/Me)^2)^{1/2}$$
$$T < 2175 \times \left(1 - (3015/4530)^2\right)^{1/2} = 1620\,kN$$
$$S2 = T/Wd = 1620/905.48 \approx 1.8$$

From above calculations it could be found that the tensile safety factor of the drill string was adequate.

The application of top quality lubricant in drill mud could reduce the rotational working frictional resistance of the drill string in the borehole, and being favourable to borehole safety.

3. **Drilling tool assembly**

Based upon the characteristics of drilling operations in CCSD-1 Well, the design of core drilling tool assembly was: PDM and hydro-hammer were adopted as the full hole drilling tool assembly for down hole power. At the shallow hole section a four stabilizer full hole drilling tool assembly with two stabilizers + two reaming shells (see Fig. 5.43a) was used. At the deep hole section a three stabilizer full hole drilling tool assembly with a jar-while-drilling + one stabilizer + two reaming shells (see Fig. 5.43b) was utilized. At oversized hole section and tool sticking hole section, a drilling tool assembly with slick drill collar or KTB drill collar + (jar-while-drilling) + PDM + hydro-hammer + diamond double tube core drilling tool was adopted.

Drilling tool assembly I (Fig. 5.43a): 157 mm sized core drill bit + 157.3 mm reaming shell + 139.7 mm core barrel + 157.3 mm reaming shell + 146 mm double tube adapter + 127 mm hydro-hammer + 120 mm (95 mm) PDM + 156 mm stabilizer + 120 mm one piece drill collar + 156 mm stabilizer + 120 mm five pieces drill collar + 88.9 mm drill rod + 89 mm kelly bar 139.7 mm core barrel + 157.3 mm reaming shell + 146 mm double tube adapter + 127 mm hydro-hammer + 120 mm (95 mm) PDM + 156 mm stabilizer + jar-while-drilling + 120 mm six pieces drill collar + 88.9 mm drill rod + 89 mm kelly bar.

Drilling tool assembly II (Fig. 5.43b): 157 mm sized core drill bit + 157.3 mm reaming shell + 139.7 mm core barrel + 157.3 mm reaming shell + 146 mm double tube adapter + 127 mm hydro-hammer + 120 mm (95 mm) PDM + 156 mm stabilizer + jar-while-drilling + 120 mm six pieces drill collar + 88.9 mm drill rod + 89 mm kelly bar.

5.3.6.2 Drilling Parameters

PDM hydro-hammer core drilling techniques are brand-new core drilling techniques which were researched and gradually developed in the construction of CCSD-1 Well. Before this, there were no existing experiences which could be used. Only by continuous exploring in drilling CCSD-1 Well the drilling parameters which were suitable to the drilling techniques were well developed.

As other drilling techniques, the main drilling parameters for PDM hydro-hammer core drilling techniques include bit pressure, rotary speed and pump delivery (P, N and Q). In drilling operations, the optimum technical and economical indexes could only be obtained from the optimum drilling speed, mainly realized by optimizing and regulating the drilling parameters. The down hole power percussive rotary core drilling techniques used were completely new and thus the drilling parameters were different from those of conventional drilling techniques. More factors would affect the drilling parameters, such as the properties of PDM, drilling mud properties, geological conditions of the borehole, the types of diamond drill bits and the working properties of hydro-hammer, etc. And the interaction of the drilling parameters (bit pressure, rotary speed and pump delivery) would become more obvious. As there were no ripe experiences to be used, the only way to determine the drilling parameters for the down hole power percussive rotary core drilling techniques was to carry on an exploration in practice, on the basis of theoretical calculation.

1. **Bit pressure**

Among the drilling parameters, bit pressure is one of the key ones. For diamond impregnated drill bit, if it does not have enough specific pressure against rock formations, the diamond cutting edge can not effectively cut into the rock, and then the penetration rate will be directly affected. In drilling medium hard to hard rocks, the recommended unit pressure for diamond impregnated drill bit to effectively cut into the rock should be $\sigma = 4.1\text{--}8.82 \times 10^6$ Pa, and from theoretical calculation that obtained the bit pressure P is 30–65 kN.

As for diamond impregnated drill bit, if the properties of bit matrix is not suitable to the rock formations to be drilled,

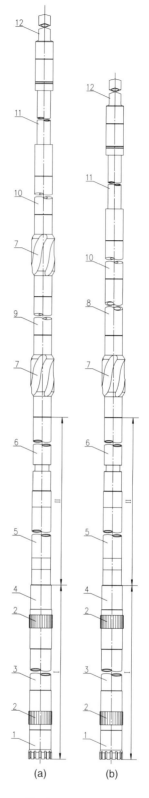

Fig. 5.43 Core drilling tool assembly for CCSD-1 Well.
1 157 mm diamond core drill bit.
2 157.3 mm diamond reaming shell. *3* 139.7 mm core barrel.
4 146 mm double tube adapter.
5 127 mm hydro-hammer.
6 C5LZ95 × 7 or 4LZ120 × 7 PDM. *7* 156 mm stabilizer.
8 120 mm jar-while-drilling.
9 120 mm drill collar single piece.
10 120 mm drill collar string.
11 89 mm drill string. *12.* 89 mm Kelly bar

or enough bit pressure can not be supplied, the bit matrix can not wear off in advance and consequently leading an automatic fall off of the diamond grains which have already lost the ability to crush the rock, and finally the self sharpening of the drill bit can not be realized. This will be manifested at hole bottom by "bit slippery" and dramatic drilling speed decrease.

However, bit pressure has a close relationship with borehole inclination. The related research indicates that even under the condition of full hole drilling tool assembly the exerted bit pressure still produces component of force which leads to hole inclination. This component of force is known as hole deviating force, which causes hole inclination. Hole deviation degree, is not only influenced by the hole deviating force, but also by the conditions of stratum, such as stratum dip angle, anisotropy degree of rocks and crustal stress, etc. On the other hand, hole inclination angle can influence hole deviating force.

Therefore, reasonable bit pressure should be determined on the basis of comprehensive consideration of a variety of variables.

The rated output torque and power of PDM must be considered in determination of bit pressure. In case that bit pressure satisfies the theoretical requirement of rock fragmentation, rock fragmentation power and torque may exceed the rated torque and power of the PDM, or the PDM often fails to work resulted from inadequate power storage, thus reducing the service life of the PDM. To solve this problem, the engineers and technicians at drill site introduced the successful experiences of the application of PDM techniques in geological exploration drilling to the down hole power core drilling in CCSD-1 Well and got a successful result. It is well known that under the action of dynamic load the drillability of rock will be considerably improved, to use hydro-hammer percussive rotary drilling method, its bit pressure can be lower than that of conventional rotary drilling, and the load of down hole power PDM drilling tool can be decreased. Based on our study in CCSD-1 Well drilling in a variety of rock formations, the recommended bit pressure should be 10–35 kN, based on different stratum conditions.

2. **Rotary speed**

For impregnated diamond drill bit, diamond is of high hardness and good wear resistance, whereas with less diamond exposure. In this case a high drilling speed can only be obtained by a high cutting speed produced by a high rotary speed. It is traditionally thought that the average linear velocity for impregnated diamond drill bit should reach to 1.5–3 m/s and in general the high value should be adopted, i.e., high rotary speed should be used, if only the stratum condition and the technical condition permit.

As for PDM + hydro-hammer + diamond double tube core drilling tool, PDM driving is the main driving while rotary table driving is auxiliary, belonging to a "double driving" or "double rotary" drilling method, with the following advantages: (i) effectively overcoming the frictional resistance of borehole wall to drilling tool and the bit pressure can be accurately controlled; (ii) reducing drill rod wear and the destroy of drill rod to borehole wall; (iii) reducing drilling tool vibration in borehole and avoiding drilling tool damage resulted from resonance; and (iv) reducing the damage to casing, especially the damage to moving casing.

Under the circumstances of adopting double rotary method with PDM driving as the main driving while rotary table driving as auxiliary, the increase of rotary speed mainly depends on PDM. The output rotary speed of PDM increases linearly with mud pump delivery, with loading characteristics that the output power and the output torque are approximate linear with pressure drop. When pressure drop reaches to a certain value, the output rotary speed decreases dramatically and the output power increases slowly, then the mechanical efficiency gets a maximum value, see Fig. 5.44.

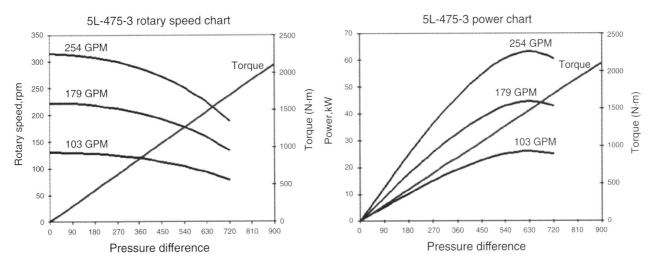

Fig. 5.44 Rotary speed and power charts of 4LZ120 PDM produced by Lilin Co. in Tianjin

The optimum output rotary speed of PDM is the output rotary speed when drill bit has adequate bit pressure and adequate working torque, and PDM has the maximum mechanical efficiency. However, as the hydro-hammer and PDM work simultaneously the large circulating pressure drop of drilling fluid restricts the increase of pump delivery and the increase of rotary speed. Concerning the PDM of 120 mm diameter, it is impossible for 157 mm sized diamond drill bit to reach a linear speed of 4 m/s (487 rpm) even if the pump delivery reaches the maximum, whereas a linear speed of 1.6 m/s (200 rpm) can be obtained under the working torque.

3. **Pump delivery**

For PDM + hydro-hammer downhole power drilling, the determination of pump delivery should first of all consider that the output rotary speed and the power of PDM guarantee the basic needs for diamond core drilling, then the need for uplifting of the cuttings, and finally the matching relationship between hydro-hammer, diamond drill bit and pump delivery.

The output rotary speed of PDM is direct linear with the pump delivery. In order to maintain high rotary speed that diamond impregnated drill bit needs, the pump delivery must be increased as high as possible, whereas excessive mud pump delivery will produce excessive mud circulating pressure, resulting in serious flushing to borehole wall, aggravated erosion to drill bit matrix and abnormal working of hydro-hammer. Under the circumstances that PDM + hydro-hammer + diamond core drilling tool is to be used, the determination of rotary speed and pump delivery should be appropriately optimized by reference to the technical parameters supplied by PDM manufacturers. On the other hand, both hydro-hammer and diamond drill bit manufacturers should, based upon this characteristic, develop their products which are suitable to CCSD-1 Well drilling, making core drilling tools, hydro-hammer and diamond drill bit compatible with the working properties of PDM, not on the contrary.

Drilling parameters for CCSD-1 Well with PDM + hydro-hammer + diamond double tube coring tool downhole power percussive rotary core drilling techniques were basically optimized through theoretical calculation and practical exploration, with the following values: bit pressure 10–35 kN, pump delivery 8.62–10.34 L/s, rotary speed 157–189 rpm (PDM) + 12 rpm (rotary table), average penetration rate over 1.1 m/h.

4. **Optimization of drilling parameters**

In comparison with conventional diamond core drilling, the working conditions of drill bit and drilling tools in borehole change a lot for downhole power percussive rotary core drilling. Available parameters in drilling regulations can no longer suit or satisfy the needs of this new drilling method, as a result of the application of downhole engine—PDM and hydro-hammer. Drilling parameters should be determined based upon the working characteristics of the downhole engine. For diamond core drilling operations, rotary speed, bit pressure and pump delivery are the most important basic parameters in drilling regulations and in conventional drilling operations these three basic parameters can be independently adjusted at surface according to requirement. However, as for downhole power core drilling techniques, downhole engine is the most basic factor to complete drilling project, especially PDM, in which pump delivery is the most important parameter to decide its working property. To ensure PDM for an optimum output property, its pump delivery must be within its reasonable limit of working flow rate, which may not be able to satisfy the requirements of conventional diamond core drilling for pump delivery. Therefore, rotary speed, bit pressure and pump delivery are interactive and must be comprehensively designed. The drilling parameters were optimized through the following ways.

i. To optimize PDM and make its output power under a certain working flow rate approach as much as possible the requirements of diamond core drilling for bit pressure, rotary speed and pump delivery.
ii. To improve hydro-hammer and make its working flow rate suitable for the working flow rate of PDM.
iii. To develop new type diamond core drill bits in order to realize a long service life and high efficient drilling for PDM hydro-hammer.
iv. To improve mud properties by tightening solid control and increasing the lubricating property of drilling mud so as to reduce as much as possible the erosion of drilling mud to diamond drill bit and the wear to PDM and hydro-hammer.

By adopting the abovementioned measures the downhole power percussive rotary core drilling techniques were gradually improved and perfected, the drilling parameters were more and more reasonable, and finally satisfactory economic and technical indexes were obtained.

5.3.7 The Application Results of Hard Rock Deep Well Core Drilling Techniques

5.3.7.1 General Situation of Core Drilling

Full hole coring was designed for CCSD-1 Well, with core recovery of no less than 80 %. Therefore, core drilling was the most important construction procedure and work load for CCSD-1 Well. For a 5000 m deep core drilling project in very hard crystalline rocks the difficulties were considerably serious, as we had not had any experiences in China, even with few experiences from foreign countries to be used. Core drilling procedure was divided up into the pilot hole core drilling and the main hole core drilling.

5.3 Down Hole Power Percussive Rotary Core Drilling System

Table 5.44 Test results of drilling tool in the second stage of the pilot hole

	PDM double tube core drilling	PDM hydro-hammer double tube core drilling	Total
Total footage drilled (m)	300.62	197.37	497.99
Total round trips drilled	130	59	189
Total time (h)	442.34	177.20	619.54
Penetration rate (m/h)	0.68	1.11	0.80
Average footage drilled per round trip (m)	2.31	3.35	2.63

1. **Pilot hole core drilling**

Based on the initial design of drilling sub-project, the pilot hole of CCSD-1 Well would be completed as long as core drilling reached to 2000 m deep, as another important task was to test the new drilling techniques to be used for the main hole. For the pilot hole of CCSD-1 Well the actual core drilling started from 101 m deep and ended up at 2046.54 m deep. According to the procedures of the core drilling method tests and the improvement and perfection of the core drilling system, the pilot hole core drilling can be divided up into four stages: the drilling stage with PDM double tube core drilling (101.00–502.06 m); the initial testing stage for PDM hydro-hammer system (502.06–1000.05 m); the improving stage for PDM hydro-hammer system (1000.05–1503.00 m) and the perfection stage for PDM hydro-hammer system (1503.00–2046.54 m).

i. The drilling stage mainly with PDM double tube core drilling

During the first stage in pilot hole core drilling (101.00–502.06 m), PDM double tube core drilling, PDM single tube core drilling and rotary table double tube core drilling were respectively tested and the results showed that PDM double tube core drilling was superior to the other two methods in main technical indexes such as penetration rate, footage drilled per round trip and core recovery. However, because rock is hard, cleavage development and easily broken, PDM double tube core drilling still had some problems such as low penetration rate (approximately 0.8 m/h) and short footage drilled per round trip (approximately 2 m).

To counter these problems which existed in PDM double tube core drilling, the drilling technicians considered to combine hydro-hammer into PDM double tube core drilling system, as hydro-hammer could increase penetration rate, reduce core blockage and improve footage drilled per round trip. The first round trip experiment for PDM hydro-hammer double tube core drilling system was carried out at the hole depth of 218.44 m and during this experiment hydro-hammer worked only for a short time, but with penetration rate obviously improved. Afterwards, experiments were carried on for 12 round trips, during which the main problems existed in hydro-hammer included unsatisfactory working reliability and short working time.

ii. The initial testing stage for PDM hydro-hammer system

In this stage (502.06–1000.05 m), PDM double tube core drilling was the main method for core drilling, at the same time hydro-hammer was improved based upon the working situation in the first stage and then PDM hydro-hammer downhole combined power drilling system was continuously tested. In this stage gneiss is the main rock, with granite eclogite and ultrabasic rocks of a certain thickness. The drilling result of PDM and the test result of hydro-hammer can be found in Table 5.44.

In this stage 189 round trips were completed, in 35 round trips more than 4 m were drilled per round trip, and among these 35 round trips 22 were completed by using hydro-hammer, amounting to 37.29 % of the number of times that hydro-hammer were employed for drilling; in 76 round trips that penetration rate exceeded 1 m/h, and among these 76 round trips 38 were completed by using hydro-hammer, accounting for 64 % of the number of times that hydro-hammer were employed. In comparison with pure PDM drilling, the average footage drilled per round trip and the penetration rate for PDM hydro-hammer drilling increased by 45 and 63 % respectively, indicating that the application of hydro-hammer played a very obvious role in improving both the footage drilled per round trip and the penetration rate.

During this stage, the problems of unsatisfactory working reliability and short service life in some parts existed in hydro-hammer.

iii. The improving stage for PDM hydro-hammer core drilling system

During this stage (1000.05–1503.00 m), the improvement of hydro-hammer was continuously made, its working reliability was obviously improved and the proportion of drilling with hydro-hammer was greatly increased, with strong points more conspicuous. PDM hydro-hammer system core drilling became a main core drilling method.

In this stage were completed 148 round trips, in which PDM hydro-hammer drilling was more employed. The working reliability of hydro-hammer was obviously improved, with an average footage of 3.94 m drilled per round trip. Among these 148 round trips, for 26 round trips the core barrels (5.6 and 8.6 m long respectively) were full

Table 5.45 Test results of drilling tool in the third stage of the pilot hole

	PDM double tube core drilling	PDM hydro-hammer double tube core drilling	Total
Total footage drilled (m)	142.66	350.77	493.43
Total round trips drilled	54	89	143
Total time (h)	200.95	300.50	501.45
Penetration rate (m/h)	0.71	1.17	0.98
Average footage drilled per round trip (m)	2.64	3.94	3.45

Table 5.46 Test results of drilling tool in the fourth stage of the pilot hole

	PDM double tube core drilling	PDM hydro-hammer double tube core drilling	Total
Total footage drilled (m)	9.55	525.91	535.46
Total round trips drilled	3	104	107
Total time (h)	7.93	483.32	491.25
Penetration rate (m/h)	1.20	1.09	1.09
Average footage drilled per round trip (m)	3.18	5.06	5.00

of core, for 6 round trips the footage drilled exceeded 8 m. The longest footage drilled per round trip reached 8.72 m and the highest penetration rate reached 2.87 m/h. In comparison with PDM drilling, the penetration rate of PDM hydro-hammer drilling increased by 65 % and the footage drilled per round trip increased by 49 %, with the application results shown in Table 5.45.

iv. The perfection stage for PDM hydro-hammer core drilling system

In this stage (1503.00–2046.54 m), 112 round trips were completed, all with PDM hydro-hammer drilling, except 5 round trips completed by rotary table drilling because of absence of PDM supply and round trips finished with PDM double tube core drilling. From the 579th round trip (hole depth 1634.38 m), to overcome the frictional resistance of drilling tool to borehole wall and let bit pressure effectively transmit to hole bottom, a "double rotary" drilling method was utilized, i.e. the rotary table was slowly rotated (8–12 rpm) while PDM rotated. In comparison with the third stage, the footage drilled per round trip was increased greatly, in 7 round trips the footage drilled exceeded 8 m and in 28 round trips the footage drilled exceeded 7 m (at the later stage the core barrel was 7.9 m in length). Especially after adding 1 % GLUB lubricant into drilling fluid a record was set that in continuous 16 round trips the core barrels were fully filled, except 2 round trips in which drill bits were worn to flat, thus the tripping time needed was dramatically reduced. The addition of lubricant greatly increased the service life and the reliability of hydro-hammer.

In general, in the fourth stage the drilling system of PDM hydro-hammer diamond double tube core drilling tools were perfected, with technical and economic indexes obviously increased (see Table 5.46). For this, the main reasons lay in (1) the continuous improvement of hydro-hammer; (2) addition of lubricant into drilling mud; (3) continuous improvement of drilling tools (core barrel) properties; (4) improvement of drilling tool application and management measures and (5) skilful operations and rich experiences gradually obtained by drillers.

2. **Core drilling in the main hole**

According to the initial design of drilling sub-project, top drive diamond wireline core drilling was to be used for main hole drilling and the hydraulic top drive and the wireline drill rod were to be aided by ICDP. At the beginning of main hole drilling, five round trips of top drive diamond wireline core drilling were tested, however, with the results of low penetration rate, short footage drilled per round trip and low core recovery. At the same time, core barrels could not be fished out for many times because there existed some problems in supplied wireline drill rods. Because of the unsatisfactory results of top drive diamond wireline core drilling test and the satisfactory results obtained from PDM hydro-hammer double tube core drilling in the pilot hole, PDM hydro-hammer double tube core drilling method was mainly used in main hole drilling. Meanwhile, the repair of wireline drill rods and the development of PDM hydro-hammer wireline core drilling system were carried out.

Main hole core drilling can be classified into three stages: the first stage CCSD-MH with the hole depth 2046.54–2982.18 m, the second stage CCSD-MH-1C with the hole depth 2974.59–3665.87 m and the third stage CCSD-MH-2C with the hole depth 3624.16–5118.2 m.

i. The first stage of core drilling in the main hole (CCSD-MH)

The first stage of core drilling in the main hole started from 2046.54 m. When drilling to 2982.18 m deep, the vertex angle exceeded 16° and then the inclination was corrected. In this stage a total footage of 935.64 m was completed.

Table 5.47 Technical and economic indexes obtained by different core drilling methods in the pilot hole

Drilling method	Round trips	Footage drilled (m)	Footage drilled per round trip (m)	ROP (m/h)	Core recovery (%)
Top drive double tube	8	5.87	0.73	0.36	5.1
PDM double tube	21	76.12	3.62	0.49	54.4
Top drive wireline	5	7.62	1.52	0.63	13.8
Top drive hydro-hammer wireline	2	8.27	4.14	0.89	99.5
PDM wireline	3	6.72	2.24	0.33	71.9
PDM hydro-hammer double tube	107	830.07	7.76	1.31	80.7

Table 5.48 Technical and economic indexes obtained in CCSD-MH-1C

Drilling method	Round trips	Footage drilled (m)	Footage drilled per round trip (m)	ROP (m/h)	Core recovery (%)
PDM hydro-hammer double tube core drilling	79	634.85	8.04	1.25	83.3

Table 5.49 Technical and economic indexes obtained in CCSD-MH-2C

Drilling method	Round trips	Footage drilled (m)	Footage drilled per round trip (m)	Drilling rate (m/h)	Core recovery (%)
PDM hydro-hammer double tube core drilling	185	1473.05	7.96	1.03	87.9
PDM double tube	1	7.01	7.01	0.67	100.0
Total	186	1480.06	7.96	1.03	88.0

In this stage six core drilling methods were adopted. Besides the methods of PDM double tube core drilling and PDM hydro-hammer double tube core drilling used in the pilot hole, other methods such as top drive double tube core drilling, top drive wireline core drilling, top drive hydro-hammer wireline core drilling and PDM wireline core drilling were employed, with the results shown in Table 5.47, from which it can be found that PDM hydro-hammer swivel type double tube core drilling technique is an effective and high quality method which can still obtain satisfactory result even in complicated borehole conditions in broken rock formations.

ii. The second stage of core drilling in the main hole (CCSD-MH-1C)

After reaching to 2770 m deep, lithology changes with gneiss as the main rock, with large dip angle and strong inclination tendency. Drilling to 2935 m deep the vertex angle of the main hole increased to 16.34°, with a buildup rate of 0.55 degree per 10 m. Inclination correction with backfill was conducted when reaching to 2982.18 m and after this core drilling for CCSD-MH-1C started from 2974.59 m. On October 2nd, 2003, when drilling to 3665.87 m deep reaming shell and drill bit dropped in the hole as a result of male thread broken of the lower reaming shell. Treatments failed for many times and then it was decided to pull out the moving casing, ream the hole, set the casing and resume core drilling. To 3665.87 m deep CCSD-MH-1C was completed. During this stage the main hole was fully completed by PDM hydro-hammer double tube core drilling system, obtaining good technical and economic indexes shown in Table 5.48.

iii. The third stage of core drilling in the main hole (CCSD-MH-2C)

Core drilling in the third stage was started at the hole depth of 3624.16 m after the second reaming and hole cementation with casing, and ended up at 5118.2 m. In this stage the main hole was completed by PDM hydro-hammer diamond double tube drilling tool system, except only one round trip finished by PDM double tube tool. The excellent results can be found in Table 5.49.

5.3.7.2 Application Results of Downhole Power Percussive Rotary Core Drilling

For PDM + hydro-hammer + diamond double tube core drilling techniques, we did not have any ripe experiences as it was a brand-new method and the rich experiences we had accumulated in the past conventional diamond core drilling could only be used for reference. Of the near 5000 m core drilling job in CCSD-1 Well, over 4000 m were completed by this method. From the phases of test, innovation, improvement to perfection, our drilling technicians threw their painstaking efforts into this brand-new drilling technique and the technical and economic indexes obtained in different phases indicated that this method came up to a high level in application, with the comprehensive technical indexes far surpassed those of conventional core drilling, with the reasons as follows:

i. Drilling in complex rock formations, downhole PDM drive can effectively reduce the disturbance of drill rods to borehole wall, reduce in-the-hole accident and improve borehole quality.
ii. The opened swivel type double tube core drilling method adopted with drilling mud lubrication can reduce the failure of core barrel swivel movement, cuttings blockage and the wear of core barrel against core.
iii. Lengthened gauging drill bit was utilized, dual reaming shell was set above core tube, and dual stabilizers were placed both above and under the first piece drill collar to avoid borehole deviation.
iv. With hydro-hammer percussive drilling, drilling with low bit pressure was realized, in which the bit pressure on a unit area was far less than that in conventional diamond drilling, thus effectively reduced the load of PDM, benefiting the increase of PDM life, the decrease of core blockage, the improvement of the footage drilled per round trip and the increase of penetration rate.
v. With the deepening of the borehole, a drilling method of "double drive" or "double rotary" was adopted, with PDM drive as main and rotary table drive as auxiliary. With this method the frictional resistance of borehole wall to the drilling tools could be effectively overcome and bit pressure could be accurately controlled.
vi. The application of PDM made it possible to assess the working situation at hole bottom from the surface according to the pressure change on the pump pressure gauge. In case of abnormality treatment could be conducted by adjusting the parameters. Even if the treatment failed the drilling tools could be lifted for further check up, so as to avoid idle drilling and reduce abnormal wear of the core.

The determination of the drilling parameters for this new method was completely different from that for conventional drilling methods; with still more influence factors, mainly depended upon the properties of PDM, mud properties, geological conditions of borehole, diamond drill bit types and the working properties of hydro-hammer, etc. The interactions among bit pressure, rotary speed and pump delivery were still more obvious. It was necessary to do well every single technical work in the whole technical system to optimum first and then conduct an overall and comprehensive study. Only by this way each single technical work could be mixed within the integrated technical system and the entire superiority of this new technique could be brought into full play.

1. **Borehole quality**

For CCSD-1 Well, borehole quality indexes included core recovery and core quality, situations of crooked borehole and oversized borehole.

i. Core quality

The purpose of scientific drilling is to obtain core, cuttings and fluids (gas and liquids) in rock formations through borehole, and obtain underground information by geophysical logging or by a long-term observation through the technical instruments set in the borehole, for instance, to understand the relationship between giant meteorite impact and the earth's surface environmental change and biological species extinction by core drilling at the bottom of an extra large meteorite impact crater; to understand global climate change in the past millions of years by continuous core drilling for sediments in the deeper lakes around the world or drilling for ice core at glaciers in polar region; to understand the constitution and the structure of mantle, and the dynamic process of collisional orogenesis by core drilling for rocks from deep mantle at the key areas of collisional orogenic belts; and to understand the mechanism of earthquake and the thermal structure of volcano system by core drilling at earthquake source and volcano source areas, so as to gradually reduce the damages caused by earthquake and volcanic explosion to mankind.

Samples such as core and cuttings are the firsthand information for geologists to make thorough research. Large funds were used for the drilling construction of CCSD-1 Well, for reason to obtain as much as possible high quality underground materials such as core, cuttings and fluids, etc. In the design of CCSD-1 Well, core recovery was requested no less than 80 %, with top quality, as scanning imagery was to be conducted on core surface, for the purpose of geological research.

PDM hydro-hammer core drilling technique was developed for CCSD-1 Well. At initial stage, however, due to the poor swivel movement of the swivel type double tube core drilling tool core recovery was low and core quality was unsatisfactory. This exerted great pressure upon the technicians at drill site. However, after continuous improvement and perfection an average core recovery of 85.82 % was obtained, with core quality greatly increased. Even in extremely broken complex formations the core recovered still maintained as much as possible the original state of the formations, thus providing high quality material samples for geological research.

The technical methods to guarantee core recovery and to increase core quality mainly included:

(i) To ensure the swivel movement of the swivel type double tube drilling tool, in order to reduce as much as possible the wear of core.
(ii) To ensure the working properties of hydro-hammer, for reducing core blockage.

(iii) To improve mud lubrication properties, for the purpose of decreasing the wear of core.
(iv) To optimize drill bit structure, so as to ensure core surface regular and smooth.
(v) To optimize drilling parameters.
(vi) To improve driller's skill, so as to ensure drilling process stable and continuous.

  ii. Hole quality

Hole quality mainly includes the intact level of borehole wall and deviation degree of borehole.

As hole quality exerts a big influence upon the quality of geophysical log data and safe operations, it is requested that borehole wall should be as regular as possible, in order to obtain still more complete and accurate logging data. On the other hand, the deviation degree of borehole is an important factor which exerts influence on borehole safety, often decides whether drilling can reach to the target safely and economically. The Cola Super Deep Well in former the Soviet Union and the KTB Project in Germany both could not reach the designed depth, resulted decisively from excessive deviation of the boreholes.

Though core, cuttings and fluid samples (gas and liquids) from rock formations are important, the basic data of the underground geological information carried by them may be lost under the influence of core recovery. The related data should be collected by using geophysical logging and be under comparative study with core and cuttings, for complement each other. Meanwhile, to make full use of the scientific borehole which is completed at great cost, the geologists will carry on a long term observation by setting instruments in the borehole to obtain different underground information. All these need a high quality passageway, either for geophysical logging or for long term observation, to repeatedly set a variety of logging instruments.

The successful application of the downhole power and the hydro-hammer greatly reduced the mechanical damage to borehole wall by high rotation drill string. This is of the utmost importance for protecting the original geometric form of the borehole in hard, brittle and broken formation, such as fault. Throughout the period of three year drilling operation, the borehole of CCSD-1 Well maintained stable, sticking of logging tools was seldom met in tens of logging operations and it never happened that the logging instruments dropped into the borehole. The successful comprehensive logging operations obtained a vast amount of valuable log data for geological research, and the routine engineering logging provided basis for making decision to guarantee normal drilling operation.

The successful application of the downhole power and the hydro-hammer made it possible to conduct a small scale drilling operation in the rock formations of high hardness, high abrasiveness and high deviating degree, so as to greatly decrease borehole deviation. At the pilot hole depth of 2028 m, the deviation was only 4.1°, thus it was realized to combine the two holes into one, and this, saved large funds and a lot of time for CCSD-1 Well.

2. **Drilling efficiency**

The successful application of hydro-hammer made it possible to conduct large diameter continuous core drilling in deep hard rock and achieve satisfactory technical and economic indexes, with the main reason that the application of hydro-hammer brought an impact load on the cutting tools. The impact load has a characteristic that the contact stress can reach to extremely high value instantaneously, with stress concentrated. In this case even though the dynamic hardness of rock is larger (generally 8–9 times larger) than the static hardness micro cracks are still easily produced. The higher the impact velocity is, the higher the rock brittleness will be, thus being favourable for crack developing. Therefore hard rock can be fragmented by using moderate impact work (tens of N m), while static indentation needs much work. Furthermore, cutting elements wear less, with the following reasons:

(i) The axial pressure required for drilling is less, and rotary speed is low.
(ii) The frictional coefficient of volumetric fragmentation is less than that of surface fragmentation, and volumetric fragmentation of rock can be easily obtained in percussive rotary drilling.
(iii) The relative wear of the cutting elements is reduced as a result of fast drilling speed.
(iv) The acting time of the cutting elements on rock is short in percussive rock fragmentation.
(v) The axial pressure exerted in percussion improves the transmission condition of impact work and then increases the impact result.
(vi) Because of the high frequency and the continuous impact load exerted on rock, cracks develop completely during the process of rock fragmentation, being favourable for fragmenting hard rock.
(vii) The continuous rotary cutting action existed in percussion improves the transmission direction of impact load, thus the percussive fragmentation and the cutting fragmentation of rock can be brought into full play.

Furthermore, core barrel bears vibration of a certain frequency during the process of core drilling, as the impact load produced by hydro-hammer is transmitted to the drill bit at hole bottom through the core barrel. It is just this vibration which can effectively reduce the probability of blockage of broken core after entering into the core barrel, and even free the core blockage, once it happens, before it is completely blocked. It is known that diamond core drill bit cuts rock annularly and if core blockage happens during the process of drilling the newly cut core can not go into the core barrel smoothly and then blocks at the entrance of the core barrel.

At this time the core will rubs directly against rock and the limited bit pressure can not effectively act on the cutting elements of drill bit for further drilling, resulting in dramatic decrease of penetration rate, abnormal wear of the inner diameter of the drill bit and excessive wear of core, and decrease of core recovery. Under these circumstances, treatment by lifting the drilling tools is the only way in case that core blockage can not be effectively freed. The application of hydro-hammer greatly increases the footage drilled per round trip, ensures the core recovery and greatly increases both the actual drilling time and the penetration rate.

3. **Borehole safety**

Although the rock formations drilled in CCSD-1 Well are located near the Tanlu Rift and the rock formations are complicated in structure, rock anisotropy develops, with strong crustal stress, resulted from the influence of violent geological structure movement of subduction and uplift and return after the collision between the Yangzi Plate and the Sino-Korean Plate, and through ultra-high temperature and ultra-high pressure metamorphism, yet each stratum maintains a state of equilibrium under the natural conditions before it is penetrated through. This original equilibrium is upset under the external interference such as drilling, and then a variety of unstable factors begin to appear, bringing about complicated phenomena in borehole. Based upon the characteristics of the strata the most important factor which affects the safety in drilling operation in CCSD-1 Well was sticking of drilling tools by falling rocks.

As down hole power was utilized and rotary table was not rotated in a high speed, the strike and the scrape of the drill string under high rotation speed against borehole wall could be avoided, reducing the damage to the original state of equilibrium of the strata, to the maximum extent. Under these circumstances rock falling and hole collapse which frequently happened in complex rock formations in conventional drilling greatly reduced, even if in case of hole collapse caused by crustal stress release the size of fallen rock pieces could be effectively limited because of the great reduction of hole collapsing degree resulted from greatly reduced disturbance of drill string against borehole wall, as it is known that large rock falling, especially in hard rock formations, is the most important reason for sticking of drill string.

Core drilling started from 101 m deep and completed at 5118.2 m deep, a total core drilling footage of 5005.87 m in CCSD-1 Well were completed, by using nine core drilling methods, basically covered all available core drilling methods in the world (Table 5.50). Among this total core drilling footage, 80.77 % was completed by PDM hydro-hammer drive double tube method, 18.16 % was finished by PDM drive double tube method, whereas only 1.07 % was finished by the other seven methods. With PDM hydro-hammer drive double tube core drilling method, a total footage of 4043.25 m were completed, with an average penetration rate of 1.13 m/h and an average footage drilled per round trip of 6.34 m, being the highest among the nine methods. By using this method, the average core recovery reached to 85.44 % and 282.73 m were completed per month at the hole depth of 3000 m.

In comparison with PDM drive double tube method, PDM hydro-hammer drive double tube core drilling method

Table 5.50 Core drilling results in CCSD-1 Well

Drilling method	Round trips	Footage drilled (m)	Actual drilling time (h)	Core length (m)	ROP (m/h)	Core recovery (%)	Footage drilled per round trip (m)
PDM drive single tube	5	6.84	12.00	2.22	0.57	32.46	1.37
Rotary table drive double tube	7	11.60	29.61	6.40	0.39	55.17	1.66
Rotary table hydro-hammer drive double tube	5	6.39	18.08	4.95	0.35	77.46	1.28
Top drive double tube	8	5.87	16.24	0.30	0.36	5.11	0.73
PDM drive wireline coring	3(8)	6.72	20.59	4.83	0.33	71.88	2.24
Top drive wireline coring	5	7.62	12.02	1.05	0.63	13.78	1.52
Top drive hydro-hammer drive wireline coring	2(3)	8.27	9.25	8.23	0.89	99.52	4.14
PDM drive double tube	397 + 1	909.31	1274.96	805.97	0.71	88.64	2.28
PDM hydro-hammer drive double tube (Down hole power percussive rotary core drilling)	636 + 2	4043.25	3563.68	3454.56	1.13	85.44	6.34
Total	1071	5005.87	4956.43	4288.51	1.01	85.67	4.67

Note 1. In the column of round trip, the number in the brackets represents the times of drilling without lifting drill string in wireline core drilling
2. In the column of round trip, the number after the plus sign represents core drilling of sidetracking
3. In the core drilling hole section of CCSD-1 Well, 71.9 m were completed by milling, tri-cone non-core drilling and drilling for deviation correction

Fig. 5.45 Comparison of drilling results by different core drilling methods in CCSD-1 Well

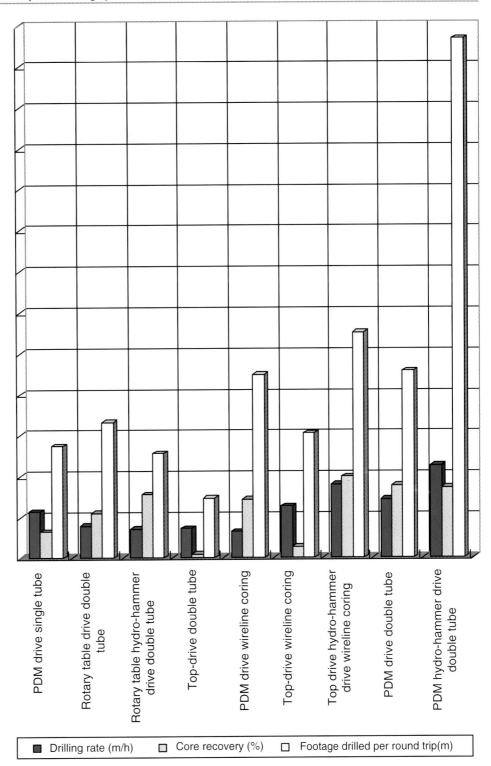

increased the penetration rate by 59 % and the footage drilled per round trip by 178 %.

Rotary table drive double tube core drilling and top drive double tube core drilling methods which are commonly used in oil drilling industry were unsuitable for hard rock core drilling in CCSD-1 Well, because of low penetration rate, extremely low core recovery, extremely short footage drilled per round trip, poor coring result and unsatisfied economy. Also, wireline core drilling techniques which are normally utilized in geological drilling industry were unsuitable for large diameter deep core drilling requested in CCSD-1 Well, as a result of low penetration rate, extremely low core

Table 5.51 Core drilling results by downhole power percussive rotary core drilling techniques at different stages

Hole section	Round trip	Footage drilled (m)	Actual drilling time (h)	Core length (m)	Drilling rate (m/h)	Core recovery (%)	Footage drilled per round trip (m)
Test stage of the pilot hole from 218.44 to 1000.05 m	72	220.01	207.11	213.63	1.06	97.10	3.06
Improvement stage of the pilot hole from 1000.05 to 1503.00 m	89	350.77	300.50	278.25	1.17	79.33	3.94
Perfection stage of the pilot hole from 1503.00 to 2046.54 m	104	525.91	483.32	464.70	1.09	88.36	5.06
Subtotal in the pilot hole	265	1096.69	990.93	956.58	1.11	87.22	4.14
Well developed and reliable application in MH section	107	830.07	631.48	669.81	1.31	80.69	7.76
First side-tracking and straightening in MH section	2	8.59	5.81	4.22	1.48	49.13	4.30
Well developed and reliable application in MH-1C section	79	634.85	506.16	528.80	1.25	83.30	8.04
Well developed and reliable application in MH-2C section	185	1473.05	1429.30	1295.15	1.03	87.92	7.96
Subtotal in the main hole	373	2946.56	2572.75	2497.98	1.15	84.78	7.90
Total	638	4043.25	3563.68	3454.56	1.13	85.44	6.34

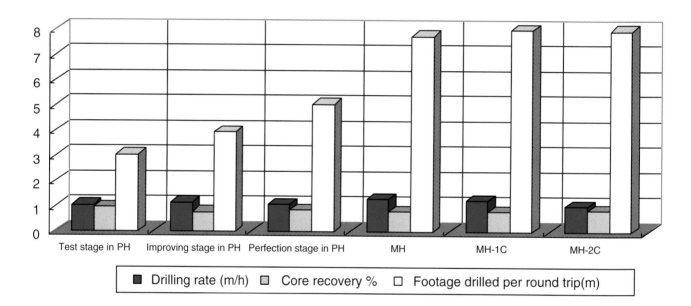

Fig. 5.46 Core drilling results by PDM hydro-hammer drive double tube core drilling techniques at different stages

recovery, extremely short footage drilled per round trip, very poor coring result and unsatisfied economy. The advantage of improving the footage drilled per round trip by wireline core drilling could not be brought into full play.

Core drilling practice in CCSD-1 Well indicates that PDM hydro-hammer drive double tube core drilling method is the best among all these drilling techniques, with the achieved technical and economic indexes obviously superior to the other methods (see Table 5.50; Fig. 5.45).

In general, downhole power percussive rotary core drilling techniques have stood the severe trials from the hard rocks and the complex formations of the Sulu Ultra-high Pressure Metamorphic Belt and achieved brilliant successes, not only creating a series of records in China but also evoking tremendous worldwide repercussions. Our drilling techniques used in CCSD-1 Well were highly appraised by world drilling circles at the Forum of Ten Years Review and Prospects of International Continental Scientific Drilling held by the Research Centre for Geosciences (GFZ) in Potsdam, Germany in March of 2005.

Core drilling results by downhole power percussive rotary core drilling techniques at different stages can be found in Table 5.51 and Fig. 5.46, from which it can be learned that penetration rate was maintained at over 1 m/h, footage drilled per round trip was gradually improved along with the tests and then stabilized at the main hole, and core recovery was stabilized over 80 %, especially in the three stages at the main hole section the average core recovery was increasingly improved along with the increase of borehole depth. It can be found from the data at different stages that this core drilling technique experienced relatively long time, from the first test to well developed. And during this period our research workers carried out a large amount of continuous study on improvement and perfection of this technique, so as to improve its reliability and its technical and economic indexes.

Diamond Core Drill Bit

Being an important part of core drilling techniques, diamond core drill bit is a basic prerequisite to improve drilling efficiency and core recovery, and its application results (drilling rate and service life) play an important role to influence the economy and the technical indexes of drilling construction.

The difficult problems of diamond drill bit application in CCSD-1 Well included the follows: (i) it was difficult for drill bit to crush rocks, as they are very hard and highly abrasive, thus drilling rate was low, with unsatisfactory gauging result of bit matrix and low service life of drill bit; (ii) as drill bit bears the percussive load from hydro-hammer, high strength drill bit matrix, top quality diamond and high holding force of matrix to diamond were required; (iii) it was decided that the diamond drill bits for CCSD-1 Well should be 157 mm OD, 96 mm ID and 240 mm height, thus bringing some difficulties for manufacture; (iv) drill bit design and manufacturing techniques should be completely changed because the drill bits for CCSD-1 Well were large in diameter and our existing manufacturing techniques for geological core drill bits were only suitable for small sizes (smaller than 91 mm). On the other hand, the available core drill bits for oil drilling were only suitable to soft sedimentary rock formations so that new structure drill bits must be designed and the related manufacturing techniques be developed; and (v) the drill bits for CCSD-1 Well should have good stability and anti-deviation property, as more than 50 % of the rocks to be drilled was gneiss, with strong deviated tendency.

6.1 The Physical and Mechanical Properties of the Rocks to Be Drilled

6.1.1 The Properties of the Rocks to Be Drilled

The hardness and abrasiveness of rocks depend mainly on the composition, grain size and compaction of rock minerals.

Translated by Geng Junfeng.

The rocks to be drilled in CCSD-1 Well were granite gneiss and felsic gneiss, with high content of quartz and feldspar, mainly medium grain size, compact crystillite and very high hardness, being difficult to be penetrated, while coarse grain sized eclogite and rich rutile eclogite are also high in hardness, difficult to be drilled, resulting in low drilling rate and excessive wear of drill bits because of their high abrasiveness. The diamond drill bits for these rocks must have the characteristics of high matrix hardness, top grade diamond and high diamond concentration.

Rocks in some borehole sections are hard, brittle and broken, so for drilling these rocks the diamond drill bits were requested to be of high hardness and wear resistance, with good gauging result.

As more than 50 % of the rocks to be drilled was gneiss, which strongly leads to hole deviation, it was very important to design and select the diamond drill bits with the function of anti-deviation.

6.1.2 The Physical and Mechanical Properties of the Rocks

Based upon the situation of the rock formations of CCSD-1 Well, felsic gneiss, biotite amphibolite, rich rutile eclogite and biotite plagiogneiss were selected as the typical rocks to be tested by Beijing Institute of Exploration Engineering, for indentation hardness, pendulum hardness, uniaxial compressive strength, abrasiveness and erosiveness. All these five testing indexes can represent the comprehensive properties of the rocks, and also play a guiding role in the design of diamond drill bits for CCSD-1 Well and in the selection of drilling parameters.

1. **Indentation hardness**

Rock indentation hardness denotes the local crushing resistant strength of the rock under the state of approximate tri-axial stress, and reflects the cementing strength of the rock. In comparison with Shore hardness and pendulum hardness, indentation hardness can still better represent

Table 6.1 Testing result of indentation hardness

No.	Rock	Sampling		Test result	Evaluated grade
		Place	Depth (m)	MPa	
1	Felsic gneiss	Donghai	1485.18	3650	VIII
2	Rich rutile eclogite	Donghai	536.40	5020	IX
3	Biotite plagiogneiss	Donghai	916.20	3150	VII
4	Biotite amphibolite	Donghai	1143.50	1750	V–VI

directly the drilling resistant property of the rock during drilling process. Indentation hardness is calculated in such a way: indenting the rock sample by using an indenter with the bottom area of 1 mm² and measuring the pressure of the indenter when it suddenly indents into the rock. For normal testing, three pieces of rock samples are needed for each type of rock and five spots are tested for each piece of rock sample, and finally the average value is adopted.

WYY-1 rock indentation hardness testing instrument was used for testing, with the results shown in Table 6.1.

2. **Pendulum hardness**

As a dynamic measuring method, the dynamic hardness of the rock is represented by bounce times of the pendulating ball on rock surface and its first spring-back angle. The harder the rock, the less energy the rock surface absorbs from the ball, the higher the ball springs back, and the more times the ball strikes against the rock surface.

WYB-64 rock pendulum hardness testing instrument was used for testing, with the results shown in Table 6.2.

3. **Uniaxial compressive strength**

The uniaxial compressive strength is a basic parameter to represent the solidness of the rock and the test of uniaxial compressive strength is also a basic way to observe the crushing process of the rock.

NYL-600 press was used for testing uniaxial compressive strength, with the results shown in Table 6.3.

4. **Rock abrasiveness test**

The wear ability of rock to rock fragmentation tools is known as rock abrasiveness. In order to actually reflect the wear state of rock to diamond drill bit, the following method is used to test rock abrasiveness: a standard diamond grinding stick is used to grind rock sample for 3 min under the pressure of 150 N and at a rotary speed of 400 rpm and then the worn part of the diamond grinding stick is used to represent rock abrasiveness.

Rock abrasiveness was tested on a modified bench drill, results shown in Table 6.4.

5. **Rock erosiveness test**

Being a new concept, rock erosiveness can reflect the wearing situation of drill bit matrix at hole bottom better than rock abrasiveness and thus it is very scientific to use it to represent the abrasion ability of rocks. Rock erosiveness is tested in such a way that a standard testing matrix sample (30 mm diameter × 5 mm, matrix hardness HRC 38–40) is eroded by a 2 mm sized spray nozzle with drilling fluid carrying 40/60 mesh rock sample powder under the pressure of 1.2 MPa and at flow rate of 9.7 L/min, for 10 min. The eroded volume of the matrix sample is used to represent rock erosiveness.

Table 6.2 Testing result of pendulum hardness

No.	Rock	Sampling		Test result		Evaluated grade
		Place	Depth (m)	Bounce times	Spring angle (°)	
1	Felsic gneiss	Donghai	1485.18	68	78	VIII
2	Rich rutile eclogite	Donghai	536.40	85	80	IX
3	Biotite plagiogneiss	Donghai	916.20	66	76	VII
4	Biotite amphibolite	Donghai	1143.50	60	70	VI

Table 6.3 Relationship between uniaxial compressive strength with Protodyakonov number and drillability grade

No.	Rock	Sampling		Test result	Protodyakonov number (f)	Drillability grade
		Place	Depth (m)	MPa		
1	Felsic gneiss	Donghai	1485.18	172	13.29	IX
2	Rich rutile eclogite	Donghai	536.40	186	14.07	X
3	Biotite plagiogneiss	Donghai	916.20	141	11.54	VIII
4	Biotite amphibolite	Donghai	1143.50	108	9.60	VII

Table 6.4 Testing result of rock abrasiveness

No.	Rock	Sampling		Test result	Evaluated grade
		Place	Depth (m)	Abrasiveness (mg)	
1	Felsic gneiss	Donghai	1485.18	8.57	High 4+
2	Rich rutile eclogite	Donghai	536.40	5.03	Medium 3+
3	Biotite plagiogneiss	Donghai	916.20	6.62	High 4−
4	Biotite amphibolite	Donghai	1143.50	3.29	Medium 3−

Rock erosiveness was tested on a self-developed instrument, with the results shown in Table 6.5.

6. **Rock drillability grade**

Drillability grade of the main rocks to be drilled in CCSD-1 Well can be obtained by synthesizing indentation hardness, pendulum hardness and uniaxial compressive strength of the rock samples, shown in Table 6.6. As these rock samples were obtained at some shallow borehole sections, though with a certain representativeness, they can not fully reflect the whole properties of all the rocks in the whole borehole, this is because that although the mineral components are basically the same for each rock, the content proportion of the minerals and their grain size are not always the same, which can affect the testing result. Taking felsic granite gneiss as an example, its quartz content varies obviously and the quartz grain size differs, which affect its hardness and abrasiveness.

Based upon drilling indexes, analysis on the rocks from the main hole and by reference to the abovementioned testing results, it was believed that the rocks in CCSD-1 Well are mainly 8–9 grade, some are 10–11 grade. However, some rocks even have lower drillability grade, such as amphibolite, 7 grade.

6.2 Selection of Diamond Core Drill Bit Types

Design and type selection of diamond core drill bit mainly depend upon the rock properties and the drilling technology, the former has already been described at above section, and the latter includes drilling method, drilling tool structure, drilling parameters and drilling fluid properties.

6.2.1 Core Drilling Technologies

PDM hydro-hammer conventional diamond core drilling technique system (by lifting drilling tool for core recovery) was the main method to be used for CCSD-1 Well. The theoretical rotary speed of PDM (C5LZ95 × 7 and 4LZ120 × 7) was from 128 to 380 rpm and the working torque reached to 2137 Nm. In applications, however, restricted by a variety of conditions the actual rotary speed of the PDM was no more than 200 rpm, in many cases between 176 and 189 rpm. For 157 mm sized drill bit, the linear velocity of outer circumference is about 1.5 m/s, being the lower limit

Table 6.5 Testing result of rock erosiveness

No.	Rock	Sampling		Test result	Evaluated grade
		Place	Depth (m)	Erosiveness (mm^3)	
1	Felsic gneiss	Donghai	1485.18	50	Strong
2	Rich rutile eclogite	Donghai	536.40	40	Medium
3	Biotite plagiogneiss	Donghai	916.20	45	Less strong
4	Biotite amphibolite	Donghai	1143.50	20	Less medium

Table 6.6 Testing report of rock drillability grade for CCSD

No.	Rock	Sampling		Test result				Evaluated grade
		Place	Depth (m)	Indentation hardness (MPa)	Pendulum hardness		Uniaxial compressive strength (MPa)	
					Times	Angle degrees		
1	Felsic gneiss	Donghai	1485.18	3650	68	78	172	VIII
2	Rich rutile eclogite	Donghai	536.40	5020	85	80	186	IX
3	Biotite plagiogneiss	Donghai	916.20	3150	66	76	141	VII
4	Biotite amphibolite	Donghai	1143.50	1750	60	70	108	V–VI

Fig. 6.1 Matrix piece drop off

(1.5–3 m/s) of the linear velocity requested by impregnated diamond core drilling.

Hydro-hammer drilling requests low bit pressure, generally 10–35 kN, while needs higher impact resistance for drill bit. Higher impact exerted an unfavorable influence on diamond drill bit, resulting in matrix piece drop-off (Fig. 6.1), even the whole matrix drop-off, especially for electro-plated diamond drill bit.

Considering rock properties and PDM properties, drill bit should be moderate in hardness and high in wear resistance, so as to maintain higher drilling speed and a fairly long service life; while in consideration of percussive rotary drilling the drill bit should be high in hardness, high in impact resistance and moderate in wear resistance. If percussive rotary drilling could be normally operated then the requirement for drill bit properties would be unitary; while in case that hydro-hammer could not work effectively at down hole for one round trip then the requirement for drill bit properties would be much more complicated, bringing about difficulties for drill bit design and manufacture.

As CCSD-1 Well is large in diameter and deep in depth, drilling fluid needs large pump delivery rate and high pump pressure, and the cuttings are high in both hardness and in abrasiveness, exerting a very strong erosive action on drill bit. Tests indicated that the erosion of the cuttings of felsic gneiss from CCSD-1 Well is 50 mm^3 and the erosion of the cuttings of rich rutile eclogite is 40 mm^3, both means that the cuttings from CCSD-1 Well are very high in erosion, causing harmful influence on diamond drill bit gauging and drill bit water-ways (Fig. 6.2).

Differing from conventional rotary drilling, scientific deep drilling requires that the diamond drill bits utilized should be suitable to the high requirements on percussive rotary drilling, i.e. the drill bits should have higher matrix hardness and wear resistance, and higher toughness against impact and erosive resistance.

Fig. 6.2 Steps caused by erosion at water-ways

6.2.2 Types of Diamond Core Drill Bits

Diamond core drill bits can be classified into two types based on diamond inserting way, surface set drill bit and impregnated drill bit.

1. **Natural diamond surface set drill bit**

High in matrix hardness, wear resistance and erosive resistance, natural diamond surface set drill bit is suitable for the drilling conditions of large pump delivery and high pump pressure in oil drilling. With high diamond exposure and coarse diamond grains, natural diamond surface set drill bit produces high drilling efficiency and long service life in drilling sedimentary rocks.

As for CCSD-1 Well, the rocks encountered are crystalline rocks, with high hardness and high abrasiveness. High drilling speed we required could only be realized by adopting high rotary speed and high bit pressure, which would inevitably increase the compound vibration to the drill bit, and for surface set drill bit this was really an unfavorable factor, which would cause the highly exposed diamond broken or dropping-off, and gradually losing its cutting ability. Our experiences have also shown that natural diamond surface set drill bit caused high drilling cost, being unsuitable for drilling hard rocks or fracture developed formations.

To improve drilling rate, hydro-hammer percussive rotary drilling was to be utilized for CCSD-1 Well, according to the design. As the surface set diamond drill bit had high diamond exposure the highly exposed diamond would be easily broken because under the action of large percussive work the reacting force of the hard rock was also large. In this case the diamond could not fully work and the service life was short. It was believed from this that natural diamond surface set drill bit was unsuitable for hard rock drilling and hard rock percussive rotary drilling.

2. **Impregnated diamond drill bit**

Impregnated diamond drill bit is suitable to medium hard to hard rocks, and the rocks with different abrasivenesses and with different intact degrees. In other words, the impregnated diamond drill bit has a wide adaptability to crystalline rocks, with good technical indexes. Most of the impregnated diamond drill bits use synthetic diamond, resulting in a low drilling cost.

The diamond used for impregnated drill bit is much smaller than that used for surface set diamond drill bit, thus with less exposed from drill bit matrix, that is to say, the diamond is enclosed with multidirectional stress and therefore is not easily broken under the action of percussive load. The impregnated drill bit is suitable for percussive rotary drilling (Fig. 6.3), from which it can be found that bit face wears normally, with fairly high diamond exposure and without diamond broken. It can also be found that the diamond has no tadpole shaped support at its tail end, which is just the characteristic of percussive rotary drilling creates for the drill bit.

With the experiences of tens of years, the design and the manufacturing technology of impregnated diamond drill bits have been well developed. The impregnated diamond drill bit is favorable to swivel type double tube core drilling, with high core recovery and good core quality obtained. Therefore, impregnated diamond drill bit was the first choice for drilling CCSD-1 Well, as it could satisfy the demands of scientific drilling.

6.3 Design and Manufacture of Impregnated Diamond Core Drill Bits

Based on the rock formations of CCSD-1 Well and the drilling techniques adopted, the overall demands for the diamond core drill bits were as following:

(i) Large overflow area of the water passages Water passages should cover 25–30 % of the bit face area and both the outer and inner water passages should be deeper than the conventional water passages, to be favorable to discharge the cuttings and cool the diamond drill bit, and to reduce the erosion of drilling fluid with high pump rate and high pressure to drill bit inside and outside gauges.

(ii) Small contacting area of diamond segments to hole bottom Bit pressure ratio can be increased so as to improve drilling speed. With small contacting area of diamond segments to hole bottom a low bit pressure can be used for drilling.

(iii) Long inside and outside gauges The outside gauges should be 40 mm while the inside gauges 30 mm, which can improve gauging effect, maintain the stability of drilling tool and reduce the radial vibration of the drill bit, with the functions of anti-deviation and anti-whirling.

(iv) High diamond working layer Under the circumstances that each meter drilled consumes the same amount of diamond, this drill bit serves longer life in comparison with the conventional drill bit.

In drilling CCSD-1 Well, impregnated diamond drill bits were produced by three different technologies (Fig. 6.4), with the designs and manufacturing methods as follows:

6.3.1 Segment Inserted Drill Bit by Twice Forming

Segments with diamond working layer are manufactured by sintering process of hot press, gauges produced by infiltration process without press and then the segments are brazed

Fig. 6.3 Diamond exposure under the percussion of hydro-hammer

Fig. 6.4 Impregnated diamond core bits. *Left* Electro-plated drill bit by twice forming method, developed by the Institute of Exploration Technologies. *Middle* Drill bit by sintering process of hot press, developed by Guilin Diamond Co. *Right* Segment inserted drill bit by twice forming, developed by the Beijing Institute of Exploration Engineering

onto drill bit body under medium temperature. To improve the brazing strength, brazing layer matrixes produced by sintering process of hot press or infiltration process without press are used.

1. **Bit matrix properties designed by drill bit expert system**

Being an advanced, scientific and reliable design method, the drill bit expert system was used for designing diamond drill bits, as it could regulate the formulas of bit matrix in time according to the variation of the rock properties and the drilling conditions, so as to make the bit matrix properties suitable to the new situations.

For felsic gneiss and rich rutile eclogite, (see Tables 6.1, 6.2, 6.3, 6.4, 6.5 and 6.6) which are difficultly penetrated, the bit matrix of high wear resistance, high erosive resistance and with high diamond concentration was adopted. Designed by the drill bit expert system, the wear resistance of the bit matrix (ML) was 0.25×10^{-5}, the erosive resistance was 28 cm^{-3}, and the diamond concentration, 85–90 %.

The characteristics that polycrystalline diamond is wear resistant, natural diamond is of high strength and carbonado diamond is of high hardness were fully utilized. A mixed set of polycrystalline diamond and natural diamond was used for drill bit inside gauging (Fig. 6.5a) while a mixed set of polycrystalline diamond, natural diamond and carbonado diamond was used for drill bit outside gauging (Fig. 6.5b)

2. **Diamond parameters**

Diamond grade has an obvious influence on drill bit quality, which directly affects drilling indexes. For this reason SDA100 + synthetic diamond (DE BEERS standard) was adopted, because this diamond can keep a long time sharpening state, without becoming dull after worn, as micro-cracks will occur when the diamond worn to a certain time. This is of benefit to increase drilling speed and to improve bit service life.

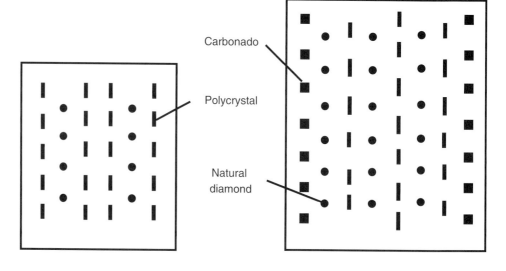

Fig. 6.5 Inside and outside gauging of segment inserted drill bit by twice forming

3. Use of ultra-fine prealloy powder

The specially developed ultra-fine prealloy powder was used, so that the time for mixing powders and the hardening of the powders during mixing were both reduced and homogeneous distribution of matrix components improved. With large specific surface and high surface energy, ultra-fine powder can not only reduce the sintering temperature and the pressure of hot press, but also improve the alloying degree, being favorable for firmly holding diamond and improving drill bit quality.

4. Drill bit manufacturing technology

(1) Manufacture of segment

Based upon the formula prepared by optimum design from the expert system, the powders were mixed, weighed, moulded, and then the moulds were sintered in a furnace automatically at the temperature of 820–900° of centigrade and the pressure of 10–20 kN. Different from the conventional techniques, the sintering technology of the segment used lower sintering temperature and higher sintering pressure to ensure that during the sintering process there were no leaks, no contraction and no clearance. In this way the dimensional accuracy of the sintered segments, the homogeneity of the matrix components and the constant quality of different batches all can be guaranteed. The whole sintering process is controlled by computer, during which temperature raise and pressure raise, temperature preservation and pressure preservation, temperature drop and pressure drop, and shut-down furnace are all unmanned operations. Lower temperature sinter is the most important factor, by which diamond strength and structure will not be damaged and diamond will not be easily worn flat or broken in drilling hard rocks, with the results of strong crushing rocks and prolonging drill bit service life.

(2) Manufacture of drill bit matrix body

Drill bit matrix body was sintered in an automatic temperature controlling furnace by using the methods of sintering process of hot press and infiltration process without press, with the produced bit matrix body of high strength, high wear resistance, strong erosion resistance and good brazing ability. On the drill bit matrix body, the positions for brazing the segments must be precisely predetermined and the width for brazing seam should be optimized, otherwise brazing will not be firm and the segments may drop off in drilling operations.

(3) Brazing of segment

Medium temperature brazing was adopted. The segments are cleaned and set into the brazing slots, with the positioning mould equipped, were brazed, and then slowly cooled to room temperature.

6.3.2 Sintered Diamond Drill Bit

1. Design of drill bit structure

Flat bit face profile was designed, with R5 transitional arc at inner and outer edges, to avoid matrix broken during drilling process.

The structure of bit gauges is shown in Fig. 6.6, from which it can be found that from bit face upwards, polycrystalline diamond and high strength diamond were used for gauging in working layer, polycrystalline diamond was for medium layer and in the top layer diamond micro-powder was used for strengthened gauging. With the characteristics of high wear resistance and erosion resistance, this gauging structure achieved very good effect, without affecting drilling result.

In the working layer, the bit face is divided up into three areas (see left of Fig. 6.6), i.e. the outer diameter area (Area A), the inner diameter area (Area B) and the medium area (Area C). The outer diameter area and the inner diameter area are strengthened areas, in which the matrix is harder, with higher diamond concentration. Besides polycrystalline

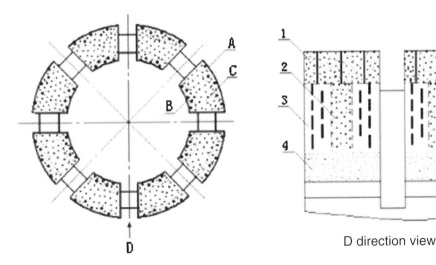

Fig. 6.6 Structure of sintered diamond drill bit. *1* Working layer. *2* Diamond gauging area. *3* Polycrystalline diamond gauging area. *4* Micro-powder strengthened gauging area

diamond gauging, polycrystalline diamond is distributed in the areas to improve the wear resistance and impact resistance of the areas, avoid an inclined wear at bit face and keep a synchronous wear of Areas A, B and C at the working layer of drill bit face, with results of longer service life and satisfactory drilling effect.

Straight slot shaped waterways are evenly distributed, with width of 10 mm, depth of 10 mm. The depths of inner and outer slots are 6.5 mm. This water flowing passageway with large flow area can satisfy the need of large pump rate and high pump pressure for CCSD-1 Well drilling.

2. **Diamond parameters**

By strict selection, high strength diamond manufactured by two anvil press was used for producing diamond drill bit. The compressive strength of the diamond should be no less than 250 N, diamond concentration was 80–95 %, and diamond size was 35/40 to 60/70 mesh. Diamond concentration at both inner and outer circular areas in matrix working layer was higher than that at the medium area, and that at the outer circular area was slightly higher than that at the inner circular area, to maintain a synchronous wear.

3. **Drill bit manufacturing technology**

(1) Material mixing and mould assembling

To prepare the materials according to the matrix formula, and then grind the mixing materials in a ball grinding mill for 24 h, with a ball to material ratio of 1:2. Store the ground powder in a bottle for use.

Place the graphite mould on a rotary table, glue the mould core with mucilage, stick the prefabricated water-way inserts onto the position of water-way, charge the working layer powder materials which contain diamond and vibrate and press for compactness, place the reinforced polycrystalline diamond into the working layer, load the powder materials for matrix non-working layer and press for compactness, insert the reinforced gauging polycrystalline diamond, assemble steel body and press, then the whole assembly is sintered in a sintering furnace.

(2) Sinter

A 160kVA intermediate frequency electric furnace was used for sintering diamond drill bit. Slowly increased temperature and relatively high sintering pressure were utilized as the sintering technology, by which the temperature was increased to 960 °C within 20 min. When the temperature was increased to 700 °C the pressure was increased to 1 MPa and then gradually increased, to 8.7 MPa when the temperature was increased to 900 °C, then maintain this pressure. Preserving the temperature for 15 min under 960 °C and then gradually decreased the temperature, unloaded the pressure when the temperature was decreased to 600 °C and the drill bit was taken out of the furnace. The sintering time-temperature curve can be found in Fig. 6.7, and the sintering time-pressure curve in Fig. 6.8.

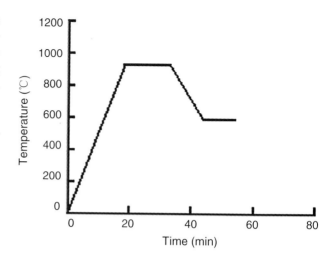

Fig. 6.7 Sintering time-temperature curve

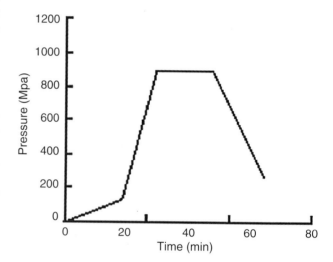

Fig. 6.8 Sintering time-pressure curve

6.3.3 Electro-plated Diamond Drill Bit by Twice Forming

The manufacturing technique of twice forming electro-plated diamond drill bit is a method that combines infiltration process without press and electroplating process, with the superiorities of good gauging effect resulted from infiltration process without press and high drilling rate caused by electroplated diamond drill bit.

1. **Gauging structure**

Based on the characteristics of electro-plated drill bit, high strength monocrystalline diamond was used for gauging in the working layer (Fig. 6.9) and polycrystalline diamond and natural diamond were used for gauging at the gauge areas (Fig. 6.9). The gauges were produced by infiltration process without press, with cast tungsten carbide as the main matrix material, which has high wear resistance and

6.3 Design and Manufacture of Impregnated Diamond Core Drill Bits

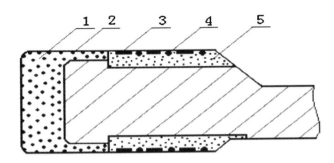

Fig. 6.9 Gauging structure of twice forming electro-plated drill bit. *1* Working layer. *2* High strength monocrystalline diamond. *3* Polycrystalline diamond. *4* Natural diamond. *5* Gauges

good holding strength to natural and polycrystalline diamond.

In the structure, the length of the gauges was increased to 42 mm, to improve the gauging effect and the stability of the drill bit.

To increase the binding strength between the gauges and the bit steel body, thread was machined and holes were drilled at corresponding position of the bit steel body (Fig. 6.10), to improve the flowability of the binding metals at the inner and outer annuluses, and to combine the bit steel body and the gauges into a firm whole. In field operations gauge drop-off never happened and the gauging effect was satisfactory.

2. **Bit matrix properties**

Nickel is the main ingredient for bit matrix. During the process of electroplating with nickel, drill bit property was adjusted by adding additives so as to suit for drilling different rock formations. The main property of this bit matrix was the high hardness, with a certain brittleness, which showed in drilling operations was not the poor diamond exposure caused by hard matrix, but a good diamond exposure and the increased drilling rate, resulted from a lower matrix wear resistance caused by its appropriate brittleness. Also, the adaptability of this bit matrix to rock formations was obviously improved; good drilling result could also be obtained even in hard rock formation with weak abrasiveness.

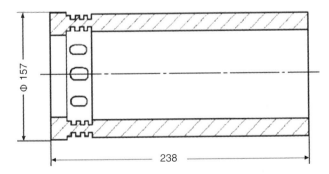

Fig. 6.10 Steel body of twice forming electro-plated drill bit

3. **Drill bit manufacturing technology**

(1) Bit gauges produced by infiltration process without press

Assembled the mould for sintering by placing the steel body in the graphite mould, filling in the inner and outer annuluses between the graphite mould and the steel body with matrix skeleton powder of fixed quantity, vibrating properly to make the matrix skeleton powder to the designed density, filling binding metal and flux, and the mould was sent into the furnace for sinter.

Infiltration temperature and heat preservation time greatly influence the quality of bit gauge. Excessively high temperature and long heat preservation time can bring harm, but inadequate temperature and short heat preservation time can not bring to infiltration, thus affecting the quality. The designed sintering temperature was 1020–1050 °C and the heat preservation time, 8–10 min.

(2) The working layer of electro-plated drill bit

Pre-treatment before electro-plating was enhanced and the following measures were adopted: de-oiling with organic solvent → cleaning with hot water → connecting wire and coating insulating paint → packing → severe erosion (acid pickling) → washing with cold water → electrochemical de-oiling (alkali boiling) → washing with hot water → washing with cold water → acid pickling (severe erosion) → live-wire treatment (weak erosion) → washing with cold water → alkali boiling → washing with hot water → washing with cold water → weak erosion → entering into the electroplating bath with electricity-on.

In the process of electroplating, the current density was set to 1.8 A/dm^2 (equal to electrodeposition speed 0.04 mm/h), the electroplating liquid temperature was set to 45 °C, and the pH value, 4.0–4.8. The addition of surface active agent (anti-pinhole agent) in the electroplating liquid could reduce the staying time of hydrogen on the surface of electroplating layer, so as to reduce the hydrogen content in the electroplating layer.

Once the height of the working layer reached the designed demand the drill bit was taken out of the electroplating bath, cleaned and placed in a constant temperature drying case for dehydrogenation (at 200 °C for 2 h) to reduce the brittleness of the working layer caused by hydrogen and to improve bit matrix properties.

6.4 Application of Diamond Core Drill Bits

6.4.1 Brief Introduction

On the basis of pre-study projects of sintered and electroplated drill bits, drill bit manufacturers all over China were invited to participate in the drill bit optimization test for

Table 6.7 Application results of the diamond drill bits used for CCSD-1 Well core drilling

Bit type		Bit quantity	Roundtrips	Footage drilled (m)	Penetration rate (m/h)	Drill bit service life (m)	Footage drilled per roundtrip (m)	Core recovery (%)
Surface set	2 kinds products in sum	12	91	171.76	0.70	14.31	1.89	96.1
Sintered impregnated	Segment inserted by twice forming	129	563	2843.65	1.05	22.04	5.05	83.4
	Sintered	46	174	1240.64	1.08	26.97	7.13	87.4
	5 types sintered	10	40	65.16	0.62	6.52	1.63	90.1
	7 kinds products in sum	185	777	4149.45	1.05	22.43	5.34	84.7
Electro-plated impregnated	Electro-plated by twice forming	27	158	451.93	0.82	16.74	2.86	88.9
	5 types electro-plated	23	45	232.73	1.14	10.12	5.17	88.5
	6 kinds products in sum	50	203	684.66	0.90	13.69	3.37	88.8
Total		247	1071	5005.87	1.01	20.27	4.67	85.7

Note Core drilling with three-in-one drilling tools in the drilling tool testing stage is excluded (footage of 3.5 m was drilled in 3 round trips)

CCSD-1 Well. Eleven drill bit manufacturers took part in the test, concerning natural diamond surface set drill bits and impregnated diamond drill bits. After two rounds of test, Beijing Institute of Exploration Engineering, Institute of Exploration Technologies and Guilin Diamond Co. were chosen as the drill bit suppliers for CCSD-1 Well.

For core drilling in CCSD-1 Well (from 101 to 5,118.2 m), 247 diamond core drill bits were used (Table 6.7), with a total footage of 5,005.87 m drilled, an average bit service life of 20.27 m, and an average penetration rate of 1.01 m/h. Among these drill bits, 12 were natural diamond surface set drill bits, 185 were sintered impregnated drill bits, and 50 were electroplated drill bits. For each drill bit, its service life and average penetration rate along with the hole depth can be found in Fig. 6.11, and the round trips each drill bit completed and its average footage drilled per round trip along with the hole depth can be found in Fig. 6.12.

There were 28 drill bits with service life exceeding 40 m, accounting for 11.3 % of the total drill bits. Of these drill bits, 3 had service life exceeding 70 m, 4 had service life of 60–70 m, 5 had service life of 50–60 m, and 16 had service life of 40–50 m. The maximum service life was 75.23 m, with this drill bit, 75.4 h drilling was completed and 70.89 m core was recovered.

The average round trip completed by each drill bit was 4.34. Two drill bits drilled 33 round trips each, with 72.06 m and 75.23 m drilled respectively. However, there were 49 drill bits which were abandoned only for one round trip, accounting for 19.8 % of the total drill bits.

1. **Application result of the drill bits used for pilot hole core drilling**

For the pilot hole (hole section from 101 to 2,046.54 m), the conventional coring technique was used, for the test and selection of core drilling method. The methods of PDM drive conventional coring (double tube, water isolated double tube, jet reverse circulation double tube and single tube), rotary table drive conventional coring (double tube, water isolated double tube and single tube) and PDM hydro- hammer drive conventional coring (double tube, water isolated double tube and jet reverse circulation double tube) were tested successively. In the light of the optimization test of the drilling methods, diamond core drill bits were tested. 111 drill bits concerning 14 kinds of three types from eleven Chinese manufacturers were tested in drilling operations (Table 6.8), with coring footage of 1,945.54 m, an average drill bit service life of 17.85 m and an average penetration rate of 0.90 m/h. For each drill bit, its service life and average penetration rate along with the hole depth can be found in Fig. 6.13, and the round trips each drill bit completed and its average footage drilled per round trip along with the hole depth can be found in Fig. 6.14.

Among the 13 drill bits which only completed one round trip (No. 77 was still used in the main hole), six drill bits were excessively worn at working layer; matrix pieces dropped for three drill bits; four drill bits were normally worn (single tube and jet reverse circulation drill bits could still be used).

6.4 Application of Diamond Core Drill Bits

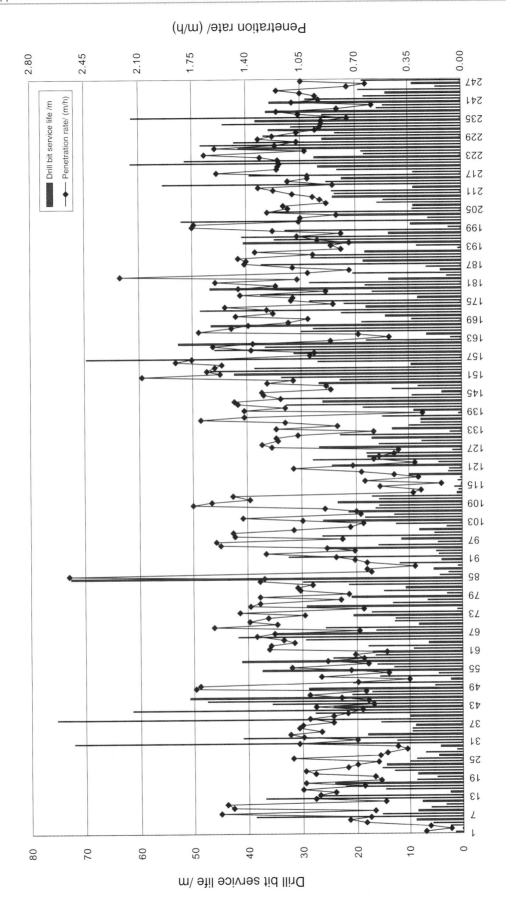

Fig. 6.11 Service life and penetration rate of the diamond drill bits used for CCSD-1 Well versus hole depth. No. 1–111 drill bits were used for the first phase core drilling for the main hole; No. 112–162 drill bits were used for pilot hole core drilling; No. 163–192 drill bits were used for the second phase core drilling for the main hole; and No. 193–247 drill bits were used for the third phase core drilling for the main hole

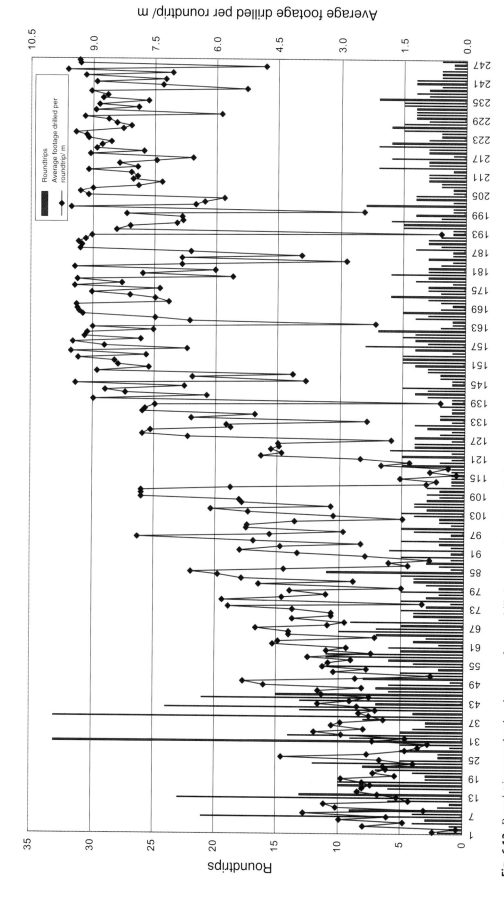

Fig. 6.12 Round trips completed and average footage drilled per round trip by each diamond core drill bit versus hole depth. No. 1–111 drill bits were used for pilot hole core drilling; No. 112–162 drill bits were used for the first phase core drilling for the main hole; No. 163–192 drill bits were used for the second phase core drilling for the main hole; and No. 193–247 drill bits were used for the third phase core drilling for the main hole

6.4 Application of Diamond Core Drill Bits

Table 6.8 Application results of diamond core drill bits in the pilot hole (PH)

Bit type		Bit quantity	Roundtrips	Footage drilled (m)	Penetration rate (m/h)	Drill bit service life (m)	Footage drilled per roundtrip (m)	Core recovery (%)
Surface set	2 kinds products in sum	12	91	171.76	0.70	14.31	1.89	96.1
Sintered impregnated	Segment inserted by twice forming	48	316	1073.11	1.01	22.36	3.40	87.1
	Sintered	4	29	85.88	0.78	21.47	2.96	80.6
	4 types sintered	9	39	62.66	0.61	6.96	1.61	93.7
	6 kinds products in sum	61	384	1221.65	0.96	20.03	3.18	87.0
Electro-plated impregnated	Electro-plated by twice forming	25	156	445.79	0.83	17.83	2.86	89.2
	5 types electroplated	11	26	106.34	1.10	9.67	4.09	94.3
	6 kinds products in sum	36	182	552.13	0.87	15.34	3.03	90.2
Total		109	657	1945.54	0.90	17.85	2.96	88.7

Note In the pilot hole a total of 111 drill bits were run into the hole. Among these drill bits No. 77 and No. 111 segment inserted drill bits by twice forming were still used in the main hole (MH), thus calculated as 0.5 for each. No. 56 and No. 105 segment inserted drill bits were still used in the main hole (MH), thus calculated as 0.5 for each

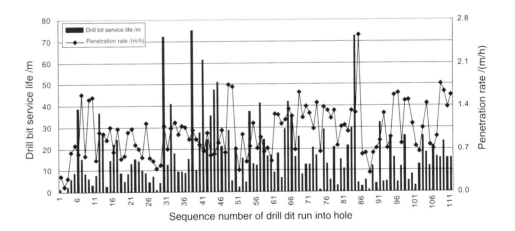

Fig. 6.13 Service life and penetration rate of the diamond drill bits used for pilot hole core drilling versus hole depth

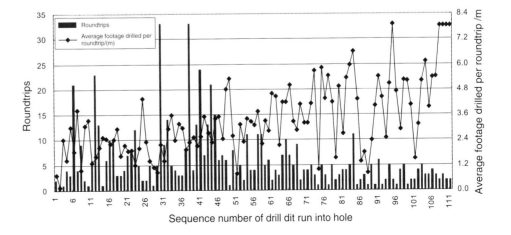

Fig. 6.14 Round trips completed and average footage drilled per round trip by each diamond core drill bit in the pilot hole versus hole depth

2. Application result of the drill bits used for the first phase of the main hole

For the first phase of the main hole (hole section 2046.54–2982.18 m), wireline core drilling techniques (conventional wireline core drilling, PDM wireline core drilling and hydro-hammer wireline core drilling) and conventional core drilling techniques (PDM conventional coring and PDM hydro-hammer conventional coring) were used, with 55 drill bits concerning 5 kinds of two types from five manufacturers (including one German manufacturer, see Table 6.9). 934.67 m were completed, with an average drill bit service life of 17.97 m and an average penetration rate of 1.11 m/h. Among these drill bits were 8 wireline core drill bits (5 kinds of two types), with footage of 22.61 m drilled, average service life of 2.83 m and average penetration rate of 0.54 m/h (Table 6.10).

For each drill bit, its service life and average penetration rate versus the hole depth can be found in Fig. 6.15, and the round trips each drill bit completed and its average footage drilled per round trip versus the hole depth can be found in Fig. 6.16.

Among the eighteen drill bits each only completed one round trip (No. 111, 105 and 77 had been used in the pilot hole), thirteen drill bits wore seriously at the working layer,

Table 6.9 Application results of diamond drill bits in the first phase of core drilling in the main hole (MH)

Bit type		Bit quantity	Roundtrips	Footage drilled (m)	Penetration rate (m/h)	Drill bit service life (m)	Footage drilled per roundtrip (m)	Core recovery (%)
Sintered impregnated	Segment inserted by twice forming	36	113	723.74	1.09	20.10	6.40	77.0
	Sintered	7	21	141.36	1.35	20.19	6.73	84.6
	German sintered	1	1	2.50	1.10	2.50	2.50	0.0
	In sum	44	135	867.60	1.13	19.72	6.43	78.0
Electro-plated impregnated	Electro-plated by twice forming	2	2	6.14	0.33	3.07	3.07	73.0
	Electro-plated	6	9	60.93	1.07	10.16	6.77	72.3
	Sum	8	11	67.07	0.89	8.38	6.10	72.4
Total		52	146	934.67	1.11	17.97	6.40	77.6

Note In the MH drilling a total of 55 drill bits were run into the hole. Among these drill bits No. 77 and No. 111 segment inserted drill bits by twice forming had been used in the pilot hole (PH), thus calculated as 0.5 for each. No. 56 and No. 105 segment inserted drill bits had been used in the pilot hole (PH), thus calculated as 0.5 for each

Table 6.10 Application results of wireline core drill bits

| Drill bit | | Well depth (m) | | Footage drilled (m) | Roundtrips | Drill bit service life (m) | Drilling speed (m/h) | Core recovery (%) | Remarks |
No.	Type	From	To						
112	Electro-plated	2047.82	2048.74	0.92	1	0.92	0.32	52.2	Top drive wireline coring
114	Sintered	2049.67	2051.22	1.55	1	1.55	0.54	27.1	Top drive wireline coring
119	Segment inserted by twice forming	2066.70	2462.70	2.65	2	2.65	0.66	5.7	Top drive wireline coring
120	German sintered	2069.19	2071.69	2.50	1	2.50	1.10	0.0	Top drive wireline coring
122	Electro-plated by twice forming	2075.63	2080.01	4.38	1	4.38	0.31	79.2	PDM wireline coring
126	Electro-plated by twice forming	2130.58	2132.34	1.76	1	1.76	0.42	57.4	Hydro-hammer wireline coring
138	Segment inserted by twice forming	2283.28	2283.86	0.58	1	0.58	0.26	60.3	PDM wireline coring
147	Segment inserted by twice forming	2442.71	2460.38	8.27	2	8.27	0.89	99.5	Hydro-hammer wireline coring
Total				22.61	10	2.83	0.54	62.4	

6.4 Application of Diamond Core Drill Bits

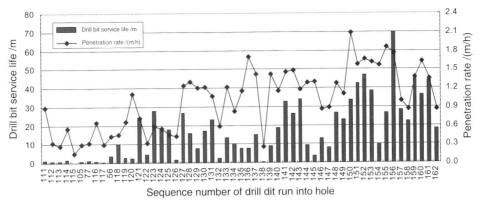

Fig. 6.15 Service life and penetration rate of the diamond drill bits used in the first phase of the main hole versus hole depth

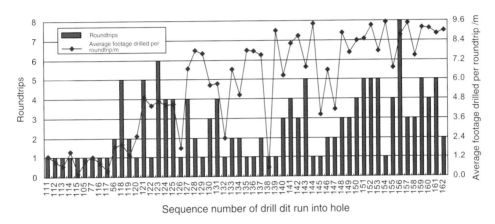

Fig. 6.16 Round trips completed and average footage drilled per round trip by each diamond drill bit in the first phase of the main hole versus hole depth

one drill bit broke in matrix and dropped off, and four drill bits wore normally (wireline core drill bits, still could be used).

In the first phase of main hole drilling, wireline core drilling techniques were to be used. Top drive double tube drilling tool (for 7 round trips) and wireline core drilling tool (for 2 round trips) were utilized for drilling continuous 9 round trips. However, abnormal phenomenon happened that all the drill bits (including 7 new drill bits and 2 drill bits had been used in the pilot hole) were completely worn out for only one round trip (Fig. 6.17).

Analysis of bit wear indicated that the drill bit had worn quickly and very smoothly at bit face, should belonging to abnormal wear, not worn by broken tungsten carbide. This abnormal wear neither resulted from unsuitability of the drill bit for rock formation, nor from unreasonable selection of drilling parameters. Analysis of the rock properties showed that in this phase the rock was broken, hole size was large (Fig. 4.16) and seriously over-sized hole section provided

Fig. 6.17 Drill bit wear in top drive core drilling

space for crook of wireline drill rods. Analysis of wireline drill rods showed that a lot of drill rods had been obviously crooked, resulted from the inadequate rigidity and the excessive flexibility, which had not only created a rotation, but also a revolution of the drilling string, which meant at the hole bottom the drill bit rotated as well as swinging and sliding, with poor stability. Swinging and sliding of drill bit were the main reason for drill bit abnormal wear and low drilling rate. From Fig. 6.18 it can be further verified that the inside diameter of the drill bit would not be worn so seriously and the core size would not be so small if the drill bit only rotated without revolution and without swinging and sliding. In Fig. 6.18, under the step of the inclined worn core was the normal sized core obtained after changing a drilling method while above the step was small sized core resulted from inclined wear.

In view of this situation, PDM drive core drilling tool was employed for drilling five round trips, to provide a good bottom hole condition for wireline core drilling, which was then tested for eight round trips of lifting drilling tool (conventional wireline coring, PDM wireline coring and hydro- hammer wireline coring) with still unsatisfactory results obtained (Table 6.10). After changing the drilling method drill bit service life, penetration rate and footage drilled per round trip were obviously increased, while the wireline drill rods and the top drive capability could not satisfy the needs of a continuous test. From the 150th drill bit (at the hole depth of 2519.64 m) 89 mm sized drill rods were used for PDM hydro-hammer conventional core drilling, with drill bit service life and footage drilled per round trip steadily increased.

3. **Application result of the drill bits used for the second phase of the main hole**

For the second phase of the main hole (hole section 2974.59–3665.87 m), the techniques of PDM hydro-hammer conventional coring by lifting drilling tools were employed, with 32 drill bits concerning 3 kinds of two types from three manufacturers (including the 163rd drill bit for

Fig. 6.18 Inner step on the drill bit and the inclined worn core

Table 6.11 Application results of diamond drill bits in the second phase of core drilling in the main hole (MH-1C)

Bit type		Bit quantity	Roundtrips	Footage drilled (m)	Penetration rate (m/h)	Drill bit service life (m)	Footage drilled per roundtrip (m)	Core recovery (%)
Sintered impregnated	Segment inserted by twice forming	17	45	356.09	1.18	20.95	7.91	80.0
	Sintered	9	28	226.50	1.35	25.17	8.09	83.9
	In sum	26	73	582.59	1.24	22.41	7.98	81.5
Electro-plated impregnated		5	9	63.01	1.30	12.60	7.00	95.1
Total		31	82	645.60	1.25	20.83	7.87	82.8

Note In the MH-1C drilling a total of 32 drill bits were run into the hole. Among these drill bits No. 158 and No. 161 segment inserted drill bits by twice forming had been used in the MH section, thus calculated as 0.5 for each. Three round trips core drilling for the first time sidetracking in the main hole was counted in MH-1C section, No. 163 and No. 161 drill bits were used, with a coring footage of 10.75 m

Fig. 6.19 Service life and penetration rate of the diamond drill bits used in the second phase of the main hole versus hole depth

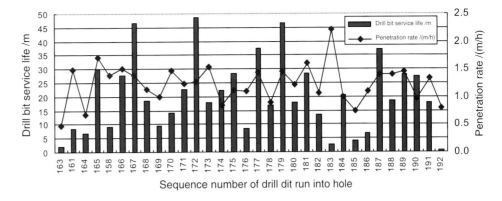

Fig. 6.20 Round trips completed and average footage drilled per round trip by each diamond drill bit in the second phase of the main hole versus hole depth

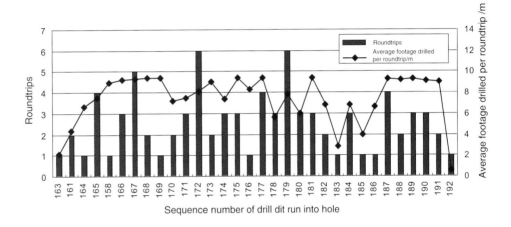

sidetracking). 645.60 m were completed, with an average drill bit service life of 20.83 m and an average penetration rate of 1.25 m/h (see Table 6.11). For each drill bit, its service life and average penetration rate versus the hole depth can be found in Fig. 6.19, and the round trips each drill bit completed and its average footage drilled per round trip versus the hole depth can be found in Fig. 6.20.

In drilling the hole section from 3475 to 3525 m, it was difficult for drill bits and drilling tool to maintain stable because of the oversized hole section, broken rock and crustal stress and under these circumstances six drill bits (No. 182, 183, 185, 186, 179 and 184) were broken and then matrix pieces dropped off, among these drill bits No. 183, 185 and 186 were abandoned only after running one round trip.

Of the eight drill bits which were run for only one round trip (No. 158 had been run in MH section, belonging to normal scrapping), three drill bits were excessively worn at the working layer, four drill bits were broken in matrix and matrix pieces dropped off and one drill bit was broken with reaming shell and dropped off at hole bottom.

4. Application result of the drill bits used for the third phase of the main hole

For the third phase of the main hole (hole section 3624.16–5118.20 m), the techniques of PDM hydro-hammer conventional core drilling were employed (only one round trip was completed by PDM conventional core drilling), with 55 drill bits concerning 3 kinds of two types from three manufacturers. Core drilling of 1480.06 m were completed, with an average drill bit service life of 26.91 m and an average penetration rate of 1.03 m/h (see Table 6.12). For each drill bit, its service life and average penetration rate versus the hole depth can be found in Fig. 6.21, and the round

Table 6.12 Application results of diamond drill bits in the third phase of core drilling in the main hole (MH-2C)

Bit type		Bit quantity	Roundtrips	Footage drilled (m)	Penetration rate (m/h)	Drill bit service life (m)	Footage drilled per roundtrip (m)	Core recovery (%)
Sintered impregnated	Segment inserted by twice forming	28	89	690.71	1.02	24.67	7.76	86.1
	Sintered	26	96	786.9	1.03	30.27	8.20	89.7
	In sum	54	185	1477.61	1.03	27.36	7.99	88.0
Electro-plated impregnated		1	1	2.45	1.75	2.45	2.45	69.4
Total		55	186	1480.06	1.03	26.91	7.96	88.0

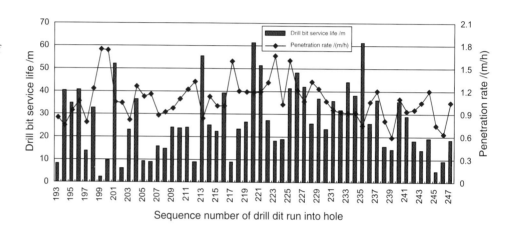

Fig. 6.21 Service life and penetration rate of the diamond drill bits used in the third phase of the main hole versus hole depth

trips each drill bit completed and its average footage drilled per round trip versus the hole depth can be found in Fig. 6.22.

Among the ten drill bits which were run for only one round trip, eight drill bits were excessively worn in working layer and two drill bits were broken in matrix and matrix pieces dropped off.

6.4.2 Application Results of Three Main Core Drill Bits

In CCSD-1 Well drilling, three main types of core drill bits including 129 segment inserted drill bits produced by twice forming, 46 sintered diamond drill bits and 27 electro-plated

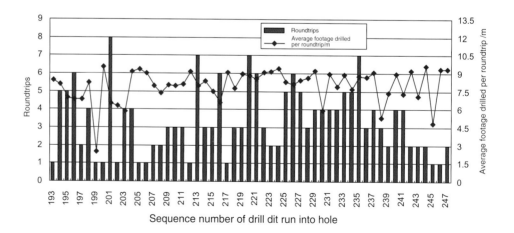

Fig. 6.22 Round trips completed and average footage drilled per round trip by each diamond drill bit in the third phase of the main hole versus hole depth

diamond drill bits manufactured by twice forming were used to complete a core drilling of 4536.22 m, which amounted to 83.6 % of the total core drilling footage of 5005.87 m, and the three main types of core drill bits accounted for 81.8 % of the total core drill bits used.

1. **Segment inserted drill bit by twice forming**

129 segment inserted drill bits produced by twice forming (Fig. 6.23) were used to complete a core drilling of 2483.65 m, amounting to 56.8 % of the total core drilling footage of 5005.87 m, and accounting for 52.2 % of the total 247 core drill bits used. The average service life of these drill bits was 22.04 m, average penetration rate was 1.05 m/h (Table 6.13); and the maximum service life of the drill bit reached 75.23 m with penetration rate of 1.00 m/h. In drilling PH hole section the average service life of the drill bits was 22.36 m, in MH hole section the average service life of the drill bits was 20.10 m, in MH-1C hole section the average service life of the drill bits was 20.95 m and in MH-2C hole section the average service life of the drill bits reached 24.67 m. For each drill bit, its service life and average penetration rate versus the hole depth can be found in Fig. 6.24, and the round trips each drill bit completed and its average footage drilled per round trip versus the hole depth can be found in Fig. 6.25.

Among the 24 drill bits which were run for only one round trip, sixteen drill bits were excessively worn at the working layer, four drill bits were broken in matrix and matrix pieces dropped off, and four drill bits (wireline core drill bits and jet reverse circulation core drill bits) were normally worn and could still be used.

The quality of the drill bits basically maintained stably, with satisfactory adaptability to the variation of the rock formations. At the initial stage of drilling, 15 drill bits were tested (Table 6.14). The segment inserted drill bit with soft matrix was unsuitable for crystalline rocks of medium abrasiveness and of strong erosiveness, mainly manifested as poor matrix wear resistance, poor matrix erosive resistance, short service life and low penetration rate. As for hand brazed segment inserted drill bit, it was only used for one round trip because of segment drop-off due to the low brazing strength. The segment inserted drill bit with medium hard matrix could basically satisfy the needs for drilling hard rock with high abrasiveness, with average service life of 31.09 m and average penetration rate of 0.98 m/h. Afterwards improvement and perfection were conducted in accordance with the characteristics of percussive rotary drilling techniques.

2. **Sintered diamond drill bit**

46 sintered diamond core drill bits (Fig. 6.26) were used to complete a core drilling of 1240.64 m, amounting to 24.8 % of the total core drilling footage of 5005.87 m, and

Fig. 6.23 Diamond segment inserted core drill bit produced by twice forming

Table 6.13 Application results of the segment inserted drill bits produced by twice forming

	Bit quantity	Roundtrips	Footage drilled (m)	Penetration rate (m/h)	Drill bit service life (m)	Footage drilled per roundtrip (m)	Core recovery (%)
Segment inserted by twice forming	129	563	2843.65	1.05	22.04	5.05	83.4
All the drill bits used in CCSD-1 Well	247	1071	5005.87	1.01	20.27	4.67	85.7
Ratio (%)	52.2	52.6	56.8	104.0	108.8	108.1	97.3

202　　6　Diamond Core Drill Bit

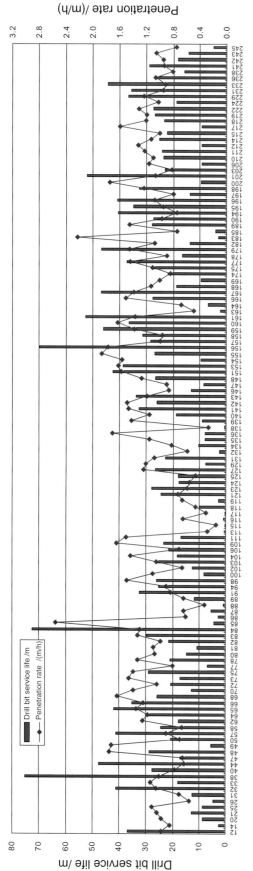

Fig. 6.24 Service life and penetration rate of the segment inserted drill bits produced by twice forming versus hole depth

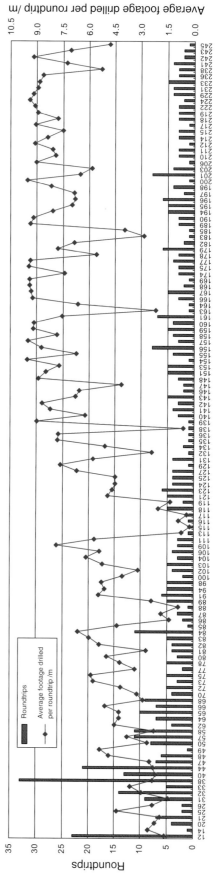

Fig. 6.25 Round trips completed and average footage drilled per round trip by each segment inserted drill bit produced by twice forming versus hole depth

6.4 Application of Diamond Core Drill Bits

Table 6.14 Drill bit testing results at the initial stage of drilling

Bit type	Bit quantity	Accumulative footage drilled (m)	Accumulative actual drilling time (h)	Average drill bit service life (m)	Average penetration rate (m/h)
Segment inserted, medium hard matrix	10	310.09	317.62	31.09	0.98
Segment inserted, soft matrix	4	43.94	34.93	8.74	1.00
Hand brazed segment inserted	1	2.53	2.55	2.53	0.99

Fig. 6.26 Sintered diamond drill bit

accounting for 18.6 % of the total 247 core drill bits used. The average service life of these drill bits was 26.97 m, average penetration rate was 1.08 m/h (Table 6.15); and the maximum service life of the drill bit reached 61.39 m with penetration rate of 1.18 m/h. In drilling PH hole section the average service life of the drill bits was 21.47 m, in MH hole section the average service life of the drill bits was 20.19 m, in MH-1C hole section the average service life of the drill bits was 25.17 m and in MH-2C hole section the average service life of the drill bits reached 30.27 m. For each drill bit, its service life and average penetration rate versus the hole depth can be found in Fig. 6.27, and the round trips each drill bit completed and its average footage drilled per round trip versus the hole depth can be found in Fig. 6.28.

Among the 6 drill bits which were run for only one round trip, two drill bits were excessively worn at the working layer, four drill bits were broken in matrix and matrix pieces dropped off.

At the beginning, the drill bit matrix did not adapt to the rock formation. Diamond grains were small, penetration rate was low, service life was short and gauging effect of the working layer was unsatisfactory. After improvement the quality of the drill bit was increased, with good adaptability for percussive rotary drilling, improved penetration rate and satisfactory service life. But the matrix hardness was higher. After drilling the main hole, another round of improvement was conducted, resulting in a perfect structure and improved matrix properties. Because of top quality diamond used, drill bit quality leapt to the first place, with long service life and high penetration rate obtained.

Table 6.15 Application results of the sintered diamond drill bits

	Bit quantity	Roundtrips	Footage drilled (m)	Penetration rate (m/h)	Drill bit service life (m)	Footage drilled per roundtrip (m)	Core recovery (%)
Sintered drill bit	46	174	1240.64	1.08	26.97	7.13	87.4
All the drill bits used in CCSD-1 Well	247	1071	5005.87	1.01	20.27	4.67	85.7
Ratio (%)	18.6	16.2	24.8	106.9	133.1	152.7	102.0

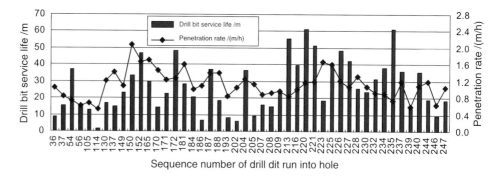

Fig. 6.27 Service life and penetration rate of the sintered diamond drill bits versus hole depth

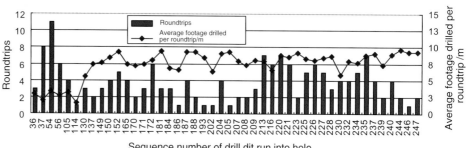

Fig. 6.28 Round trips completed and average footage drilled per round trip by each sintered diamond drill bit versus hole depth

3. Electro-plated diamond drill bit produced by twice forming

27 electro-plated diamond drill bits produced by twice forming (Fig. 6.29) were used to complete a core drilling of 451.93 m, amounting to 9.0 % of the total core drilling footage of 5005.87 m, and accounting for 10.9 % of the total 247 core drill bits used. The average service life of these drill bits was 16.74 m, average penetration rate was 0.82 m/h (Table 6.16); and the maximum service life of the drill bit reached 72.06 m with penetration rate of 1.07 m/h. In drilling PH hole section the average service life of the drill bits was 17.83 m, in MH hole section the average service life of the drill bits was 3.07 m (wireline core drilling). For each drill bit, its service life and average penetration rate versus the hole depth can be found in Fig. 6.30, and the round trips each drill bit completed and its average footage drilled per round trip versus the hole depth can be found in Fig. 6.31.

Among the 6 drill bits which were run for only one round trip, two drill bits were excessively worn at the working layer, three drill bits were broken in matrix and matrix pieces dropped off and one drill bit was normally worn (wireline core bit still could be used).

During the initial period of the test the drill bits showed good properties, with long service life and fast penetration rate. However, along with the change of core drilling method, the drill bits became unsuitable to percussive rotary drilling, with the working layer dropped off. Under these circumstances drill bit properties could not be brought into full play and the drill bits could only complete one or two

Fig. 6.29 Electro-plated diamond drill bit produced by twice forming

6.4 Application of Diamond Core Drill Bits

Table 6.16 Application results of the electro-plated diamond drill bits produced by twice forming

	Bit quantity	Roundtrips	Footage drilled (m)	Penetration rate (m/h)	Drill bit service life (m)	Footage drilled per roundtrip (m)	Core recovery (%)
Electro-plated drill bit by twice forming	27	158	451.93	0.82	16.74	2.86	88.9
All the drill bits used in CCSD-1 Well	247	1,071	5005.87	1.01	20.27	4.67	85.7
Ratio (%)	10.9	14.8	9.0	81.2	82.6	61.2	103.7

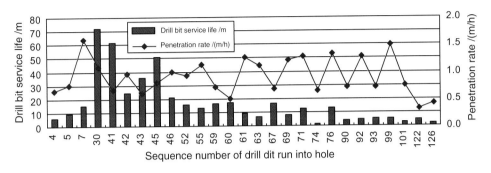

Fig. 6.30 Service life and penetration rate of the electro-plated diamond drill bits produced by twice forming versus hole depth

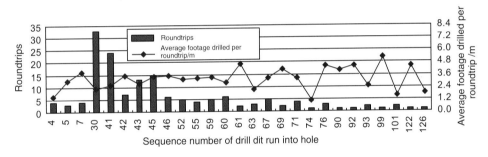

Fig. 6.31 Round trips completed and average footage drilled per round trip by each electro-plated diamond drill bit versus hole depth

round trips. During the main hole drilling only two wireline core drill bits were used.

6.4.3 Application Results of Other Type Core Drill Bits

1. **Natural diamond surface set drill bit**

Twelve natural diamond surface set drill bits from two manufacturers were used to complete 171.76 m, with an average drill bit service life of 14.31 m and an average penetration rate of 0.7 m/h (see Table 6.17). For each drill bit, its service life and average penetration rate versus the hole depth can be found in Fig. 6.32.

The natural diamond surface set drill bit produced by Chuanke Chrida Diamond Bit Co. Ltd. (Fig. 6.33) was narrow and shallow in water hole, and in inner and outer water passages, with small water flow cross section, being unfavourable for effectively cooling drill bit and quickly discharging the cuttings. During drilling operations pump suffocation often happened, affecting drilling efficiency and drill bit service life.

Beijing Institute of Exploration Engineering produced a natural diamond surface set drill bit (Fig. 6.34), however, without getting satisfactory result.

Because of low penetration rate, natural diamond surface set drill bits did not get a satisfactory drilling result. Especially after drilling for a period of time drilling efficiency dramatically decreased and it was discovered after lifting the drilling tool that a part of natural diamond had been broken in cutting edges and most of the diamond had been worn to dull (Fig. 6.35). From the picture it can be seen that a complete natural diamond grain is hardly found.

2. **Impregnated diamond drill bit**

Ten impregnated diamond core drill bits by hot pressing (including the wireline core drill bit provided from Germany) supplied from other five manufacturers completed

Table 6.17 Application results of the natural diamond surface set drill bits

Drill bit		Roundtrips	Well depth (m)		Footage drilled (m)	Penetration rate (m/h)	Drill bit service life (m)	Footage drilled per roundtrip (m)	Core recovery (%)
Sequence number	Code number		From	To					
6	1013	21	110.11	254.44	38.65	0.61	38.65	1.84	96.8
8	1015	9	162.97	189.53	8.52	0.58	8.52	0.95	51.5
11	TGS-TB1	6	189.53	252.45	7.81	0.51	7.81	1.30	112.3
13	1607857	13	233.40	277.94	26.60	0.94	26.60	2.05	99.2
15	1607856	6	259.94	279.95	14.60	1.05	14.60	2.43	100.6
16	1607858	10	279.95	326.45	22.26	0.65	22.26	2.23	95.5
17	1607855	10	284.85	322.60	24.24	1.03	24.24	2.42	101.1
18	1607883	3	326.45	852.68	8.77	0.54	8.77	2.92	98.7
19	1607884	3	331.32	336.24	4.92	0.58	4.92	1.64	97.0
24	1607893	5	380.12	1021.64	10.02	0.56	10.02	2.00	95.9
51	1607892	3	1021.64	1164.71	2.34	0.35	2.34	0.78	86.8
79	1607898	2	1413.95	1416.98	3.03	0.75	3.03	1.52	83.2
Total		91			171.76	0.70	14.31	1.89	96.1

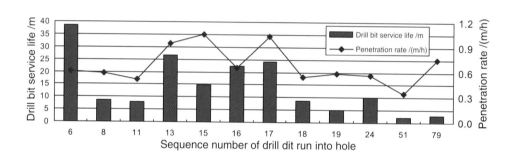

Fig. 6.32 Service life and penetration rate of the natural diamond surface set drill bits versus hole depth

Fig. 6.33 Surface set drill bit produced by Chuanke Chrida Diamond Bit Co. Ltd.

Fig. 6.34 Surface set drill bit produced by Beijing Institute of Exploration Engineering

6.4 Application of Diamond Core Drill Bits

Fig. 6.35 A worn surface set drill bit produced by Chuanke Co. Ltd.

Fig. 6.36 An impregnated drill bit by hot pressing from Chuanke Chrida Diamond Bit Co. Ltd.

Fig. 6.37 An impregnated drill bit by hot pressing from the Tool Institute, Drilling Research Academy of Shengli Oil Field

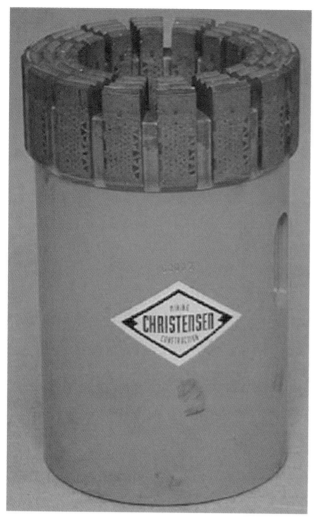

Fig. 6.38 An impregnated drill bit by hot pressing from Germany

Fig. 6.39 An impregnated drill bit by hot pressing from the Industrial and Commercial Co. China University of Geosciences (Wuhan)

Table 6.18 Application results of other impregnated core drill bits

Manufacturer	Bit quantity	Roundtrips	Footage drilled (m)	Penetration rate (m/h)	Drill bit service life (m)	Footage drilled per roundtrip (m)	Core recovery (%)
Hunan No. 409 brigade	2	10	11.27	0.47	5.64	1.13	93.7
Mingzhu Co., Wuxi	2	2	3.50	0.25	1.75	1.75	38.9
Chrida Co. (Fig. 6.36)	1	4	10.07	0.85	10.07	2.52	107.9
Institute of Tool, Shengli Drilling Academy (Fig. 6.37)	4	23	37.82	0.72	9.46	1.64	94.9
German made (Fig. 6.38)	1	1	2.50	1.10	2.50	2.50	0
Total	10	40	65.16	0.62	6.52	1.63	90.1

Table 6.19 Application results of other electro-plated impregnated core drill bits

Manufacturer	Bit quantity	Roundtrips	Footage drilled (m)	Penetration rate (m/h)	Drill bit service life (m)	Footage drilled per roundtrip (m)	Core recovery (%)
Hunan No. 409 brigade	2	3	1.58	0.21	0.79	0.79	75.9
Zhongda Co., University of Geosciences	1	2	6.11	1.50	6.11	6.11	95.9
Qinglongjian Co., University of Geosciences	2	7	19.08	1.00	9.54	9.54	100.3
Lanqiao Co., Fujian	1	1	4.68	1.60	4.68	4.68	39.5
Industrial and commercial Co., University of Geosciences (Fig. 6.39)	17	32	201.28	1.19	11.84	11.84	88.4
Total	23	45	232.73	1.14	10.12	5.17	88.5

a coring footage of 65.16 m, with an average drill bit service life of 6.52 m and an average penetration rate of 0.62 m/h (Table 6.18).

Twenty three electro-plated impregnated diamond core drill bits supplied from another five drill bit manufacturers completed a coring footage of 232.73 m, with an average drill bit service life of 10.12 m and an average penetration rate of 1.14 m/h (Table 6.19).

Reaming Drilling Techniques of Hard Crystalline Rock

The main problems encountered in the reaming drilling of CCSD-1 Well were as follows:

1. Hard formation was difficult for drilling: according to the cores recovered from the pilot hole drilling, the formation encountered by reaming drilling was crystalline rock (mainly eclogite and gneiss), with drillability grades of 8–9 in most cases, the major was grade 9, and the drillability of a few rocks was even up to grade 10–11, and the penetration rate per hour was lower. If diamond reaming bit was to be used, since low rotary speed of the rotary table, the penetration rate would be very low; while using rock bit (mainly with volumetric fragmentation form) to ream the hole would still lead to low penetration rate since the serious bit bouncing and rough drilling and the bit pressure couldn't be exerted to the drill bit.
2. Serious bit bouncing would happen: according to the geological information obtained from the pilot hole drilling, the rock properties of the formation to be drilled were hard and brittle, and a great amount of broken zones would be penetrated through, with soft and hard interbedded formations. When rock bit was used to ream hole, the hole bottom shape was irregular, which was easy for rock bit to bounce in the process of rolling, and in the meantime, the rock bit was frequently obstructed by broken face of the rock in the process of rotation in the hole and thus the drill bit was easy to bounce, so that the drilling tool and the drill bit would bear larger impact load over a long period of time, very easily resulting in fatigue and damage for the drilling tool and the drill bit, being detrimental to drilling safety.
3. Rock blocks could frequently fall from hole wall: the formation rocks to be drilled were very hard and broken, rock blocks often fell off from the hole wall, the rocks had no permeability, very difficult for drill mud to form mud cake to effectively protect hole wall, and thus the unstable rock in the hole wall was in a free state; the releasing of the crustal stress in the formations promoted the new fracture to form and the unstable rock block to fall after the hole wall was produced; in the process of drilling, the bouncing of the drilling tool could produce disturbance and beat to the hole wall, easily causing rock broken and falling. Since the fallen rock has high hardness and strength, drilling tool sticking would very easily happen.
4. Drilling tool could wear seriously: the rock drilled is crystalline rock, contains a lot of quartz minerals and thus has high hardness and very high abrasiveness. Under these circumstances the side face of the pilot bit body, the outer and gauge teeth of rock bits, the leg gauge and bit gauge, the stabilizers, the reamers, the collars and so on, would be worn off very seriously, leading to a lower bit footage, shorter bit service life, the undersized hole diameter and hole accidents. Since the drill bit wore quickly, the hole section drilled by each drill bit would obviously become conical, when a new drill bit was changed in the process of drilling, much time was needed to drill off, and it was easy for the hole to become a wedge shape and thus the drilling tool be stuck. This drilling tool sticking or drill bit deformation was detrimental to hole safety.
5. Threat from falling objects: in the process of drilling, pilot body, tools, cones, tungsten carbide inserts and other objects would probably fall into the hole. These falling objects could easily produce an obstruction to the drilling tool in the process of drilling and stick the bit cones and the bit pilot body so that drilling couldn't normally advance, and also would bring a great threat to safe reaming drilling.

7.1 Development of Pilot Reaming Bits

The rock bits for oil drilling are mainly used for softer sedimentary rocks, precedents that the rock bits used for reaming drilling in so hard rock have seldom been found.

Translated by Zhang Yongqin.

The conventional reaming rock bits have a poor adaptability to the rock formations in this drilling project. At the same time, the hole size designed for this drilling project was special, and there was no any standard rock bit size both at home and abroad, so specially designed reaming rock bits were needed. Since the assembled rock bit was adopted, the welding strength for combination, the type option of the cone and the arrangement of the cones etc. still needed further study, and the load exerted to the bit pilot body and the strength design were one of the main difficult points. The state of force borne by the cone bearings of the pilot reaming drilling was different from that of the conventional non-core drilling, when the rock fragmentation ring was narrow, the direction of the axial force of cone bearings would be changed, the axial outward force was smaller and easily caused cone falling.

Some problems that needed to be solved in the application of the reaming rock bits were as follows:

1. The abrasiveness and impact resistance of the cutters: since the rock formation was hard, broken and highly abrasive, the main requirements to the bits included the abrasiveness of the bit cutters so that a longer service life and a higher drilling speed could be guaranteed. At the same time, because of frequent bouncing of the drilling tool in the process of drilling, the teeth would break off as the single cone or tungsten carbide cutter bore a tremendous load instantaneously.

2. Bit gauge protection: poor gauge protection would lead to a decrease of the hole diameter as the drill bit worn and that could cause an increase of drilling-off work when a new drill bit was used in the next roundtrip, easily leading to a wedge-shaped sticking of the drilling tool in the meantime, and thus bringing serious damage to the follow-up drill bits.

3. Bit strength: drill bit would bear large impact load because of hard and broken rock formations and the falling rock blocks and other falling objects. In the process of design and manufacture of the drill bit, it would be necessary to prevent the deformation, breaking-off, falling-off of the cone or the pilot body from the drill bit so that in the hole accidents would be avoided.

4. Good pilot performance: since the rock formations were broken, the problems of soft and hard non-homogeneous formations may probably exist. For the sake of preventing deviating from the axel line of original hole drilled as the pilot hole in the process of reaming drilling, a good pilot ability of the pilot drill bit would be needed, the long pilot body would have a good pilot ability, however, it would bear bigger torque. A high strength pilot body was needed.

5. Rock fragmentation of pilot bit: to avoid protruding rocks, falling rock blocks or broken debris in the small diameter pilot hole, the end face of the pilot bit body would have rock cutting ability in order to avoid the obstruction to the drill bit and non-advancing.

6. The abrasiveness of pilot body: since the reaming drilling was to be carried out in the φ 157 mm hole which had been completed before, and during core drilling φ 89 mm drill rods and φ 120 mm collars were used and when reaming drilling φ 127 mm drill rods and φ 177.8 and φ 203.2 mm collars were used, so the rigidity of the two drilling tools was different. Because the rigidity of the drilling tool in the process of reaming drilling was bigger, after assembling the stabilizer which coincided with axel line of the pilot hole, the drilling tool assembly produced a bending and made the down-hole drilling tool bear bigger bending moment, easily causing the problems of drilling tool bouncing and sticking, big friction obstruction during pulling-up and running-in, wearing of the bit pilot body, etc. Wear of the pilot body would decrease its diameter and cause φ 311.1 or φ 244.5 mm drill hole deviating from φ 157 mm pilot hole axis line. Under these circumstances the afterward run-in bit pilot body would bear bigger bending moment and damage very easily. However, if the rigidity of the reaming drilling tool was too low, the strength of the drilling tool would be low and the hole dog-leg angle would be big, and then leading to a difficulty to run casing.

In order to meet the urgent demands of the drilling project, the CCSD-1 headquarters cooperated respectively with the Institute of Exploration Techniques and the Jianghan

Fig. 7.1 KZ Series of rock bit developed by IET

7.1 Development of Pilot Reaming Bits

Fig. 7.2 KHAT series of rock bit developed by Kingdream

Kingdream Engineering Drilling Tool Company to research and manufacture the KX and KHAT types of the reaming drilling bits. The reaming drilling bits used in CCSD-1 reaming drilling were just these two types (Figs. 7.1 and 7.2).

7.1.1 KZ157/311.1 Type Reaming Bit

1. **Design and Improvement of KZ157/3111.1 Type Reaming Bit**

   (1) Option of the bit cutter

   Cone tooth breaks the rock in ways of impact, indentation and shearing; in hard rock formations these fragmentation ways can produce volumetric fragmentation effect, especially in reaming drilling phase. Since the free surface of the slim hole had been produced, under the action of the impact way, the free surface can be effectively used and big volumetric fragmentation effect can be produced. Rock bits are widely used in oil drilling industry and are very perfect; in comparison with other cutters rock bits have the advantages of long service life and high drilling speed. If diamond drill bit was used for drilling, the cutting volume per revolution would be very little and high revolution would be needed for cutting, which just was what the rotary table drill rig at the drill site couldn't satisfy. If the PDC bit was used, and then since the rock formation was extremely hard, it would be difficult for the cutters to cut into the rock formation and the scraping effect couldn't be realized, the non-homogeneousness of the rock formation could also lead to PDC cutter breaking in the process of revolution because of the impact load.

   The load bearing state of drill bit cutters in reaming drilling was very bad, so the bigger diameter bit leg would be needed as far as possible. But if the bit leg was too big, it would affect the cross-section area for connection between the bit pilot body and bit body and thereby further affect the bit strength.

   The reaming drilling bits can be made into tri-cones, four cones, six cones and so on. The stability of four-cone reaming drilling bit is poor and easily vibrates in drilling. The stability of tri-cone bit and six-cone bit is good, but if the number of the cone is much more and the bigger bit leg is selected in a limited arrangement space, it will irresistibly affect the connecting strength between the pilot body and the bit body and it will be detrimental for increasing the cutter and bit body strength.

   The KZ reaming drilling bit used the cone leg of the φ 215.9 mm insert rock bits as the cutters, φ 215.9 mm insert rock bit is the most common bit size in petroleum drilling industry, with the cone cutters perfectly developed. The bearings of the rock bit have the pressure-balance lubrication system and is suitable for the deep well drilling; the cone bearings use the metal floating sealing as the sealing structure, with very good performance; the ability against the impact load is very strong and is suitable for high linear velocity drilling; the spare part models are all ICDC537 and 547 (or equal to 537 and 547), the blade height of the tungsten carbide cutters is not high, with better toughness, not easy to break off under higher impact load when drilling in hard rock formations. The arrangement of tungsten carbide teeth on the cone is closer, the outer diameter teeth and gauge protection teeth are strengthened, and this arrangement of the cutter teeth is suitable for drilling in highly abrasive crystalline metamorphic rock (eclogite, gneiss). The bit leg used tungsten carbide to strengthen the gauge protection and to increase the abrasiveness resistance of the leg.

   (2) Enhancing the integral strength of the bit

   An integral structure design (Fig. 7.3) was used for drill bit. The bit body, the large diameter stabilizer and the extended pilot-body were integrated and had high strength; forged low-carbon alloy steel, which was tempered in heat treatment, was selected for the material for main bit body.

   In the early time, 35CrMoA high strength alloy steel was selected for the bit body. Although the weldability of 35CrMoA steel was acceptable, it couldn't meet the demands

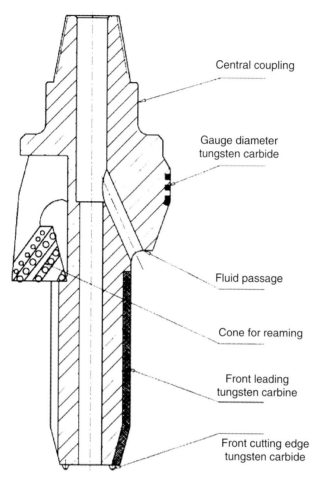

Fig. 7.3 The structure scheme of KZ reaming bit

in practical drilling. In the process of drilling operations, since the formation was brittle and broken so that an irregular hole bottom would be easily formed, which made a single cone leg bearing the impact load of the whole drilling tool instantaneously, and thus the welding seams frequently fractured. Among the drill bits used in drilling applications, KZ157/311-05 bit deformed after the welding seam fractured in drilling, the bit deformed, the drilling tool was stuck in the hole, and when the drilling tool was pulled up with strong force, two bit legs fell off in the hole bottom. Although some improvement methods were taken, such as preheating, heat preservation, increase of welding slope slot and welding seam height and use of high strength welding rod, the problems of welding seam breaking still frequently occurred.

In the later period, a batch of forged low carbon steel 20CrNiMoA was ordered as the bit material from a special steel plant, and its weldability and welding strength were obviously improved. After the new material was used, the welding seam between the bit leg and the bit body never broke again, the problem of welding seam strength was solved and then the welding reliability of the bit was ensured.

(3) Enhancing the bit pilot performance

The bit used its steel body as the pilot, which was 350 mm long at the beginning. In order to further decrease the turning torque of the drilling tool and the breaking of the pilot body, the effective length of the pilot body was reduced to 250 mm subsequently. Tungsten carbide was inserted onto the outer cylindrical surface of the pilot section and thus the wear-resistance was increased. Then the drill bit had a good and stable pilot performance in a long term.

In order to prevent larger rock blocks from falling off into φ 157 mm pilot hole during reaming drilling, a tungsten carbide bit or tri-cone rock bit was designed in the front end of the pilot body as the pilot bit.

The pilot bit with the inserted tungsten carbide cutters: the tungsten carbide cutters were integrated with the bit body, i.e., high blade tungsten carbide cutters were coldly inserted onto the lower end of the pilot body; the cutters could clean up or break the falling blocks simply. The strength of this bit was weaker and the cost was lower, therefore, this bit structure was mainly used in reaming drilling.

The cone pilot bit: a 152.4 mm rock bit was jointed in the lower end of the pilot body (Fig. 7.4), with the main purpose to clean up the settlings in the pilot hole bottom, such as the tungsten carbide and the collapsing rocks from the pilot hole wall. This structure was mainly used for the final three reaming drill bits.

(4) Strengthening gauging and stabilizing effects

Besides the gauge protection by cones and tungsten carbide cutters on bit legs, integral gauging metal blocks were set onto the bit body and were arranged in crisscross with three cone legs, with the outer diameter of 309 mm. Columnar gauging tungsten carbide cutters were set by fully tight and cold- setting method on the surface of bit body (Figs. 7.1, 7.3 and 7.4). Practical applications proved that the gauge added onto the bit body played an important role for preventing hole diameter undersize and also effectively reduced hole deviation tendency. However, once a long footage completed by the drill bit and the rock formation was very high in abrasiveness, bit gauge still wore quite seriously.

Fig. 7.4 KZ series pilot reaming bit with rock bit

(5) Design of drill bit fluid passage

A four passageway design was adopted, with the drilling fluid equally distributed for each cone, to fully satisfy the demands of cooling cones and cleaning the hole bottom; the pilot part of the drill bit was also distributed with partial drilling fluid thus a certain uplift drilling fluid velocity was produced in the annulus between the bit pilot body and pilot hole wall, this could prevent the rock debris produced in reaming drilling from settling to the pilot hole bottom, and at the same time, prevent the broken cuttings and rock blocks from aggregating in the lower end of the pilot body, which could be easily worn off.

Nozzle structure was added in the first drill bit KZ01. But drilling practice showed that the addition of the nozzle could lead to an overpressure of the pump. Drilling in hard and brittle rock formations, the function of hydraulic rock fragmentation was not obvious, without evident role of increasing drilling speed. In the meantime, considering no bit balling problem existed, too high nozzle pressure drop could easily cause "pricking leakage" and other problems, and so from the KZ 02 bit on, nozzle was canceled, because with nozzle the disadvantage was more than advantage.

2. **The technical parameters and indexes of KZ157/311.1 reaming bit**

The technical parameters and indexes of KZ157/311.1 reaming bit can be found in Table 7.1.

3. **Process control of KZ157/311.1 reaming bit manufacture**

The decisive factor affecting the service life of rock bit is bearing seal, so during the process of the KZ type rock bit manufacture, besides choosing high quality tri-cone rock bit, the temperature in the process of manufacture must be strictly controlled. In the process of disassembling the tri-cone rock bit, a special technology was adopted, and the welding temperature was strictly controlled in the welding process to prevent the bearing rubber seal and lubricating oil storage capsule from high temperature damage.

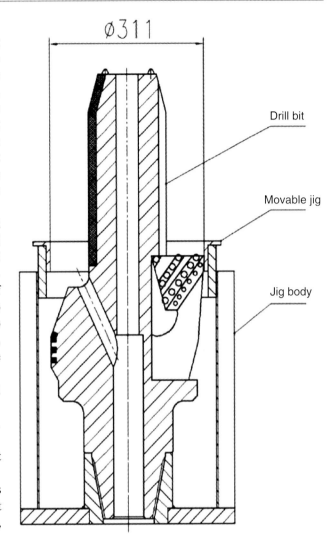

Fig. 7.5 Jig for reaming bit welding

Table 7.1 The technical parameters and indexes of KZ157/311.1 reaming bit

Item	Parameter
Bit outer diameter	311.1 mm
Bit pilot diameter	The cone was 152.4 mm, bit steel blank was 156 mm
Bit pilot length	350 mm (later changed to 250 mm)
Bit service life	Under normal drilling conditions, the service life of bit bearing was no less than 60 h, with the maximum over 70 h
Drilling speed	Based upon different rock formations, drilling speed would reach 0.5–1.5 m/h
Bit mass	170 kg

The machining precision of rock bit is also an extremely important factor that affects the application of the rock bit. In order to guarantee the machining precision, all the bit parts were machined before assembling the bit. In addition, a special jig for bit production (Fig. 7.5) was designed to guarantee the concentricity between the thread and the cone pilot body of the bit and guarantee the normality of the bit diameter; the welding technology was strictly controlled in the process of welding to avoid an over deformation which would cause abandonment before use.

7.1.2 KHAT 157/311.1 Reaming Bit

With the basic structure and cone selection the same as KZ series bit, KHAT 157/311.1 reaming bit (Fig. 7.2) also has its own following features:

1. The length of the pilot section was short, the spherical tungsten carbide teeth were used for gauge protection in the outer cylinder surface. There was no cutting tooth at the end surface;
2. The cone legs were designed and manufactured by the original drill bit plant and thus cutting, separation and other procedures were unnecessary. The cone legs could be used directly for welding into the reaming bit;
3. Rich experiences in rock bit assembling and welding techniques were accumulated, the selection of materials and process was correct and thus the welding seams didn't break off.

7.1.3 Development and Improvement of KZ157/244.5 Reaming Bit

KZ157/244.5 reaming bit was manufactured on the base of successful application of KZ157/311.1 reaming bit and keeping design conception and structural advantages of KZ157/311.1 reaming bit. Since the structure space was more narrow, the drilling depth was more deeper and the factors of unstable hole wall increased, the successful development of KZ157/244.5 reaming bit (Fig. 7.6 left) was resulted from the careful design. At the initial stage, following attempts were made for the selection of the cones and the ways of gauging.

1. **Retrofit of cone leg**

Since the structure space of KZ157/244.5 bit body was small, in order to guarantee the structural strength, the cone legs of φ 165.1 mm rock bit were used for the first six rock bits. However, since the cones were small, with small and less tungsten carbide teeth to break rock, and in the meantime, the linear velocity of the bearings was higher, the bit service life was unsatisfactory, as the average actual drilling time was only 16.87 h, the average footage drilled was 14.69 m, and the average penetration rate was 0.87 m/h. Because the gauging teeth were less, the cone diameter wore quickly; basically, the tungsten carbide teeth wore flat in each drilling roundtrip (Fig. 7.7). Penetration rate was lower, drilling speed was lower, and application cost of the bit was fairly high.

After repeated study, starting from the seventh bit, the cone legs of HJ series 215.9 mm (8½ in.) rock bit with metal seal were used. Since the diameter of reaming bit was small, the use of large bit cone legs made the bit body structural strength very weak, especially the strength of juncture area between bit body and pilot body (shadow area of Fig. 7.8), the front pilot body more easily broke off under large twisting. So retrofit of cone leg greatly increased the difficulty in design and manufacture (Fig. 7.8).

In order to solve the problem of bit strength, firstly, the pilot body length and the pilot diameter were appropriately decreased so as to reduce the force borne; secondly, the modification to the cone structure was carried out. Under the precondition that bit application would not be affected, the front point tip of the cone which doesn't break rock was cut off with the air-blasting or electrical cutting so that the front size of the cone was reduced to prevent the interference of the bit body positions from each other and to provide a larger space for increasing the bit body strength; thirdly, while considering the space of the cone positions, the cone body was precisely manufactured by reference to the cone

Fig. 7.6 157/244.5 reaming rock bits

Fig. 7.7 The 6½ in cones of the reaming bit were worn flat

Fig. 7.8 Structure comparison of KZ157/244.5 reaming rock bit with different cone legs

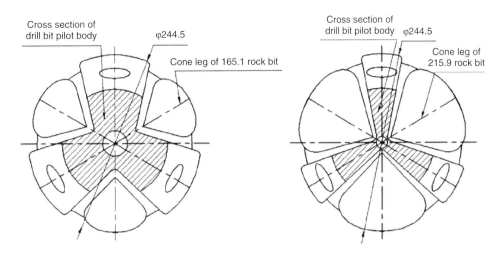

Table 7.2 Comparison of effects on cone leg improvement of KT157/244.5 reaming bit

Statistics range	Bit quantity/piece	Average footage drilled (m/bit)	Average actual drilling time (h/bit)	Average penetration rate (m/h)	Remarks
Before KT157/244.5-6#	6	14.69	16.87	0.87	7 roundtrips
After KT157/244.5-7#	15	64.75	57.89	1.12	17 roundtrips
Contrast of the improvement (after improvement/before improvement)		4.41	3.43	1.29	

shape and the outside space of the cone was fully used to increase the cross-section area for jointing and reserve the solid body parts of the bit as much as possible in order to enhance the structural strength of the bit body.

Application practice proved that the improvement of the rock bit leg obtained a good result, the integral strength of the bit was guaranteed, with the service life of the bit greatly increased (the average drilling time was 57.89 h, the average footage drilled was 64.75 m), the drilling speed increased obviously (Table 7.2). However, when serious sticking or bit bouncing happened with the front pilot body at hole bottom, the pilot body still easily deformed. In MH-1K-27 roundtrip the pilot body of KZ157/244.5-22 bit broke off (Fig. 7.9).

2. **Improvement of gauging methods**

On the base of the successful application of the reaming bit in the first phase, KZ157/244.5 reaming rock bit was also strengthened in gauging. Firstly, the quantity of the inserted tungsten carbide for gauging was increased; secondly, the whole cone legs were inserted with tungsten carbide under the condition that cone leg strength would not be affected. The wear to the bit was improved somewhat, but the gauging ability of the bit did not match the ability of down hole rock fragmentation, undersize of the hole was still serious. Under these circumstances, diamond reamer was used in the drilling tool assembly.

The polycrystalline diamond inserts were attempted for guaranteeing the bit diameter; the polycrystalline inserts,

Fig. 7.9 Falling-off of the pilot body of KZ157/244.5-22 bit

made by the Beijing Institute of Exploration Engineering, were inserted with silver welding onto the gauging section of the bit body. But in practice, the polycrystalline diamond insert was not as good as the tungsten carbide in gauge protection. The reasons included that the hardness and strength of the polycrystalline diamond inserts were lower and the

Fig. 7.10 Wearing on gauge of the reaming bit

working area for gauge protection was small, in the process of cutting rock in lower rotation and larger torque, the cutters cut the rock with powerful squeezing-pressing and percussive actions, the diamond inserts could not bear such a large impacting force and then deformed and partially fell off, with diamond grains falling off together with the matrix, so that polycrystalline diamond insert could not act for gauge protection (Fig. 7.10).

7.2 Design of Drilling Tool

7.2.1 Strength Check of Drilling String

1. **The calculation equations of the drill string strength**
1. WOB
   Refer to the section of optimization of drilling parameters in this chapter.
2. The total length of the drill collars

$$\text{the needed collar weight} = \frac{\text{designed maximum WOB} \times \text{saftey factor}}{\text{buoyancy factor of drilling fluid}} \quad (7.1)$$

$$\text{the total length of collar} = \frac{\text{the needed collar weight}}{\text{collar weight per meter}} \quad (7.2)$$

3. The total length of drill rods

$$\text{the total length of drill rods} = \text{hole depth} - \text{the total length of collar} \quad (7.3)$$

4. Tensile stress of the upper drill string
   While the drill string is hanged up, the tensile stress of the drill rods in drilling fluid is:

$$\sigma_{L1} = \frac{QK}{A} \quad (7.4)$$

in which

σ_{L1} tensile stress of the drill rods hanged up in drilling fluid, MPa
Q the weight of the drill rods in air beneath the hole mouth, N
K the buoyancy factor
A cross section of the drill rod at hole mouth, mm^2

In drilling, the tensile stress:

$$\sigma_{L2} = \frac{QK - P}{A} \quad (7.5)$$

where

Q the weight of the whole drill string in air, N
P WOB, N

During tripping, the tensile stress is:

$$\sigma_{L3} = 1.5 \times \frac{QK}{A} \quad (7.6)$$

5. Compressive stress of the lower drill string

$$\sigma_Y = \frac{P}{A} \quad (7.7)$$

in which

σ_Y the compressive stress of the lower drill string, MPa
P WOB, N
A calculated cross section area, mm^2

6. Shear stress

$$\tau = \frac{9.8 \times 10^6 \cdot N \cdot k \cdot \eta}{n \cdot W} \quad (7.8)$$

where

τ the shear stress, MPa
N the power of power engine, kW
k the possible overload factor of power engine
η transmission efficiency form power engine to rotary table

n   rotary speed of drill string, r/min
W   the modulus of torsional resistance cross-section of the checked drill string, mm³

7. Bending stress

$$\sigma_W = \frac{100 \cdot M}{W} \quad (7.9)$$

in which

σ_W bending stress, MPa
M   the maximum bending moment, N·m
W   the modulus of the bending resistance cross section of the drill string, mm³

8. the compound stress in drilling

The compound stress of the upper drill string:

$$\sigma_1 = \sqrt{(\sigma_{L2} + \sigma_W)^2 + 4\tau^2} \quad (7.10)$$

where

σ compound stress of the upper drill string, MPa
σ_{L2} tensile stress of the upper drill string, MPa
σ_W bending stress, MPa
τ shear stress, MPa

The compound stress of the lower drill string:

$$\sigma_2 = \sqrt{(\sigma_Y + \sigma_W)^2 + 4\tau^2} \quad (7.11)$$

in which

σ compound stress of the lower drill string, MPa
σ_{L2} compressive stress of the lower drill string, MPa
σ_W bending stress, MPa
τ shear stress, MPa

2. **Check of drill string strength**

With help of the above-mentioned strength check methods of the drilling tools used in oil drilling, substituting the input power and transmission efficiency of the rotary table, WOB, rpm, total drilling tool weight, cross-section area of the drill rods, parameters of drill rod mechanical performances, drilling fluid density and other basic parameters of ZJ70D drill into the formulas, the drilling tool strength was checked, with the result that the φ 127 mm oil drill rod of G105 steel grade, which met the strength demands in the drilling, was suitable to 3800 m hole depth. However, considering the larger vibration of the drilling tool and more frequent stress alteration in reaming drilling, the drill rod of higher grade steel was still more suitable for reaming drilling. Therefore it was decided that S135 grade steel should be used for the drill rod for reaming drilling, with data shown in Table 7.3.

7.2.2 Selection of Drilling Tools

In reaming drilling, in order to increase the stability of the drilling tool in the hole, appropriate numbers of stabilizers were added to the drilling tool assembly. Considering the frequent and serious drilling tool bouncing, shock absorber was added to the downhole drilling tool assembly. In order to deal with the possible sticking accident, jar-while-drilling was added to the drilling tool assembly so that the sticking of the drilling tool could be released. After serious wear of the bit gauge protection occurred, diamond reamer was added above the drill bit.

1. **Impregnated diamond reaming shell**

Because of the strong abrasiveness of the rock formations, the gauge protection ability of the reaming rock bits was not enough, during reaming drilling, the bit gauge protection wore quickly with the outer diameter of the drill bit and the hole size undersized. The new drill bit could not be run into the hole bottom in the next roundtrip and redressing would be conducted.

The above case, especially after three HAKT reaming rock bits were continuously used in φ 311.1 mm reaming hole section, became more serious. The follow-up new bits wore off at outside diameter when reaching to the hole bottom. To solve this, it was decided that an impregnated diamond reamer would be added on the reaming rock bit. The diamond reamer was made by the sintering processing with hot pressing and twice forming, and impregnated diamond matrix was used as the cutting element.

In φ 311.1 mm reaming-hole, it was not until the MH-1K-42 Roundtrip that the diamond reamer was used. At first, a long steel bar brazed with impregnated diamond matrixes was welded onto the drilling tool, so as to decrease the cost. In the late phase, considering the quality problems of the welding, the integral diamond reamers were made. The two ends of the diamond reamer wings were conical shape, with diamond matrixes welded at the conical section. The diamond matrix was made of natural diamond and SDA100 + synthetic diamond. The diamond layer thickness in gauge protection was 3.5 mm, and in the lower conical section was 5 mm, with the diamond concentration of 100 % and the matrix hardness of HRC40 ± 3. The size of the impregnated diamond matrix was 30 mm × 10 mm × 10 mm. For each reamer, there were 8 reaming strips, on each of which there were 21 impregnated diamond matrixes, with the total of 168 matrixes. The length of the diamond reaming strip was 200 mm (Fig. 7.11a).

In φ 244.5 mm reaming drilling, the integral structure reamer was used. Differing from φ 311.1 mm reaming drilling, a lot of polycrystalline diamonds were inserted into the impregnated diamond matrixes and the gauge protection ability of the diamond reamers was further increased. Six diamond reamers were totally used in this phase (Fig. 7.11b).

Table 7.3 Data for checking drilling tool strength in reaming drilling

Reaming size and hole section			φ 157/311.1 (101.00–2033.00 m)			φ 157/244.5 (2028.00–3625.18 m)		
Drilling tool			Drill rod	Collar	Collar	Drill rod	Collar	Collar
Size/mm			127	177.8	203.2	127	177.8	203.2
Steel grade			S135			S135		
Inner diameters (mm)			108.6	71.4	71.4	108.6	71.4	71.4
Unit weight (N/m)			284.7	1606	2190	284.7	1606	2190
Length (m)			1888	27	54	3500	54	27
Weight per section (kN)			538	43.5	118	997	87	59
WOB (kN)			100			58		
Rpm (r/min)			85			70		
Mud density (g/cm³)			1.06			1.06		
Tensile force margin (kN)			2420			2006		
Tensile strength (kN)			3107			3107		
Torsion yield (kN·m)			100			100		
Safety factor	Tension	Design requirement	1.8			1.8		
		Actuality	3.65			2.72		
	Torsion	Design requirement	1.25			1.25		
		Actuality	17			17		

After using the diamond reamers, the situation of hole undersize was obviously improved, and the redressing time was reduced during running-in the drilling tool.

2. **Double direction shock absorber**

Since the formations drilled were very hard and the rock in some hole sections was broken, in reaming-drilling, the annular area of reaming was bigger and thus the vibration of drilling tool was unavoidable, and sometimes it could affect the reaming drilling effect, especially in the hole sections with severe deviation change. The shock absorber could effectively absorb various vibrations caused in drilling and improve bouncing of the drilling tool, thus the normal reaming drilling could be effectively guaranteed. The double direction shock absorber is a kind of drilling tool which can alleviate or eliminate the vertical and peripheral vibrations of the drill string, to keep normal WOB and torque so as to reduce the vibration damages to drilling tools and equipment while to increase drilling speed and decrease drilling cost. After the double-direction shock absorber was used in reaming drilling, bit bouncing was obviously improved, and the down hole accidents were evidently reduced. Large WOB could be exerted to the drill bit during drilling and thus the drilling speed was increased.

3. **Stabilizer**

The purpose to set stabilizer (Fig. 7.12) was to centralize drill collar and stabilize drilling tool, and to reduce the wear of drilling tool. Meantime, when inflection point occurred in the hole, the stabilizer could rectify the hole wall to some extent.

Fig. 7.11 Diamond reaming shells for reaming drilling. **a** φ 311.1 mm diamond reamer, **b** φ 244.5 mm diamond reamer

7.2 Design of Drilling Tool

Fig. 7.12 Stabilizer used in reaming drilling

The main factors that should be considered for setting the stabilizer included outer diameter, length, number and setting position of the stabilizer. Usually, two stabilizers were set in the drilling tool assembly, under the collar. The two stabilizers were set at a distance of one or two pieces of drill collar.

The outer diameters of the stabilizers used in φ 311.1 mm reaming drilling were φ 309 and φ 305 mm. Since the outer diameter of the reaming bits was not standard, especially for the reaming bits in preliminary phase, the outer diameter was less than φ 311.1 mm in most cases, being only φ 309 mm, and the outer diameter of a few reaming bits was even φ 308 mm, after using φ 309 mm stabilizer, it was easy to cause bit bouncing and even cause rough drilling. Therefore,

when reaming drilling was conducted in MH-1K-17 roundtrip, φ 305 mm stabilizer was used instead of φ 309 mm stabilizer, and by adjusting the position of the stabilizer, reaming drilling was more stable.

The outer diameter of the stabilizers used in φ 244.5 mm reaming drilling was φ 214 mm.

4. **Double direction jar while drilling**

The jar-while-drilling was used to release the vibration force (upward or downward) and free the sticking when the drilling tool was stuck in the hole. During the second reaming drilling, the hole condition was complicated, with falling fish and falling rocks caused by formation collapse. Thus it was necessary to set the jar onto the drilling tool. According to practical conditions and drilling tool assembly used in this hole section, drilling tool sticking point was mainly at the drill bit, therefore, the jar-while-drilling was set at the lower end of the collar and the upper end of shock absorber. In the operations of releasing drilling tool sticking, the jar-while-drilling played a very important role.

5. **Hydro-hammer**

The development of hydro-hammer was one of the pre-research projects of CCSD, during the period of φ 311.1 mm reaming drilling, hydro-hammer originally designed for non-core drilling was run in hole for testing eight roundtrips in all. KSC-203 hydro-hammer was run in hole for testing three roundtrips in all, and SYZX-273 hydro-hammer (Fig. 7.13) was run in hole for testing five roundtrips with the longest continuous working time of only 17 h. Hydro-hammer obviously increased drilling speed. Since the testing of the hydro-hammers was arranged in the middle of the production drilling, there was no enough time to solve the problems found in the testing, so, in the whole process of reaming drilling, hydro-hammer didn't play a proper role.

The influence of the percussive power of hydro-hammer to drilling effect and rock bit was obvious. For instance, when SYZX-273 hydro-hammer was run in the MH-1K-47 roundtrip for testing, the bit used was φ 311.1 mm ($12^1/_4$ in) rock bit, and the rock in this hole section was amphibolite and eclogite. The hydro-hammer only worked for 4 h, with the footage drilled of 53 m in this roundtrip. The teeth of the bit cone broke and seriously wore, the bearings of three cones all loosened and lost efficacy (Fig. 7.14). The matching between the parameters of the hydro-hammer and

Fig. 7.13 SYZX-273 hydro-hammer and the affiliated 311 mm button bit

Fig. 7.14 Rock bit lost efficacy

the working properties of the rock bits still needed further study, otherwise, the service life of the bit would be affected.

7.2.3 Design of Drilling Tool Assembly

1. **φ 157/311.1 mm reaming drilling tool assembly**

According to the principle of drill collar grading and the result of drill string strength checking, φ 157/311.1 mm reaming drilling tool assembly can be found in Table 7.4, and the schematic drawing of the drilling tool assembly in Fig. 7.15.

Pilot reaming drilling tool assembly was mainly used for conventional reaming drilling, being a main drilling tool assembly form in reaming drilling. Pilot reaming drilling tool without stabilizer was mainly used in the period that drilling tool rotation torque needed to be reduced, by using this drilling tool the rotation resistance could be reduced, and meantime rock block dropping and drilling tool sticking could be avoided. This drilling tool was mainly used in the late phase of reaming drilling. Reaming drilling tool assembly with reamer was mainly used for rectifying the hole wall and drilling-off for setting casing in the final reaming drilling phase.

The starting phase of reaming drilling was mainly for exploring the drilling bit structure and drilling technology, so the footage drilled per roundtrip was short. Reaming drilling went into normal after solving the problems of welding seam strength of the drill bits. Since the long delivery term of vibration absorber, the vibration absorber was not used to the drilling tool assembly until the MH-1K-17 roundtrip; the vibration absorber alleviated the bouncing of the drilling tool to a certain extent. After that, the drilling tool assembly was adjusted mainly according to the hole bottom conditions, which are described as follows.

The stabilizer was disassembled from the drilling tool assembly from the MH-1K-27 roundtrip, the torque of the drill bit, which was caused due to the dog-leg degree, was decreased in borehole so that the drilling tool accident would be averted in deep hole section. Diamond reamer was jointed on the drill bit for reaming from the MH-1K-42 roundtrip, at the same time of drilling, previous hole wall was rectified and the drilling-off workload was reduced to the follow-up new bit and the hole undersize phenomenon was improved.

From the MH-1K-44 roundtrip, considering the obvious variation of the hole vertex angle at the hole bottom in this section and large resistance to the big rigidity reaming drilling tool in drilling operation, flexible pendulum drilling

Table 7.4 φ 157/311.1 reaming drilling tool assemblies

Drilling tool	Drilling tool assembly
Conventional pilot reaming drilling tool	φ 157/311.1 mm reaming bit + vibration absorber + φ 203.2 mm collars × 18 m + φ 309 mm stabilizer + φ 203.2 mm collar × 9 m + φ 309 mm stabilizer + φ 203.2 mm collar × 27 m + φ 177.8 mm collar × 27 m + φ 127 mm drill rod
Hydro-hammer drilling tool assembly	φ 157/311.1 mm reaming bit + φ 273 mm or φ 203 mm hydro-hammer × 3.2 m + vibration absorber + φ 203.2 mm collar × 9 m + φ 309 mm stabilizer + φ 203.2 mm collar × 8 m + φ 309 mm stabilizer + φ 203.2 mm collar × 27 m + φ 177.8 mm collar × 27 m + φ 127 mm drill rod
Pilot reaming drilling tool assembly without stabilizer	φ 157/311.1 mm reaming bit + vibration absorber + φ 203.2 mm collar × 54 m + φ 177.8 mm collar × 27 m + φ 127 mm drill rod
Reaming drilling tool assembly with reamer, without pilot	φ 311.1 mm tri-cone rock bit + diamond reamer + φ 203.2 mm collar × 9 m + φ 309 mm stabilizer + φ 203.2 mm collar × 9 m + φ 309 mm stabilizer + φ 203.2 mm collar × 36 m + φ 177.8 mm collar × 36 m + φ 127 mm drill rod

7.2 Design of Drilling Tool

Fig. 7.15 157/311.1 reaming drilling assemblies

tool (φ 203.3 mm collar between two stabilizers was changed to φ 158.8 mm collar) and non-core tri-cone rock bit were used and the gauge protection diamond reamer was disassembled, the pilot capacity of the drilling tool was decreased in hope that the follow-up drilling would not be affected by original hole trace, the hole dog-leg degree and vertex angle would be decreased in order to lay a solid foundation for the follow-up drilling (however, since the pilot hole had already formed a slim hole as a free fragmentation face, it was impossible to completely separate from the slime hole, but it would be possible to partially alleviate the hole deflection intensity). In the MH-1K-47 roundtrip, after the redressing was carried out again for rectifying the hole wall, cementing bag was drilled, then the reaming drilling was accomplished in this hole section.

2. **φ 157/244.5 mm reaming drilling tool assembly**

In φ 157/311.1 mm reaming, diamond reamer was designed in order to rectify the hole size. The drilling tool assembly for φ 157/244.5 mm reaming (Table 7.5) was basically the same as that for φ 157/311.1 mm reaming, with the schematic drawing in Fig. 7.16.

Based upon the experiences that using tri-cone rock bit didn't deviate off the axis of the original slim hole in φ 157/311.1 mm reaming drilling, and considering that pilot reaming drill bit had large torque, pilot body was easily to stick, and the whole body strength was less than the conventional tri-cone rock bit under complicated down hole condition and large torque condition, tri-cone rock bit reaming drilling tool assembly was mainly used in reaming drilling. The flexible tri-cone rock bit reaming drilling tool assembly was sometimes used in order to avoid sticking and breaking of the pilot body of the pilot reaming drill bit, when the down-hole condition was relatively complicated and the spring leaf of the elastic stabilizer of the moving casing might fall off.

In reaming drilling of this phase, there were a lot of broken spring leaves and broken pilot body of the drill bit at the hole bottom, a slim hole milling tool assembly was used to mill the falling objects and to drill off the slim hole for many times in order to prevent the damage to the drill bits and hole accidents. In the drilling process after 3447 m, there were not only "falling fish" and debris, but also the falling rock blocks from the hole wall resulted from hole oversize, and then milling shoe and fishing-debris drilling tool assemblies were used to fish and mill the debris at the hole bottom, shown in Table 7.6 and Fig. 7.17.

7.3 Optimization of Drilling Parameters

7.3.1 WOB

WOB for reaming rock bit should be less than that for conventional tri-cone rock bit with same size.

The load bearing condition of reaming bit was inferior to that of integral tri-cone rock bit, the pilot body bore very large bending torque, the vibration was large while drilling, and bouncing easily happened to produce large damage force to the cone leg welding seam. In addition, the cone body of the reaming rock bit was smaller; excessive WOB could cause an early damage to the cone bearings.

According to the foreign experiences and on-the-spot experimental research, the specific pressure of reaming drilling was 0.25–0.65 kN/mm (hole diameter-pilot hole diameter). WOB for φ 157/311.1 mm reaming drilling was 38.5–100 kN, and WOB for φ 157/244.5 mm reaming drilling was 22–57 kN.

The practical WOB should be adjusted according to the conditions of the formations to be drilled. If bit bouncing occurred in drilling, small WOB should be taken for further drilling. Low WOB was to be used in the initial phase of each roundtrip, after the wear of tungsten carbide teeth, WOB was gradually increased to ensure drill bit to cut the rock effectively.

7.3.2 Rotary Speed

The rotary speed of drill bit mainly depends on the time used for bit teeth breaking the rock, the formation strength drilled and the bearing capacity of the bearings. According to the

Table 7.5 φ 157/244.5 mm reaming drilling tool assemblies

Drilling tool	Drilling tool assembly
Conventional pilot reaming drilling tool	φ 157/244.5 mm reaming bit + φ 244 mm diamond reamer + φ 203 mm vibration absorber + φ 203.2 mm collar × 27 m + φ 178 mm vibration absorber + φ 214 mm stabilizer + φ 177.8 mm collar × 54 m + φ 127 mm drill rod
Flexible pilot reaming drilling tool	φ 157/244.5 mm reaming bit + φ 244 mm diamond reamer + φ 203 mm vibration absorber + φ 178 mm vibration absorber + φ 214 mm stabilizer + φ 177.8 mm collar × 72 m + φ 127 mm drill rod
Flexible tri-cone rock bit reaming drilling tool	φ 244.5 mm tri-cone bit + φ 244 mm diamond reamer + φ 203 mm vibration absorber + φ 178 mm vibration absorber + φ 214 mm stabilizer + φ 177.8 mm collar × 90 m + φ 127 mm drill rod

7.3 Optimization of Drilling Parameters

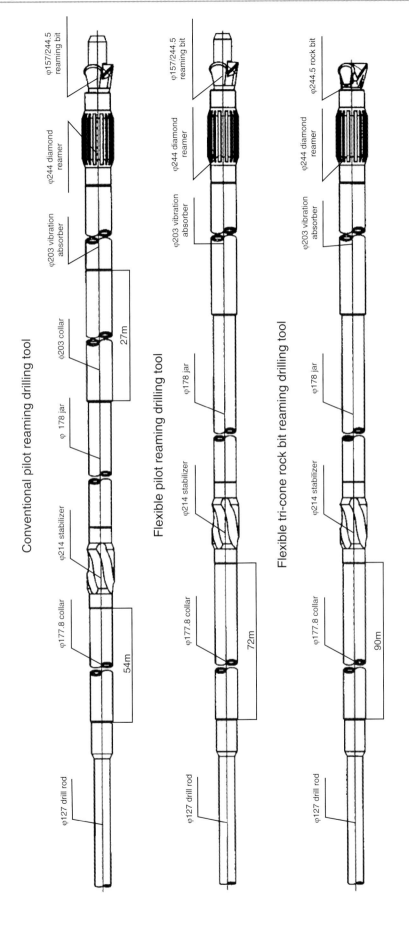

Fig. 7.16 157/244.5 reaming drilling assemblies

Table 7.6 φ 157/244.5 mm drilling tool assemblies for milling, drifting and fishing debris

Drilling tool	Drilling tool assembly
Slim hole drilling tool assembly for milling hole	φ 150 mm milling shoe + SK153 stabilizer + SK146 wire-line coring collars × 54 m + SK139.7 wire-line drill rod × 27 m + φ 178 mm jar + φ 214 mm stabilizer + φ 177.8 mm collar × 54 m + v 127 mm drill rod
Milling shoe + fishing-debris drilling tool assembly	φ 158.8 mm rock bit + φ 89 mm drill rod + φ 193.7 mm × 4.38 m special 1# fishing cup + φ 193.7 mm × 6.52 m special 2# fishing cup + φ 203 mm vibration absorber + φ 178 mm jar + φ 214 mm stabilizer + φ 177.8 mm collar × 90 m + φ 127 mm drill rod

research on rock fragmentation by tungsten carbide teeth from Biryukov, a scholar of the former Soviet Union, the contact time (t) between the rock and teeth is no less than 0.02–0.03 s, if less than 0.02–0.03 s, the acting effect that teeth exert the pressing force onto the rock will decrease sharply. According to above theory, the rotary speed range of φ 311.1 mm reaming rock bit could be obtained.

Suppose the big end diameter of the cone is d, the teeth number of the big end is z, the bit diameter is D, than an arc length between the teeth of the big end of the cone L = πd/z, the linear velocity of rotation of big end of the cone V_G = L/t, and then V_G = πd/zt. Under pure rolling, n/n_G = d/D, and then n_G = nD/d, in which, n is bit rotary speed, n_G is the rotary speed of the cone.

Considering that the teeth of reaming rock bit do not roll purely at hole bottom, the rotation speed decreases slightly, therefore, the above formula should be corrected to n_G = knD/d, in which, k = 0.95 is the thumb experience factor of the velocity loss.

From $V_G = \frac{\pi d n_G}{60}$, $n_G = \frac{60 V_G}{\pi d} = \frac{60 \pi d}{\pi d z t} = \frac{60}{zt} = k\frac{nD}{d}$

Obtain:

$$n = \frac{60d}{Dkzt} \quad (7.12)$$

For φ 157/311.1 mm rock bit, to put k = 0.95, t = 0.02–0.03 s, z = 19, d = 130 mm, D = 311.1 mm into the formula (7.12), and then we can obtain n = 46.3–69.5 r/min.

For φ 157/244.5 mm rock bit, to put k = 0.95, t = 0.02–0.03 s, z = 19, d = 130 mm, D = 244.5 mm into the formula (7.12), and then we can obtain n = 59–88 r/min.

In addition, while the reaming rock bit drilling in hard formations, in the arrangement design of the cones, big axial and radial slip shift existed, excessive high revolution could lead an early damage, and meantime, the journal bearing structure of the cone could not bear such high revolution. So, the rotary speed of the reaming rock bit should be limited to a reasonable scope of 40–70 r/min.

In the process of drilling, the upper limit of WOB and rotary speed ranges could not be used at the same time, otherwise, the severe bouncing of drilling tool would take place while drilling and it would be easy to cause hole accidents, rotary speed should be adjusted often in the light of the bouncing of the drilling tool at hole bottom. Therefore, rotary speed and WOB used for reaming drilling were lower, the practical rotary speed in drilling was 45–60 r/min.

7.3.3 Pump Displacement

In theory, the larger the pump displacement is, the better the effect of cleaning out the cuttings at hole bottom will be, the minimum uplift flow rate of the mud should be no less than 0.5 m/s:

In φ 311.1 mm reaming drilling, the pump displacement $Q = v\pi(D^2 - d^2)/4 = 0.5 \times 10 \times 3.14 \times (3.111^2 - 1.27^2)/4$ = 32L/s.

According to mud performance and pump conditions at the drill site, the pump displacement range for φ 157/311.1 mm reaming drilling was 30–35 L/s.

The pump displacement in φ 244.5 mm reaming drilling $Q = v\pi(D^2 - d^2)/4 = 0.5 \times 10 \times 3.14 \times (2.445^2 - 1.27^2)/4 =$ 17L/s.

According to mud performance and pump conditions at the drill site, the pump displacement range for φ 157/244.5 mm reaming drilling was 25–30 L/s.

7.4 Effect of Reaming Drilling

7.4.1 General Drilling Conditions

The reaming drilling of CCSD-1 Well was mainly divided into two times:

1. **The first reaming drilling of the main hole**

After the pilot hole was finished, since the quality of the hole body was relatively ideal, with the maximum deviation of only 4.1°, which was far less than the 14° deviation of the original design, in accordance with the double-hole design program, a decision was made to directly ream the φ 157 mm pilot hole to φ 311.1 mm (12^1/$_2$ in.). In May of 2005, after φ 157 mm core drilling to 2046.54 m depth in the pilot hole, the upper φ 244.5 mm (9^5/$_8$in.) moving casings were pulled out, the hole section from 101.00 to 2033.00 m was reamed to φ 311.1 mm in diameter, and then φ 273.1 mm technical

7.4 Effect of Reaming Drilling

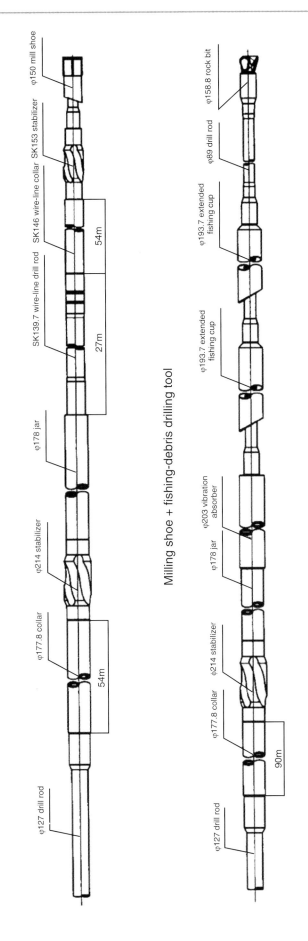

Fig. 7.17 157/245 drilling assemblies for milling, drifting and fishing cuttings

casings and φ 193.7 mm moving casings were set in, afterwards, coring drilling was carried out in the main hole. Reaming drilling lasted 122 days.

2. The second reaming drilling in the main hole

In October 2003, when core drilling in the main hole reached to 3665.87 m, the lower reamer of the core drilling tool broke down and lead to an accident in the hole. The fishing operations were carried out repeatedly for many times but without any success. And meantime, in the process of handling the accident, sticking of the drilling tool also occurred because of the falling rock blocks from the upper hole wall, besides, a casing program had been planned in the design in this hole section. For safe drilling and preventing a serious hole accident in follow-up drilling, and meantime, for bypassing the "fallen fish", a decision was made to carry out the second reaming drilling. The upper φ 193.7 mm moving casings were pulled out, the hole section from 2028 to 3525.18 m was reamed to φ 244.5 mm diameter. After sidetracking drilling, φ 193.7 mm technical casings were set in, and then core drilling in the main hole continued. It took 141 days for reaming drilling in this hole section.

The situation of the two reaming drillings can be found in Table 7.7, the technical index of the reaming drilling in Table 7.8, the variations of the service life and the rate of penetration of the reaming bits with hole depth changes for the first and the second reaming drillings respectively in Figs. 7.18 and 7.19.

7.4.2 Application of Pilot Reaming Bits

1. KZ157/311.1 reaming bit

From the start of reaming drilling on May 7th, 2002 to the end of the reaming on September 5th, 2002, twenty seven KZ reaming bits were used in succession and satisfactory

Table 7.7 The situations of the reaming drilling

Reaming drilling	Drilling date	Day	Hole section (m)	Diameter of reaming hole (mm)	Reaming bit				
					Piece	Total footage (m)	Footage (m/bit)	Service life (h/bit)	Drilling speed (m/h)
I	2002.05.07–2002.09.05	122	101–2033	311.1	34	1926.89	56.67	54.13	1.05
				215.9	1	5.00	5.00	3.43	1.46
				Subtotal	35	1931.89	55.20	52.68	1.05
II	2003.10.29–2004.03.14	141	2028–3525.18	244.5	28	1497.18	53.47	50.18	1.07
Total		263			63	3429.07	54.43	51.57	1.06

Note Milling drilling for 0.11 m in the first reaming drilling, φ 215.9 mm rock bit was still usable

Table 7.8 The technical statistics of the reaming drilling

Reaming drilling method		Bit/piece	Roundtrip	Total footage (m)	Actual drilling time (h)	Rate of penetration (m/h)	Footage drilled per roundtrip (m)
The first reaming drilling	157/311.1 reaming rock bit	30	42	1770.37	1684.85	1.05	42.15
	311.1 tri-cone rock bit	3	4	156.01	153.79	1.01	39.00
	311.1 flat face bit for hydro-hammer	1	1	0.51	1.67	0.31	0.51
	215.9 tri-cone rock bit	1	1	5.00	3.43	1.46	5.00
	Subtotal	35	48	1931.89	1843.74	1.05	40.25
The second reaming drilling	157/244.5 reaming rock bit	22	25	1145.46	1145.46	1.09	45.82
	244.5 tri-cone rock bit	6	9	351.72	349.34	1.01	39.08
	Subtotal	28	34	1497.18	1404.98	1.07	44.03
Total of two reaming drilling operations		63	82	3429.07	3248.72	1.06	41.82

7.4 Effect of Reaming Drilling

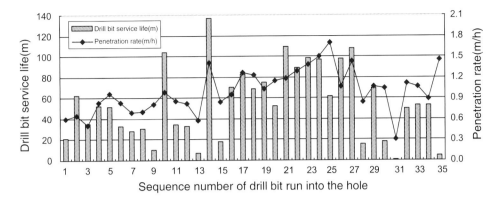

Fig. 7.18 Variation of service life and the rate of penetration of the reaming bits versus hole depth change for the first reaming drilling No. 1–18 and No. 22–30 bits were KZ pilot reaming bits; No. 19–21 bits were KHAT pilot reaming bits; No. 3 bit was spherical teeth full face percussive bit; No. 32–34 bits were φ 311.1 mm tri-cone rock bits; No. 35 bit was φ 215.9 mm tri-cone rock bit

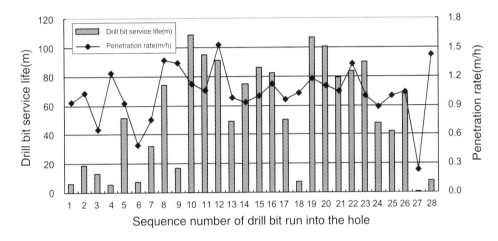

Fig. 7.19 Variation of service life and the rate of penetration of the reaming bits versus hole depth change for the second reaming drilling No. 1–2, No. 4, No. 6–9, No. 12–13, No. 16–27 bits were KZ pilot reaming bits; No. 15 bit was KHAT pilot reaming bit; No. 3, No. 5, No. 10, No. 11, No. 14, No. 28 bits were φ 244.5 mm tri-cone rock bits

results were obtained in the rock formations with drillability of Grade 7–11. The total reaming footage was 1534.05 m, the average rate of penetration was 1.04 m/h, the highest penetration rate was 1.7 m/h. The longest service life of a single bit was 105.94 h, and the highest footage was 136.53 m (Table 7.9).

According to the testing results, it can be found that the bits with vibration absorber had better application effect than those without vibration absorber. In the initial stage of bit application, the bit was not mounted with vibration absorber so that WOB and rotary speed could not be fully exerted to the bit, the drilling tool bounced seriously. After the vibration absorber was added to the bit, drilling became normal, WOB could be increased to 7–9 t, the rotary speed was increased to 50–60 r/min. However, when hydro-hammer was added on the bit, the abnormal damage of the bearings or the tungsten carbide teeth of the bit was serious.

2. KHAT 157/311.1 reaming bit

Three KHAT reaming bits were used, with the total reaming footage of 236.32 m, the total drilling time of 209.90 h, the average drilling speed per hour of 1.126 m/h, the highest rate of penetration of 1.19 m/h. The longest service life of a single bit was 91.86 h, and the highest footage drilled was 109.16 m (Table 7.10).

As the delivery of this type of reaming bit to the drill site was relatively late, only three drill bits were used in succession. Since KHAT drill bit lacked of gauge protection design on the bit body, the gauging capacity of this series of drill bit was affected. The contacting points between the bit and hole wall at hole bottom were only three points, when whirl of rock bit happened in the process of drilling, it would be easy to form a multi-petal hole profile, the actual hole diameter was slightly smaller.

Table 7.9 The statistics of KZ157/311.1 reaming bits

Serial num Dec of drill bit	Roundtrip	Footage drilled (m)	Drilling time (h)	Drilling speed (m/h)	Serial number of drill bit	Roundtrip	Footage drilled (m)	Drilling time (h)	Drilling speed (m/h)
KZ157/311.1-01	1	20.96	35.10	0.60	KZ157/311.1-15	2	136.53	97.00	1.41
KZ157/311.1-02	5	62.42	97.79	0.64	KZ157/311.1-16	1	70.23	73.99	0.95
KZ157/311.1-03	2	35.23	68.68	0.51	KZ157/311.1-17	1	82.77	65.08	1.27
KZ157/311.1-04	1	52.35	62.82	0.83	KZ157/311.1-18	1	68.59	55.82	1.23
KZ157/311.1-05	1	51.83	54.05	0.96	KZ157/311.1-19	1	89.00	68.93	1.29
KZ157/311.1-06	2	33.16	39.82	0.83	KZ157/311.1-20	1	62.14	36.46	1.70
KZ157/311.1-07	2	28.00	40.64	0.69	KZ157/311.1-21	1	96.65	64.49	1.50
KZ157/311.1-08	2	30.26	43.29	0.70	KZ157/311.1-22	1	98.71	70.95	1.39
KZ157/311.1-09	1	10.07	12.48	0.81	KZ157/311.1-23	1	97.66	91.31	1.07
KZ157/311.1-10	2	103.80	105.94	0.98	KZ157/311.1-24	1	107.74	75.54	1.43
KZ157/311.1-11	2	34.56	40.36	0.86	KZ157/311.1-25	1	18.13	17.38	1.04
KZ157/311.1-12	1	32.50	39.75	0.82	KZ157/311.1-26	1	69.57	64.76	1.07
KZ157/311.1-13	1	7.29	12.68	0.57	KZ157/311.1-27	1	15.60	18.35	0.85
KZ157/311.1-14	1	18.30	21.49	0.85	Total 27	38	1 534.05	1 474.95	1.04

Table 7.10 The statistics of KHAT157/311.1 reaming bit

Serial number of drill bit	Roundtrip	Footage drilled (m)	Actual drilling time (h)	Rate of penetration (m/h)
2002-A031	2	74.91	72.57	1.03
2002-A032	1	52.25	45.47	1.15
2002-A033	1	109.16	91.86	1.19
Total	4	236.32	209.90	1.13

When this type of bit was used, after a new bit was lowered to hole, it would need long time for redressing. After 236.32 m were drilled by three drill bits of this type and KZ type drill bit was used instead, hole section for redressing was even much longer, and it made the outer diameter of the new bit wear seriously when the new bit redressed to hole bottom, if drilling continued, and the wear of the outer diameter of the drill bit would increase continuously, then the outer diameter was far less than the standard outer diameter and lead to a conical-shape hole profile in the hole section drilled, therefore in the follow-up drilling, a longer distance redressing would be needed if a new bit was used again. This situation lasted to the end of reaming drilling. Finally, diamond reamer was added to the drilling tool, in combination with hole rectification in drilling or in special running-in for many times, an obvious effect was obtained.

3. **KZ157/244.5 reaming bit**

From the start of reaming drilling on October 29th, 2003 to the end of reaming drilling on March 14th, 2004, twenty one KZ 157/244.5 reaming rock bits were used. The total reaming footage drilled was 1059.34 m, the average rate of penetration was 1.09 m/h, and the highest rate of penetration was 1.53 m/h. The longest service life of a single bit was 91.85 h, and the highest footage was 107.06 m (Table 7.11).

Since the dog-leg degree in the small diameter hole trace in this reaming phase was big, and the bending moment borne by the pilot body of the reaming bit was large during reaming drilling, and at the same time, due to the tungsten carbide blocks, elastic steel pieces and other hard materials

7.4 Effect of Reaming Drilling

Table 7.11 The statistics of KZ157/244.5 reaming bit

Series number of drill bit	Roundtrip	Footage drilled (m)	Drilling time (h)	Drilling speed (m/h)	Series number of drill bit	Roundtrip	Footage drilled (m)	Drilling time (h)	Drilling speed (m/h)
KZ157/244.5-01	1	6.53	6.99	0.93	KZ157/244.5-13	1	82.10	73.63	1.12
KZ157/244.5-02	1	19.12	18.60	1.03	KZ157/244.5-14	1	78.98	75.01	1.05
KZ157/244.5-03	2	5.69	4.64	1.23	KZ157/244.5-15	1	100.59	91.85	1.10
KZ157/244.5-04	1	31.90	42.56	0.75	KZ157/244.5-16	1	107.06	90.58	1.18
KZ157/244.5-05	1	7.63	15.58	0.49	KZ157/244.5-18	1	67.65	64.75	1.04
KZ157/244.5-06	1	17.25	12.83	1.34	KZ157/244.5-19	1	0.44	1.87	0.24
KZ157/244.5-07	1	73.85	54.01	1.37	KZ157/244.5-20	1	42.00	42.06	1.00
KZ157/244.5-08	1	91.00	59.39	1.53	KZ157/244.5-21	1	83.39	62.53	1.33
KZ157/244.5-10	2	49.22	50.21	0.98	KZ157/244.5-22	2	47.56	53.30	0.89
KZ157/244.5-11	1	50.03	51.90	0.96	KZ157/244.5-23	1	89.72	89.78	1.00
KZ157/244.5-12	1	7.64	7.41	1.03	Total 21	24	1 059.34	969.48	1.09

Fig. 7.20 Seriously worn pilot body in MH-2K-21 roundtrip

Fig. 7.21 Seriously worn pilot body in MH-2K-21 roundtrip

at hole bottom, the pilot position of the drill bit wore seriously (Fig. 7.20). In follow-up drilling, the pilot body of the bit run in again also bore big bending moment, with the same wear appeared. However, along with the increase of hole depth, the drilling tool gradually returned to the original hole axis under the guidance of the pilot body, with the wear disappeared.

4. **KHAT157/244.5 reaming bit**

The structure of KHAT157/244.5 bit basically followed the original structure of KHAT157/311.1, with specially designed cone legs adopted. In strengthening gauge protection capacity, tungsten carbides were inserted on the top of cone legs, however, due to inadequate tungsten carbides which were inserted at the same position direction in the circumference of the cones, the effect of gauge protection and rectifying well wall was unsatisfactory. As there was no cutting capacity at bottom, drilling could not be realized when foreign articles existed in the pilot hole. Another big problem was that the connecting area between bit pilot body and bit body was small in the design. Limited by this small size, the pilot body of the first drill bit broke and fell down to the hole bottom (Fig. 7.21), although no tungsten carbides

Table 7.12 Statistics of the application results of the pilot reaming drilling bits in the whole borehole

Reaming diameter (mm)	Type	Bit used	Total footage drilled (m)	Total drilling time (h)	Average footage drilled (m/bit)	Average service life (h/bit)	Average ROP (m/h)
157/311.1	KZ KHAT subtotal	27	1534.05	1475.95	56.82	54.66	1.04
		3	236.32	209.90	78.77	69.97	1.13
		30	1770.37	1685.54	59.01	56.16	1.05
157/244.5	KZ KHAT subtotal	21	1059.34	969.48	50.44	46.17	1.09
		1	86.12	86.16	86.12	86.16	1.00
		22	1145.46	1055.64	52.07	47.98	1.09
Total		52	2915.83	2740.49	56.07	52.70	1.06

Note Some KZ bits still could be used

were inserted at the front end and more space obtained. But the consideration to the strength in the design was still not enough (see right of Fig. 7.6). Other two drill bits were not used. The first bit drilled 86.12 m, drilling time was 86.16 h and the average drilling rate was 1.00 m/h.

5. **Summary of pilot reaming bits**

During two reaming drilling, four kinds of bits of two sizes and two types were used; statistics of the application effect of the bits can be found in Table 7.12. Among the bits used, most bits were KZ bits, mainly because the KZ bits had a stable performance, better gauge protection and safety in design, thus played an important role for the smooth accomplishment of the reaming drilling. Even though the average footage drilled and service life of the two kinds of KHAT reaming rock bit in the statistics were better, big problems still existed in gauge protection and safety, needing further improvement.

Well-Deviation Control Techniques for Strong Dipping Strata

The factors influencing hole deviation are extensive, they are generally divided into three kinds, the geological factor, the technical method factor and operation factor. The geological factor mainly includes the rock structures, anisotropy, the schistosity and bedding, soft and hard interbeds and occurrence and so on. The technical method factor mainly includes the deflection in hole opening, the well wall annular space, the rigidity of the bottom drilling tool assembly and deflection state and so on. The operation factor mainly is the feeding WOB. During drilling CCSD-1 Well, the strata factor and features of bottom drilling tool assembly were the important factors. Since the strata factor was unchangeable, so the reasonable bottom drilling tool assembly was mainly taken with help of the WOB control technique to prevent the hole deviation.

The formations drilled in CCSD-1 Well mainly were granitic-gneiss, plagiogneiss, eclogite, amphibolite and ultrabasic rock, among these rocks, the gneiss was typical strongly dipping formation. The bottom hole power (hydrohammer + PDM) was mainly taken to recover core by pulling drill string, with core drilling diameter of $\varphi 157\,mm$. The bottom drilling tool assembly with stabilizers was used for preventing hole deviation, when the hole deviation exceeded the limitation or the hole accident was very difficult to deal with, the side-tracking drilling technology was used.

8.1 Summary

According to hole structure, drilling method and logging requirement and so on, the well-deviation index (see the Sect. 2.8 of Chap. 2) of the preliminary design of CCSD-1 Well was decided. A flexible double hole program was taken, at first, a 2000 m pilot hole would be drilled, and then the hole position was moved to another place for drilling the main hole to 5000 m deep. If the actual deviation of the pilot hole would not exceed the design index in the 0–2000 m section, and the casing program of the pilot hole could meet the requirement of the main hole drilling, the pilot hole position could be used for the main hole drilling, and preliminary double hole program could be changed into the single hole program (double holes would be combined into one hole). So, to guarantee the pilot hole deviation index was not only a single technical problem, but also had important economical significance at the same time.

8.1.1 The Formation Conditions

The forecasting formations of the CCSD-1 Well were divided into five units (see the Sect. 2.2 of Chap. 2), between each of the five rock structure units was separated by the toughness shear belt. The rocks to be drilled were mainly divided into gneiss, eclogite, serpentinized peridotite, schist rock, amphibolite, mylonite and cataclastic rock and so on. The actual rock histogram can be found in Fig. 8.1.

1. **The strong natural deviation tendency of rock strata**

Since the metamorphic depth of rocks to be drilled was not identical, there were big difference of hardness, density and porosity of the rocks, rock beddings and natures frequently changed and stratification developed very much and gneiss structure, so the formations to be drilled were strong deviation strata. The tendency angle of rock beddings was 85–185°, the main tendency was nearly eastward (100° or so), the bedding inclination angle was 25–60°, the upper was steep and the lower was gradual. The fracture was developed, most fractures above the 3000 m hole section were identical with the bedding tendency and inclination angle; the tendency of the most fractures below the 3000 m hole section was southwestern and the inclination angel was gradually steep. Several pieces of mylnite belts would be drilled in the pilot and the main holes, besides the developed planar joints, mineral grained size would be fine and hardness become higher.

Translated by Yongqin Zhang.

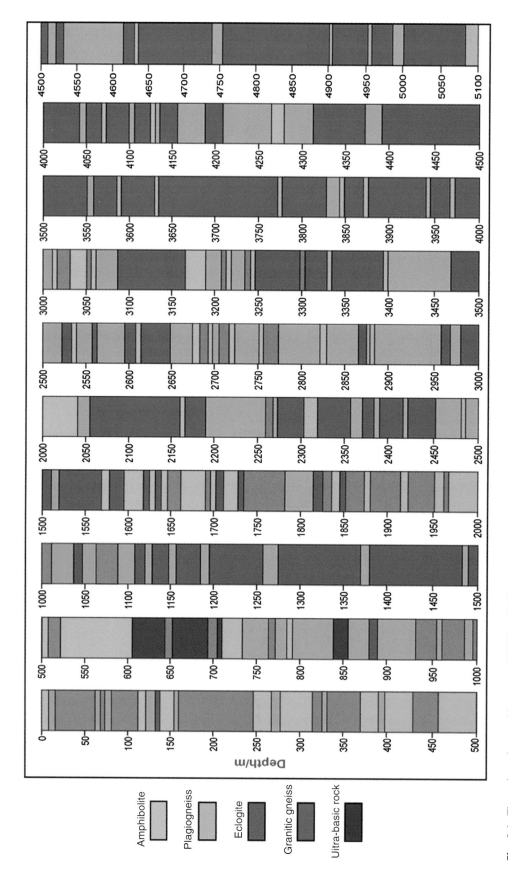

Fig. 8.1 The main rock strata histogram of CCSD-1 Well

The No. 6 Geological Prospecting Brigade of Jiangsu Provincial Bureau of Geology had ever drilled several hundred holes in the near of the drill site of CCSD-1 Well, the average vertex angle build-up rate reached 3.72(°)/100 m. In CCSD-PP2 hole which was 382 m far away from the CCSD-1 Well, although many hole deviation prevention measures were taken, the hole vertex angle still reached 4.7° at 88 m depth, when hole depth was 737.25 m, the hole vertex angle was 20° at measured depth of 700 m, the deviation-correction and backfill sidetrack deviation-correction were carried out respectively and then the final hole vertex angle was kept 14.7° at 1028.68 m.

2. **The rock aeolotropism and ground stress**

The rock aeolotropism was caused mainly by the high angle fractures, structure stresses non-equilibrium or both. The aeolotropism development in the high angle and developmental fracture was intense, and when the fracture strike was identical with the direction of the rock aeolotropism, the measurement of the aeolotropism got stronger. The aeolotropism in the developed collapse well wall and fracture sections got stronger, and the aeolotropism in the compact well section got weaker (Niu Yixiong et al. 2006. Report of logging sub-project for CCSD project, pp. 348–365).

   i. The strong aeolotropism well sections: the well wall stress collapse got serious and there was no natural fracture mark in 1450–1515 m, 2090–2110 m, 2580–2700 m, 3660–3758 m, 4330–4390 m sections.

   ii. The less strong aeolotropism well sections: there were no well wall stress collapse and high angle fracture got developed in 523–600 m, 1111–1287 m, 1515–1560 m, 1620–1720 m, 1783–1810 m, 2015–2045 m, 2255–2285 m, 2410–2580 m, 3620–3660 m, 3948–4010 m, 4480–4625 m, 4750–4950 m; and there was evident well wall stress collapse and the high angle fracture got developed in 1303–1450 m, 1810–2015 m, 2045–2055 m, 2130–2145 m, 2160–2255 m, 2330–2410 m, 3758–3948 m, 4010–4330 m, 4390–4480 m, 4625–4750 m, 4950–5060 m.

   iii. The weak aeolotropism well section: the rock formations were compact and there were no well wall stress collapse and no developed fractures in 600–700 m, 1287–1303 m, 1600–1620 m, 2055–2090 m, 2150–2160 m.

8.1.2 The Well Deviation Control Technology

In the 1950s, an American scholar Lubinski researched the deformation of drill string under force and multi-bending in the straight well, and created a new situation of the research on drill string mechanics and provided a theory base for the application of the pendulum drilling tool. In the 1960s, Hoch presented the well deviation prevention theory with double stabilizers on the basis of the counteraction force from the well wall limitation when the drill string was pushed down and bent in the well, and then provided a theory base for the development of full-hole drilling tool. At the moment, in the oil drilling industry, the differential equation method (Lubinski), finite element method (Millheim), the vertical and horizontal bending method (Bai Jiazhi) and energy method (Walker) are mainly used for analyzing and calculating the force holding and deformation of the drilling tool assembly at hole bottom and explored the side force and dip angle of the bit (Su Yinao 2003).

1. **Deviation prevention techniques widely used in petroleum drilling industry**

The common deviation prevention and correction drilling tools mainly include full-hole and pendulum drilling tools. The full-hole drilling tool can control the variation rate of the hole deviation, but it cannot effectively control the magnitude of the hole deviation angle. The pendulum drilling tool is an effective means to control the hole deviation and is used most widely at the moment, the pendulum drilling tool includes the slick collar drilling tool, tapered drilling tool and stabilizer drilling tool and so on. Besides the full-hole and pendulum drilling tools, there are still square collar, eccentric collar, flat collar and HCY anti-deviation drilling tool assemblies, etc.

Currently, new deviation prevention techniques such as eccentric axis drilling tool assembly, flexible drilling tool assembly, pilot drilling tool, reverse pendulum, automatic-vertical drilling system, etc. are under development.

2. **The deviation prevention techniques in scientific drilling**

The vertical drilling in scientific drilling includes the passive vertical drilling system and the initiative vertical drilling system. The passive vertical hole drilling system means that the drilling tool assembly has only the functions of deviation prevention, the deviation correction and keeping-straightness in drilling process, there is no MWD instrument in the drilling tool assembly and cannot measure and rectify while drilling, such as pendulum and full-hole drilling tools. The initiative vertical drilling system means that the drilling tool assembly is equipped with MWD and deviation correction system; it can measure and rectify the hole deviation while drilling.

The passive vertical drilling system was mainly used for scientific drilling project in the former Soviet Union. In the early Cola SG-3 Well, a turbine drilling tool assembly (pendulum drilling tool) with a special stabilizer was used in 2100 m well depth. The full-hole drilling system with two or more turbine motors arranged side by side was used, for drilling from the surface to 4000 m deep under 11–15 r/min revolution and a good drilling effect with 1° deviation was achieved. The common pendulum passive vertical hole drilling tool was used in the hole section beneath 4000 m.

The pendulum passive vertical drilling tool was used with φ 69 mm coring rock bit in the upper section of the pilot hole in German KTB, a hole deviation prevention method of the geological exploration core drilling was used for 152.4 mm wire-line core drilling.

The initiative vertical hole drilling system and MSS control system and fully-stable well-deviation control system with measurement system were used in the KTB main hole. 3978.2 m (124 roundtrips) were drilled with the initiative vertical hole drilling system, among the 3978.2 m, 3204.8 m (100 roundtrips) were drilled with VDS vertical hole drilling system, 773.4 m (24 roundtrips) were drilled with the ZBE vertical hole drilling system. 1099.9 m (49 roundtrips) were drilled with the MSS motor control system. When the well depth reached to 7000 m, the well deviation angle was basically kept within 1°, the horizontal deflection distance was less the 15 m.

8.1.3 The Basic Conditions of Well Deviation Control in CCSD-1 Well

For φ 157 mm diamond core drilling, the basic geological exploration core drilling method was used, the outer diameter of the upper and lower reaming shells was 0.3 mm larger than that of the diamond bit. For preventing well deviation, the length of the outer gauge protection of the diamond core bit was extended to 40 mm, the reaming wing length of the reaming shell was extended to 110 mm, the bottom collar pulling-down method was used and the stabilizer was installed in the drilling tool assembly, a full-hole and keeping-straight drilling tool assembly was formed (Fig. 8.2), and meantime the outer diameters of the drill bit and reaming shell were measured after each pulling-up of drilling tool in order to guarantee the performance of the drilling tool assembly unchanged.

During core drilling, when the well deviation surpassed the deviation angle limitation required or when the well accident was very difficult to handle, the single bend PDM and MWD tool (wire or wireless) was to be used to rectify the hole deviation (hole bottom correction) by side-tracking or to bypass the drilling obstacle by side-tracking. During CCSD-1 Well drilling, 2 times of the backfill side-tracking, 3 times of hole bottom deviation-correction and 1 time of continuous deflector deviation-correction testing were carried out. At the hole depth of 2749 m, the φ 120 mm single bend PDM and the wire MWD tool (DST) were used for φ 157 mm diameter side-tracking drilling and deviation correction; at 3400 m hole depth, φ 172 mm single bend PDM and the wireless MWD were used for φ 244.5 mm diameter side-tracking drilling to bypass the drilling obstacle; in 3127.54–3253.33 m hole section (MH-1C), the φ 120 mm single bend PDM and the wire MWD tool (DST) were used for 3 times of hole bottom deviation correction drilling (3127.54–3139.39 m, 3171.28–3191.28 m, 3244.98–3254.33 m); in 5129.36–5134.66 m hole section, the φ 120 mm PDM driving continuous deflector and wire MWD tool (DST) were used for the hole bottom deviation correction test.

The pilot hole ended at 2946.54 m, and the well deviation angle was 4.1° at 2028 m, the dog-leg degree was 0.4(°)/10 m (Fig. 8.3) at 1950 m, the technical program which the pilot hole and the main hole were put together was fully completed, the cost to remove the derrick to other position was saved to avoid spending a huge amount of expenditure to import foreign VDS vertical drilling system (used to drill the upper 2000 m hole section of the main hole), drilling time was shortened.

Although 5 times of directional drilling all got successful, the lower formation of the hole almost was gneiss, the well deviation became very serious. After each stage of directional drilling the well deviation angle quickly grew (Fig. 8.4), the well deviation angle was 17.18° at the depth of 4000 m. According to the geologists' main viewpoint that well deviation would not exert great influence to the geological achievement of CCSD-1 Well in the homogeneous and thick geologic body and the lessons from German KTB that a complicated hole shape had been formed by a number of deviation correction operations and many hole accidents in the KTB pilot hole emerged so that the hole had to be ended up ahead of schedule, the CCSD-1 Center reported to the leading group and after getting an approval, decided to give up one more time deviation correction drilling. The final hole depth of CCSD-1 Well is 5158 m, the final hole deviation angle is 23.5° (wire MWD), the vertical depth is 5027 m, and the closing distance 396 m (Fig. 8.4).[1]

8.2 Deviation Prevention Drilling Technology

In recent years, in petroleum drilling industry a dynamic deviation prevention theory appeared in relation to the traditional deviation prevention theory by small WOB and drilling tool assembly (Gao Deli et al. 2004), Gao Deli thought that the drill string is in the state of whirling in the borehole (the drill string rotates around the its axis, and at the same time it also rotates around the axis of the borehole), the drilling tool of a certain kind of combination deforms in spiral and crooked forms under heavy WOB (usually 200 kN WOB is exerted to φ 215.9 mm rock bit), the revolution eccentric force of whirling makes the bit produce a force toward the lower wall of the borehole by deformation and then get the effects of

---

[1] Engineering logging to 5075 m depth, the logging data of the DST wired MWD instrument were used after 5100 m.

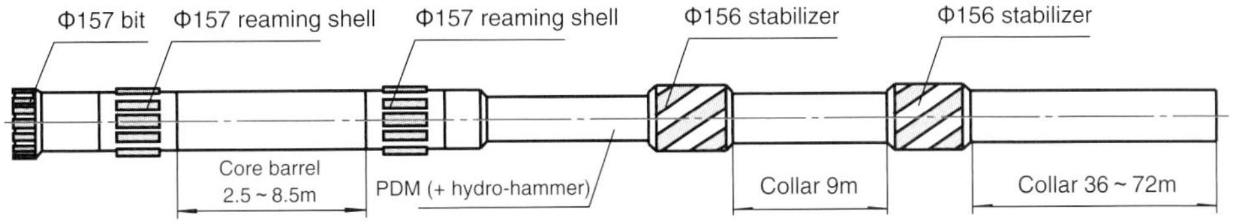

8.2 Deviation Prevention Drilling Technology

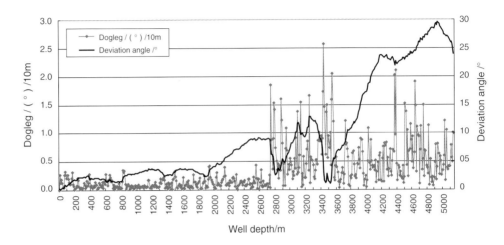

Fig. 8.2 Sketch map of the core drilling tool assembly in the hole bottom

Fig. 8.3 Variation curve of deviation angle and dog leg degree versus well depth of CCSD-1 Well

deviation prevention and deviation correction. But CCSD-1 Well differed from the petroleum drilling industry:

i. In core drilling, drill bit is connected directly with the core barrel by reaming shell, the pendulum force of the drill collar above the core drilling tool can't be effectively transferred to the drill bit.

ii. The strength, rigidity and weight of diamond drill bit, reaming shell and core barrel are all far more less than those of drill collar, the strength of thread connection is even weaker, the heavy WOB is not allowed to exert to the drilling tool.

iii. For CCSD-1 Well, PDM hydro-hammer down-hole drive conventional core drilling (by lifting drilling tool for core recovery) was mainly used, the output torque of the PDM decided that bit could not bear the over-heavy WOB.

So, the deviation prevention of CCSD-1 Well still adopted the traditional theory of deviation prevention as the basis.

8.2.1 The Well Deviation Control in Core Drilling

1. **The technical measures of deviation prevention in core drilling**

(1) Rigid, straight, full and heavy drilling tool assembly

Rigid: the petroleum casing with the size of $\varphi 139.7\,mm \times 10.54\,mm$ and the steel grade of N80 (API standard) was adopted as the core barrel, and the outer tubes of the down hole power (PDM and hydro-hammer) were all high quality steel tube with the wall thickness not less than 10 mm. They could fully guarantee against bending under given WOB.

Straight: the core barrel was carefully selected and then the concentricity and coaxiality of the connecting thread of the large diameter drilling tool could be guaranteed.

Full: the diameter of the reaming shell was slightly larger than the bit diameter, and the large diameter drilling tool always fill in the borehole. Rational quantity of stabilizers was installed on the collars in order to prevent the large diameter drilling tool from tilting as much as possible (Fig. 8.2).

Heavy: $\varphi 120\,mm$ collars of 45–81 m were connected with the drilling tool to make the drill string in a tensile condition.

The calculation on the well deviation force (Table 8.1) of core drilling tool assembly was carried out by China Petroleum University in a research project of Research on Drilling Hole Deviation Prevention. In different lengths of core barrel assemblies, only short core barrel assembly had certain deviation correction force, but after all the value of this deviation correction force was too small, which was not enough to contend with the formation kick-off force.

According to in-the-hole conditions and the requirement of drilling technology and so on in core drilling of CCSD-1

Fig. 8.4 Projection diagram of the deviation angle, azimuth and well trajectory of CCSD-1 Well

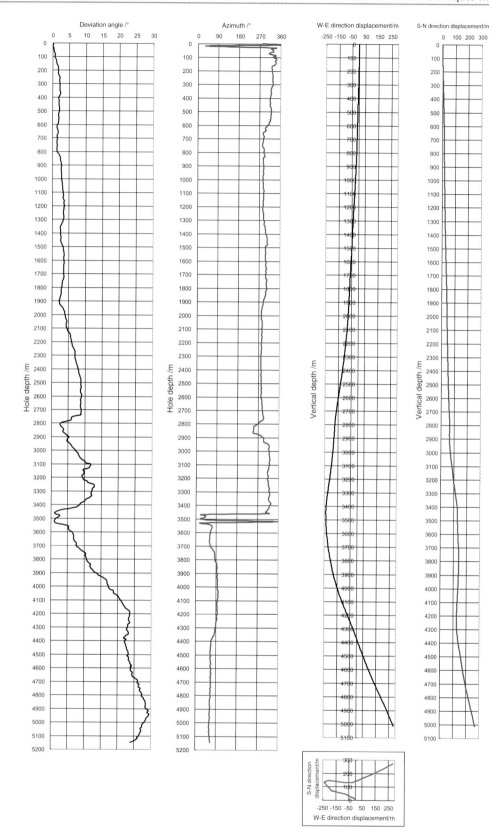

Table 8.1 Calculation on deviation correction force of the coring drilling tool assemblies

Type of PDM	Core barrel length (m)	Well deviation angle (°)	WOB/kN 10 — Well deviation force (N)	20 — Well deviation force (N)	30 — Well deviation force (N)	40 — Well deviation force (N)
LZ120	3.0	1	−12.6	−12.7	−12.8	−12.9
		2	−25.1	−25.3	−25.5	−25.6
	4.5	1	3.8	3.7	3.6	3.6
		2	7.7	7.6	7.4	7.3
	6.0	1	24.8	24.9	24.9	25
		2	49.7	49.8	49.9	50.1
	8.5	1	69.7	70.5	71.3	72.1
		2	139.5	141.1	142.7	144.3
LZ95	3.0	1	−16.3	−16.5	−16.6	−16.8
		2	−32.5	−32.8	−33.1	−33.4
	4.5	1	0.5	0.3	0.1	0
		2	1.0	0.7	0.4	0.1
	6.0	1	21.8	21.8	21.7	21.7
		2	43.7	43.6	43.6	43.5
	8.5	1	67.3	67.9	68.6	69.3
		2	134.6	135.9	137.3	138.7

Well, the down hole drilling tool assemblies used can be found in Fig. 8.5, where,

P1 shows the PDM core drilling tool assembly used for changing hole diameter;
P2 shows slick PDM core drilling tool assembly;
P3 shows the PDM core drilling tool assembly with one stabilizer;
P4 shows the PDM core drilling tool assembly with two stabilizers;
P5 shows the PDM hydro-hammer core drilling tool assembly with two stabilizers;
M1 shows the core drilling tool assembly of SK wire-line drill string driven by top drive;
M2 shows the PDM core drilling tool assembly of SK wire-line drill string;
M3 shows the PDM hydro-hammer drilling tool assembly of SK wire-line drill string;
M4 shows slick PDM hydro-hammer core drilling tool assembly;
M5 shows slick jar-while-drilling PDM hydro-hammer core drilling tool assembly;
M6 shows SK wire-line collar jar-while-drilling PDM hydro-hammer core drilling tool assembly;
M7 shows upper stabilizer jar-while-drilling PDM hydro-hammer core drilling tool assembly;
M8 shows lower stabilizer jar-while-drilling PDM hydro-hammer core drilling tool assembly.

(2) Percussive rotary drilling with small WOB

The small WOB drilling was an important theory of Lubinski on deviation prevention. At the same time, since the PDM drilling tool was used in CCSD-1 Well, the larger WOB was not possible for PDM driving. Therefore, the WOB range of core drilling was controlled in 10–35 kN, with the maximum WOB no more than 40 kN. According to the calculation on the bit face area of the diamond drill bit to be used and rock hardness to be drilled, the WOB of core drilling should be 60–85 kN, so, although the smaller WOB drilling was beneficial to deviation prevention, the drilling efficiency was greatly affected. It not only created lower penetration rate, but also core blocks often happened since smaller WOB could not overcome the squeezing force caused by the broken core to the inner barrel, the footage drilled per roundtrip was very short. The pulling-up and running-in speed of petroleum drill rig was far lower than that of core drill, the non-drilling time at well head of large diameter core drilling was much longer than that of small diameter drilling. According to the statistics of the relation between non-drilling time and the well depth shown in Fig. 8.6 (in the regress curve $t = \Delta t + ah$, $\Delta t \approx 1.5$ h is the non-drilling time at well head, $a \approx 0.42$ is the tripping time per meter in the well), it can be found that the influence of the footage drilled per roundtrip to the drilling efficiency is still larger than penetration rate in deep hole drilling. In order to increase penetration rate and footage drilled per roundtrip

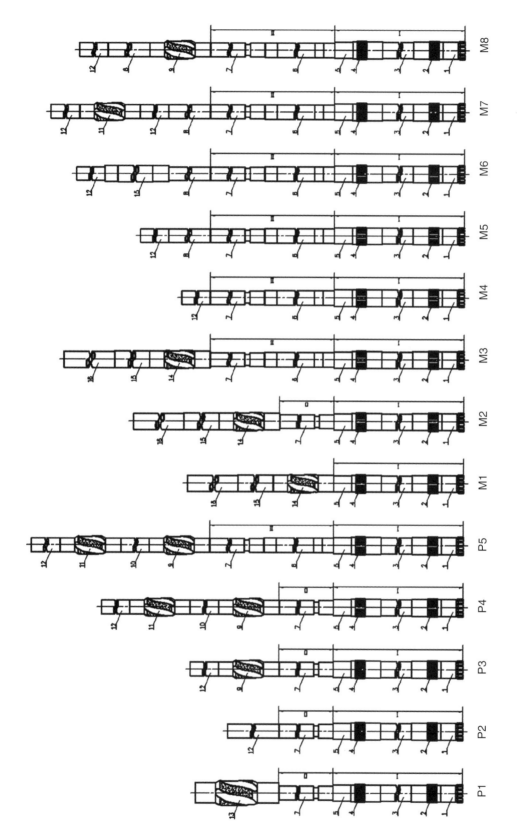

Fig. 8.5 Down hole drilling tool assemblies for CCSD-1 Well. *1* φ157 mm coring bit *2, 4* φ157 mm reamer *3* φ139.7 mm core barrel *5* φ146 mm swivel type sub *6* φ127 mm hydro-hammer *7* φ120 mm PDM *8* φ120 mm jar-while-drilling *9, 11* φ156 mm stabilizer *10.* φ120 mm single collar *12* φ120 mm collar string *13* φ215.9 mm stabilizer *14* SK157 stabilizer *15* SK146 wireline drill collar string *16* SK139.7 wireline drill rod string

8.2 Deviation Prevention Drilling Technology

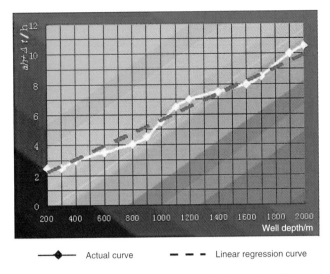

Fig. 8.6 The relationship between non-drilling time and well depth

Fig. 8.7 Serious one-side wearing of the stabilizer when drill string turning

under the small WOB condition, hydraulic percussive rotary drilling technique must be considered. From 2.18.44 m deep, PDM drive hydro-hammer drilling technique was tested, through two turns of improvements of the hydro-hammers and the mud performance, a very good drilling effect was obtained from this technique, penetration rate was increased to 1.13 m/h from 0.71 m/h of pure rotary drilling, the average footage drilled per roundtrip was increased to 6.34 m from less 2.28 m of pure rotary drilling, thereby higher drilling efficiency was effectively guaranteed and deviation prevention drilling was realized under small WOB.

(3) Deviation prevention action driven by down-hole PDM

In conventional rotary drilling, the rotary force of drill bit is transferred from the surface through the drill string, it is unavoidable for the long drill string to turn around the hole axis (revolution). The turning of the drill string around the hole axis (revolution) causes detrimental effect to well stabilization in following three aspects:

i. Either in traditional deviation prevention theory or in dynamic deviation prevention, when calculating the bottom drilling tool deformation and side force on drill bit, Δr, the well visible radius would be used as an important parameter in all aspects. Δr is the difference value between wellbore radius and bit radius (collar, stabilizer, reaming shell, etc.), which demonstrates a constraint of well wall to the drilling tool, and the bigger the Δr is, the bigger the angle of the drilling tool axis deviating from wellbore axis is. The eccentric force caused by revolution makes the reaming shell or the tungsten carbide stabilizer above the large diameter drilling tool produce an additional side force to the well wall, this side force makes the drilling tool wear in one side in hard rock formations and the well diameter enlarge in soft formations, with an overall effect that the tipping degree of drilling tool enhances with the increase of Δr. Figure 8.7 shows the damage by revolution of drill string driven at surface to stabilizer, from which, it can be judged the influence produced by drill string revolution to well wall.

ii. In slant well, the axis of the drill string revolution above the large diameter drilling tool descends to the wellbore axis because of gravity action, so, a lifting-up force should be exerted to the drill bit with the help of the upper reaming shell as a supporting point.

iii. If the dynamics nature of the rock formation is anisotropic, so, when the wellbore above the large diameter drilling tool is enlarged, even the rigid drilling tool is in a vertical well section, straightness of the wellbore still cannot be kept.

For PDM down hole driving form, there was no rotation action above the thick diameter drilling tool, this not only got rid of the enlargement action to the well while the drilling string turned around the well axis, and at the same time, made the angle between the thick diameter drilling tool axis and well body axis unchangeable under certain axial force, since strict measures of keeping rigidity and fullness of the drilling tool was taken, the angle between the thick diameter drilling tool axis and well body axis was controlled within very small range.

2. **Well deviation control conditions in core drilling**

Thirteen types of down hole drilling tool assemblies, shown in Fig. 8.5, were used to drill 4954.52 m with the conventional core drilling method (lifting drill string to recover core), the footage completed by each core drilling tool assembly can be found in Fig. 8.8, and the concrete conditions of drilling and well deviation shown in Table 8.2.

The drilling tool assembly with dual stabilizers and dual reamers (P4, P5) was mainly used in the pilot hole section; the long and full-hole (SK wireline core drilling string) drilling tool assembly with one stabilizer and dual reamers

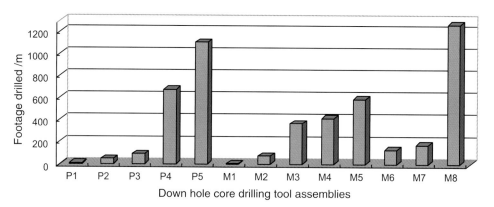

Fig. 8.8 Footage drilled by different down hole drilling tool assemblies

(M3), the drilling tool assembly with slick collar dual reamers (M4, M5), the drilling tool assembly with one stabilizer and dual reamers (M8) were mainly used in the main hole.

In spite of the formation conditions, the well deviation angles of the P4 and P5 drilling tool assemblies were basically stable, the well deviation angle of the other drilling tool assemblies were greatly variable. In the main hole section, in consideration of the need of controlling the probability of drilling tool sticking, the dual stabilizer drilling tool assembly (P5) was not used in the main hole, with the well deviation greatly variable.

8.2.2 Well Deviation Control in Non-core Drilling and Reaming Drilling

1. **Non-core drilling for hole opening**

In opening hole, a drilling tool assembly (φ 444.5 mm rock bit) with φ 177.8 mm collar (Fig. 8.8a) and φ 444.5 mm stabilizer, the collar was poured with lead for aggravation and centering to φ 406.4 mm, was used to drill to 28.62 m. From 28.62 to 60.78 m, a full-hole drilling tool was used (Fig. 8.8c). From 60.78 to 100.36 m, a pendulum drilling tool was used (Fig. 8.8d). During drilling, WOB was controlled to 40–100 kN, rotary table revolution was 30–60 r/min, drilling fluid pump displacement was 50 L/s. The φ 444.5 mm bit was used for non-core drilling at first hole diameter, the well deviation angle was less than 1° (refer to the Sect. 4.2 of Chap. 4).

2. **The first reaming drilling for the main hole**

In the pilot hole from 1720 to 1900 m, the well deviation angle decreased from 3.67° to 2.32°; from 1900 to 2000 m, the well deviation angle increased to 4.09° (Fig. 8.4). When guided reaming drilling to 1871.48 m, a rigid and flexible pendulum drilling tool was used for decreasing well deviation, by using its function of natural decrease of well deviation.

(1) The combination design on rigid and flexible pendulum drilling tool

In the pre-research project of "the research on hole deviation prevention technology", 3-D dynamic finite element method was taken to analyze and calculate the "deviation decreasing rate" of different rigid and flexible pendulum drilling tool assemblies, the recommended rigid and flexible pendulum drilling tool assembly consisted of φ 311.1 mm drill bit + two pieces of φ 203.2 mm collar + φ 311.1 mm spiral stabilizer + φ 158.8 mm single collar + φ 311.1 mm spiral stabilizer + six pieces of φ 177.8 mm collar + φ 158.8 mm collar. The anticipated result of the deviation decreasing rate of this drilling tool assembly can be found in Table 8.3.

(2) Situation of deviation decreasing reaming drilling with rigid and flexible pendulum drilling tool

From 1871.48 to 1871.99 m, SYZX-273 hydro-hammer rigid and flexible pendulum drilling tool with flat bottom button bit was used to drill 0.51 m, with WOB of 30 kN, rotary speed of 40 r/min, and pump displacement of 33 L/s; from 1871.99 to 1921.99 m, the rigid and flexible pendulum drilling tool with tri-cone bit was used to drill 20 and 30 m separately, with WOB of 35–60 kN, rotary speed of 35–55 r/min, and pump displacement of 31.5–35 L/s; from 1921.99 to 1975.00 m, KSC-203 hydro-hammer rigid and flexible pendulum drilling tool with rock bit was used to drill 53.01 m, with WOB of 35–55 kN, rotary speed of 40–65 r/min, and pump displacement of 30.8–33.9 L/s; from 1975.00 to 2028.00 m, SYZX-273 hydro-hammer rigid and flexible pendulum drilling tool with rock bit was used to drill 53.00 m, with WOB of 40–60 kN, rotary speed of 50–60 r/min and pump displacement of 31.8–35.4 L/s. The concrete situations of the drilling tool assemblies can be found in Sect. 4.2 of Chap. 4.

Because of reaming drilling in the existed small diameter borehole, hard rock, smaller pendulum force in slim and deviated hole section and many other reasons, deviation decreasing reaming drilling with rigid and flexible pendulum drilling tool was not successful.

8.2 Deviation Prevention Drilling Technology

Table 8.2 Core drilling footage and well deviation by different down hole drilling tool assemblies

Well section	Drilling tool	Well depth/m From	Well depth/m To	Footage drilled (m)	Initial well deviation (°) Well deviation angle	Initial well deviation Azimuth	End well deviation (°) Well deviation angle	End well deviation Azimuth	Well deviation conditions
PH	P1	101.00	112.95	11.80	0.68	325	0.94	343	The well deviation angles steadily increased. Drilling 0.15 m with rotary table
	P2	112.95	152.37	46.84	0.94	343	1.34	336	The well deviation angles steadily increased
		763.53	770.95		1.53	286	1.47	289	The well deviation angles were steady
	P3	152.37	233.27	92.47	1.34	336	1.87	328	The well deviation angles steadily increased and tended to be steady in the later period. 0.98 m was drilled by hydro-ham
		730.48	742.05		1.41	286	1.47	284	The well deviation angles were steady
	P4	233.27	730.48	673.94	1.87	328	1.41	286	The well deviation angles steadily increased and then slightly decreased. 0.24 m was drilled with rotary table and 82.75 m drilled by hydrohammer(P5)
		742.05	763.53		1.47	284	1.53	286	The well deviation angles were steady
		770.95	396.10		1.47	289	2.64	300	The well deviation angles were basically steady, but variable. 9.52 m were drilled by rotary table and 380.39 m by hydrohammer(P5)
	P5	1396.10	2046.54	1102.50	2.64	300	4.58	286	The well deviation angles were basically steady, but variable. 8.08 m were drilled by rotary table and 3 m by pure PDM(P4)
MH	M1	2046.54	2055.40	5.87	4.58	286	4.56	287	The well deviation angles were basically steady. 2.47 m were drilled by wire-line core drilling and 0.82 m drilled by milling shoe
	M2	2055.40	2144.05	76.12	4.56	287	5.57	282	The well deviation angles steadily increased. 8.96 m were drilled by wire-line core drilling. 0.30 m drilled by top drive, and 3.27 m drilled by hydro-hammer (M3)
	M3	2144.05	2519.64	367.53	5.57	282	8.83	283	The well deviation angles steadily increased and tended to be steady in the later period. 11.18 m were drilled with wireline core drilling and 0.15 m by milling shoe
	M4	2519.64	2936.90	417.26	8.83	283	15.73	307	The well deviation angles were steady and then sharply increased in later period
	M5	2936.90	2982.18	45.28	15.73	307			
MH-1C	M5	2974.59	3127.54	152.95	6.33	318	11.5	319	The well deviation angles sharply increased
	Correction 1	3127.54	3139.39		11.5	319	9.7	314	
	M5	3139.39	3171.28	31.89	9.7	314	9.87	312	The well deviation angles were basically steady
	Correction 2	3171.28	3191.28		9.87	312	9.25	316	
	M5	3191.28	3244.98	53.70	9.25	316	12.13	317	The well deviation angles sharply increased
	Correction 3	3244.98	3253.33		12.13	317	12.87	317	
	M5 3	253.33	3536.00	266.44	12.87	317	8.62	331	The well deviation angles gradually decreased. 15.39 m were drilled by rock bit non-core drilling and 0.84 m by milling shoe
	M6	3536.00	3665.87	129.87	8.62	331	7.4	330	The well deviation angles were basically steady
MH-2C	M5	3624.16	3664.69	40.53	6.92	62	7.15	60	The well deviation angles were
	M7	3664.69	3846.00	173.90	7.15	60	11.73	94	The well deviation angles sharply increased. 7.41 m were drilled by rotary table
	M8	3846.00	5118.20	1265.63	11.73	94	26.1	63	The well deviation angles sharply increased, then tended to be steady and slightly decreased at 4184 m deep, the well deviation angles gradually increased at 4408 m and then gradually decreased at 4947 m. 6.12 m were drilled by rock bit non-core drilling, and 0.45 m by milling shoe

Note 10.75 m which were drilled with core drilling in the first side-tracking drilling for deviation correction, 22.61 m drilled with wire-line core drilling, 17.99 m completed with rotary table core drilling, and 3.5 m by core drilling for test of "three-in one" drilling tool are not added up in this table

Table 8.3 The anticipated deviation decreasing rate of the rigid and flexible pendulum drilling tool

WOB (kN)	Rock threshold pressure (kN)	Well deviation angles (°)	Deviation decreasing rate (°/30 m)	Drilling fluid density (g/cm^3)
100	1.5	1	−1.80	1.12
		2	−1.86	
		3	−1.93	

3. **Non-core drilling in the core drilling section**

From 3516.66 to 3516.96 m in MH-1C hole section, because of the complicated hole conditions, a drilling tool assembly with φ158.8 mm rock bit and slick collar was used for drifting for 0.3 m. From 3520.91 to 3536.00 m, since the hole wall was not stable, φ158.8 mm rock bit was used. In the drilling tool assembly, besides φ120 mm collar, wire-line coring φ157 mm stabilizer and φ146 mm × 108 mm collar (Fig. 8.19d) were added to the assembly above the rock bit, and then the well deviation angle was basically steady and decreased slightly (Fig. 8.20).

In the MH-2C hole section from 3707.84 to 3712.55 m, φ158.8 mm rock bit was used for drifting for 4.71 m; from 3843.30 to 3846.00 m, φ158.8 mm rock bit was used for drifting for 2.7 m; from 4398.47 m to 4401.49 m, φ158.8 mm rock bit driven by PDM was used for drifting for 3.02 m; from 5072.06 m to 5075.16 m, φ158.8 mm rock bit was used for drifting for 3.10 m. 4 times of rock bit non-core drilling were taken as a technical measure because of the abnormality in the borehole, each time of drilling was short. As for the drilling tool assembly adopted, the core drilling tool assembly used in last roundtrip was adopted (refer to the Sect. 4.2 of Chap. 4).

In addition, in the process of drilling tool testing, φ158.8 mm rock bit was used for the non-core drilling for 3 roundtrips. The first roundtrip for drifting was for testing the "three-in-one" drilling tool and 7.66 m were completed with φ158.8 mm rock bit and slick drilling tool. The second and third roundtrips were completion-well drilling after the test of deviation correction with continuous deflector, φ158.8 mm rock bit and slick drilling tool assembly driven by PDM were used to drill 18.92 and 4.42 m respectively, the well deviation angle appeared a continuous decreasing trend.

4. **Non-core drilling in the hole section of side-tracking to avoid obstacle**

Well depth reached 3445.62 m after the second side-tracking in the main hole, approximately 220 m away from the coring depth of MH-1C (3665.87 m), for the purpose of reducing repeated core drilling, φ244.5 mm rock bit was used to drill to 3200 m deep and then φ193.7 mm technical casing was set.

From 3445.62 to 3542.73 m, a drilling tool assembly (Fig. 8.21e) with φ244.5 mm tri-cone bit + φ172 mm PDM + φ178 mm jar while drilling + φ214 mm stabilizer + ten pieces of φ177.8 mm collar + φ127 mm drilling rods was used to drill 97.11 m, with WOB of 55–70 kN, rotary speed of 149–166 r/min (PDM) + 15–25 r/min (rotary table), and pump displacement of 25.59–28.52 L/s. Before setting casing, a comprehensive logging indicated (Fig. 8.22) that well deviation angle decreased to 1° from 2° and then increased to 3°, and again decreased to 1° and then increased again to 2.5°, the azimuth increased in clockwise and varied within about 125°.

From 3542.73 to 3576.68 m, a drilling tool assembly with φ244.5 mm tri-cone bit + φ172 mm PDM + φ214 mm stabilizer + φ178 mm jar while drilling + ten pieces of φ177.8 mm collar + φ127 mm drilling rod was used to drill 33.95 m, with WOB of 55–65 kN, rotary speed of 149–161 r/min (PDM) + 12–15 r/min (rotary table), and pump displacement of 25.59–27.65 L/s, the well deviation angles increased to 5.5° from 2.5° and then tended to steady, the azimuth was steady at about 70° (Fig. 8.22).

From 3576.68 to 3600.00 m, a φ244.5 mm tri-cone bit drilling tool assembly (Fig. 8.2f) with slick collar was used to drill 23.32 m, with WOB of 100–110 kN, rotary speed of 45 r/min (rotary table), and pump displacement of 27.15–28.52 L/s. The well deviation angle increased 0.5°, the azimuth maintained basically steady (Fig. 8.22).

From 3600.00 to 3623.91 m, a φ215.9 mm tri-cone bit drilling tool assembly with fishing cup and slick collar (Fig. 8.21g) was used to drill 23.91 m, with WOB of 90–100 kN, rotary speed of 45 r/min (rotary table), and pump displacement of 27.15–27.65 L/s. The well deviation angle and the azimuth were both basically steady (Fig. 8.22).

From 3600.00 to 3620.00 m, a φ244.5 mm tri-cone bit drilling tool assembly with slick collar was used for reaming drilling for 20.00 m, with WOB of 30–70 kN, rotary speed of 45 r/min (rotary table), and pump displacement of 25.92 L/s.

8.3 Drilling Techniques for Deviation Correction

A drilling technology of advanced open hole was used in CCSD-1 Well, when necessary, it still needed to ream hole for casing. So in deviation correction, it was not allowed to leave any hidden trouble in the borehole. According to the borehole deviation and formations conditions, backfill side-tracking drilling and down hole deviation correction methods with dynamic drilling tool (PDM and MWD instrument) were used in CCSD-1 Well.

8.3.1 Side-Tracking Deviation-Correction Techniques

1. **Main technical difficult points**
   i. The formations mainly are gneiss, eclogite and amphibolite, with strong natural deviation build-up intensity, the rock is hard with drillability grade from 8 to 9.
   ii. The cement plug strength is much lower than the formation strength.
   iii. No well-developed technology and experiences for hard rock side-tracking drilling for reference.
2. **Side-tracking drilling deviation-correction technology**

(1) Selection of the side-tracking point
   i. Under the prerequisites for satisfying side-tracking drilling, side-tracking point should be as deep as possible in order to reduce repeated drilling.
   ii. To keep away from the very irregular and long larger-diameter well section, otherwise, it will be difficult to guarantee a good quality of cement plug and then the side-tracking drilling effect would be affected.
   iii. To select the rock formations with lower drillability by reference to the drilling and geological information in the old wells.
   iv. Selection of the side-tracking point should be beneficial to the control of the new well trajectory after the side-tracking drilling, to avoid azimuth changing again or large variation of the well trajectory.
   v. To fully use the beneficial conditions of the old well, the position with large borehole diameter should be selected as the side-tracking point in the regular well; and this can reduce the shoulder-building difficulty in the initial period of side-tracking drilling and will be advantageous to produce a new hole. On the top of the well section with large trajectory change, side-tracking drilling to the opposite direction of the well trajectory will be easily divorced from the old well.

(2) Selection of PDM

The PDM for kick-off has different forms of single bend, same direction double bend, reversed double bend and surface adjustable single bend house structure and so on, also the bend joint can be installed on the upper end of the straight PDM for kick-off purpose. To kick-off in hard rock formation, for selection of the kick-off PDM, the first thing that has to be considered is running-into guarantee that drilling tool can be run into the side-tracking position, and the second is the comprehensive consideration for kick-off intensity and the structure performance and so on. In the two times of side-tracking for CCSD-1 Well, a single bend PDM with simple structure and reliable performance was selected.

There was a structure bend angle (usually less than 3°) in the universal axis case of the single bend PDM, there was a nearly full hole lower reaming shell on the transmission shaft case, and a drill string stabilizer was installed above the bypass valve. The side-direction force of the drill bit increased remarkably as the bend angel increased, while the bit inclination angle decreased as the bend angle increased. The side-direction force of the drill bit decreased as the bend point position went up, the bit inclination angle increased as the bend point position went up. As the position of lower stabilizer moved up, both the side-direction force of the drill bit and bit inclination angle decreased remarkably. As the outer diameter of the lower stabilizer wore, the side-direction force of the drill bit remarkably decreased and the bit inclination angle increased. The increase of the upper stabilizer's diameter could make the side-direction force of the drill bit decrease, and bit inclination angle increased, but the influence of the upper stabilizer's diameter change was not bigger than the lower stabilizer's (Su Yinao 2001).

The 0.5–1° single bend 5LZ120 × 7 PDM was taken without installing the lower stabilizer, but the structure of welded cushion block was adopted.

(3) The MWD instrument

Deviation-correction of side-tracking drilling requires decreasing the well deviation angle, so the drilling tool face must be directionally controlled. In CCSD-1 Well a directional method of wire MWD instrument + directional key marking was used.

When the wire MWD instrument worked for directional drilling, the accessory equipment (the winch, cable passing swivel and oil pump, etc.) was complicated, and the working procedure was also complicated. In the shallow well section deviation-correction, the Kelly bar must be used to balance the reverse torque from the PDM, after drilling a single pipe, the MWD instrument must be pulled out of the well and single pipe would be connected and the direction would be fixed again.

(4) The selection for drill bits

The side cutting ability of drill bit is the main basis for selecting the drill bits for side-tracking deviation-correction, in hard rocks, the drill bits for side-tracking drilling deviation-correction available for selection include natural diamond surface set drill bits, impregnated diamond drill bits and rock bits. In side-tracking drilling for deviation-correction, abovementioned three kinds of drill bits were all used, and finally small diameter (φ140 mm) impregnated diamond drill bit was successful in the side-tracking drilling.

The side cutting edge of natural diamond surface set drill bit was sharper, the bit face and side surface could be designed into a large spherical transition shape so that the drilling tool could be easily run in, but under small WOB (small side force), the diamond grains at the side cutting edge would be easily polished.

The side cutting ability of impregnated diamond bit was weaker than that of surface set drill bit, but under small side force, the fine diamond grains could better grind hard rock. In order to easily drill a new hole, the bit bottom face and side surface needed a small spherical transition or rectangular transition, but the PDM with bend house could make the drill bit tend to one side, and thus in irregular well section slack-off would easily happen in running-in.

When rock bit was used for side-tracking drilling deviation-correction in hard rock, the gauge cutter (the outer row cutter) would be easy blunt and made the side cutting ability decrease.

(5) The side-tracking drilling diameter

Full gauge side-tracking drilling can drill out a new well once that will satisfy the follow-up core drilling, but in the practical applications, it may be restricted by the shape of the original old well. If the old well can continuously keep an even well diameter, the running-in will be smooth. But if there exists an abrupt changing diameter section in the old well, the drill bit will set on the rock at which the well diameter is oversized. In geological exploration, the slim diameter deviation-correction may encounter the above situation, to turn the drill string by manpower at the hole mouth can change the drilling tool bend direction in order for continuous running-in, but for oil drilling, it is difficult for the large diameter drill string be turned around by manpower. When slack-off is encountered during running-in, Kelly bar has to be connected and the rotary table operated for redressing, and this not only wastes time and power, but also cannot guarantee the drilling tool to go through after breaking out the Kelly bar. If the bottom side of the drill bit is designed into big spherical transition, the running-in difficulty will be decreased, but the sharpness of the bottom side edge is also relatively weakened, therefore, this will only be suitable for surface set diamond drill bit; since the side cutting ability of impregnate diamond bit is weaker, the bottom side should be nearly rectangular or small spherical transition, but this drill bit form will be difficult to run into the hole bottom. In addition, in the full gauge side-tracking drilling, drill bit tends to slide down along the old well trace. So in hard rock, it will be much difficult for the full gauge side-tracking drilling to drill out a new hole.

The bit diameter of small diameter side-tracking drilling should be one grade smaller than old well diameter. Bigger angle single bend PDM can be selected for drilling, and at the same time this would relatively increase the bit inclination angle and be apt to drill out a new hole. However, after side-tracking drilling, reaming drilling had to be carried out, and this not only made the working procedure complicated, but also brought about the risk of damaging cement wedge.

In side-tracking for deviation-correction, the full gauge side-tracking drilling program was selected, through 6 running-in trips (3 trips encountered slack-off in the hole), no successful result was obtained. Then the drill bit was changed into the small diameter φ 140 mm bit, side-tracking drilling was successful only in one running-in trip.

(6) WOB control

   i. The drillability grade of rock differs obviously from the drillability grade of the cement plug, to guarantee to drill a new hole with side-tracking drilling, WOB and feeding speed must be strictly controlled, especially in the initial period of side-tracking the drilling tool should be hanged up without WOB for drilling out the shoulder.
   ii. The relationship between WOB and ROP should be monitored strictly while drilling, before confirming that drill bit cut into the rock, WOB should not be exerted blindly onto the drill bit. When the drill bit cuts into the rock in certain proportion, ROP gradually slows down under certain WOB, and the sensitivity of pump pressure to WOB change becomes low.
   iii. Before confirming that area proportion of drill bit into rock reaches to a certain degree, the drilling tool should absolutely not be allowed to lift up in order to prevent WOB change from causing the change of the drilling tool face, which can make the hole with small rock shoulder already formed slips back into the old hole.
   iv. The proportion of drill bit into the rock can be judged from the appraisal result to the solid content of returned drilling fluid.

(7) Analysis to drilling chips

In side-tracking drilling deviation-correction, the MWD instrument was placed at 9 m above drill bit, since the 9 m footage for side-tracking drilling in hard rock would need 20 h, it would be very difficult to depend on the readings of the MWD instrument to judge if the new hole was formed. When the drill string didn't rotate and drilling under small WOB reached a certain time, since the friction force from the well wall, the display of ROP and surface WOB couldn't really reflect if the drill bit cut into the rock, and before the formation of new hole was confirmed the lowering-down and lifting-up of the drilling tool for eliminating the formation friction resistance should be avoided.

After the drilling fluid returned from well mouth conduit and was processed by oscillating screen, the solid grains over 175 mesh could be separated out, microscope analysis to the solid content of the samples could obtain the proportion of the cement chips and rock chips from the samples. So the analysis of the drilling chips (rock chips and cement chips) from the returned drilling fluid could become an important measure to judge if the drill bit cut into rock and the proportion into rock.

(8) Points for attention

  i. The readings of MWD instrument must be accurate, before the instrument was run into the well, it must be proofread.

  ii. The readings of the drilling tool face controlled within the design permission scope (±30°)must be guaranteed, once it exceeded the scope, the drilling tool face must be adjusted.

  iii. Since the drill string did not rotate, the axial resistance from the well wall to the drill string could not be eliminated in time, so after drilling for some time, WOB could not be transferred to the drill bit because of the well wall friction resistance, with the manifestation of slow ROP or even no penetration. Even though WOB was greatly increased, the pump pressure and the reverse torsion angle still did not show a change. When encountering this situation, it would need to lift-up and lower-down the drilling tool, after the well wall friction resistance was eliminated, drilling could continue.

  iv. After side-tracking with rock bit for a certain time, the side cutting edge would be worn flat, the kick-off ability would become greatly weak and even disappear. In the strong deviation building-up formations, if above situation was encountered and drilling still continued, not only a good effect could not be obtained, but also the hole trace would develop along the formation deviation building-up trend, at this time, the drilling tool had to be pulled out to change the drill bit at once.

8.3.2 Situation on Side-Tracking Drilling for Deviation-Correction

When the main hole was drilled to 2982.12 m, the well deviation angle increased from 8.29° (azimuth 294°) at 2760 m deep to 16.35° (azimuth 308°) at 2935 m deep, the well deviation quickly increased (Fig. 8.9), the deviation-correction was imperative. Three deviation-correction schemes were put forward: core drilling well-bottom deviation-correction, non-coring well-bottom deviation- correction and backfill side-tracking drilling for deviation-correction.

  i. Diamond core bit, short core barrel and PDM were used to directly correct the well deviation at the bottom through core drilling, this way had some advantages that core would not be lost and would be continuous. But this method also had some disadvantages that the deviation-correction ability was weak and experience was not ripe, under such strong formation deviation building-up trend, this method was not effective to correct the well deviation; furthermore, after the core drilling tool equipped with the bend house (bend couple) PDM, the bit axis was much off from the well center axis, it was difficult to run in the drilling tool, and at the same time, the well body quality of the drilled well could not be guaranteed.

  ii. Diamond non-core bit or rock bit was used to directly correct the well deviation at the well bottom; this method had better economy, but could lose some cores, and the well-body quality of the drilled well could not ensured.

  iii. Side-tracking drilling deviation-correction with backfill had some advantages which had higher reliability and no core lost and the well-body quality of the drilled well could not ensured, but it could make some repeated drilling work and thus economy was unsatisfactory.

Through comprehensive consideration, the method of side-tracking drilling deviation-correction with backfill was adopted.

1. **Selection of the side-tracking drilling point**

In the 2500–2750 m well section, the well deviation angle was basically steady between 8° and 9° (Fig. 8.9), the azimuth slowly increased, and the well diameter basically steady. In the 2740–2760 m well section, the well deviation angle naturally decreased (Fig. 8.10), and in the 2753–2756 m well section, the well diameter was oversized, with the maximum of 319 mm. But below 2760 m the well deviation angle naturally increased. The rock near 2749 m was garnet-plagiogneiss, in the MH120 (2742.43–2751.62 m) and MH121 (2751.62–2761.13 m) well sections, the average penetration rate of the two roundtrips completed by PDM hydro-hammer conventional core drilling were respectively 1.72 and 1.96 m/h, the core recovery were respectively 100.0 and 99.7 %, and the rock was basically intact (Fig. 8.11). Therefore, the side-tracking drilling point was selected at 2749 m depth.

Fig. 8.9 The variation curve of well deviation angle and dog-leg with the well depth in the MH well section

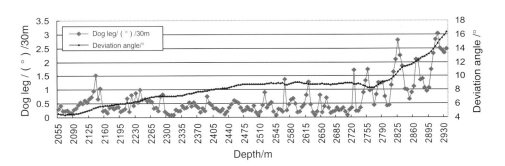

Fig. 8.10 The well deviation angle, well diameter and azimuth near the side-tracking drilling point

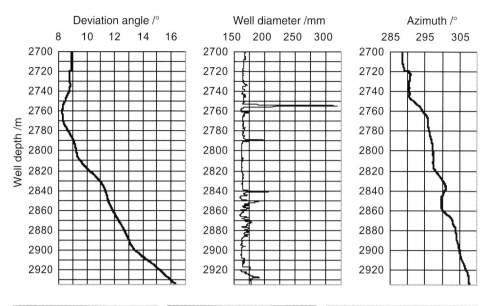

Fig. 8.11 Partial cores near the side-tracking drilling point

 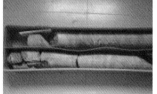 

2. **Building-up the manmade well-bottom**

The slick drilling tool assembly was used for injecting cement slurry (Table 8.4), after curing, rock bit was used for drill-off to 2748.00 m, core drilling was adopted again to recover the cement plug samples (Fig. 8.12), and the quality of cement plug in the side-tracking drilling point was checked. The details of the manmade well bottom building-up can be found in Sect. 4.2 of Chap. 4.

3. **Situation of side-tracking drilling for deviation-correction**

(1) Full gauge side-tracking drilling in 2749.00–2758.00 m

At first, φ 157 mm natural diamond surface set drill bit (Fig. 8.13a) was used for the full gauge side-tracking drilling; with the drilling tool assembly shown in Table 8.5. At beginning, smaller WOB (5 kN) was taken in order to deviate a new borehole, in the late period the WOB reached

Table 8.4 The basic data for injecting cement plug

Item	Parameter	Item	Parameter	Item	Parameter
Drilling fluid density in the well (g/cm^3)	1.04	The depth under the cement plug (m)	2982.18	The depth over the cement plug (m)	About 2750
Outside diameter of drill rod (mm)	89	Inside diameter of drill rod (mm)	70.3	The inner diameter of well wall (mm)	169
Cement quantity (t)	9.7 t	Thickening time (min)	300	Ratio of water to cement	0.38
Cement slurry density (g/cm^3)	1.97	Cement strength (MPa)	40	Cement type	JHG grade cement

Fig. 8.12 The cement plug sample in the side-tracking drilling point

Fig. 8.13 Natural diamond surface set drill bit. **a** Before use. **b** After drilling 3.05 m

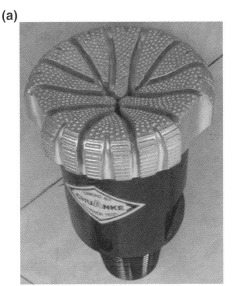

40 kN, but ROP was very low, even hardly advanced and then drill string was pulled up. The roundtrip footage was 3.05 m, penetration rate was 0.45 m/h, both the side cutting edges and outer edges of the drill bit wore seriously, but bit bottom edges hardly wore (Fig. 8.13b).

158.8 mm rock bit was used and the stabilizer was removed from the drilling tool assembly, the 1° single bend PDM (the drilling tool assembly shown in Table 8.5) was selected for drilling. In drilling operations, although WOB was lower, penetration rate was fairly fast. After drilling 4.29 m, the drill string was pulled up.

The φ 157 mm surface set diamond drill bit (impregnated bit was not delivered to the drill site) was again used, with the collar (the drilling tool assembly shown in Table 8.5) was removed from the drilling tool assembly to drill for one roundtrip.

In the former three drilling trips, the solid grains returned from the mud at the well mouth were basically cement grains, no new borehole was drilled out with side-tracking drilling (Table 8.6). Then 158 mm impregnated diamond drill bit (Fig. 8.14) was used with the side-tracking drilling tool assembly (shown in Table 8.5) for running-in, however, slack-off was encountered in all the three running-in at 2187 2118 and 2358 m respectively (Table 8.6).

(2) The small diameter side-tracking drilling in 2758–2771.35 m

φ 140 mm impregnated diamond drill bit (Fig. 8.15a) was used for small diameter side-tracking drilling, and the drilling tool assembly can be found in Table 8.7. In the initial drilling, WOB was strictly controlled and WOB feeding was steady and basically maintained to about 5 kN (Fig. 8.16). After drilling 6.7 m (the well depth was 2764.70 m, actual drilling time 25 h and 50 min), WOB was increased to

Table 8.5 The drilling tool assembly and the drilling parameters of full gauge side-tracking drilling in 2749.00–2758.00 m

Serial number	Drilling tool assembly	WOB (kN)	RPM (r/min)	Pump displacement (L/s)
1	φ 157 mm natural diamond surface set bit + 0.75° single bend 5LZ120 × 7 PDM + φ 120 mm directional sub + φ 104 mm non magnetic collar + φ 156 mm spiral stabilizer + φ 120 mm collar × 7 + φ 89 mm drilling rod	5–40	109–125	9.31–10.69
2	φ 158.8 mm rock bit + 1° single bend 5LZ120 × 7 PDM + φ 120 mm directional sub + φ 104 mm non magnetic collar + φ 120 mm collar × 7 + φ 89 mm drilling rod	5–8	111	9.48
3	φ 157 mm natural diamond surface set bit + 1° single bend 5LZ120 × 7 PDM + φ 120 mm directional sub + φ 104 mm non magnetic collar + φ 89 mm drilling rod	3–10	95–97	8.10–8.27
4	φ 158 mm impregnated diamond bit + 0.75° single bend 5LZ120 × 7 PDM (adding 11 mm thick welded pad block) + φ 120 mm directional sub + φ 104 mm non magnetic collar + φ 156 mm spiral stabilizer + φ 120 mm collar × 7 + φ 89 mm drilling rod			
5	φ 158 mm impregnated diamond bit + 0.75° single bend 5LZ120 × 7 PDM (adding 11 mm thick welded pad block)			
6	φ 120 mm directional sub + φ 104 mm non magnetic collar + φ 89 mm drilling rod			

Table 8.6 Situations of the first six side-tracking drilling roundtrips

Serial number	Deviation-correction bit	Well section (m)	Footage drilled (m)	Actual drilling time (h)	ROP (m/h)	Deviation-correction results
1	φ 157 mm surface set diamond bit	2749.00–2752.05	3.05	6.83	0.45	Old well
2	φ 158.8 mm tri-cone bit	2752.05–2756.34	4.29	5.55	0.77	Old well
3	φ 157 mm surface set diamond bit	2756.34–2758	1.66	19.99	0.08	Old well
4	φ 158 mm impregnated diamond bit	Encountered slack-off at 2187 m				The drilling tool could not be run into the well bottom
5	φ 158 mm impregnated diamond bit	Encountered slack-off at 2218 m				The drilling tool could not be run into the well bottom
6	φ 158 mm impregnated diamond bit	Encountered slack-off at 2358 m				The drilling tool could not be run into the well bottom

Fig. 8.14 158 mm impregnated diamond bit

20 kN, the ROP didn't increase quickly (Fig. 8.16), and this showed that the drill bit had most cut into the rock. In the later period of drilling, the solid grains in the returned mud from well mouth were basically cuttings. When drilling to 2769.59 m, WOB was increased, however, without any penetration, the drilling tool was lifted up for a certain height and could not be lowered to the original position, and then the drilling tool was pulled out. The dead point of the drill bit center enlarged (Fig. 8.15b), and a blind hole of φ 20 mm × 20 mm was worn out, the diamond layer of the outer side cutting edge was basically worn flat. The footage drilled was 11.59 m; and penetration rate was 0.28 m/h.

The MWD instrument was used, the well deviation angle began to decrease from 8.2° at the beginning of side-tracking drilling to 6.2° (Fig. 8.17), the azimuth was steady between 291–294° (the old well azimuth in this well section was 292–296°), the overall angle building-up deviation rate was 0.3(°)/m. The total horizontal displacement of the new and old wells is shown in Fig. 8.18, distance between the new and

Fig. 8.15 Drill bits for small diameter side-tracking drilling. **a** φ 140 mm impregnated bit. **b** φ 140 mm impregnated bit after side-tracking drilling. **c** φ 157 mm reaming bit with guide

8.3 Drilling Techniques for Deviation Correction

Table 8.7 The drilling tool assembly and the drilling parameters of the small diameter side-tracking drilling in 2758.00–2771.35 m

Serial number	Drilling tool assembly	WOB (kN)	RPM (r/min)	Pump displacement (L/s)
7	φ 140 mm impregnated diamond bit + 1° single bend 5LZ120 × 7 PDM (adding 6 mm thick welded pad block) + φ 120 mm directional sub + φ 104 mm non magnetic collar + φ 156 mm spiral stabilizer + φ 89 mm drill rod	3–25	97–122	8.79–10.34
8	φ 157 mm diamond bit with φ 110 mm pilot + 4LZ120 × 7 PDM + φ 120 mm collar × 3 + φ 89 mm drill rod φ 157 mm coring bit + φ 157 mm reamer + φ 139.7 mm	3–10	161–173	8.9–9.48
9	core barrel × 4.59 m + φ 157 mm reamer + upper joint + 4LZ120 × 7 PDM + φ 120 mm collar × 3 + φ 89 mm drill rod	15–25	173–183	9.48–10.00

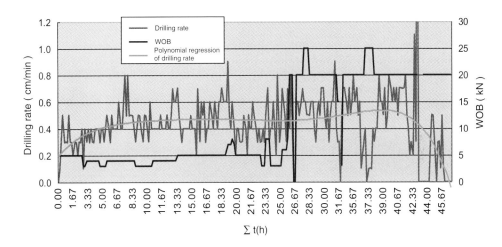

Fig. 8.16 Curve of side-tracking drilling process of φ 140 mm impregnated drill bits

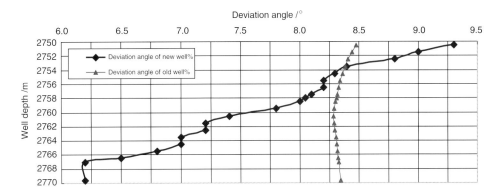

Fig. 8.17 The well deviation angles of the new and old wells (the old well deviation angle data was from the logging and the new well deviation angle data was from the MWD)

old well axis was 253.4 mm at the depth of 2769.59 m, the rock wedge thickness between the new and old wells was about 101 mm (the old well diameter was 164.7 mm at this point); distance between the new and old wells axis at 2767 m depth was about 157.5 mm, the new and old wells were basically separated (157.5 − 164.5/2 − 140/2 = 5.25 mm).

The φ 157 mm impregnated bit (Fig. 8.15c) with pilot was used for reaming drilling from 2758.00 to 2769.19 m, with the average penetration rate of 1.24 m/h. The drilling tool assembly is shown in Table 8.7.

The φ 157 mm impregnated diamond drill bit was used for core drilling from 2769.19 to 2771.35 m, with the drilling tool assembly shown in Table 8.7, the footage drilled was 2.16 m, penetration rate was 0.48 m/h, and core recovery was 81.5 %. The upper core recovered in the core drilling is shown in Fig. 8.19, from which we can see that the central positions of the non-core bit and the coring bit had deviated much at the end point of side-tracking drilling, the core drilling decreased the trend that well continued to deviate and reached the purpose of alleviating the dog-leg degree.

(3) Angle dropping drilling with rock bit in 2771.35–2942.11 m

φ 158.8 mm rock bit and 0.5° single bend PDM were used for angle dropping drilling at 2771.35–2797.01 m (the drilling tool assembly is shown in Table 8.8), with the footage drilled of 25.66 m and penetration rate of 0.94 m/h.

Fig. 8.18 Comparison of horizontal displacement of new and old wells

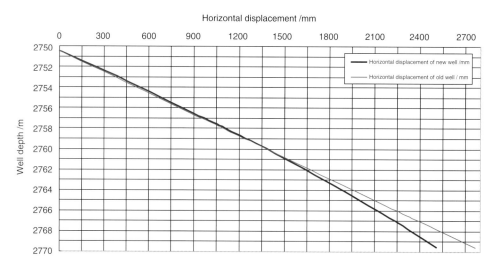

Fig. 8.19 The core recovered at the end point of side-tracking drilling

The MWD instrument in working indicated that there was no much change of the well deviation angle, and the azimuth angle decreased by about 15° (Fig. 8.20).

φ158.8 mm rock bit and 0.75° single bend PDM (adding 16.5 mm thick pad block) were used for angle dropping drilling at 2797.01–2812.03 m (drilling tool assembly is shown in Table 8.8), with the footage drilled of 15.02 m and the penetration rate of 1.09 m/h. The MWD instrument in working indicated that the well deviation angle decreased by 2° and the azimuth angle decreased by about 5° (Fig. 8.20).

φ158.8 mm rock bit and 0.5° single bend PDM were used for angle dropping drilling at 2812.03–2850.01 m (the drilling tool assembly is shown in Table 8.8), with the footage drilled of 37.98 m and the penetration rate of 1.01 m/h. The MWD instrument in working indicated that the well deviation angle began to decrease 1° and gradually increased to 3.3° at the late period, the azimuth began to decrease 15° and tended to be steady at the late period (Fig. 8.21).

φ158.8 mm rock bit and 0.75° single bend PDM (adding 11 mm thick pad block) were used for double rotary drilling (PDM with 121 r/min + the rotary table with 20 r/min) at 2850.01–2866.40 m, with the footage drilled of 16.39 m and the penetration rate of 1.04 m/h.

φ158.8 mm rock bit and 0.75° single bend PDM (adding 11 mm thick pad block) were used for angle dropping drilling at 2866.40–2888.79 m (drilling tool assembly is shown in Table 8.8), the footage drilled was 22.39 m and the penetration rate was 1.10 m/h. The total footage drilled was 38.78 m and the average penetration rate was 1.07 m/h. The MWD instrument in working showed that the well deviation angle was steady at the beginning and gradually increased to 4.8° at the late period, the azimuth was also basically steady at the beginning and gradually increased to 280° (Fig. 8.21) at the late period.

φ158.8 mm rock bit and 0.75° single bend PDM (adding 11 mm thick pad block) were used for angle dropping drilling (the drilling tool assembly is shown in Table 8.8) at both the two well sections of 2888.79–2907.70 m and 2907.70–2922.23 m, with the footages drilled of 18.91 and 14.53 m respectively, and the penetration rate of 0.76 and 0.85 m/h respectively. The MWD instrument in working showed that the well deviation angle increased instead of decrease, the azimuth angle increased (Fig. 8.22).

φ158.8 mm rock bit and 1° single bend PDM (adding 17 mm thick pad block) were used for angle dropping drilling (the drilling tool assembly is shown in Table 8.8) in 2922.23–2942.11 m, with the footages drilled of 19.88 m and the penetration rate of 0.93 m/h. The MWD instrument

8.3 Drilling Techniques for Deviation Correction

Table 8.8 The drilling tool assembly and drilling parameters at the second stage of side-tracking drilling in 2758.00–2771.35 m

Serial number	Drilling tool assembly	WOB (kN)	RPM (r/min)	Pump displacement (L/s)
10	φ158.8 mm rock bit + 0.50° single bend 5LZ120 × 7 PDM + φ120 mm directional sub + φ104 mm non magnetic collar + φ120 mm collar × 7 + φ89 mm drill rod	15–20	111	9.48
11	φ158.8 mm rock bit + 0.50° single bend 5LZ120 × 7 PDM (adding 16.5 mm thick pad block) + φ120 mm directional sub + φ104 mm non magnetic collar + φ120 mm collar × 7 + φ89 mm drill rod	10–30	111	9.48
12	φ158.8 mm rock bit + 0.50° single bend 5LZ120 × 7 PDM + φ120 mm directional sub + φ104 mm non magnetic collar + φ120 mm collar × 7 + φ89 mm drill rod	5–25	109–111	9.31–9.48
13	φ158.8 mm rock bit + 0.50° single bend 5LZ120 × 7	10–40	109–121	9.31–10.34
14	PDM (adding 11 mm thick pad block) + φ120 mm directional sub + φ104 mm non magnetic collar + φ120 mm collar × 7 + φ89 mm drill rod	10–35	109–111	9.31–9.48
15	φ158.8 mm rock bit + 0.50° single bend 5LZ120 × 7	5–20	111–121	9.48–10.34
16	PDM (adding 17 mm thick pad block) + φ120 mm directional sub + φ104 mm non magnetic collar + φ120 mm collar × 7 + φ89 mm drill rod	5–20	109–111	9.31–9.48

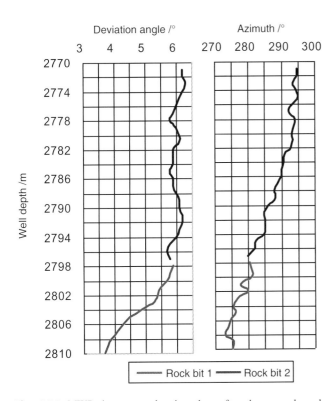

Fig. 8.20 MWD instrument logging data after the second angle dropping drilling with rock bit

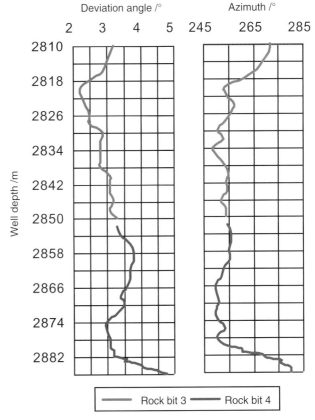

Fig. 8.21 MWD instrument logging data after the fourth angle dropping with rock bit

in working showed that the well deviation angle slightly deceased and the azimuth was basically steady (Fig. 8.22).

(4) The angle dropping drilling of impregnated diamond drill bit in 2942.11–2974.59 m

φ158 mm impregnated diamond drill bit and 0.75° single bend PDM (welded with 17 mm thick pad block) were used for running in. When drilling tool was run into 2350 m deep, slack-off was encountered, drilling-off was carried out and pump blockage often occurred, and then the drilling tool was pulled out when drilling off to 2364.55 m.

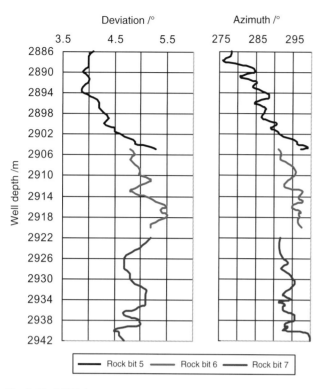

Fig. 8.22 MWD instrument logging data after angle dropping drilling

The φ158.6 mm hole rectifier (Fig. 8.23) drilling tool assembly (Table 8.9) was used for rectifying-hole drilling from 2374.22 to 2942 m.

φ158 mm impregnated diamond drill bit (Fig. 8.24) and 0.75° single bend PDM (welded with 17 mm thick pad block) were used for angle dropping drilling (the drilling tool assembly can be found in Table 8.9) from 2942.11 to 2952.96 m, when adjusting the drilling tool face and lifting-

Fig. 8.23 φ158 mm hole rectifiers

up with 90 t force (the drilling tool weighs 60 t) and pressing-down with 6 t couldn't release the sticking. Lifting-up with 100 t for many times and lowering-down with whole drilling tool hanging weight couldn't release the sticking, rotating the drilling tool and mud circulation was normal. Lifting-up with 102 t and pressing-down with 8 t in a trice and restoring the original whole drilling tool hanging weight, the bit sticking was released. The footage drilled was 10.85 m, with penetration rate of 0.41 m/h. The MWD instrument in working showed that the well deviation angle didn't decrease but increased, and the azimuth increased by 10° (Fig. 8.25).

158 mm impregnated diamond drill bit (Fig. 8.24) + 1° single bend PDM (welded with 17 mm thick pad block) + 156 mm upper stabilizer (the drilling tool assembly can be found in Table 8.9) were used for running in, when the drilling tool reached to 2 m from the well bottom, the resistance force began to increase, but the pump pressure was normal. The drilling tool was lifted up repeatedly with 4–17 t force and lowered down repeatedly with 4–8 t force to 1 m from the well bottom, but it was very difficult to run to the well bottom, and then the drilling tool was pulled out.

158 mm impregnated diamond drill bit + 1° single bend PDM (welded with 17 mm thick pad block) were used for angle dropping drilling (the drilling tool assembly can be found in Table 8.9, the upper stabilizer was removed from the drilling tool assembly) from 2952.96 to 2956.91 m, when drilling to 2956.91 m, to lift up and lower down the drilling tool encountered the sticking, and the pump pressure was over–pressured, and then the drilling tool was pulled out. The small core (Fig. 8.26) was recovered out with the drilling tool; the footage drilled was 3.95 m, with penetration rate of 0.34 m/h.

140 mm impregnated diamond drill bit (Fig. 8.15a) + 1° single bend PDM were used for the angle dropping drilling (the drilling tool assembly can be found in Table 8.9) from 2956.91 to 2961.91 m, when drilling to 2961.91 m, to lift up and lower down the drilling tool encountered sticking, and the pump pressure was over–pressured, and then the drilling tool was pulled out. The half matrix of the drill bit fell off to the well bottom (Fig. 8.27), the footage drilled was 5.00 m, with penetration rate of 0.48 m/h.

157 mm impregnated diamond drill bit with pilot (Fig. 8.15c) was used for reaming drilling from 2956.91 to 2761.03 m, the drilling tool assembly can be found in Table 8.9, with the average penetration rate of 1.03 m/h.

158.5 mm hole rectifier (Fig. 8.23) drilling tool assembly (Table 8.9) was used for rectifying-hole drilling from 2495 to 2961.38 m.

157 mm impregnated core bit (the drilling tool assembly is shown in Table 8.9) was used for running in, the drilling-off started at 6 m from the well bottom and the blockage and drill pipe sticking happened at 0.37 m from the well bottom. Redressing was carried out repeatedly to 2961.85 m and then

Table 8.9 The drilling tool assembly and drilling parameters of second side-tracking drilling in 2758.00–2771.35 m

Serial number	Drilling tool assembly	WOB (kN)	RPM (r/min)	Pump displacement (L/s)
17	158 mm impregnated diamond bit + 0.75° single bend 5LZ120 × 7 PDM (welded with 17 mm thick pad block) + 120 mm directional sub + 104 mm non magnetic collar + φ 120 mm collar × 7 h 89 mm drill rod	When running into 2350 m, the resistance force was encountered, and then redressed to 2364.55 m, the drilling tool was pulled out		
18	158.6 mm hole rectifier + 139.7 mm core barrel × 3.92 m + 158.6 mm hole rectifier + 139.7 mm core barrel × 3.92 m + 158 mm reamer + 4LZ120 × 7 PDM + 120 mm collar × 7 h 89 mm drill rod		160–183	8.76–10.00
19	158 mm impregnated diamond bit + 0.75° single bend 5LZ120 × 7 PDM (welded with 17 mm thick pad block) + 120 mm directional sub + 104 mm non magnetic collar + 120 mm collar × 7 h 89 mm drill rod	10–30	119	10.17
20	158 mm impregnated diamond bit + 1° single bend 5LZ120 × 7 PDM (welded with 17 mm thick pad block) + 120 mm directional sub + 104 mm non magnetic collar + 156 mm spiral stabilizer + 120 mm collar × 7 h 89 mm drill rod	The resistance force increased at 2 m from well bottom, the drilling tool was lifted up and lowered down repeatedly at 1 m from well bottom, but it was difficult for the drilling tool to run in, and then the drilling tool was pulled out		
21	158 mm impregnated diamond bit + 1° single bend 5LZ120 × 7 PDM (welded with 17 mm thick pad block) + 120 mm directional sub + 104 mm non magnetic collar + 120 mm collar × 7 h 89 mm drill rod	5–20	119	10.17
22	140 mm impregnated diamond bit + 1° single bend 5LZ120 × 7 PDM + 120 mm directional sub + 104 mm non magnetic collar + 120 mm collar × 7 h 89 mm drill rod	10–15	121	10.34
23	157 mm diamond bit with 110 mm pilot + 4LZ120 × 7 PDM + 120 mm jar while drilling + 120 mm collar × 7 h 89 mm drill rod		142–192	7.76–10.51
24	158.5 mm hole rectifier + 139.7 mm core barrel × 3.92 m + φ 158.6 mm hole rectifier + 139.7 mm core barrel × 3.92 m + 157.3 mm reamer + 4LZ120 × 7 PDM + 120 mm jar while drilling + 120 mm collar × 7 h 89 mm drill rod		173–186	9.48–10.17
25	157 mm coring bit + 157 mm reamer + 139.7 mm core barrel × 4.59 m + 157 mm reamer + swivel type joint + YZX127 hydro-hammer + 4LZ120 × 7 PDM + 120 mm jar while drilling + 120 mm collar × 7 h 89 mm drill rod	When redressing off to 0.37 m from the well bottom, the pump pressure was over-pressured and sticking happened. Redressing repeatedly to 2961.85 m, and then the drilling tool was pulled out		
26	156 mm milling shoe with fishing cup + 120 mm jar while drilling + 120 mm collar × 7 h 89 mm drill rod	20–30	30	5.69–6.04
27	157 mm coring bit + 157 mm reamer + 139.7 mm core barrel × 4.59 m + 157 mm reamer + swivel type joint + YZX127 hydro-hammer + 4LZ120 × 7 PDM + 120 mm jar while drilling + 120 mm collar × 7 h 89 mm drill rod	10–15	186	10.17
28	157 mm coring bit + 157 mm reamer + 139.7 mm core barrel × 4.59 m + 157 mm reamer + swivel type joint + YZX127 hydro-hammer + 4LZ120 × 7 PDM + 120 mm jar while drilling + SK157 stabilizer hSK146 wire-line coring collar × 15 h 89 mm drill rod	10–20	173	9.48
29	140 mm impregnated diamond bit + 1° single bend PDM + 120 mm jar while drilling + 120 mm directional sub + 104 mm non magnetic collar + 120 mm collar × 7 h 89 mm drill rod	Starting the pump when drilling tool was 5 m from well bottom, the drilling tool was lifted up and lowered down but the pump pressure was still higher. The drilling tool below the directional sub was blocked by sunk sand		
30		10–25	101–111	8.62–9.48

the drilling tool was pulled out. Ten pieces of the bit matrix fell off to the well bottom.

156 mm milling shoe with fishing cup was used to mill the well bottom (the drilling tool assembly is shown in Table 8.9), from 2961.68 to 2961.91 m, circulating mud and static-fishing were conducted respectively for four times. After milling from 2961.91 to 2961.95 m, the drilling tool was pulled out. The outer diameter was 155.5 mm; the small pieces of blocks fished out were 1.5 kg.

157 mm impregnated core bit was used for 2 roundtrips (the drilling tool assembly is shown in Table 8.9) from 2961.91 to 2970.54 m, with the footage drilled of 5.00 and 3.59 m respectively, penetration rate of 2.02 and 1.08 m/h respectively, and core recovery of 52.0 and 45.1 % respectively.

Fig. 8.24 158 mm impregnated bit

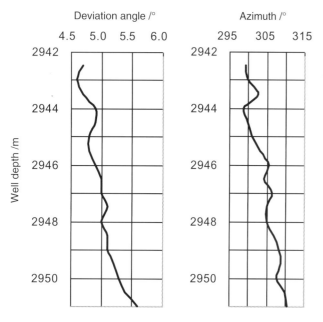

Fig. 8.25 MWD logging data after angle dropping drilling

Fig. 8.26 The small core recovered

140 mm impregnated diamond drill bit (Fig. 8.15a) + 1° single bend PDM (the drilling tool assembly is shown in Table 8.9) were used for running in, the pump was started at 5 m from the well bottom, the pump pressure rose quickly to 8.5 MPa, the drilling tool was lifted up and lowered down, but the pump pressure didn't return to the normal pressure, and then the drilling tool was pulled out. The part of the drilling tool under the directional sub was blocked by the sunken sand.

140 mm impregnated diamond drill bit (Fig. 8.15a) + 1° single bend PDM (the drilling tool assembly is shown in Table 8.9) were used for angle dropping drilling from 2970.54 to 2974.59 m, when drilling to 2974.59 m deep, the electricity power was broken because of the thunder storm, the deviation survey instrument broke down and then the drilling tool was pulled out. The footage drilled was 4.05 m, with penetration rate of 0.3 m/h.

Till then, the operation of the first side-tracking drilling (deviation correction) of the main hole was completed, the well deviation angles and azimuth angles of the new and old wells before and after the side-tracking drilling can be found in Fig. 8.28.

4. **The results of side-tracking drilling for deviation-correction**

In side-tracking drilling for deviation-correction 15 roundtrips were drilled (in another 6 roundtrips slacking-off was encountered when running in the drilling tool), the total footage completed was 215.20 m, with the average penetration rate of 0.69 m/h. Three types of single bend 5LZ120 × 7 PDM with 0.5°, 0.75°, and 1° angle were used, without lower stabilizer on the transmission axis, some PDM were welded with pad blocks with different thickness (6, 11, 16.5, and 17 mm), and upper stabilizer was seldom used. Eight structure types of single bend 5LZ120 × 7 PDM were used in the 15 drilling roundtrips (see Table 8.10; Fig. 8.29).

The bits used were 157 mm natural diamond surface set drill bit, 158.8 mm rock bit and impregnated diamond drill

8.3 Drilling Techniques for Deviation Correction

Fig. 8.27 140 mm impregnated drill bit with matrix fell off

bit (Table 8.11; Fig. 8.30), among which, the impregnated drill bits had 3 sizes of 158 and 140 mm. For the impregnated diamond drill bits, two structures of crushing face and center hole were adopted at drill bit bottom face center to overcome the central "dead point". At drill bit bottom face and side cutting edge, small (nearly square) and large spherical arcs transition structures were designed.

The successful drilling tool assembly for side-tracking drilling was 140 mm impregnated diamond drill bit + 1° single bend 5LZ120 × 7 PDM (welded with 6 mm pad block) + φ120 mm directional sub + 104 mm non magnetic collar + 156 mm spiral stabilizer + 89 mm drill rod (2758.00–2769.59 m well section).

The successful drilling tool assembly for angle dropping was 158.8 mm rock bit + 0.75° single bend 5LZ120 × 7 PDM (welded with 16.5 mm pad block) + 120 mm directional sub + 104 mm non magnetic collar + 120 mm collar × 7 + 89 mm drill rod (2797.01–2812.03 m well section).

From the hard rock and 157 mm diameter side-tracking drilling (deviation correction) we can see that: (i) the slim diameter impregnated diamond drill bits should be used for side-tracking drilling new well; (ii) the structural bend angle of the single bend PDM should not be small to meet the requirement of running-in, for installing the lower stabilizer, the addition of welded pad block structure was more reasonable; (iii) when well deviation angle is decreased to the small deviation angle, it would be inadvisable to continue decreasing the deviation angle, and the measures of stabilizing the deviation should be taken; and (iv) in the hard and strongly creating-deviation formations, it would not be suitable to correct the well deviation in a long well section.

8.3.3 Deviaton Correction at the Well Bottom of MH-1C Well Section

After the completion of the first side-tracking drilling (deviation correction), the well deviation angle developed quickly with core drilling from 2974.59 to 3127.54 m, the deviation angle increased from 6.33° to 11.51° (the maximum 11.88°) and the azimuth angle was basically steady at 320°. So, the single bend PDM (DST wire MWD instrument) was used for well bottom deviation correction drilling for three times.

1. **Drilling for deviation correction from 3127.54 to 3139.39 m**

The drilling tool assembly (Fig. 8.19b) with 140 mm impregnated diamond drill bit + 1° single bend 5LZ120 × 7 PDM + 120 mm jar while drilling + the directional sub + 104 mm non magnetic collar × 1 + 120 mm collar × 7 was used for deviation correction drilling in granite-gneiss rock for 11.85 m, with WOB of 10–40 kN, RPM of 111–129 r/min, and pump displacement of 9.48–11.03 L/s.

157 mm diamond drill bit with 110 mm guide driven by PDM was used for reaming drilling to 3138.99 m, with WOB of 10–12 kN, RPM of 176–198 r/min, and pump displacement of 9.65–10.86 L/s.

The logging data showed that (Fig. 8.20) the well deviation angle decreased from 11.51° to 9.70°, the azimuth angle decreased from 318.8° to 313.8°, and the full bending angle was 2.03°.

2. **The deviation correction drilling from 3171.28 to 3191.28 m**

The drilling tool assembly (Fig. 8.19c) with 158.8 mm rock bit + 1° single bend 5LZ120 × 7 PDM + the directional sub + 104 mm non magnetic collar × 1 + 120 mm collar × 7 was used for deviation correction drilling in amphibolite and plagiogneiss for 20.00 m, with WOB of 25–35 kN, RPM of 115–125 r/min, and pump displacement of 9.82–10.69 L/s.

The logging data showed (Fig. 8.20) that the well deviation angle decreased from 9.87° to 9.25°, the azimuth angle increased from 312.3° to 316.1°, and the full bending angle was 0.88°.

3. **The deviation correction drilling from 3244.98 to 3253.33 m**

The drilling tool assembly (Fig. 8.19b) with 140 mm impregnated diamond drill bit + 1° single bend 5LZ120 × 7 PDM + 120 mm jar while drilling + the directional sub + 104 mm non magnetic collar × 1 + 120 mm collar × 7 was used for deviation correction drilling in amphibolite and granite- gneiss for 8.35 m, with WOB of 10–75 kN, RPM of 117–132 r/min, and pump displacement of 10–11.2 L/s.

157 mm diamond drill bit with 110 mm guide driven by PDM was used for reaming drilling to 3252.86 m, with WOB of 10–15 kN, RPM of 183 r/min, and pump displacement of 10 L/s.

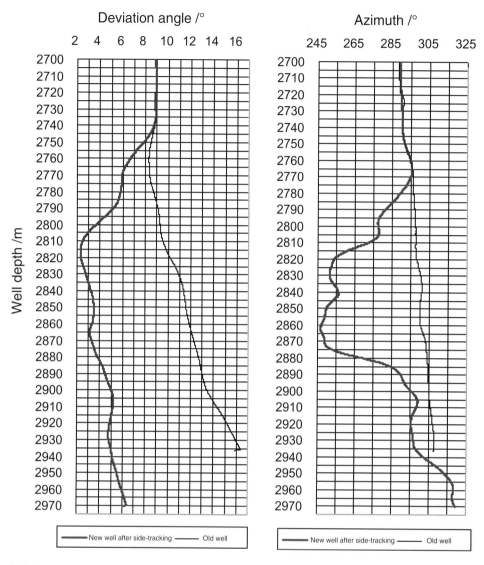

Fig. 8.28 The well deviation angles and azimuth angles of the new and old wells after the side-tracking drilling for deviation correction (engineering logging data)

Table 8.10 The roundtrip numbers and footage situations of different structures of the single bend PDM for the first side-tracking drilling (deviation correction) in the main hole

PDM structure	Roundtrip	Footage (m)
1° single bend 5LZ120 × 7 PDM	4	15.00
1° single bend 5LZ120 × 7P DM (welded with 6 mm pad block)	1	11.59
1° single bend 5LZ120 × 7 PDM (welded with 17 mm pad block)	2	23.83
0.75° single bend 5LZ120 × 7 PDM	1	3.05
0.75° single bend 5LZ120 × 7 PDM (welded with 11 mm pad block)	3	72.22
0.75° single bend 5LZ120 × 7 PDM (welded with 16.5 mm pad block)	1	15.02
0.75° single bend 5LZ120 × 7 PDM (welded with 17 mm pad block)	1	10.85
0.5° single bend 5LZ120 × 7 PDM	2	63.64
Total	15	215.2

8.3 Drilling Techniques for Deviation Correction

Fig. 8.29 The footage and roundtrip numbers drilled by single bend 5LZ120 × 7 PDM

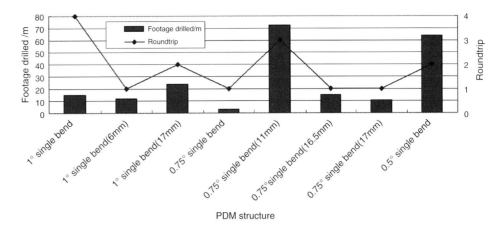

Table 8.11 The situations of drill bits application of the first side-tracking drilling (deviation correction) in the main hole

Bit type		Roundtrip	Footage drilled (m)	Average footage drilled per roundtrip (m)	Actual drilling time (h)	ROP (m/h)
157 mm natural diamond surface set drill bit		2	4.71	2.36	26.82	0.18
158.8 mm rock bit		8	175.05	21.88	184.01	0.95
Impregnated diamond bit	140 mm impregnated bit	3	20.64	6.88	65.31	0.32
	158 mm impregnated bit	2	14.80	7.40	37.86	0.39
	Sub total	5	35.44	7.09	103.17	0.34
Total		15	215.20	14.35	314.00	0.69

Fig. 8.30 Drill bit application conditions of the first side-tracking drilling (deviation correction) in the main hole

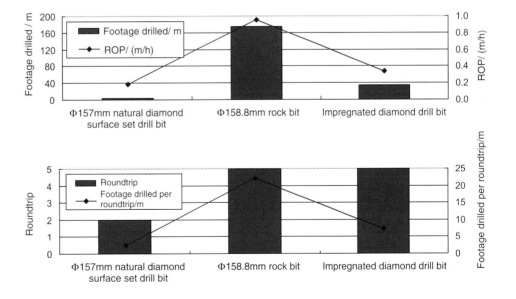

The logging data showed (Fig. 8.20) that the well deviation angle increased from 12.13° to 12.87°, the azimuth angle was steady at 317°, and full bending angle was 0.74°. Although the logging data showed that the well deviation angle during deviation correction drilling increased, yet in core drilling in follow-up granite-gneiss rock, the well deviation angle gradually decreased, the deviation correction drilling interfered the natural bending regular-pattern of the well and changed the well deviation trend.

8.4 Side-Tracking Drilling for Bypassing Obstacles

When core drilling of the second phase (MH-1C) in the main hole advanced to 3665.87 m, the upper pin thread of the lower reaming shell broke, and the diamond drill bit and the reaming shell fell to the well bottom, through many times of fishing and milling, the fallen fishes were not treated very well, and at the same time, the formation collapsed and fell off seriously, thus the accident handling was more difficult. Under these circumstances, reaming drilling to the top of the fallen fishes was planned to cover the fallen fishes, but in the process of pulling up the 193.7 mm moving casing, the 72 pieces of the elastic steel sheet of the elastic stabilizer of the casing fell off in the well, and during reaming drilling, three guide heads all broke and fell off, and well condition became more complicated. So, side-tracking drilling to bypass the fallen fishes was taken to go through the complicated well section (the reaming depth of 244.5 mm diameter was 3525.18 m).

8.4.1 Selection of Side-Tracking Drilling Tool

The diameter of the old well for side-tracking drilling to bypass the obstacle was 244.5 mm, according to the experience of side-tracking drilling for deviation correction, 215.9 mm diameter impregnated diamond drill bit was to be used for side-tracking drilling the new well. After the new well was successfully formed by side-tracking drilling, then 215.9 mm rock bit could be used to decrease the well deviation.
1. **Single bend PDM**
   The 1.75° single bend 5LZ172 × 7 PDM, with the working torque of 5 grades from 4350 to 7548 Nm, was used with 208 mm straight stripe and asymmetric lower stabilizer, with the theoretical building-up rate of 13.3(°)/30 m in 215.9 mm well.
2. **MWD instrument**
   The wireless MWD instrument (Fig. 8.31) was used. When the wireless MWD instrument worked for the directional drilling, the working procedure was much simpler, but the direct cost was more than that of the wire MWD instrument.

3. **Impregnated diamond drill bit**
   215.9 mm impregnated drill bit (Fig. 8.32a) was used for side-tracking drilling, this was favorable for the drill bit to incline at the well bottom and thus would increase the reliability and the speed of building the shoulder for side-tracking drilling, and at the same time, the small diameter side-tracking drilling was favorable for the large angle PDM side-tracking drilling tool to run in. The inner conical angle of the bit bottom face was 160°, the center water hole at the bottom face was 14 mm in size (overcoming the center "dead point"), there were eight 14 mm sized center water holes at bottom face, the outer side-kerfs were R5 mm transitional spherical arc, the large-size natural diamonds were set around the center water holes and the outer side-kerfs transitional spherical arc.

   244.5 mm impregnated diamond drill bit (Fig. 8.32c) with 180 mm guide was used, the outer diameter of the bottom face was 244.5 mm, the inner diameter was 185 mm, the guide body was 180 mm in diameter, the length was 240 mm, the guide head shape was artillery head, the center water hole was 50 mm in size.

8.4.2 Drilling Conditions of Side-Tracking Drilling to Bypass Obstacles

1. **The selection of side tracking point**
   In 3400–3470 m well section, the well diameter was relatively stable, the well deviation angle was steady and tended to decrease naturally and the azimuth was steady (Fig. 8.33). The oversized well diameter was 369 mm below this well section at the depth of 3485 m. Meanwhile, this well section was near the well depth of 3525.16 m, the cement backfill height was proper. From 3400 m depth down there was a short well section with diameter slightly increased, which was favorable for 215.9 mm diamond kick-off bit to side-track and to build the shoulder to form a new hole. The average penetration rate of the three roundtrips of MH-1C-R46 (3390.54–3400.04 m), MH-1C-R47 (3400.04–3404.13 m) and MH-1C-R48 (3404.13–3413.53 m) were 1.03, 1.62 and 1.40 m/h respectively, with the core recovery of 96.1, 111.5 and 100.0 % respectively. The rocks were basically intact (Fig. 8.34), felsic gneiss and amphibolite in the 3395.24–3400.46 m well section, amphibolite plagiogneiss with stavrite and epidote stavrite in the 3400.46–3408.70 m well section. In that case, the side-tracking point was selected at 3400 m depth.
2. **The construction of manmade well bottom**
   The drilling tool with SK wire-line collar was used for grouting cement (Table 8.12), rock bit was used for drill-off to 3350 m for curing, to drill out cement plug to 3395 m and then 157 mm core drilling tool was used for recovering the

8.4 Side-Tracking Drilling for Bypassing Obstacles

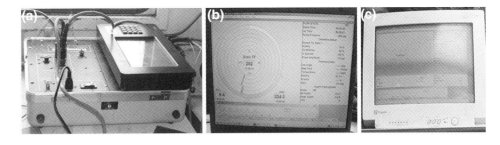

Fig. 8.31 Mud pulse MWD instrument. **a** Data processing instrument, **b** computer operating interface, **c** display for driller, **d** down hole instrument assembly

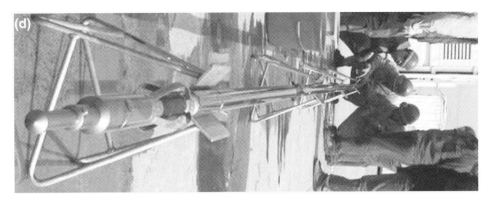

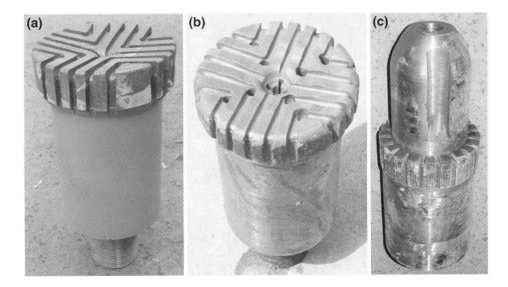

Fig. 8.32 Drill bit for small diameter side-tracking drilling. **a** 215.9 mm impregnated diamond bit, **b** impregnated bit after drilling 17.37 m, **c** reaming bit with guide

cement plug samples (Fig. 8.35), and finally, 244.5 mm rock bit was used for ream-off to 3400 m well depth. The details of the construction of manmade well bottom can be found in Sect. 4.2 of Chap. 4.

3. **Drilling operations of side-tracking drilling to bypass the obstacles**

(1) Side-tracking drilling at 3400.00–3417.37 m

The drilling tool assembly for small diameter side-tracking drilling consisted of 215.9 mm impregnated diamond drill bit (Fig. 8.32a) + 1.75° single bend 5LZ172 × 7 PDM + 160 mm non-magnetic sub (setting the MWD instrument within the sub) + 162 mm non-magnetic collar + 177.8 mm collar × 10 + 127 mm drill rod.

During the side-tracking drilling, drilling speed (Table 8.13) was strictly controlled, which was subjected to the drilling tool face angle. The face readings of the MWD allowed to change within regulated ±30°, when the tool face angle exceeded the allowance, the drilling tool was to be lifted up and lowered down to adjust the tool face angle. When the side-tracking drilling started, MWD was used to measure the well deviation and in the process of drilling, the well deviation was measured advisably in order to control the side-tracking drilling (Table 8.14). After drilling to 3417.23 m, the drilling speed gradually decreased and the pump pressure decreased, to increase the WOB from 30 to 55 kN was not obviously effective, and then drilling tool was

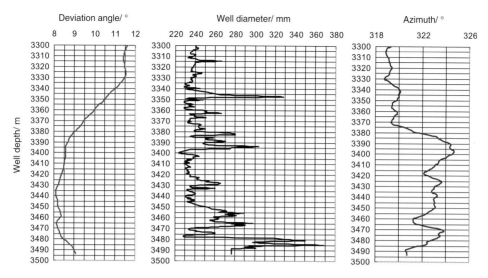

Fig. 8.33 Well deviation angle, well diameter and azimuth near the side-tracking point

Fig. 8.34 Core recovered near the side-tracking point. **a** Core from 3398.68 to 3401.46 m, **b** core from 3401.46 to 3401.96 m, **c** core from 3401.96 to 3403.56 m

Table 8.12 The basic data of grouting cement plug

Item	Parameter	Item	Parameter	Item	Parameter
Mud density in well (g/cm³)	1.19	Cement plug bottom depth (m)	3511.26	Cement plug top depth (m)	About 3400
Drilling rod OD (mm)	127	Drilling rod ID (mm)	108	Well wall ID (mm)	245
Cement amount (t)	27	Thickening time (min)	250	Well flushing time (h)	2.5
Time for grouting cement (min)	25	Slurry displacement time (min)	18	Ratio of water to cement	0.38
Cement slurry density (g/cm³)	1.97	Cement strength (MPa)	40	Cement type	JHG grade cement

pulled out. The outer diameter of the drill bit (Fig. 8.32b) wore from 215.9 to 214 mm, all the diamonds set in the periphery and around the center hole of the drill bit fell off, the center hole wore into a 50 mm concentric concave, there was no thin core in the drill bit, other places of the drill bit wore normally. The footage drilled was 17.3 m, with the average penetration rate of 0.32 m/h.

After 2 h of the side-tracking drilling (with one mud circulation), each half hour the samples were taken from the shaker of the solid controlling system for binocular microscope analysis (Fig. 8.36), the ratio of rock cuttings to the cement particles was 65:35 when pulling out the drilling tool.

After pulling out the drilling tool, engineering logging was carried out, the logging instrument encountered the resistance many times at 3407 m depth and didn't reach the well bottom (3417.37 m), the well deviation angle decreased to 7.56° (Fig. 8.37) from 8.47° at the start of side-tracking drilling, the azimuth angle was steady at 325° (the azimuth angle of the old well was 322–324° at this well section).

In order to calculate the horizontal offset distance between the new and old wells, the deviation angle and the azimuth angle of 3408–3417 m section of the new well were calculated sparingly according to 7.56° well deviation angle and 325° azimuth angle at the 3407 m depth. The total horizontal displacement of the new and old wells can be found in

8.4 Side-Tracking Drilling for Bypassing Obstacles

Fig. 8.35 The cement plug samples recovered near the side-tracking point

Table 8.13 Drilling time control during side-tracking drilling

Well section (m)	Drilling time control (cm/min)	WOB (kN)	RPM (r/min)	Mud displacement (L/s)
3400.00–3402.34	2–3	10–15	145	24.82
3402.34–3404.61	3	10–17	149	25.59
3404.61–3405.72	3	5–10	167	28.52
3405.72–3410.65	3–6	7–30	151	25.92
3410.65–3416.12	6–13	10–20	158	27.15
3416.12–3417.37	12–13	29–40	141	24.20

Table 8.14 MWD measurement data during side-tracking drilling

Well depth (m)	Depth of measurement point (m)	Well deviation angle (°)	Azimuth angle (°)	Well depth (m)	Depth of measurement point (m)	Well deviation angle (°)	Azimuth angle (°)
3400.00	3386.95	9.7	324.0	3410.77	3397.72	8.7	325.2
3406.11	3393.06	9.1	325.3	3415.25	3402.20	8.6	324.5
3408.74	3395.69	8.9	325.1	3417.37	3404.32	8.4	325.2

Fig. 8.38, at 3417 m depth, the distance between the new and old well axes was about 237.6 mm, the rock wedge thickness between the new and old wells was at least 14 mm (the old well diameter was 232 mm at this depth). The engineering logging data after the rock bit angle dropping showed that the distance between the new and old wells axes at 3417 m depth reached 510.5 mm, the rock wedge thickness was 287 mm (the old well diameter was 232 mm at this depth), the new and old wells separated off at 3411 m depth.

(2) Angle dropping drilling from 3417.37 to 3445.62 m

The drilling tool assembly consisted of 215.9 mm rock bit + 1.75° single bend 5LZ172 × 7 PDM + 160 mm non-magnetic sub (set a MWD instrument in the assembly) + 162 mm non-magnetic collar + 177.8 mm collar × 10 + φ127 mm drill rod.

28.25 m were completed by angle dropping drilling, with penetration rate of 1.31 m/h, WOB of 20–40 kN, RPM of 149 r/min, and mud displacement of 25.59 L/s. When the angle dropping drilling started, MWD measurement was conducted and the well deviation was measured properly during drilling operations in order to keep abreast of the angle dropping conditions (Table 8.15).

After pulling out the drilling tool, engineering logging was carried out, the well deviation angle decreased to 0.41° (Fig. 8.39) from 5.22° at the beginning of angle dropping drilling, the azimuth angle decreased to 297° from 317° at the beginning of angle dropping drilling, the total horizontal displacement of the new and old wells is shown in Fig. 8.40.

(3) Pilot reaming drilling from 3400.00 to 3443.31 m

From 3400.00 to 3407.42 m, the drilling tool assembly with pilot 244.5 mm impregnated diamond drill bit (Fig. 8.32c) + 4L120 × 7 PDM + 178 mm jar while drilling + 177.8 mm collar × 10 + 127 mm drill rod was used for reaming drilling, with the footage drilled of 7.42 m, the average penetration rate of 0.97 m/h, RPM of 237–248 r/min, and mud displacement of 12.96–13.57 L/s.

From 3407.42 to 3443.31 m, the drilling tool assembly with pilot 244.5 mm impregnated diamond drill bit + 4L172 × 7 PDM + 178 mm jar while drilling + 177.8 mm collar × 10 + 127 mm drill rod was used for reaming drilling, with the footage drilled of 35.89 m, the average penetration rate of 2.02 m/h, RPM of 132–146 r/min (PDM) + 15–20 r/min (rotary table), and mud displacement of 23.27–25.06 L/s.

Fig. 8.36 Ingredient of the cuttings of side-tracking at 3400 m

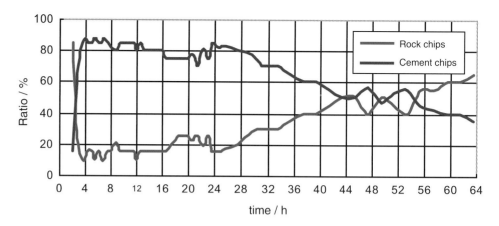

Fig. 8.37 The deviation angles of the new and old wells (engineering logging)

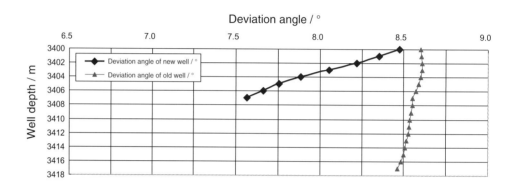

Fig. 8.38 Comparison of horizontal displacement of the new and old wells

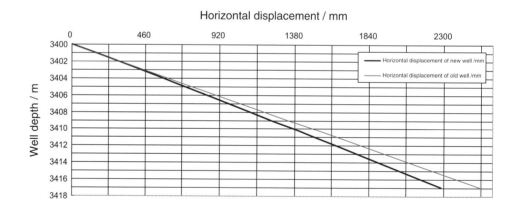

Table 8.15 MWD measured deviation data during angle dropping drilling

Well depth (m)	Measuring point depth (m)	Well deviation angle (°)	Azimuth angle (°)	Well depth (m)	Measuring point depth (m)	Well deviation angle (°)	Azimuth angle (°)
3417.33	3415.28	8.4	324.7	3442.41	3429.36	4.2	310.7
3428.25	3415.20	7.5	321.2	3445.45	3432.40	3.3	307.2
3436.88	3423.83	5.6	315.1				

8.4 Side-Tracking Drilling for Bypassing Obstacles

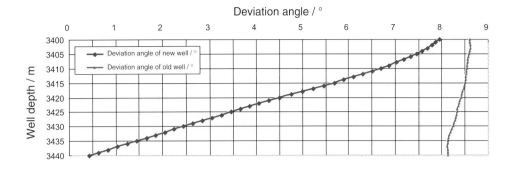

Fig. 8.39 Well deviation angles of the new and old wells (engineering logging)

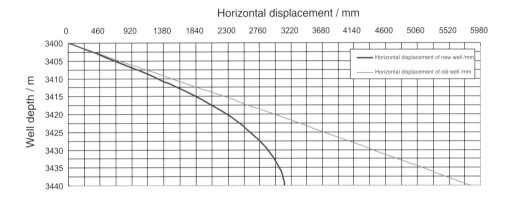

Fig. 8.40 Comparison of horizontal displacement of the new and old wells

On the base of the first side-tracking drilling for deviation correction, this side-tracking drilling for bypassing the obstacle was smooth, side-tracking drilling the new well and angle dropping operations were completed only in two times of running-in.

8.5 Development of PDM Drive Continuous Deflector

The working principle of the continuous deflector is different from that of the single bend PDM, when it bears the axial force, the slide wedge block near the drill bit extends radically and pushes the well wall, and at the same time, it transforms the radial restricting force from the well wall to the drill bit and completes the kick-off. When the axial force disappears, the slide block retrieves, and the drilling tool keeps a state of same diameter and same axis, therefore, the running-in problems in the full gauge kick-off basically do not exist. But the rock-cutting torque of existing continuous deflector depends on the drill string, in the process of kicking-off, the whole drill string rotates except the stator, from this the problem of directional reliability appears, therefore, since the PDM directional drilling technology was popularized, the application of the continuous deflector decreased greatly. The purpose of this project was to develop a new drilling tool which combines the characteristics of well-bottom PDM drive with the characteristics of same diameter and same axis from continuous deflector under free state, in order to realize the full gauge kick-off in large diameter well with diamond drill bit.

8.5.1 Working Principle of the Drilling Tool

The drilling tool was designed into two parts, i.e. driving and executing mechanisms. The driving part was rebuilt from the PDM from the factory; the executing part was the designed key part, i.e. continuous deflector. The whole structure of the drilling tool can be found in Fig. 8.41. The upper and lower ends of the upper couple 2 are connected respectively with the output shaft of the PDM rotor and spline case 4, which is jointed with the rotor key of the continuous deflector, for PDM transmitting the torque; the outer tube 1 is connected with the thread of the PDM stator, for transmitting WOB; as a part of the stator of the deflector, the slide block is concentric with the drilling tool under the free state, and slides out along the radial direction under the axial force and then exerting the side force to drill bit by limitation of the well wall; when the drilling tool is connected, the datum line of the slide block is correspondence with the directional sub of the upper end of the PDM, with orientation principle the same as PDM directional drilling.

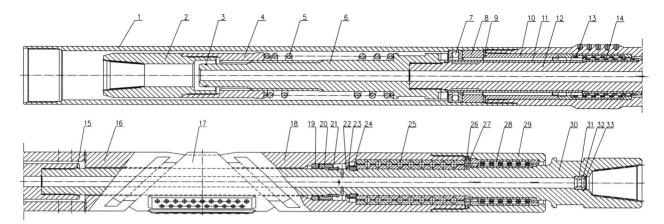

Fig. 8.41 LZ-157 continuous deflector driven by PDM. *1* Outer tube *2* Upper couple *3* O seal ring *4* Spline case *5* Restoration spring *6* Spline shaft *7* Thrust bearing *8* Adjusting washer *9*. Flat key *10* Upper bearing outer fixing ring *11* Upper bearing inner restricting ring *12* Upper shaft *13* Upper alloy bearing non-moving ring *14* Alloy bearing moving ring *15* O seal ring *16* Upper half wedge *17* Slide block *18* Lower half wedge *19* Metal seal still ring *20* Metal seal moving ring *21* Hook head key *22* Elastic restricting ring for shaft *23* Jaw case *24* Jaw *25* Ball thrust bearing series *26* Outer restricting ring for lower bearing *27* Inner restricting ring for lower bearing *28* Lower alloy bearing still ring *29* Lower pushing lid *30* Driving shaft *31* Adjusting ring *32* O seal ring *33* Elastic restricting ring for hole

8.5.2 Practical Drilling Test at Drill Site

The prototype drilling tool was tested for two rounds in practical drilling, the first round of test was conducted in the testing hole in order to check the working performances and safety of the prototype drilling tool in the hole; and the second round of test was conducted in the main hole of CCSD-1.

1. **The test in the testing hole**

(1) The basic conditions of the testing

The prototype drilling tool was tested for six roundtrips in the testing hole, with the testing conditions shown in Table 8.16, the actual drilling time of the prototype drilling tool was 64 h and during which the breakdown of the prototype drilling tool never happened, and also any part was never changed.

(2) The condition analysis and explanation

i. From the deviation survey results of the first and the fourth roundtrips, the kick-off effects of the drilling tool were good, after the two roundtrips, conventional core drilling were carried out, through observing the upper core recovered from the kick-off hole section, it also showed that the drilling tool got a good kick-off effect (Fig. 8.42). Although the kick-off effect of diamond drill bit could not be measured, it could be speculated that the kick-off result would exceed that of rock bit.

ii. To analyze the deviation survey data and core it was found that kick-off core drilling also got some kick-off effect (it was impossible and could not be allowed for this kick-off effect to be so obvious as kick-off drilling).

iii. The fourth and fifth tests encountered the obstacles at 2.5 and 4.89 m from hole bottom respectively in the hole. The 15–18 kN WOB and 8.62 L/s pump displacement were adopted respectively to drill off to the hole bottom with 0.6 and 2.3 h, with the pump pressure of 2.5 MPa. During drilling off, a large WOB was once attempted, but up to 22 kN, the pump pressure increased to 5 MPa, the attempt was given up for sake of worrying that the torque reaction of the PDM would have a big fluctuation and thus damage the wedge block.

iv. In comparison with the kicking-off drilling by single bend PDM, whatsoever rock bit or diamond bit was used, the ROP of the prototype would be more than one time lower than that of the single bend PDM, the key reason should be that the prototype consumed a lot of axial force to three bracing points in the hole wall. The hole wall drilled by rock bit was rough, and the rock bit consumed large axial pressure, since the cutting area of the diamond kicking-off bit was originally large and the axial pressure was offset by the hole wall, the bit could not get the needed unit pressure for breaking the rock and thus the drilling efficiency was greatly decreased. In addition, since the kicking-off regular pattern of the drilling tool was not ever mastered, a conservative and small WOB was taken in the testing process, thus the drilling speed was also affected to a certain extent.

v. Before the side-tracking drilling, a manmade hole bottom was built from 63.51 m hole depth with J Grade oil well cement to approximate 40 m. After curing and probing the bottom, the cement plug was drilled off to 43 m. To submit to the need of the main hole drilling (since the testing hole and the main hole used the same pump unit) the side-tracking drilling tool was pulled

8.5 Development of PDM Drive Continuous Deflector

Table 8.16 The testing situations of the continuous deflector in the testing hole

Serial number	Hole section (m)	Testing contents	Footage drilled (m)	ROP (m/h)	Vertex angle change (°)	Deflecting intensity (°/m)	Wear of bit outer diameter (mm)	Explanations
1	81.12–85.06	The swivel, lubrication and safe performance of die drilling tool	3.94	0.47	0.4 → 0.8	0.10		WOB 25–30 kN, pump displacement 8.62 L/s
2	130.62–131.68	Feasibility of of coring kicking-off	1.06	0.64			0.15	WOB 20 kN, pump displacement 8.62 L/s, coring 1m. Since core block and little footage drilled, deviation survey was not carried out
3	131.68–132.21	Diamond bit kicking-off	0.53	0.06				WOB 20–50 kN, pump displacement 8.62 L/s, since a little footage drilled, deviation survey was not carried out
4	132.21–137.14	Rock bit kicking-off	4.93	0.17	1.6 → 2.0	0.08		WOB 35–39 kN, pump displacement 8.62 L/s
5	137.14–138.83	Coring kicking-off drilling	1.69	0.27			0.10	WOB 25–35 kN, pump displacement 8.62 L/s, coring 1.6 m. Since core block and little footage drilled, deviation survey was not carried out
6	43.00–46.49	Side-tracking drilling	3.94	0.39	0.5 → 0.6			WOB 10 kN, pump displacement 9.48 L/s, drilling the cement hole bottom. The side-tracking drilling was not successful

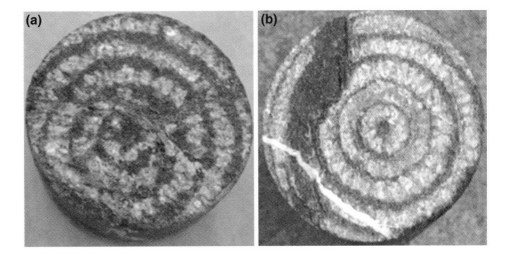

Fig. 8.42 The core (the upper end face of the core drilling) after the end of rock bit kicking-off. **a** After the first testing, **b** after the fourth testing

out after only drilling 3.94 m. Through analysis, the causes of side-tracking failure included three points: the part of the mixed cement slurry of the upper cement plug was not completely drilled off, thus side-tracking drilling could not be started at the hard cement plug (the comparison of the core recovered above the side-tracking point and at the end of the side-tracking drilling can be found in Fig. 8.43, it can be seen that the quality of the two sections of the cement plug was quite different); the footage of the side-tracking drilling was too short, it was impossible to drill a new φ 157 mm branch-hole even using very successful single bend PDM side-tracking drilling method; and lacking the experience of using oil well cement.

Fig. 8.43 Comparison of the cores recovered before and after the side-tracking drilling at the manmade hole bottom. **a** The cement core recovered before side-tracking drilling, **b** the cement core recovered after side-tracking drilling

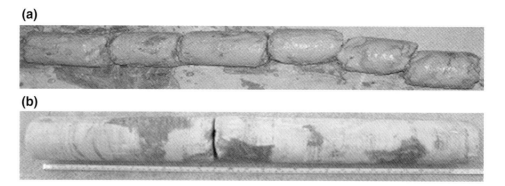

2. **The drilling test in the main hole**

Form 5129.36 to 5134.66 m in the main hole, the directional deviation-correction test of the prototype was carried out for the second time (φ158.8 mm rock bit + continuous deflector + wired MWD instrument), the total footage drilled was 5.30 m, the actual drilling time was 13.81 h, and ROP was 0.38 m/h. After drilling to 5153.58 m with non-coring rock bit, hole deviation survey was carried out, with the result shown in Fig. 8.44, from which it can be seen that the drilling tool got very good directional deviation correction effect.

8.5.3 Test Result Commentary

1. The whole structure design of the drilling tool was reasonable and reliable; during the whole testing, the connection of the PDM with the continuous deflector, the respective moving of the stator and rotor, the extending and returning and limiting of the sliping block, the bracing point of the sliping block and abrasion-proof of upper bracing point, TC bearing and cluster bearing lubrication

Fig. 8.44 Well deviation angle and azimuth angle (the wired MWD instrument) of directional deviation correction test in the main hole

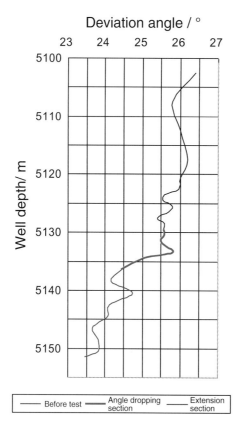
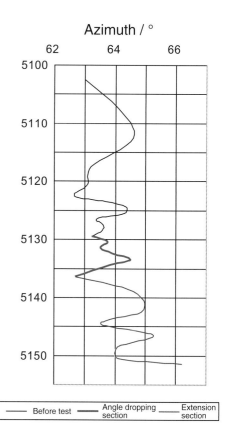

8.5 Development of PDM Drive Continuous Deflector

Table 8.17 The situations of well deviation control of CCSD-1 Well

Well section	Well depth (m) From	Well depth (m) To	Footage drilled (m)	Drilling method	Well deviation situations
PH	6.87	101.00	94.131	φ444.5 mm rock bit	Well deviation angle was less than 1°
	101.00	2046.54	945.54	φ157 mm coring	Well deviation angle was less than 4.1°, azimuth angle was between 280–325°
MH	2046.54	2982.18	935.64	φ157 mm coring	In 2046.54–2760 m well section deviation angle was between 4.42–9.08°, azimuth angle was between 280–294°; beneath 2760 m well section, well deviation angle quickly increased, azimuth angle was between 294–309°, at 2935 m the well deviation angle was up to 16.35°
MH-1X	2749.00	2974.59	225.59	Deviation correction drilling	
	2749.00	2758.00	9.00	φ157 mm diameter side-tracking drilling (no new hole coming up)	The well deviation angle 7.10° → 6.29°, azimuth angle 293° → 295°. The MWD instrument showed that well deviation did not change, the side-tracking drilling was not successful
	2758.00	2769.59	11.59	φ140 mm impregnated bit side-tracking drilling	The well deviation angle 6.29° → 6.07°, azimuth angle 295° → 292°. The MWD instrument showed: the well deviation angle 8.2° → 6.2°, the azimuth angle 291° → 294°
	2769.19	2771.35	2.16	φ157 mm coring	The well deviation angle 6.07° → 6.06°, azimuth angle 292° → 291°
	2771.35	2942.11	170.76	φ158.8 mm rock bit deviation correction	The well deviation angle 6.06° → 2.57° → 5.15°, azimuth angle 291° → 249° → 305°, the minimum well deviation angle was at 2810 m
	2942.11	2956.91	14.80	φ158 mm impregnated bit deviation correction	The well deviation angle 5.15° → 5.80°, the azimuth angle 305° → 318°
	2956.91	2961.91	5.00	φ140 mm impregnated bit deviation correction	The well deviation angle 5.80° → 6.01°, azimuth angle 318° → 320°
	2961.95	2970.54	8.59	φ157 mm coring	The well deviation angle 6.02° → 6.20°, azimuth angle 320° → 319°
	2970.54	2974.59	4.05	φ140 mm impregnated bit deviation correction	The well deviation angle 6.20° → 6.33°, azimuth angle 319° → 318°
MH-1C	2974.59	3665.87	691.28	φ157 mm coring	In 2974.59–3127.54 m the well deviation angle increased from 6.33° to 11.51°, the azimuth angle was between 318–322°; beneath 3127.54 m well section, the well deviation angle gradually decreased after 3 deviation corrections, and at 3653 m well depth, the well deviation 7.40°, azimuth angle was between 313–341°
	3127.54	3139.39	11.85	φ140 mm impregnated bit deviation correction	The well deviation angle 11.51° → 9.70°, the azimuth angle 319° → 314°
	3171.28	3191.28	20.00	φ158.8 mm rock bit deviation correction	The well deviation angle 9.87° → 9.25°, the azimuth angle 312° → 316°
	3244.98	3253.33	8.35	φ140 mm impregnated bit deviation correction	The well deviation angle 12.13° → 12.87°, azimuth angle 317°
	3516.66	3516.96	0.30	φ158.8 mm rock bit	
	3520.91	3536.00	15.09	φ158.8 mm rock bit	The well deviation angle 9.13° → 8.62°, azimuth angle 333° → 331°

(continued)

Table 8.17 (continued)

Well section	Well depth (m) From	Well depth (m) To	Footage drilled (m)	Drilling method	Well deviation situations
MH-2X	3400.00	3445.62	45.62	φ215.9 mm side-tracking drilling to bypass obstacle	Well deviation angle 8.71° → 2.06°, azimuth angle 323° → 307°
	3445.62	3624.16	178.54	φ244.5 mm rock bit	Well deviation angle 2.06° → 1.07° → 2.90° → 0.98° → 6.92°, azimuth angle 307° → 341° → 44° → 0.98° → 62° (corresponding well depth: 3445.62 m → 3462.5 m → 3478 m → 3524.5 m → 3624.16 m)
MH-2C	3624.16	5118.20	1494.04	φ157 mm coring	Well deviation angle 6.92° → 23.68° → 21.84° → 29.45° → 27° → 26.2°, azimuth angle 62° → 96° → 75° → 62° → 61° → 63° (corresponding well depth: 3624.16 m → 4192.5 m → 4380 m → 4925 m → 5075 m → 5118.2 m)
	3707.84	3712.55	4.71	φ158.8 mm rock bit	Well deviation angle 7.83° → 8.08°, azimuth angle 63° → 66°
	3843.30	3846.00	2.70	φ158.8 mm rock bit	Well deviation angle 11.79° → 11.73°, azimuth angle 94°
	4398.47	4401.49	3.02	φ158.8 mm rock bit	Well deviation angle 22.40° → 22.17°, azimuth angle 69°
	5072.06	5075.16	3.10	φ158.8 mm rock bit	Well deviation angle 27.00°, azimuth angle 61°
MH-T	5118.20	5158.00	39.80	Drilling tool test	
	5118.20	5125.86	7.66	φ158.8 mm rock bit	Well deviation angle 26.2° → 25.8°, azimuth angle 63° → 64°
	5125.86	5129.36	3.50	Core drilling tool testing with the wire-line, PDM and hydro-hammer	Well deviation angle 25.8° → 25.6°, azimuth angle 64°
	5129.36	5134.66	5.30	The deviation correction testing of the continuous deflector	Well deviation angle 25.6° → 25.1°, azimuth angle 64°
	5134.66	5158.00	23.34	φ158.8 mm rock bit	Well deviation angle 25.1° → 23.5°, azimuth angle 64° → 66°

8.5 Development of PDM Drive Continuous Deflector

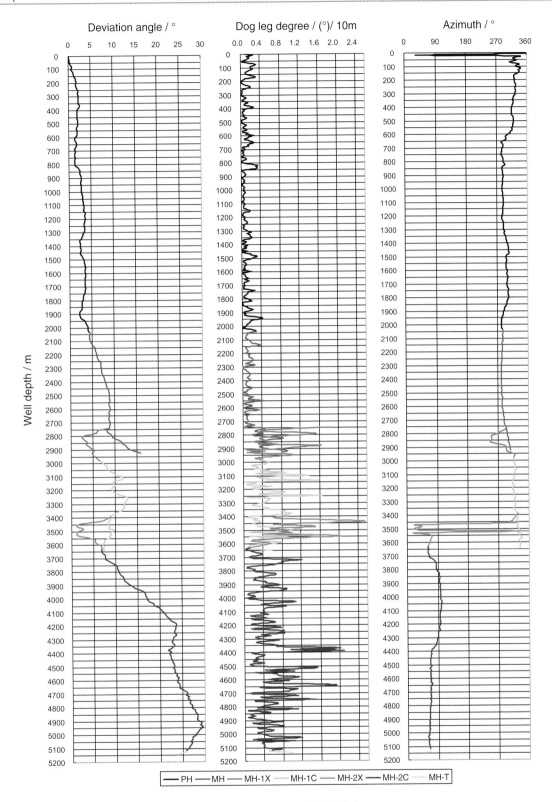

Fig. 8.45 The well deviation angle, dog-leg degree and azimuth angle in CCSD-1 Well

and other mechanisms did not go any wrong. The drilling tool connection was safe and reliable, the thread looseness did not appear; all the parts of the drilling tool were not changed during more than 70 h drilling (not including drilling-off and other auxiliary work).

2. The kicking-off effect of the drilling tool with the rock bit reached $0.1°/m$. Even though the kicking-off effect of diamond bit was not measured since the drilling footage was too short, it could be concluded that the kicking-off intensity of diamond bit would be more than that of the rock bit at the same formation conditions.
3. The drilling tool could be used for drill-off under WOB of less than 18 kN, but we would suggest avoiding drilling-off as much as possible.
4. The main reason of lower ROP was that high friction force of the drilling tool bracing points with the hole wall made the WOB not be able to be transferred effectively along the axis direction to the bit. How to decrease the friction force of the bracing points and increase the drilling speed and trip footage (especially the drilling speed and trip footage of diamond kicking-off bit) would be the further research trend for the drilling tool. The tentative idea of the kicking-off coring drilling could only be realized by increasing drilling speed and trip footage.
5. Side-tracking drilling not only requires that drilling tool can drill with long travel, but also requires that drilling tool can drill with a small WOB, controllable drilling speed, small vibration and positive pressure to hole wall in borehole, otherwise the drilling tool may slide down along the cement plug and shake and crush the cement wedge face formed by side-tracking drilling. Further research and testing work are needed for improvement of the drilling tool performances.

8.6 The Analysis on Well Deviation Control Effect

The formations drilled in CCSD-1 Well mainly are strong deviation building formations, the well deviation control was a key technique in the drilling engineering. The overall tactics taken for well deviation control was that prevention was a priority, when the well deviation was over the regulated well deviation value, the deviation correction would be carried out. The CCSD-1 Well needed to be cored in the whole hole section, the drilling program adopted an advanced open hole drilling method, so the deviation prevention mainly was the deviation prevention for core drilling.

The main technical measures for core drilling deviation prevention were: the rigid, straight, full and heavy drilling tool assembly would be used for drilling tool; and the down hole power percussive rotary core drilling method would be used for core drilling. For the pilot hole PDM hydro-hammer with double stabilizes or PDM conventional core drilling were adopted as the main drilling method, the depth of the pilot hole was 2046.54 m, the maximum well deviation angle was $4.1°$ and thus the technical requirement that "two holes combined into one" was satisfied; PDM hydro-hammer conventional core drilling was mainly used in the main hole, the well deviation angle steadily increased in most well sections, one time of side-tracking drilling deviation correction, one time of side-tracking drilling to bypass obstacle for deviation correction and three times of well bottom deviation corrections were totally carried out, the final hole depth was 5158 m, the vertex angle of the final hole was $23.4°$ and the demands of the well body quality was satisfied. The well deviation situations of CCSD-1 Well can be found in Table 8.17; Fig. 8.45 respectively.

Analysis of the well deviation control techniques of CCSD-1 Well can obtain some experiences as follows:

1. The deviation prevention techniques taken for the core drilling was successful, the rigid, straight, full and heavy drilling tool assembly was adopted for drilling tool and the down hole power percussive rotary core drilling method was used for core drilling.
2. The side-tracking drilling was feasible for deviation correction in hard rock and large diameter deep hole, single bend PDM was adopted for side-tracking deviation correction drilling and wired or wireless MWD instruments were used for MWD. For side-tracking in hard rock, large diameter and deep well, small diameter drill bit is a suitable tool to start a new hole, and then followed with reaming drilling.
3. Down hole full gauge deviation correction was feasible for hard rock, large diameter and deep well; the continuous deflector taken for deviation correction was also feasible.
4. Although coring in fractured zone was difficult, well deviation angle tended to decrease.
5. Side-tracking drilling for deviation correction was easy to cause a dog-leg well section; the repeated deviation corrections would be easy to make the well trace shape complicated; the complicated well trace shape would lead to the increase of down hole accidents and even cause the hole abandoned. Therefore, a prudent deviation correction tactics should be considered in hard rock deep well.

9 Drilling Fluids and Solids Control Technology

Core drilling, expanding drilling (reaming) and non-core drilling were the main drilling methods employed for CCSD-1 Well. Different drilling methods needed drilling fluids with different performances to ensure implementation of the drilling technologies. Roller bit was employed for expanding drilling and non-core drilling with bigger debris granule (2–5 mm) produced while diamond bit was employed for core drilling with very fine debris (5–100 μm) produced. Rotary table drive with low rotation speed of 30–60 r/min was employed for non-core drilling and expanding drilling while downhole PDM and hydro-hammer with high rotation speed of 160–200 r/min were employed for core drilling. The pump capacity of drilling fluid was 25–35 L/s for non-core drilling and 9–11 L/s for core drilling. So, discrepancy was a little wide for the performance requirements of drilling fluid to different kinds of drilling methods. Carrying capability of mud was required to be good to clean away bigger debris granule produced in non-core drilling and expanding drilling, and then improve drilling efficiency finally, while core drilling required mud with the performance of not only be good at carrying, but also suspending debris while circulation was stopped, what's more, the requirement of driving medium as a bottom power system should be satisfied. Simple for preparation, easy regulation of performance and convenient management were all required for drilling fluid in consideration of field preparation and management. For saving cost and reducing discharge, it was best to use a single drilling fluid system at drilling field, thus when drilling technology changed, the requirement of the changed drilling technology could be satisfied without changing drilling fluid system but with only simple regulation of its performance. The LBM-SD drilling fluid system specially designed for CCSD-1 Well was employed to accomplish core drilling, expanding drilling and non-core drilling.

9.1 Requirements of Scientific Drilling for Drilling Fluid

Besides satisfying the basic requirements of scientific drilling construction, drilling fluid should satisfy the requirements of scientific test, well logging and borehole log, etc. Detailed requirements can be found as follows:

1. No chemical composition which may affect scientific test, well logging and borehole log;
2. Small influence to environment safety;
3. Good lubrication and low cost;
4. Effectively relieving complex downhole situation, including high temperature and high pressure;
5. With strong weighted bearing capacity, including inverting into salt drilling fluid system;
6. Low solid content, with less interference to well logging, acoustic transmission and visual reception, avoiding jam of drilling tool;
7. With strong capability of carrying debris (cuttings), and can produce thin and good tenacity mud cake.

9.1.1 Strata Encountered and Requirements of Well Structure

Main strata in CCSD-1 Well are eclogite, paragneiss and orthogneiss. Lithology of these strata are not of water sensitivity, and their permeability are extremely tiny, so there was no special requirement for inhibition capability of drilling fluid to water sensitive layers except some broken borehole sections with collapses and block dropping. Hardness and abrasiveness of those strata are high, the cuttings produced in drilling were extremely tiny and with large density from 3.5 to 3.8 g/cm^3, so good lubrication of drilling fluid was required

Translated by Duan Longchen.

to alleviate the wear of borehole wall to drilling tool. Because of the high density and tiny size of the cuttings, good carrying and suspension capabilities were required for drilling fluid to carry the cuttings produced in drilling out of hole in time, and then to be removed by solids control equipments, so as to improve drilling efficiency and avoid bit bouncing and bit burying. Cuttings produced in diamond drilling can not be removed by normal solid control equipment for oil drilling, so high efficiency solid control equipment and related methods must be employed for solid removal.

Though most borehole sections encountered in drilling were integral and there was no formation pressure in those integrated crystalline rock strata, several broken zones and leakage zones were still encountered in drilling. Leakage problem was solved by sealing with inert material, and circulation of drilling fluid could be maintained, however, repeated leakage might be easily happened in drilling because leakoff pressure in those sealed borehole section was still small. So density and rheological parameters of drilling fluid should be strictly controlled to decrease the circulation pressure of drilling fluid.

In drilling construction, performance of drilling fluid and parameters of drilling technology are related to borehole structure. Flow state of drilling fluid should be kept constantly and abrupt change of local flow state should be avoided in drilling. Usually, "one sized casing extends to hole bottom" is impossible, several steps of casing need to be run in hole, so drilling fluid can easily form turbulence at the area between open borehole and casing, leading to an accumulation of cuttings. Well accident such as bit burying and bouncing caused by abrupt slumping of the cuttings accumulated to a certain amount will happen. Influence of drilling fluid flow state to drilling status is very obvious, too, not only related to suspension and carry of the cuttings, but also to the stability of borehole wall, circulating pressure loss of drilling fluid and operation conditions of downhole power drilling tools.

During the period of core drilling in the pilot hole, in order to ensure the carry of the cuttings, 244.5 mm ($9\frac{5}{8}$ in.) moving casing was run in the 339.7 mm ($13\frac{3}{8}$ in.) casing, with the main purposes of improving the flow state of drilling fluid and improving uplift velocity. Even so, because borehole diameter of core drilling was 157 mm, the inner diameter of 244.5 mm moving casing was 222.4, and 89 mm ($3\frac{1}{2}$ in.) drill stem was employed for core drilling, so the uplift velocities of drilling fluid in two borehole diameters differed 2.5 times. During the period of core drilling in the main hole, though 193.7 mm ($7\frac{5}{8}$ in.) moving casing was set in the borehole above 2019 m, the difference of uplift velocities was still comparatively obvious. So, carry of cuttings in different hole diameters must be considered when designing the properties of drilling fluid.

9.1.2 Requirements of Core Drilling

PDM and hydro-hammer downhole power drive core drilling method was the main way employed for CCSD-1 Well. Drilling fluid not only just worked as normal mud, but also provided working medium and power to the downhole engine (PDM and hydro-hammer). The working parameters of those downhole power drilling tool assemblies were determined by discharge rate of drilling fluid, at the same time, quality of drilling fluid properties would directly affect the working quality and service life of the downhole power system.

There is a close relationship between operating characteristics of PDM and circulation pressure and discharge rate of drilling fluid. The output torque is proportional to pressure difference between motor inlet and outlet, and output rotary speed is related to the discharge rate through the motor but no relation with pressure difference basically. So, when designing drilling fluid parameters, especially for drilling parameters or discharge rate, it was very important and should be the key element to determine the discharge rate of drilling fluid according to rotation speed of diamond drilling at first. Uplift velocity and circulating pressure drop are determined by discharge rate, and carrying capability is determined by uplift velocity and rheological parameter of drilling fluid. After PDM core drilling roundtrip is finished, circulation for hole flushing is forbidden before lifting the drilling tool in case of core loss, and 10 min gel strength value of drilling fluid should be increased to improve its suspension capability just in case of cuttings sedimentation.

Percussive power and percussive frequency of hydro-downhole-hammer are determined by the pressure difference between inlet and outlet of the hammer and the discharge rate of drilling fluid. Conventional valve type hydro-hammer is very susceptive of quality and performance of drilling fluid, so, when viscosity of drilling fluid is high, there will be a comparatively big loss of pressure between the two ends of the hydro-hammer, and this will affect single stroke percussive power of the hydro-hammer. Solid content, especially sand content of drilling fluid is very unfavorable to hydro-hammer, especially to valve type and fluidic type hydro-hammer. Drilling fluid with high sand content can seriously erode valve body and fluidic element in high speed of flow behavior, causing a premature wear or even abandonment of the hammer.

Allowable working discharge rate of hydro-hammer in drilling is limited. In hydro-hammer drilling, the range of this discharge rate should be fully considered to ensure effective work of the hammer and the carrying capacity of the cuttings. The range of working discharge rate of PDM should be considered, too when PDM hydro-hammer drilling is employed. In addition that PDM and hydro- hammer both have high speed moving parts, so lubrication property of drilling fluid is very important to improve efficiency and service life of drilling tool. Thus, PDM hydro-hammer core drilling has more requirements for drilling fluid in comparison with conventional rotary drilling.

9.1.3 Requirements of Non-core Drilling and Expanding Drilling

The size of the cuttings produced in non-core drilling and expanding drilling is comparatively large with diameter of 2–5 mm as a result of roller bit is employed. Especially in some broken borehole sections, even larger cuttings are usually produced because of dropping blocks from borehole wall, so yield point value of drilling fluid should be improved in drilling fluid treatment.

Rotary table drive is employed for non-core drilling and expanding drilling, all the drilling tools will be rotated in borehole. When hole deviation is large, the wear of hole wall to drilling tool will be rather serious because no mud cake is produced on hole wall as a result of extremely small permeability of strata. Thus, lubrication property of drilling fluid should be improved in drilling fluid treatment.

The time used for non-core drilling and expanding drilling in each roundtrip is comparatively long. The temperature of drilling mud at surface will be higher than 40 °C and the temperature in hole is even higher because of the long time continuous circulation of drilling fluid in deep well, especially in summer. At that time, the polymer in drilling fluid will be degraded, not only producing a large quantity of foam, but also changing the performance of drilling fluid. Thus, problem of anti-corrosion should be solved in the process of drilling fluid treatment.

9.1.4 Requirements of Borehole Log

Accurately and completely acquiring the core, liquid and aeriform samples released in drilling with modem scientific and technological means is one of the main purposes of CCSD-1 Well project. And providing the basic data for geological research, long term observation and other subject study is the task of borehole log. Geological logging data and follow-up monitoring of compound logging instrument can also provide technical support for drilling project and ensure that the project can be successfully implemented.

Borehole log methods used in CCSD-1 Well included core logging, cuttings logging, drilling fluid logging and follow-up monitoring with compound logging instrument, all those methods were related intimately with the performance and components of drilling fluid. Variations of every kind of ion and gas in drilling fluid could be detected by different liquid and gas analytical and detecting instruments in time. Testing methods and technical descriptions of drilling fluid samples are shown in Table 9.1.

Table 9.1 Testing methods and technical descriptions of drilling fluid samples

Instrumentation	Testing components	Testing range (mg/L)
ICP-MS	Mg, Li, Ca, Na, K, Sr, Fe, Al, Si, P, Zr, Mn, Ti, Sc, Ni, V, Cu, Pb, Zn, Co, Mo, Be, Ba and 15 rare earth elements	1×10^{-2} to 1×10^3
LC	Organic addition agent and kinds of polymers	1×10^{-2} to 1×10^3
IC	Anion such as F^-, Ci^-, Br^-, I^-, NO_3^-, SO_4^{2-} and CO_3^{2-}, etc.	1×10^{-2} to 1×10^3
GC-MS	Gas such as N_2, O_2, Ar, CO_2, CH_4, H_2 and He, etc.	1×10^{-6} to 1×10^{-5}
GC	Gas such as CH_4, C_2H_6, C_3H_8 and C_4H_{10}, etc.	1×10^{-6} to 1×10^{-5}

GC-MS, GC, IC and LC were employed for accurate surveying of microvariations to each component in drilling fluid.

After full hydration, components of drilling fluid in drilling process are basically steady and will be the background data of borehole log. Drilling fluid should be adjusted many times as a requirement of drilling project. So, drilling fluid components after adjusting need to be tested before running in hole. While drilling fluid is run in hole, element metathesis and chemical combination or even new crystalline minerals may occur after actuating with the rocks in open hole section in circulation. When abnormal phenomena happen in borehole log, it means that something of stratums has intruded into drilling fluid.

Thus, following requirements must be satisfied when drilling fluid is prepared and disposed to ensure the veracity of borehole log and explanation of abnormal phenomena.

1. Components of drilling fluid are known;
2. Components of drilling fluid should be as simple as possible;
3. Treatment times of drilling fluid should be as least as possible;
4. Drilling fluid system should have high temperature stability and higher resistance to fouling;
5. No components which may influence borehole log;
6. Intimate coordination with loggers in treatment progress of drilling fluid.

9.1.5 Requirements of Environmental Protection

Along with the improvement of human living standard and advance of science and technology, the awareness of mankind to environmental protection is increasing, too. Usually,

Table 9.2 Main technical descriptions of the drilling fluid for core drilling

Item	Parameter	Item	Parameter
Density (g/cm^3)	1.05–1.07	Fann-viscosity φ_{600}	10–16
Funnel viscosity (s)	28–32	Fann-viscosity φ_{100}	3.5–5.0
Apparent viscosity (mPas)	≤8	Sand content (%)	<0.1
API filter loss (ml)	10~12	YP/PV (ks^{-1})	0.2–1.0
Initial gel (Pa)	0.5–1.0	Cuttings transport ratio λ	≥0.5
10-min gel strength (Pa)	2.0–4.5	pH	9–10

Table 9.3 Main technical descriptions of the drilling fluid for non-core drilling and expanding drilling

Item	Parameter	Item	Parameter
Density (g/cm^3)	1.06–1.20	Funnel viscosity (s)	34–45
Apparent viscosity (mPas)	≤22	Plastic viscosity (mPas)	≤15
Initial gel (Pa)	1.5–4.5	10-min gel strength (Pa)	7–12
API filter loss (ml)	<20	Sand content (%)	<0.3
Mud cake thickness (mm)	≤0.5	pH	9–10

drilling construction will cause a certain influence to ambient, such as noise pollution, groundwater fountain pollution, discharge pollution of construction waste liquid and domestic pollution and so on, in which, discharge pollution of construction waste liquid is more serious. Drilling fluid treatment of CCSD-1 Well should satisfy the following requirements for environmental protection:

1. During construction, discharge of waste water and trash must be strictly controlled, waste drilling fluid from field construction, debris from hole and trash from well site need to be innocuously treated.
2. When the project is finished, all the waste drilling liquid must be left in mud pit and all with debris will be exported and buried after evaporation and concentration under natural conditions.

9.1.6 Requirements of Drilling Fluid Design

Based upon the above-mentioned situations, following requirements to the drilling fluid system of CCSD-1 Well was proposed:

1. Should not contain chemical materials which might influence well logging, borehole log and scientific experiment;
2. With the performances of low viscosity, low shear strength and shear-thinning property to decrease the circulating pressure drop;
3. Stability of temperature resistance should be larger than or equal to 150 °C;
4. With good lubrication property to reduce drilling torque, and to improve the service life and reliability of the down hole engine;
5. With good carrying capacity, especially for the upper borehole section where the uplift velocity in the annulus is rather slow;
6. Should be a good carrier to transport energy for down hole power system besides acting as normal drilling fluid;
7. With good fluidity and low flow resistance to meet the working requirement of PDM and hydro-hammer;
8. With sand content as low as possible to decrease erosion and wear of drilling fluid to PDM and hydro-hammer;
9. With good suspension capacity to avoid cuttings depositing;
10. Non-corrosive to metals;
11. Be favor of solid control;
12. With strong capability of invasion resistance.

Based upon above-mentioned requirements, drilling fluid system with low solid and low molecular polymer was to be adopted for CCSD-1 Well drilling, in accordance with the method of core drilling to be used, strata to be encountered and the hole structure designed, with the main technical descriptions shown in Table 9.2.

Artificial sodium bentonite or LBM-SD was to be added in the drilling fluid for core drilling to regulate its performance, so as for non-core drilling and expanding drilling (Table 9.3).

9.2 Drilling Fluid System

According to the requirements of scientific drilling, drilling fluid system should have performances of low viscosity, low gel, low solid content, low filter loss, and with good lubrication property.

9.2.1 Selection of Drilling Fluid System

Along with the development of petroleum drilling technology, drilling fluid technology has made considerable progress. According to dispersed medium in drilling fluid, drilling fluid can be divided into oil base, water base and gas base drilling fluid. In oil base drilling fluid there are a lot of organic hydrocarbon substances, which may seriously interfere borehole log, being unsuitable for scientific drilling. Gas or foam base drilling fluid is mainly applied to drill the strata with low formation pressure, and rotary blowout preventer needs to be installed at well mouth for safe drilling which may cause many troubles to drilling construction. Moreover, as down hole power is employed, drilling fluid is needed to work as the transmission and working medium for down hole power engine, but because of the compressibility of the gas base drilling fluid the power cannot be effectively transmitted to ensure a normal running of the down hole power drilling tool. Therefore, gas base drilling fluid is unsuitable for scientific drilling either. Only water base drilling fluid is the best choice of CCSD-1 Well and it is the most common type of drilling fluid in oil drilling and in geological exploration drilling. With good dispersion property to drilling materials, the properties of water base drilling fluid, especially rheological and weighted bearing capacity can be easily regulated; besides, the influence of drilling fluid to borehole log can be reduced as much as possible by optimizing mud making materials and mud conditioners. So, water base drilling fluid can fully satisfy the requirements of scientific drilling.

Organic drilling fluid treating chemicals are the most common in drilling, and the organic polymers for drilling fluid are different substances composed of repeated or homologous cells or monomers. Organic colloid materials can be applied to drilling fluid for declining filter loss, stabilizing clay soil, flocculating cuttings, increasing carrying capacity and can work as emulsifier and lubricant. With a variety of polymers, low solids non-dispersed mud system (LSND) can be widely used.

According to fabrication methods, organic polymer can be classified into three kinds, namely: natural polymers which can be used with simple treatment, such as amylum and guar; semi-artificial polymers such as sodium carboxymethyl cellulose (CMC) and derivant of amylum and balata; the most popular petroleum chemical derivants such as polyacrylamide (PAM) and polyacrylate are full-artificial polymers.

Based upon the types of monomer, polymer can be classified into single, double or copolymer with manifold monomers. According to the structure, copolymer can be of linear style or branched chain style, but cross-linking can take place in both styles by connecting of covalent bonds. Different compositions of a variety of monomers produce polymers with different structures, so, with different performances when working as drilling fluid treating agents. And this provides broad space for developing drilling fluid treating agents with different drilling purposes.

9.2.2 LBM-SD Composite Drilling Fluid Material

LBM-SD composite drilling fluid material was developed and manufactured by Beijing Institute of Exploration Engineering according to the 5000 m continuous core drilling situation of CCSD-1 Well, to satisfy the requirements of single property of drilling fluid, long term in drilling construction and high quality. LBM-SD is mixed and epurated with organic copolymer (LPA) and artificial sodium bentonite. According to the requirement of drilling operation, a satisfactory performance of drilling fluid can be obtained without adding other treating agents except water with a certain ratio. Thus, field mud making technology can be simplified and drilling fluid performance can be easily controlled and regulated.

LBM-SD is a kind of composite mud powder with low viscosity, low gel strength and low fluid loss, with strong capacities of salt resistance, calcium-magnesia resistance and temperature resistance. LBM-SD is a kind of selective flocculating and non-beneficiated mud making material. The main component of LBM-SD is LPA, which is a kind of liquid low polymer copolymerized at low temperature with crylic acid, acrylonitrile and acrylamide, mainly contains groups such as $-CONH_2$, $-COONa$, $-CN$ and $(-COO)_2Ca$, etc.

9.2.3 Drilling Fluid Mechanism and Composition of LPA Polymer

One of the most obvious innovations of modern drilling fluid technology is the introduction of superpolymer flocculant and the development of low solids non-dispersed and clay-free superpolymer mud system. Superpolymer flocculant can make the cuttings and inferior soil in drilling fluid in a non-dispersed flocculating state, which then can be cleaned by solids control equipment, and thus the problems of cuttings dispersing and accumulation exist in dispersed drilling fluid system can be well solved. The polymer in drilling fluid not only mainly works for flocculating, but also for restraining, lubricating for drag reduction, cross-linking for sealing and dilution shearing. One superpolymer molecule is adsorbed on several grains and as a bridge connecting each grain, then the adsorbed grains was agglomerated and coagulated by curling of superpolymer.

With different molecular weights and degrees of hydrolysis, polymer can be divided into beneficiated and non-

Fig. 9.1 Beneficiated and non-beneficiated active mechanism of polymer

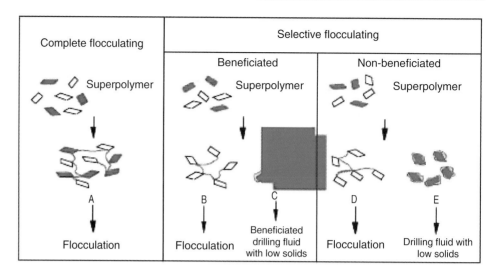

beneficiated selective flocculants according to its functions. Both the flocculants can flocculate the cuttings and inferior soil in drilling fluid but do not flocculate the colloidal grains of bentonite. The difference of the two flocculants is whether they can improve the viscosity of drilling fluid. The sketch of beneficiated and non-beneficiated active mechanism of polymer is shown in Fig. 9.1. Non-beneficiated selective flocculation is employed for diamond drilling while beneficiated selective flocculation is employed for oil drilling. LPA in LBM-SD composite mud material developed especially for CCSD-1 Well belongs to non-beneficiated selective flocculant.

1. **Active mechanism of function group**

The function group of organic macromolecular treating agent is the key for drilling fluid to give full play to its main function, and it is also the base for developing new treating agent.

(–COO)$_2$Ca is the main function group of CPA (calcium polyacrylate) superpolymer. CPA is a non-water-soluble macromolecular compound, which cannot be directly used as drilling fluid treating agent but can be coordinated with solutizer and dissolved in water, then used as treating agent. Generally sodium carbonate is used as solutizer but its addition has big influence upon the properties of the product. According to the test results, the optimal proportion of CPA to sodium carbonate should be 4:1 (mass) more or less (Pingquan and Shiliang 2003). If the addition of sodium carbonate is not enough, the hydrotrope of CPA is not good and the corresponding treatment effect will be poor. On molecular structure, there is an optimal ratio range of (–COO)$_2$Ca/–COONa, in this range, almost 43 % mol of (–COO)$_2$Ca are inverted into –COONa, and the ratio of Ca/Na is 3:1 more or less. If the addition of sodium carbonate is too much, most of the (–COO)$_2$Ca with weak hydratability will be inverted into –COONa with strong hydratability, then the antipollution capability of CPA will be reduced while filter loss increases as a result of net structure destroyed. With the most important advantage of strong antipollution, CPA is an excellent filtrate reducer of calcium salt drilling fluid. The fluid loss effect of CPA is better than Na-CMC. Filter loss cab be decreased from 50 ml to 4.5 ml by adding 0.6 % CPA into salt water drilling fluid, while Na-CMC can only decrease to 20 ml with the same addition. When salt water drilling fluid is treated by CPA, filter loss will be declined precipitously with the increase of the addition of CPA, while the viscosity will be dramatically increased, thus easily causing a thickening.

–CONH$_2$ is the main function group of polyacrylamide (PAM). Molecular weights of PAM produced under different ways and reaction conditions are extremely different, some with average molecular weight of just thousands, and some can be more than tens of millions. The contacting area of long chain PAM and water increases, flow resistance increases, too, so the viscosity of water solution is increased, and this viscosity varies with different molecular weights, densities and degrees of hydrolysis of PAM. Generally, if the other conditions are identical, the bigger the molecular weight and the longer the chain is, the higher the viscosity of water solution will be. The backbone of PAM molecule is C–C key connection, being reliable, thus with good heat stability and without obvious degradation reaction at high temperature of 130 °C. When PAM molecule is adsorbed on grain, drill stem or borehole wall surface, there will be produced a film with certain lubricity which can change the friction between solids surface into the friction between macromolecular chains, thus, friction resistance will be greatly decreased, viscous factor of filter cake will be declined, thus with good performance of lubrication and drag reducing.

The most distinct characteristic of –CONH$_2$ is that it has flocculating capability in comparison with propenyl drilling fluid. Flocculating capability of PAM is greatly related to its molecular weight, and the bigger of the molecular weight is, the stronger of the flocculating capability will be. Long molecular chain is good at bridge connecting more clay grains and the flocculating effect will be better, ordinarily, chain length of molecule is required to be six to ten times of clay grain. For PAM, the flocculating capability will be notable if molecular weight is bigger than one million, and will be weak if smaller than one million. There is no evident viscosity variation when pH value is five to nine, but usually viscosity of solution will be increased as a result of the increase of $-COO^-$ because of hydrolysis reaction and molecular chain is assumed to outspread under the influence of electrostatic repulsion in solution with high pH value. Besides, viscosity of PAM decreases with the increase of temperature and mineralization degree of the solution, thus, its antipollution capacity is weak.

$-CN$ is the main function group of polyacrylonitrile (PAN), and can be manufactured into hydrolyzed polyacrylonitrile (HPAN) under alkaline water reaction by hydrolization. Hydrolysis degree of HPAN is closely related to the product property and fluid loss function of drilling fluid. Test results indicated that the fluid loss effect would be the best when the content of carboxyl group was between 70 and 80 %, too much or too less was not beneficial (Pingquan and Shiliang 2003). Controlling hydrolysis condition is very important because too large hydrolysis degree will affect the adsorption of clay grains while too small degree will reduce hydratability.

A certain content of $-CN$ can improve salt resistance based upon different hydrolysis degrees. HPAN is a linear water soluble superpolymer, its water solubility is related to hydrolysis degree. There are three groups in HPAN molecular chain, $-CN$, $-CONH_2$ and $-COONa$. The first two are adsorbing groups while the latter is hydrating group. Adsorbing group determines the adsorptive capacity of superpolymer and clay, more adsorbing groups indicate strong adsorption of superpolymer to clay grains, i.e., lots of clay grains are adsorbed onto superpolymer, thus flocculating structure will be formed and this is unfavourable to the reaction of HPAN and clay grains; too less adsorbing group is unfavourable to clay adsorbing onto the molecular chain of superpolymer, thus the reaction effect is not good either. So the quantity of adsorbing groups should be appropriate. The amount of hydrating groups determines the effect of fluid loss of HPAN in drilling fluid, and the content of $-COONa$ is usually selected to 70–80 %. The molecular chain internodes of HPAN is $-C-C-$ structure, so its high temperature stability is satisfactory, to 200–230 °C, salt resistance capability is stronger, too; however, calcium resistance capability is weak and flocculence sediment will be produced when calcium contamination encountered.

The characteristics of the active mechanism of the hydrating and adsorbing groups in drilling fluid can be found as follows:

1. Low viscosity—low molecular weight;
2. Low solid phase—with some $-CONH_2$ (with non-beneficiated selective flocculating capability);
3. Low fluid loss—with some $-COONa$ (the effect of low fluid loss can be realized);
4. Salt resistance—with part of $-CN$ formed (certain of salt resistance capability achieved);
5. Calcium resistance—with part of $(-COO)_2Ca$ formed (calcium and magnesium resistance capability of drilling fluid can be improved).

A superpolymer can be produced by combining the three polymers through adopting their strong points to offset their weaknesses, and the requirements of mud making performance for scientific drilling can be realized by regulating the ratio of each function groups. Then a composite drilling fluid material can be produced by mixed refining this superpolymer with high quality sodium bentonite, and it can not only satisfy the requirements of CCSD for drilling fluid, but also greatly simplify mud technology at drill site. LPA is just a low ternary copolymer made according to the above-mentioned principle.

LPA contains four groups ($-CONH_2$, $(-COO)_2Ca$, $-CN$ and $-COONa$). Besides the function of fluid loss, LPA has the functions of selective flocculation and strong salt, calcium and magnesium resistance. Through reasonable proportion of the four groups in synthesis process, an ideal performance of drilling fluid can be achieved. Flocculation of LPA results from the lumpy flocculation produced by hydrogen bond adsorption and bridging between hydrogen of $-CONH_2$ adsorbing group on molecular chain and oxygen of clay surface, and then the lumpy flocculation causes floccule because of unstable subsidence on dynamics.

There are a certain amount of $-COONa$ hydrate groups in LPA molecular chain. $-COONa$ can be ionized into $COO^- + Na^+$ in water, makes the chain to be negative charge and hydrated. Thus, LPA has stronger hydration besides the functions of adsorbing and flocculation. To different kinds of clay, it's capabilities of adsorbing and flocculation are different. For example, bentonite grain is minute, cation exchange capacity is high (CEC = 80–150 meg/100 g), permanent negative charges on grain surface are much, electrical double layer is thick, ζ potential is high, hydrated film is thick and repulsion is large, so, adsorbing and flocculation capability of LPA to bentonite grain is low. Even some of bentonite grains are adsorbed and bridged, a stable subside on dynamics is hardly to be achieved. While inferior clay grains are coarse, cation exchange capacity is low (CEC = 0–40 meg/100 g), permanent negative charges on grain surface are less, electrical double layer is thin, ζ potential is low, hydrated film is thin and repulsion is small, so, the adsorbing, bridging and flocculation capabilities of LPA to inferior clay are stronger. To bentonite and inferior clay (for example, cuttings), LPA shows a selective flocculation effect.

Table 9.4 Main raw materials for composing LPA

Number	Raw materials	Code	Molecular formula	Physical property
1	Crylic acid	AC	$CH_2 = CH-COOH$	Molecular weight is 72.06, achromatic liquid, melting point is 13 °C, boiling point is 141.6 °C, density is 1.038 g/cm^3 under 30 °C
2	Acrylonitrile	AN	$CH_2 = CH-CN$	Molecular weight is 53, achromatic liquid, dissolved in water
3	Acrylamide	AM	$CH_2 = CH-CONH_2$	Molecular weight is 71, white powder, dissolved in water
4	Ammonium persulfate	APS	$(NH_4)_2S_2O_8$	Molecular weight is 228.2
5	Potassium persulfate	KPS	$K_2S_2O_8$	Molecular weight is 270.3
6	Sodium hydroxide		NaOH	Molecular weight is 40
7	Calcium hydroxide		$Ca(OH)_2$	Molecular weight is 74

In LPA molecular chain there are amount of $(-COO)_2Ca$, which can improve the capacities of fluid loss and calcium and magnesium resistance for drilling fluid.

In LPA molecular chain there are amount of –CN, which can improve the capacities of salt resistance and high temperature resistance for drilling fluid. Thus drilling fluid can have better stability under high temperature.

2. **Main raw materials for composing LPA Polymer**

LPA is a multicomponent copolymer polymerized with crylic acid (AC), acrylamide (AM) and acrylonitrile (AN) monomers under low temperature. The main raw materials of LPA and their physical properties are shown in Table 9.4.

changed into nano-carboxyl and then causing a decline of polymer's resistance to fouling. Quantity of $Ca(OH)_2$ relates to the resistance to fouling of the polymer, too much quantity will cause the increase of structural strength and filter loss, and the decline of solubility. As the type and quantity of initiating agent, and polymerization time directly affect the molecular weight of the polymer, so all must be strictly controlled otherwise the performances of the products cannot achieve the expectant results. The polymerization reaction process of ternary polymer with potassium persulfate as initiating agent is described as follows:

$$CH_2=CH + CH_2=CH + CH_2=CH + NaOH + Ca(OH)_2 \longrightarrow$$
$$\quad\quad | \quad\quad\quad\quad | \quad\quad\quad\quad |$$
$$\quad\quad CN \quad\quad COONH_2 \quad COOH$$

$$[-CH=C-]_x \cdots [-CH=C-]_y \cdots [-CH=C-]_z \cdots [-CH=C-]_z + NH_3$$
$$\quad\quad | \quad\quad\quad\quad\quad | \quad\quad\quad\quad\quad | \quad\quad\quad\quad\quad |$$
$$\quad\quad CN \quad\quad\quad\quad COONa \quad\quad\quad COOCa_{1/2} \quad\quad CONH_2$$

3. **Polymerization technology**

Pouring AC, AN, and AM proportionally into a container with NaOH water solution, agitating slowly, heating appropriately till the mixture can calorify by itself. Thickness is increasing while agitating and degree of polymerization can be controlled by thickness. When a certain thickness is achieved, that means that the three monomers have changed into a ternary polymer—LPA.

The ratio of each monomer, degree of polymerization and hydration, initiating agent and polymerization rate all can directly affect the performance of the polymer. In polymerization technology of LPA, the ratio of the three kinds of amphoteric organic monomer will directly affect not only the performance of composed polymer, but also the performance of LBM produced. The quantity of NaOH determines the hydration degree of the polymer and in case that the quantity is too much, nearly all the carboxyl will be

It is a heat radiation reaction in above-mentioned polymerization process. Polymerization temperature should be advertent in reaction process. Reaction process may intensify or even explode under too high temperature. Reaction progress can be controlled according to the viscosity variation of the product. After achieving a certain viscosity, reaction should be stopped for blending with artificial sodium bentonite in time and fabricating into LBM-SD, otherwise some larger molecular weight polymer will be produced after a certain time.

9.2.4 Manufacture Technology of LBM-SD

LBM-SD is blended mainly with artificial sodium bentonite (NV-1) and LPA polymer, with the main technological process of compounding—blending—extrusion—desiccation—comminuting—packing (Fig. 9.2).

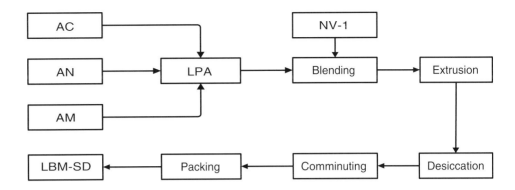

Fig. 9.2 Production technological process of LBM-SD

Blending and extrusion are very important in production technological process of LBM-SD.

Blending makes the two components mixed more symmetrically. Extrusion is completed at higher temperature of 120 °C more or less. Under high temperature and high pressure, through shear action, clay grains finish the organic combination with polymer under unhydrated condition, thus creating conditions for full-hydration and shortening hydration time in mud making, also avoiding agglomeration of the polymer usually occurred in mud making. LBM-SD can be directly used in borehole soon after agitation, so mud making technology at drill site is greatly simplified.

9.2.5 Evaluation Procedure of Drilling Fluid

1. **Test condition of drilling fluid performance**

   (1) Make-up medium: fresh water, 10 % NaCl water solution and artificial seawater. All prepared with distilled water.

   (2) Test temperature: room temperature of (24 ± 3) °C and high temperature of 150–210 °C

   (3) Test instrument: six-speed rotary viscosimeter, API FL press, HTHP FL press under pressure of 3.45 MPa and temperature of 260 °C, Type 50C HTHP rheometer, high temperature roller oven and high-speed agitator of 12,000–15,000 r/min.

   (4) Test standard: general routine of API RP 13B

   (5) Test items: rheological parameters (apparent viscosity (AV), plastic viscosity (PV), yield point (YP), gel strengths of G_{10s} and G_{10min}), API FL, HTHP FL with temperature of 150 °C and pressure of 3.45 MPa, weighted mud test (density of 1.1–2.0 g/cm^3), HTHP rheological property curve under temperature of 150 °C and pressure of 3.45 MPa.

2. **Routines and methods of drilling fluid test**

   (1) Preparing base mud for drilling fluid with fresh water, 10 % NaCl water solution and artificial seawater respectively.

   (2) High temperature performance tests, include filter loss (HTHP FL) and HTHP rheological property tests under high temperature. Sample is rolled for 16 h in a high temperature aging roller oven under constant temperature of 150 °C, then cooling to room temperature.

   (3) In HTHP FL testing, put the high temperature aged and cooled sample into filter loss press, testing the filter loss in 30 min under pressure of 3.45 MPa and constant temperature of 150 °C. Testing results are all listed into the datasheets of drilling fluid regular properties.

   (4) Weighting test. High density drilling fluid is mainly employed for balancing the pressure between formations and the stability of borehole. Though there was nearly no abnormal high formation pressure in CCSD-1 Well, for being ready for all eventualities, it was required that all the drilling fluid systems employed should have certain weighted bearing capability, reaching to 2.0 g/cm^3 according to the design requirements.

9.2.6 Performance of LBM-SD Drilling Fluid System

1. **Performance of fresh mud**

The dosage of LBM-SD in distilled water are 3, 4, 5, 6, 7, 8 % respectively in conventional performance tests. According to API RP 13B: 20 min high speed agitation, 16 h aging in room temperature of (24 ± 3) °C, and 5 min high speed agitation. Conventional performance testing includes rheological parameter and API filter loss. Conventional performances of fresh mud with different dosage of LBM-SD can be found in Table 9.5.

According to Table 9.5, fresh mud has less filter loss of 9.6 ml with LBM-SD dosage of 3 %, even when dosage increases to 8 %, yield point-plastic viscosity ratio is just 0.34 ks^{-1} and viscosity is just 14.5 MPas. This characteristic of low viscosity and low yield point can improve the fluidity of the drilling fluid and decrease circulating pressure loss, being essential to diamond core drilling.

Table 9.5 Performance of LBM-SD fresh mud

LBM dosage (%)	Apparent viscosity AV (mPas)	Yield-plastic ratio YP/PV (ks^{-1})	API filter loss (ml)	HTHP filter loss (ml)
3	4.0	0.14	9.6	40
4	5.5	0.15	8.0	36
5	7.0	0.17	7.8	30
6	9.5	0.19	7.6	26
7	11.8	0.26	7.4	22
8	14.5	0.34	6.8	21

Table 9.6 Performance of LBM-SD salt mud

LBM dosage (%)	Apparent viscosity AV (mPas)	Yield-plastic ratio YP/PV (ks^{-1})	API filter loss (ml)	HTHP filter loss (ml)
3	6.3	0.50	26	96
4	8.5	1.10	22	80
5	12.5	1.50	18	63
6	16.3	1.50	17	58
7	19.5	1.78	16	36
8	25.0	2.50	15	32

Table 9.7 Formula of artificial seawater

Component	NaCl	CaCl$_2$	MgCl$_2$	MgSO$_4$	KCl
Dosage (g/L)	41.54	1.69	3.69	4.92	2.24

2. **Performance of salt water mud**

Salt pollution is caused by different salts intruding into drilling fluid while drilling salt stratums. Usually the soluble salts encountered in drilling can be divided into monovalent and bivalent salts. The commonest monovalent salt encountered is NaCl, and KCl is less seen. Apparent viscosity, yield value, gel strength and filter loss of the drilling fluid are increased as a result of salt pollution. While drilling aquifer which contains salt, intruding of saline can cause the same pollution. Thus, salt resistance of drilling fluid is an important index in evaluation of drilling fluid for scientific drilling.

Preparing 10 % NaCl water solution with distilled water, and making salt drilling fluid with LBM-SD dosages of 3–8 % respectively. According to API RP 13B, drilling fluid performances with different dosage of LBM-SD are shown in Table 9.6.

LBM-SD salt mud prepared with 10 % NaCl water solution still has low viscosity and shear force, even LBM-SD dosage increases to 8 %, yield point-plastic viscosity ratio is only just 2.5 ks^{-1} (API standard is less than 3). Usually, drilling fluid will cause flocculation and viscosity increase when salt contamination is encountered. Based upon the salt resistance test it can be found that this LBM-SD system has satisfactory salt resistance stability, which is very important in scientific drilling, for during drilling process strata which contain soluble salt are likely encountered, and if drilling mud does not have good salt resistance, a series of complex borehole problems will arise.

3. **Performance of artificial seawater mud**

The main purpose of seawater resistance testing is to detect the comprehensive capability such as salt, calcium and magnesium resistance of drilling fluid. Bivalent metallic ions which mainly contained in artificial seawater, especially calcium and magnesium ions can obviously influence the drilling fluid properties. Aimed at detecting invasion resistance capability of LBM-SD, high concentration artificial seawater (Table 9.7) was prepared. Dosages of each ion in artificial seawater are shown in Table 9.8. Performances of artificial seawater mud with different dosages of LBM-SD can be found in Table 9.9.

In scientific drilling construction, many unknown factors will be encountered in drilling rock strata. A variety of soluble minerals which will greatly affect drilling fluid performance are likely encountered in drilling. Great risks exist if drilling fluid does not have good invasion resistance. Therefore, drilling fluid used for scientific drilling should not only be easily regulated according to construction requirements, but also have stronger antipollution capacity.

The comparison of apparent viscosities of fresh water, salt water and seawater with different dosages of LBM-SD is shown in Fig. 9.3.

Table 9.8 Dosages of each ion in artificial seawater

Component	Na^+	Ca^{++}	Mg^{++}	K^+	Cl^-	SO_4^-
Artificial seawater (mg/L)	16,330	610	1910	1170	30,120	3940

Table 9.9 Performance of LBM-SD artificial seawater mud

LBM dosage (%)	Apparent viscosity AV (mPas)	Yield-plastic ratio YP/PV (ks^{-1})	API filter loss (ml)	HTHP filter loss (ml)
3	2.5	0.25	28.6	88
4	3.5	0.28	23.0	82
5	5.3	0.31	17.0	48
6	6.3	0.50	14.6	32
7	8.3	0.70	12.8	32
8	9.5	0.90	11.0	28

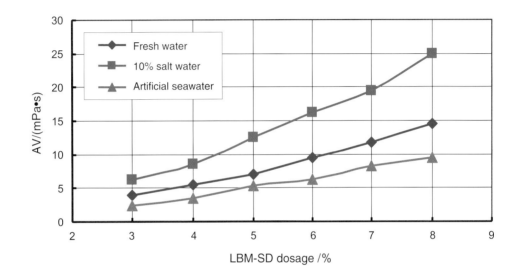

Fig. 9.3 Comparison of LBM-SD salt and seawater resistance properties

According to Fig. 9.3, it can be found that LBM-SD drilling fluid system has stronger calcium and magnesium resistance capability. Total amount of calcium and magnesium resistance reaches to more than 2500 mg/l. Thickening phenomenon can be found in salt water but the drilling fluid still has good fluidity.

4. **Property of temperature resistance**

The influence of high temperature to drilling fluid includes two aspects: on the one hand, with the increasing of temperature, drilling fluid properties change at the same time, thus, there is a great difference of the drilling fluid properties between under bottom hole high temperature and at wellhead low temperature, and this difference can only be detected by using the instrument which simulates bottom hole actual condition of high temperature and high pressure. On the other hand, each component of drilling fluid itself and some variations that may not be occurred under low temperature will all be excited under high temperature, then the rock formation contamination will become worse.

Generally speaking, bottom hole temperature of a 5000 m well may reach to 150–180 °C or even higher. Such high temperature must be a great influence to drilling fluid system. To water base drilling fluid, influence from pressure may be small while influence from high temperature will be a main one.

High temperature and high pressure influence drilling fluid properties in following ways:

(1) Physical aspects: along with the increase of temperature, liquid viscosity decreases; while with the increase of pressure, liquid density increases and then viscosity increases, too.

(2) Chemical aspects: all hydroxides will react with clay under temperature higher than 100 °C more or less. To low alkaline drilling fluid, the influence of temperature to rheological property is not serious, while it will become serious for high alkaline drilling fluid. This influence depends upon temperature and types of hydroxide metallic ions, when temperature is approximately higher than 150 °C, high solid mud treated with lime will produce hydrated aluminosilicate, and coagulate to be with rheology property similar to cement.

Fig. 9.4 Comparison of apparent viscosity variation between LBM-SD and HPAN with temperature changing

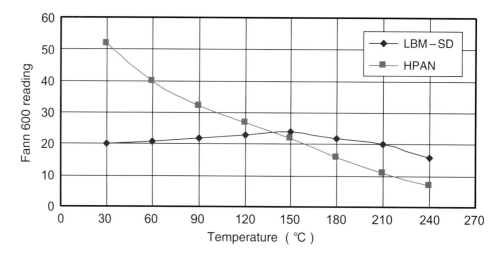

(3) Electrochemistry aspects: ionic active capability of each electrolyte will increase along with the increase of temperature. The exchange of alkaline ions changes the balance between grain gravitational and repulsion force. It also changes the balance between decentralization and flocculation at the same time. The degree and orientation of these changes and their influence to drilling fluid rheology vary along with the electrochemistry characteristics of the drilling fluid.

Usually, if drilling fluid is not of a floccular type, its plastic viscosity and yield point will decrease when temperature increases to 177 °C. If drilling fluid is floccular type, its plastic viscosity will decrease while yield point increase dramatically when temperature reaches to over 100 °C. Aimed at detecting the high temperature resistance property of LBM-SD drilling fluid system, ammonium hydrolyzed polyacrylonitrile (NH_4–HPAN) or ammonium salt with moderate price was chosen for comparison test. NH_4–HPAN was hydrolysate of waste acrylic fibers under high pressure of 2 MPa and high temperature of 200 °C, or called as high pressure hydrated polyacrylonitrile with hydrolysis degree of approximate 50 % and molecular weight of no more than one hundred thousand. NH_4^+ has the similar inhibiting capability as K^+, with the properties of anti-sloughing, high temperature resistance and non-increase of viscosity.

Based upon performance-price ratio, usual dosage of NH_4–HPAN in low solids drilling fluid is 1–2 %. Dosage of NH_4–HPAN in laboratory mud property comparison test was 2 %, in comparison with the high temperature resistance performance of LBM-SD drilling fluid.

High temperature property tests include high temperature high pressure filter loss (HTHP FL) and HTHP rheology testing. Samples are rolled for 16 h in high temperature aging roller oven under constant temperature of 150 °C, then, cooled to room temperature for use. In HTHP FL testing, the sample which has been aged under high temperature and then cooled is poured into HTHP filtration instrument, with pressure of 3.45 MPa and constant temperature of 150 °C maintained, and then the filter loss in 30 min is detected. HTHP FL of LBM-SD can be found in Tables 9.5, 9.6 and 9.9.

HTHP rheology testing was conducted in 50 °C HTHP rheometer, with the rotation speed of the rotor of the high temperature rheometer was set at a constant of 600 r/min and starting temperature set at 30 °C with heating rate of 1 °C/min. Readings on the viscometer at nine measuring points of 30, 60, 90, 120, 150, 180, 210 and 240 °C were respectively collected, at each measuring point the constant temperature time was 30 min. Rheological properties of LBM-SD and HPAN mud systems under different temperatures are compared in Fig. 9.4.

It can be found from Fig. 9.4 that LBM-SD has better high temperature stability than HPAN, and its viscosity is still stable even when temperature increases to 210 °C. The viscosity of HPAN decreases gradually with the increase of temperature, showing its sensitivity to high temperature.

Yield value of drilling fluid, also called as yield point, is the measurement of attraction force between solid grains when drilling fluid is under flowing state. Usually, yield value of drilling fluid increases with the increase of temperature. From the test results in Fig. 9.5 it can be found that though the yield values of both the treating agents increase with the increase of temperature, yield value of LBM-SD is always lower than that of HPAN.

5. **Bearing capacity**

To ensure the safety in drilling construction, the drilling fluid system used for scientific drilling must have better bearing capacity or suspending capacity for weighting materials, then mud can be weighted to enough density and should have better rheological property once the abnormal borehole pressure happened. Drilling fluid performance of 8 % LBM-SD fresh mud weighted to density of 2.0 g/cm³ is shown in Table 9.10. According to Table 9.10 it can be

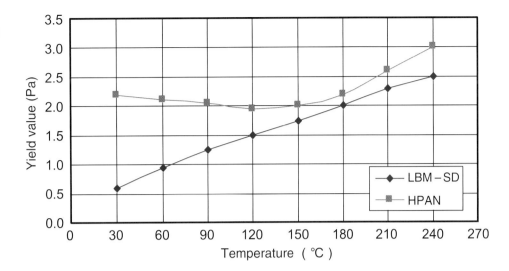

Fig. 9.5 Comparison of yield value variation between LBM-SD and HPAN with temperature changing

Table 9.10 Performance of LBM weighted drilling fluid

LBM-SD content (%)	AV (mPas)	YP (Pa)	YP/PV (ks^{-1})	API FL (ml)	HTHP-FL (ml)
8	64	19	0.3	4	30

found that even when the 8 % LBM-SD drilling fluid system is weighted to density of 2.0 g/cm³, its filter loss can still be kept at 4 ml.

According to above-mentioned testing results it can be found that LBM-SD drilling fluid system has the characteristics of low viscosity, low yield value, low filter loss and low solid phase, and with good capabilities of salt resistance, invasion resistance and temperature resistance and good bearing capacity, thus can fully satisfy all the requirements for mud performance in scientific drilling.

9.3 Drilling Fluid for Core Drilling

9.3.1 Properties

To prepare drilling mud at drill site, LBM-SD composite mud powder can be added directly into the water used only for mud making and then agitation. Other materials are not need to be added and thus field mud preparation technology was greatly simplified. To keep rheological parameters steady in using, a certain amount of LBM-SD can be supplemented according to drilling requirements to control mud performance easily. The pH value of drilling fluid may be decreased as a result of weather change or invasion of minerals and groundwater from rock formation, so pH value should be detected in time in application, and NaON is employed for controlling pH value in the range of 8–9.

Because shear force of LBM-SD drilling fluid system was low and the drift diameter of the upper Φ244.5 mm moving casing was as big as Φ218.4 mm, the problem of low annulus uplift velocity occurred in pilot hole core drilling, so artificial sodium bentonite and HV-CMC were properly added to improve suspending and carrying capacities.

1. **Density**

3 % LBM-SD single composite material was employed for preparation of core drilling fluid with density of 1.05 g/cm³. Adopting this low density and solid phase mud was because of the following factors:

(1) Because PDM hydro-hammer drilling was applied and low density mud could improve work efficiency of those down hole power drilling tools.
(2) Low solid phase could reduce the wear of mud to drilling tools, improve service lives of the drilling tools, and maintain longer steady working time and reliability of working.
(3) All the rock formations drilled were metamorphic rock with tight structure, subminiature permeability and there was almost no formation pressure, thus the problem of using mud density to balance formation pressure did not exist.
(4) Low density mud could improve efficiency of diamond core drilling, being good at solid control and improving lubricating effect of lubricant.
(5) The cost of mud preparation was low.

2. **Viscosity**

Viscosity is a ratio between shear stress and shear rate and is a measurement of flow resistance inside a fluid. To improve drilling efficiency, mud viscosity must be controlled. The viscosity of an ideal mud should be decreased with the increase of shear rate viewing from the angle of viscosity, and water base drilling fluid belongs to this kind. Being different from density, the influence of viscosity to rotation speed is not direct because it influences ROP through affecting the elimination of the cuttings at hole bottom. When mud becomes thinner, its circulation pressure loss is decreased and the pressure loss through drill bit is increased, so that the fluidic velocity for eliminating the cuttings is improved and a larger lashing force to hole bottom can be produced.

The influence of bottom hole cleanliness on diamond drilling efficiency is very obvious. Usually, the circulating passage between core drilling tool and down hole power tools is confined, the viscosity of drilling fluid for diamond core drilling cannot be too high, or the increase of viscosity will cause larger circulating pressure loss in the process of circulation and jamming may occur as a result of higher viscosity. Purely seeing from cuttings removal and drill bit cooling, fresh water is the best drilling fluid for diamond drilling, but because its weak carrying capacity, quick settling of the cuttings may cause bit burying after circulation is stopped.

In view of the drilling methods employed and the hole structure designed in CCSD-1 Well, better fluidity of drilling fluid was required to decrease the circulating pressure drop, increase the suspension capability for cuttings and prevent deposition of the cuttings. In pilot hole core drilling, because the inner diameter of upper moving casing was bigger, uplift velocity of drilling fluid was as low as 0.2–0.25 m/s, thus shear force could be properly improved in preparation of mud according to the condition of cuttings carrying in drilling. Considering that Φ157 mm drill bit and Φ89 mm drill rod were employed for diamond drilling in open hole section, annular uplift velocity could achieve 0.76 m/s under delivery capacity of 10 L/s and cuttings were distributed between 5 and 100 μm in size, so 3 % LBM-SD drilling fluid with Fann θ600 value of 10–12 and funnel viscosity of 28–32 s more or less could satisfy all the requirements of carrying and suspending the cuttings.

3. Filter loss

In drilling operations, to avoid formation fluid entering into borehole, the hydrostatic fluid column pressure of drilling fluid is usually needed to be higher than the pressure of fluid in formation pore, then drilling fluid trends to intrude into permeable strata. During filtration, solid phase in drilling fluid will produce a layer of mud cake at borehole wall; the permeability of the mud cake is closely related to drilling fluid quality. Mud cake with high permeability will make mud cake thicker and thus the diameter of borehole is decreased, then a variety of problems will arise, such as too large torque when drill rod rotation, tight pull, too high swabbing pressure and surge pressure in lifting drilling tools, and so on. To maintain the stability of the borehole wall and decrease the intrusion of filtrate into strata, drilling fluid must be treated to ensure the permeability of the mud cake as low as possible.

Filtration property of drilling fluid manifests as (1) the capability of solid phase in drilling fluid producing mud cake; and (2) permeability magnitude of mud cake. For drilling fluids with similar solid phase concentration, the lower the permeability of mud cake, the thinner the mud cake and the less the filter loss will be. Filtration property is determined by content and physical properties of gel material in drilling fluid. Drilling fluid with adequate gel material can decrease in-the-hole accidents to a minimum. Two filtrations exist when drilling fluid is in borehole: static filtration exists when drilling fluid circulation is stopped and mud cake will become thicker as time goes on; when drilling fluid is being circulated, mud cake will be eroded by flowing drilling fluid and thicker mud cake is restricted, it is called dynamic filtration. Dynamic filtration rate is far more than static filtration rate. API filter press is usually employed though it is only an envelope test and cannot indicate the filtration property of drilling fluid actually in borehole, only for reference.

After filtration, the prerequisite of drilling fluid to produce a layer of mud cake is that the stratum must be permeable. In CCSD-1 Well, all the strata encountered are metamorphic rocks with extremely low permeability and nearly non-permeable except some broken or lost circulation borehole sections, furthermore, there is no water-sensitive stratum either. So, to drilling fluid, high or low filter loss is not obvious in protecting borehole. If the filter loss of drilling fluid is decreased too much, the rheological property of drilling fluid will be affected and the cost of drilling fluid will be increased, what's more, it is unfavorable to diamond core drilling.

According to the laboratory test of the LBM-SD drilling fluid material used in CCSD-1 Well, drilling fluid prepared with 3 % LBM-SD has API filter loss of no more than 10 ml. In actual drilling construction, a small amount of PAL should be added into drilling fluid to make sure that the filter loss would be less than 12 ml, which could satisfy the requirements of drilling.

4. **Rheological property**

Rheological property of drilling fluid is directly related to its flow rate and circulation pressure, and affects the flow behavior of drilling fluid, too. Laminar flow is the main in low rate flowing, with regular current, and its relationship between circulation pressure and flow rate is a function of fluid viscosity. Turbulent flow is the main in high rate flowing, with aberrant current and mainly restricted by inertia of fluid motion, and its flow equation is empirical. Drilling fluid usually presents three flowing patterns, i.e., Newtonian fluid, Bingham plastic fluid and pseudoplastic fluid.

Flow behavior of drilling fluid is directly related to pressure loss and cuttings carrying capacity and can influence drilling process at the same time. Drilling fluid without satisfactory flow behavior may cause a series of problems, such as hole sloughing, sand settling, decrease of penetration rate, borehole enlargement, sticking, lost circulation, and so on. The phenomenon of effective or apparent viscosity (shear stress/shear rate) decreasing with the increase of shear rate is called shear thinning effect which is very useful to

Table 9.11 Lubrication test result of GLUB

Style of drilling fluid	Dosage of GLUB (%)	3 % LBM-SD		6 % LBM-SD	
		Decrease of friction coefficient (%)	Decrease of torque (%)	Decrease of friction coefficient (%)	Decrease of torque (%)
Water drilling fluid	0.5	66.82	53.33	63.35	50.00
	1.0	77.25	63.33	71.20	57.14
	1.5	83.89	70.00	78.53	64.29
	2.0	90.52	76.67	85.86	71.43
Drilling fluid with 10 % NaCl	0.5	24.51	17.65	71.57	57.89
	1.0	39.92	29.41	76.59	63.16
	1.5	62.85	50.00	81.61	68.42
	2.0	62.85	50.00	81.61	68.42
Artificial seawater drilling fluid	0.5	27.06	18.75	46.64	35.29
	1.0	51.76	37.52	66.40	52.94
	1.5	68.24	50.00	78.26	64.71
	2.0	68.24	50.00	78.26	64.71

drilling fluid. Effective viscosity of drilling fluid is comparatively smaller under the condition of high shear rate in drill rod, and then pump pressure is declined. While in annular space, cuttings carrying capacity is improved as a result of the comparatively high effective viscosity of drilling fluid under the condition of low shear rate.

The hole structure designed and the core drilling tool assembly (with the length of 5000 m and 89 mm drill rod has 70.2 mm inner diameter) employed for CCSD-1 Well determined that the drilling fluid must have better shear thinning effect to decline circulating pressure. Meanwhile, drilling fluid with lower yield value was required to reduce the pump starting pressure and decrease the pump starting pressure surge.

5. Oiliness

In scientific drilling, especially in ultradeep well drilling, high requirements are put forward for lubricating technology of drilling fluid. In diamond drilling high speed rotation is required for drilling tool. A rotating linear velocity of 1.5–3 m/s is required for drill bit, especially when synthetic diamond impregnated drill bit is employed. Because the annular clearance between drilling tool and borehole wall is small, the resistance force produced by high speed drilling tool rotation is huge, thus the drilling fluid used should have good oiliness. When borehole is of large deflection, large frictional drag between drilling tool and borehole wall will be produced during tripping and key slots on borehole wall will be drawn out. It is very necessary to improve the oiliness of the drilling fluid and reduce the frictional drag to ensure the success of 5000 m core drilling in CCSD-1 Well.

GLUB lubricant was prepared especially for the drilling requirements of CCSD-1 Well by Beijing Institute of Exploration Engineering. It is a kind of emulsion lubricant with vegetable oil and mineral oil as the base oil, with strong capability of calcium, magnesium, salt and high temperature resistance, and without fluorescent display. The main composition of GLUB lubricant includes soy oil, white oil, ABS, SP-80 and stearin etc., with recommended dosage of 0.5–2.0 %.

Based upon the Extreme Pressure Lubricant Test Method Used for Drilling Fluid, China oil and gas professional standards SY/T5662-94, reduced rate of friction coefficient and reduced rate of relative torque of GLUB lubricant were tested, and at the same time the compatibility performance of GLUB lubricant with LBM-SD drilling fluid system was tested. The base mud of drilling fluid was prepared with LBM-SD dosage of 3 and 6 % respectively, with the lubrication test results shown in Table 9.11.

The compatibility performance of GLUB and LBM-SD was good. There was no obvious influence to rheological properties of drilling fluid, according to test, no flocculated aggregation was produced based upon observation and there was no obvious variation in filter loss, shear force, invasion resistance capability of the drilling fluid after being prepared. The comparison test results of the compatibility performance of 3 % LBM-SD base mud with 1 % GLUB can be found in Table 9.12.

Table 9.12 Compatibility performance test result of LBM-SD and GLUB

GLUB (%)	Φ_{600}	Φ_{300}	Φ_{200}	Φ_{100}	Φ_6	Φ_3	G_{10s}/G_{10min}	API filter loss (ml)
0	17.0	9.5	7.0	3.5	1.0	0.5	0.5/0.5	12
1	16.0	9.0	6.0	3.0	0.3	0	0/0	12

Through the comparison, it can be found that there is no obvious influence to rheological properties after 1 % GLUB is added into LBM-SD base mud.

9.3.2 Circulating Pressure Drop

Pressure drop and distribution of each location of pressure drop circulation in the circulation system need to be accurately forecasted in the processes of design and inspection of drilling constructions. Those pressures should be understood in optimizing core drilling system to transfer energy to down hole power system to the maximum limit. Especially in some fractured formations, annular pressure drop should be understood in time to reduce the overbalanced pressure of wellbore to the minimum. The circulating pressure drop of the drilling fluid in CCSD-1 Well is huge, if the shear force of the drilling fluid cannot be effectively decreased, not only the circulating resistance force will be increased, but also the working horsepower of the down hole power system will be decreased at the same time, what's more, drilling fluid is hard to discharge from drill rod in the progress of running out of hole, which may cause drilling fluid ejection at wellhead. There is a close relationship between high-quality, fast and safety drilling and regulating, controlling and optimizing rheological parameters of drilling fluid according to the characteristics of strata and the requirements of drilling technology. Optimization of the rheological properties of drilling fluid plays an important role in solving problems such as cuttings carrying and suspension, and reducing circulating pressure drop.

The value of drilling fluid circulating pressure drop in drill rod will directly affect the efficiency of down hole power system. The frictional pressure drop of drilling fluid in annular space (between drill rod and borehole wall) is intimately related to the working behavior of down hole power system, too. Besides, the increase of drilling fluid circulating pressure drop will increase the load of mud pump.

1. **Selection of rheological model**

The selection of rheological model is the basis of rheology calculation, only if the rheological model which is proximal to field mud is chosen, reliability of the calculation can be guaranteed. Usually, the simplest Binghanm model is chosen to describe the flowing properties of drilling fluid, but it cannot accurately describe the rheological properties of drilling fluid. Generally speaking, it is believed that Power model, Casson model and Hershel-Bulkley model all can describe the properties of drilling fluid fairly well.

Standpipe pressure in drilling progress is composed of circulating pressure inside drill rod, circulating pressure in annulus and circulating pressure of down hole drilling tool. Accurate and real-time calculation of circulating pressure drop of drilling tool provides reliable basis for correctly judging the behavior of down hole drilling tool, also provides reference for drilling technology design, especially for drilling fluid design.

According to the tour report at the construction site of CCSD-1 Well, a total of 196 groups of data from well depth of 101–1000 m were continuously acquired. Based upon the principle of data statistics, small probability data with deviation larger than 3 were eliminated. The average readings of $\Phi600$–$\Phi3$ on Fann viscosimeter of the drilling fluid in this hole section were 9.25, 6.83, 4.44, 2.86, 0.80 and 0.65.

Binghanm, power law and Casson models were primarily determined as the objective models. For easy to deal with, the models were changed into linear form, namely y = ax + b type.

Binghanm flow pattern, $y = \tau_b + \eta_b \gamma$

Power law flow pattern, $\tau = \kappa \gamma^n$, changed into the pattern of $\tau_m = \tau_k + n\gamma_m$

where, $\tau_m = \log \tau$, $\tau_k = \log \tau$, $\gamma_m = \log \gamma$

Casson flow pattern, $\tau^{1/2} = \tau_C^{1/2} + \eta_C^{1/2} \gamma^{1/2}$

changed into the pattern of $\tau_{1/2} = \tau_{C1/2} + \eta \tau_{C1/2} \gamma_{1/2}$

where, $\tau_{1/2} = \tau^{1/2}$, $\tau_{C1/2} = \tau_C^{1/2}$, $\eta_{C1/2} = \eta_C^{1/2}$, $\gamma_{1/2} = \gamma^{1/2}$

Calculation results by linear regression are shown in Table 9.13.

Through regression analysis, three rheological models were educed as follows:

Binghanm model, $\tau = 0.58 + 0.0044\gamma$
Power law model, $\tau = 0.134\gamma^{0.5}$
Casson model, $\tau^{1/2} = 0.526 + 0.051\gamma^{1/2}$

According to the statistics principle and statistical results in Table 9.13, three models all achieved the conditions of F < F $_{0.05,(1,4)}$ = 7.71 and ρ > ρ $_{0.05,4}$ = 0.8114. Casson model had the smallest total deviation quadratic sum (Q) and F checkout, and

Table 9.13 Regression analysis results of the three rheological models

Rheological model		Binghanm	Power-law	Casson
Rheological equation		$\tau = \tau_b + \eta_b\gamma$	$\tau = k\gamma^n$	$\tau^{1/2} = \tau_C^{1/2} + \eta_C^{1/2}\gamma^{1/2}$
Regression result		$\tau_b = 0.58$	$n = 0.5$	$\tau_C^{1/2} = 0.526$
		$\eta_b = 0.0044$	$k = 0.134$	$\eta_C^{1/2} = 0.051$
Total deviation quadratic sum (Q)		0.482	0.410	0.173
F checkout		0.0085	0.0027	0.0015
Correlated checkout (P)		0.983	0.995	0.999

9.3 Drilling Fluid for Core Drilling

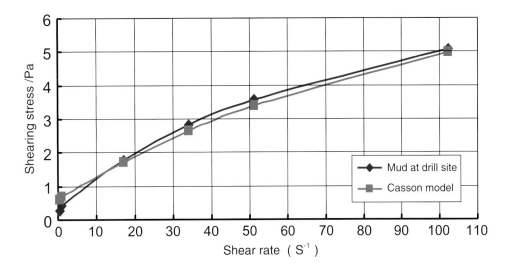

Fig. 9.6 Curve comparisons of CCSD-1 well site mud and Casson rheological model

with the largest correlated checkout, thus, Casson model could be optimized as the rheological model of CCSD-1 Well.

Mud rheological performance was directly related to pressure loss and capacity of carrying cuttings in drilling progress, also related to drilling efficiency at the same time. According to the drilling fluid tour report at drill site of CCSD-1 Well, 96 groups of data within the well depth from 4500 to 5000 m were continuously acquired. On the base of mathematical statistics principle, small probability data with deviation larger than 3 were eliminated. The average readings of $\Phi 600$–$\Phi 3$ on Fann viscosimeter of the drilling fluid in this hole section were 9.94, 7.00, 5.50, 3.47, 0.81 and 0.53. Rheological performance of the mud at drill site drawn based upon the statistical results is shown in Fig. 9.6.

The swirling flow formula of Casson fluid which flows in rotary viscosimeter can be expressed as:

Casson viscosity, $\eta_c^{1/2} = 0.03779(\phi_{600}^{1/2} - \phi_{100}^{1/2})$

Casson shearing stress, $\tau_c^{1/2} = 0.4775[(6\phi_{100})^{1/2} - \phi_{600}^{1/2}]$

Through calculation, it could be found that the Casson rheological model of the drilling mud at drill site was $\tau_w^{1/2} = 0.673 + 0.0487 \gamma_n^{1/2}$.

In comparison with Binghanm and power law models, Casson model has a fine description of drill mud in a large range of shear rate. Most of the flowing behavior of drill mud coincides well with Casson curve; even under low shear rate the calculation result is satisfactory. Casson plastic viscosity is the measurement of internal friction property of drill mud, which is only determined by types, content and fineness of solid, viscosity of liquid and content of macromolecular substances. Different from Binghanm plastic viscosity, Casson plastic viscosity is hardly affected by structure property of drill mud. Casson yield value is the measurement of coupling force of drill mud under dynamic state, and also the measurement of polarity or dynamic structural strength of mud grains. So every kind of substance which can influence the electrochemical property of fluid, such as dispersing agent, electrolyte, organic flocculent, solid content and dispersion degree, all can influence Casson yield value.

2. **Calculation of circulating pressure drop**

Circulating pressure drop in the process of circulation mainly includes the circulating pressure drop inside drilling pipe, circulating pressure drop in annulus and down hole power pressure drop (including dispersed flow on PDM, hydro-hammer, core drilling tool and bit face). When performance of drilling fluid is constant, circulating pressure drop in annulus and inside drill pipe will increase with well depth, while down hole power pressure drop will not change, thus, pressure drops of down hole power affected by variation of discharge capacity can be measured by wellhead test.

(1) Calculation of circulating pressure drop in annulus and inside drill pipe

In drilling fluid rheology, after pipe wall shear stress τ_w is determined, pressure drop along pipe length L can be acquired by the following formula.

$$P = \frac{4\tau_W L}{D} \quad (9.1)$$

In which, P is pressure drop, Pa. To annular flow, $D = D_H - D_P$. To inside pipe flow, $D = D_0$.

Under the condition of laminar flow, the rheological parameters of Casson fluid in annular flowing is given by

Pipe wall shear stress in annular flow

$$\tau_W = \eta_c \gamma_n \cdot \frac{1}{1 - 2.4\phi^{1/2} + 1.5\phi - 0.1\phi^3} \quad (9.2)$$

Annular flow, unit rate of flow kernel,

$$\phi - (1 - 2.4\phi^{1/2} + 1.5\phi - 0.1\phi^3)\Pi = 0 \quad (9.3)$$

where, $\Pi = \frac{\tau_c}{\eta_c \gamma_n}$

Table 9.14 Calculation of circulating pressure drop in each part of borehole circulation passageway

Location	ID (m)	OD (m)	Length (m)	D (m)	V (m/s)	$R_n^{0.25}$	f	P (MPa)
Drill pipe inside casing	0.070		5000	0.070	2.46	16.9	0.00468	4.84
Annular drill pipe (in casing)	0.175	0.089	3600	0.086	0.56	12.2	0.00648	0.22
Annular drill pipe (in borehole)	0.157	0.089	1307	0.068	0.76	12.36	0.00640	0.15
Annular drill collar	0.157	0.120	81	0.037	1.24	11.94	0.00662	0.09
Annular drill string	0.157	0.140	12	0.017	2.52	11.77	0.00672	0.06

Annular flow

$$\gamma_n = \frac{12\bar{v}}{D_H - D_P} \quad (9.4)$$

Under the condition of laminar flow, the rheological parameters of Casson fluid of inside pipe flow is given by
Pipe wall shear stress of conduit flow

$$\tau_w = \eta_c \gamma_n \cdot \frac{21}{21 - 48\phi^{1/2} + 28\phi - 0.1\phi^4} \quad (9.5)$$

Conduit flow, flow kernel of unit rate,

$$21\phi - (21 - 48\phi^{1/2} + 28\phi - \phi^4)\Pi = 0 \quad (9.6)$$

where, $\Pi = \frac{\tau_c}{\eta_c \gamma_n}$

Conduit flow

$$\gamma_n = \frac{8\bar{v}}{D_0} \quad (9.7)$$

To Newtonian fluid, it is laminar flow if Re is less than 2100, while it is turbulent flow when Re is larger than or equal to 2100. In core drilling of CCSD-1 Well, no matter in annulus or inside pipe flowing conditions of drilling fluid were both of non-Newtonian fluid, namely turbulent flow. To simplify calculation, Fanning equation was employed for the calculation of circulating pressure drop

$$P = f \cdot \frac{2\rho v^2}{D} L \quad (9.8)$$

where, f is Fanning friction factor which is a dimensionless number. P is pressure drop, Pa.
To annular flow, $D = D_H - D_P$. To inside pipe flow, $D = D_0$.
Simplified Blasius equation is usually employed in engineering construction, to Casson liquid, Blasius equation is given as

$$f = 0.0791/R_n^{0.25} \quad (2100 \le R_n \le 100000) \quad (9.9)$$

In which, $R_n = \frac{\rho D \bar{v}}{\eta_c}$

When 89 mm drill rod and Φ157 mm diamond core drill bit were employed for drilling, the circulating discharge rate reached 10L/s with mud density of 1.07 g/cm³, and the hole depth reached to 5000 m, the circulating pressure drop of inside pipe and in annulus at different hole sections was calculated by Casson flow model, with the calculated parameters and results shown in Table 9.14.

From the calculating results in Table 9.14 it is known that circulating pressure drop of inside drill pipe is 4.84 MPa and sum of annular circulating pressure drop is 0.52 MPa.

(2) Circulating pressure drop of double tube core drilling tool

Generally speaking, circulating pressure drop of double tube core drilling tool does not change with the increase of well depth. As a result of the structure limitations of swivel type joint of swivel type double tube drilling tool and the narrow (5 mm) annular overflow area between inner and outer core barrels, large circulating pressure drop will be caused in mud circulating progress. When the discharge rate is 10 L/s more or less, circulating pressure drop of double tube drilling tool will be approximately 2 MPa, and at well depth of 1800–2000 m, it accounts for about 20 % of the total circulating pressure drop. Circulating discharge rate and circulating pressure drop of core drilling tool tested at drill site are shown in Fig. 9.7.

(3) Circulating pressure drop of PDM

Circulating pressure drop of 4LZ120 × 7 PDM tested at drill site can be found in Fig. 9.8. The circulating pressure drop of PDM is mainly affected by the perfectness of the PDM (old or new) and well depth. Circulating pressure drop will be fluctuated within a certain range in drilling progress because of the variation of rotary torque.

(4) Circulating pressure drop of hydro-hammer

When discharge rate was 10 L/s, circulating pressure drop of YZX127 hydro-hammer tested at drill site was 2.5 MPa, shown in Fig. 9.9.

3. Pressure distribution of circulation system

Under the conditions of well depth of 5000 m, discharge rate of 10 L/s, 4LZ120 PDM and YZX127 hydro-hammer employed, the pressure drop distribution at each part of mud circulating path can be found in Table 9.15 and Fig. 9.10.

Meter readings of standpipe pressure (SPP) at each roundtrip in core drilling from 4991.50 to 5050.77 m are listed in Table 9.16. In eight roundtrips listed, the average

9.3 Drilling Fluid for Core Drilling

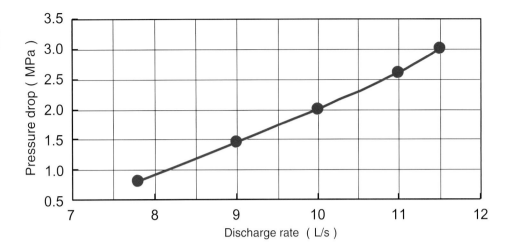

Fig. 9.7 Circulating discharge rate and circulating pressure drop of core drilling tool tested at drill site

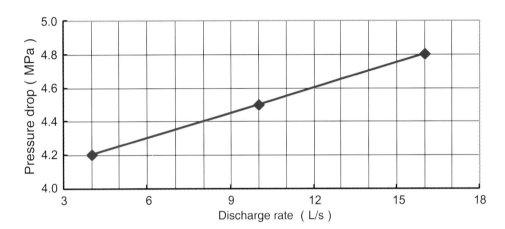

Fig. 9.8 Circulating pressure drop of 4LZ120 × 7Y PDM tested at drill site

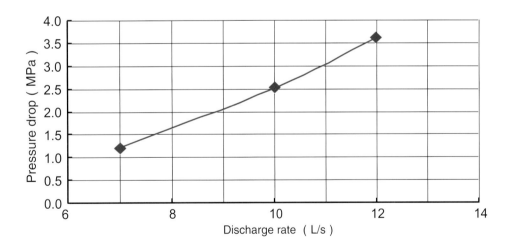

Fig. 9.9 Circulating pressure drop of YZX127 hydro-hammer tested at drill site

Table 9.15 Pressure drop distribution in mud circulating progress

Item	Inside pipe	Annulus	Double tube drilling tool	PDM	Hydro-hammer	Total pressure drop
Pressure drop (MPa)	4.84	0.52	2.00	4.50	2.50	14.36
Ratio (%)	34	3.6	14	31	17.4	100

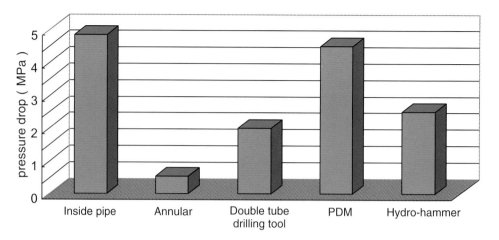

Fig. 9.10 Pressure drop distribution in mud circulating progress

standpipe pressure is 14.45 MPa, which is very close to the theoretical calculation result.

The theoretical calculation results were lower than the values tested at drill site because in theoretical calculation the factors such as pressure drop produced by cross flow on bit bottom, well deviation and measurement error were not taken into consideration. Besides, with the variation of strata, pressure change, reading accuracy of pressure meter and variation of temperature at hole bottom all would cause the pressure drop variation of PDM. Even in the same roundtrip, circulating pressure drop would vary continuously. Under normal conditions, circulating pressure drop is low at the beginning of a roundtrip and constantly increases in the whole roundtrip. This variation range is no more than 1.0 MPa under normal drilling conditions.

9.3.3 Lubrication Effect of Drilling Fluid

Lubricity of drilling fluid is very important to the behavior of down hole drilling tools, especially in deep hole section when deviation angle is larger than 20°, the lubricity has a nonreplaceable effect to decreasing drilling torque, stabilizing drilling parameters and deceasing the wear of drilling tools, etc. The detected results of lubricity of drilling fluid samples at drill site before and after added with lubricant are shown in Table 9.17.

From the detected results it can be found that the friction factor of drilling fluid was obviously decreased after added with lubricant. The average reduced rate of friction factor was 82.5 %.

In the early stage of pilot hole construction, no lubricant was added into the drilling fluid in core drilling in the hole section upper than 1837.09 m in consideration of down hole power was to be employed and drill rod would not rotate. However, the bearing of the double tube swivel type core drilling tool was seriously worn and had to be changed nearly in every one or two roundtrips. Swivel action failed as a result of the serious wear of the bearing, it had considerable large destroy to core and seriously affected the core recovery and the footage drilled per roundtrip. After adding with 1.3 % lubricant in the hole section under the depth of 1837.09 m, drilling efficiency was obviously improved. To compare the effect of adding lubricant, a statistical result of the core drilling data in the hole sections of 210 m before and after adding GLUB lubricant is shown in Table 9.18.

Table 9.16 Mud standpipe pressure of each roundtrip in well depth from 4991.50 to 5050.77 m of MH2C hole section

Roundtrip	MH-2C-171	MH-2C-172	MH-2C-173	MH-2C-174	MH-2C-175	MH-2C-176	MH-2C-177	MH-2C-178
SPP/MPa	15.8	14.8	14.4	14.4	14.3	14.8	12.7	14.4

Table 9.17 The detected results of lubricity of the drilling fluid at CCSD-1 Well site

Sample	Instrument reading	Friction factor	Reduced rate of friction factor (%)
1# (field mud without lubricant)	61.5	0.6638	
2# (office mud with lubricant)	10.2	0.1101	83.4
3# (field mud with lubricant)	10.9	0.1177	82.3

Note Sample 1# was the circulated mud acquired on-site. Sample 2# and 3# were laboratory and production field mud after added with 1.33 % GLUB respectively

9.3 Drilling Fluid for Core Drilling

Table 9.18 Comparison of core drilling data before and after adding lubricant

Lubricant	Drilling section (m)	Footage drilled (m)	Roundtrip			Average footage drilled per roundtrip/m	ROP (m/h)	Core recovery (%)
			Total roundtrip	Lifting with full barrel	Ratio (%)			
None	1626.11–1837.09	210.98	51	13	25.5	4.14	0.96	84.8
GLUB	1837.09–2046.54	209.45	31	20	64.5	6.76	1.24	96.9

From Table 9.18 it can be understood the effect of using lubricant. The using of lubricant played an important role in improving drilling efficiency, as in the 31 roundtrips added with lubricant 20 roundtrips with full barrel. Above-mentioned comparison indicated that when PDM hydro-hammer core drilling method was employed, though drill pipe were not rotated, down hole power engine still need a good lubrication environment, especially the swivel action components of hydro-hammer and double tube drilling tool were very hypersensitive to the lubricity of drilling fluid, and poor lubricity could seriously influence drilling efficiency and service life.

Drilling fluid with good lubricity can provide high quality working medium for PDM and hydro- hammer and supply good lubrication condition for swivel type assembly of double tube core drilling tool, and obviously improve the swivel action performance of the drilling tool, it also reduces the resistance of core entering into core barrel and avoids core jam.

The application result of lubricant could be confirmed by examination and repair of the drilling tools at drill site and by checking the consumption of bearings. After lubricant was used, in one hole section in which 191 m were drilled, during which no bearing was changed. It was found through several inspections at drill site that the bearing looked like as new one and obvious wear was hardly seen. Before lubricant was employed, the service life of swivel action bearing of core drilling tool was about 3–5 roundtrips, with corrosion and scaling phenomenon on the components of hydro-hammer and with serious wear; after lubricant was employed, corrosion phenomenon disappeared, and oil film could be found on the surface of the components, thus the service life improved. In conclusion, the application of GLUB lubricant provided an important guarantee for overall improving the technical indexes for core drilling.

9.4 Solid Control Technique of Drilling Fluid

The solid phase in drilling fluid can be divided into two kinds according to its function: one is usable solids such as bentonite, chemical treating agents and ground barium sulfate and so on; while the other one is hazardous solids such as cuttings, poor bentonite, grit, etc. The aim of solid control is to eliminate the hazardous solids and preserve the usable solids so as to satisfy the requirement of drilling technology to drilling fluid performance. Cuttings are the main pollutions of drilling fluid; mainly influence the physica properties of drilling fluid, which can increase the density, viscosity, yield value, filter loss, mud cake, abrasiveness, glutinousness and flow resistance of drilling fluid. Solid control has become the important factor which can directly influence safety, high quality and high-speed drilling. Solid control is one of the important ways to realize optimized drilling. Statistics from oil drilling field indicate that when density of drilling fluid is in low range, the solid contents in drilling fluid decreases each 1 % (be equal to that the density of drilling fluid decreases 0.016 g/cm^3) the penetration rate will be generally improved by approximate 8 per cent in soft formation. So, it can be found that the benefit of good solid control is very significant.

To the PDM hydro-hammer diamond core drilling techniques used in CCSD-1 Well, cuttings in drilling fluid are the main reason for wearing of the whole circulating system, especially the wear of down hole power engine, cuttings are also the key factor of causing drill bit burying (circulation flushing cannot be started before lifting drilling tool for PDM core drilling, otherwise core may be lost). In diamond core drilling, most cuttings produced by grinding are smaller than 170 meshes (91 μm), a small amount of larger cuttings are produced by volumetric fragmentation under the impact of hydro- hammer, and there are still very few larger rock pieces dropped from borehole wall. So the distribution range of cuttings size in drilling fluid is comparatively wide, if those cuttings cannot be eliminated in time, it will not only influence the mud performance, but also seriously affect the drilling efficiency and wear of drilling tool.

A high requirement for sand content in drilling fluid (less than 0.1 %) was put forward in drilling fluid design based upon the drilling technique employed. In drilling construction, strict, advanced and high efficiency solids control technology must be employed to fulfill the requirement of design and ensure the success of drilling technology.

In geological core drilling, the performance of drilling fluid is generally regulated by increasing or decreasing the clay content, adding chemical treating agents, and water diluting. Though by using those methods certain of effects can be achieved, the cost of drilling fluid maintenance is too high. If the formation mud making is encountered in drilling, the content of cuttings and inferior clay in drilling fluid will increase, which causes the increase of viscosity, shear force

	Centrifuge	Cyclon	Desilter	Desander	Super oscillating screen
5μ	12μ	25μ	40μ	74μ	
Colloid	Clay				Grit

Fig. 9.11 The cuttings size eliminated by solids control equipment

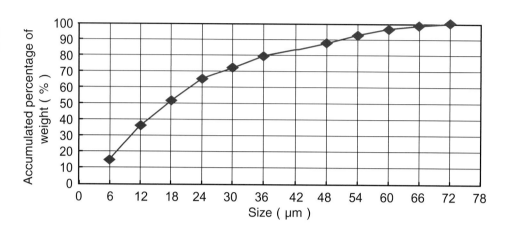

Fig. 9.12 An accumulated distribution curve of grain size of the cuttings produced by diamond core drilling in CCSD-1 Well

and filter loss of the drilling fluid and decrease of fluidity. Water dilution can just change the solids content in drilling fluid, but cannot change the distribution of the grain size. What's more, this method decreases the content of high quality bentonite and chemical treating agents in drilling fluid, causing enormous waste. Once the thinner is degraded, the viscosity, shear force and filter loss of drilling fluid will increase at once, then thinner and fluid loss reducer need to be added again. Such repeated addition will increase the volume of drilling fluid, which adds to the difficulty of waste mud treatment and may cause environment pollution. The most common method to regulate the performance of drilling fluid is to eliminate the hazardous solids in drilling fluid, mainly cuttings and inferior clay.

The development of drilling fluid solids control technique offered a suit of more efficient method to regulate the performance of drilling fluid. Common solids control equipment mainly includes oscillating screen, desander, desilter, super cyclone and centrifuge. Those solids control equipment can effectively eliminate cuttings and inferior clay, maintain the content of high quality clay and thus achieve the aim of regulating the performance of drilling fluid. When the viscosity of drilling fluid is low, the range of solid grains which can be eliminated by solids control equipment can be found in Fig. 9.11.

9.4.1 Cuttings Size Analysis

The designed well depth of CCSD-1 Well was 5000 m and continuous coring was conducted in the whole well. The inner and outer diameter of drill bit was 96 and 157 mm respectively. Not taking the cuttings produced by expanding drilling and core breaking into account, the cuttings produced by core drilling amounted to approximately 60 m^3, weighed more than 160 t, being three to four times of the bentonite used for drilling fluid in actual core drilling. To PDM hydro-hammer core drilling, the down hole power engine is very susceptive to cuttings, especially to the cuttings with high content of quartz, high hardness and strong abrasiveness such as gneiss and eclogite, which will seriously abrade down hole power engine under the comparatively high speed of drilling fluid, thus the reliability and service life of the down hole power engine will be reduced. With rotary drilling as the main and percussive drilling as the auxiliary, PDM hydro-hammer percussive rotary drilling produces fine sized cuttings, with an average specific surface area of 0.79 m^2/g and average grain size of 17 μm. The distribution of cuttings is illustrated in Fig. 9.12.

From Fig. 9.12 it can be found that the cuttings with grain size smaller than 12 μm accounts for about 37 % of the total. From comparison of Fig. 9.11 with Fig. 9.12 it can be understood that most of the cuttings produced by PDM hydro-hammer diamond core drilling are smaller than 74 μm, to which oscillating screen can hardly act for solids control, while desander can only eliminate about 12 % of the cuttings, desilter can eliminate just 35 % of the cuttings, even the super cyclone can only eliminate 85 % of the cuttings, and centrifuge can eliminate about 90 % of the cuttings. Thus, centrifuge was chosen as the main solids control equipment for CCSD-1.

9.4.2 Requirement of Solids Control Equipment to Drilling Fluid

When the properties of drilling fluid are to be designed, the separating capacity of centrifuge must be considered. Grain size which can be separated out by centrifuge can be calculated by following formula (according to the compilation group of a Chinese-English Dictionary of Petroleum 1994):

$$d_{\min} = \sqrt{\frac{18\mu \ln(r_2/r_1) V_s}{\pi h \omega^2 (\rho_s - \rho_f)(r_2^2 - r_1^2)}} \quad (9.10)$$

where,

- $d_{\min}$ is the minimum grain size which can be separated out by centrifuge, m
- μ is the dynamic viscosity of drilling fluid, kg s/m²
- r_1 is the distance between grain and rotation axle, m
- r_2 is the inner radius of the drum tumbler of a centrifuge, m
- ρ_s is the density of cuttings, kg/m³
- ρ_f is the density of drilling fluid, kg/m³
- V_s is the volume flow rate, m³/s
- H is the length of tumbler, m
- ω is the angular speed of drum tumbler, rad/s

It can be found from the formula (9.10) that the grain size separated out by centrifuge is direct proportion to the square roots of viscosity and flow rate of drilling fluid, and in inverse proportion to the angular speed of the drum tumbler of a centrifuge. Once the type of a centrifuge is determined, the lower the viscosity of drilling fluid is, the smaller the grains can be separated out by the centrifuge and the better the effect will be. On the other hand, the higher the viscosity is, the larger the grains can be separated out by the centrifuge, thus the worse the separating capability of the centrifuge, and the higher the solids content in drilling fluid will be. Therefore, the density and gel strength of drilling fluid must be reduced as much as possible so as to effectively eliminate the cuttings in drilling fluid.

The minimum grain size separated out by centrifuge is inversely as square root of the density contrast between solid grain and drilling fluid. If centrifuge type and viscosity of drilling fluid are invariable, the lower the density of drilling fluid is, the smaller solid grain can be separated out by centrifuge and the better the solids control efficiency will be. Otherwise, the higher the density of drilling fluid is, the larger solids grain be separated out by centrifuge and the worse the solids control efficiency will be.

Circulation of drilling fluid will be stopped when making a pipe connection, running in/out of hole or a breakdown happened for equipment. At that time, it is required that drilling fluid system should quickly produce a space grid structure to suspend cuttings or let the cuttings subside in a very slow speed. When starting pump, pump pressure should not be too high otherwise formation leakage may be produced. The suspension capability of drilling fluid mainly depends upon its gel strength. If the gel strength is too low, the suspension capability will be weak, then once pumping is stopped, lots of cuttings will deposit at hole bottom, drilling tool cannot be run down to hole bottom, or even causing bit burying.

The minimum gel strength of drilling fluid required to suspend cuttings can be obtained approximately by Stokes formula. Assuming that cuttings are in globular shape, the minimum gel strength can be obtained according to the relationship between cuttings gravity and buoyancy from drilling fluid (Moore 1974):

$$\frac{\pi d_s^3 \rho_s g}{6} = \frac{\pi d_s^3 \rho_f g}{6} + \pi d_s^2 \tau \quad (9.11)$$

So,

$$\tau_{\min} = \frac{d_s (\rho_s - \rho_f) g}{6} \quad (9.12)$$

where,

- d_s is the diameter of cuttings grain, m
- $\tau_{\min}$ is the gel strength of drilling fluid, Pa
- g is the acceleration of gravity, g = 10 m²/s

From formula 9.12 it is understood that the minimum gel strength of drilling fluid is in direct ratio to cuttings grain diameter and in direct ratio to the density difference between cuttings and drilling fluid. If the densities of cuttings and drilling fluid are constant, the smaller the cuttings are, the smaller the gel strength needed to suspend solids will be while the bigger the cuttings are, the bigger the gel strength is needed. Or if the cuttings density and diameter are constant, the smaller the density of drilling fluid is, the bigger the gel strength needed to suspend solids will be while the bigger the density of drilling fluid is, the smaller the gel strength is needed.

It can be found from formulas (9.10) and (9.12) that the requirements of cuttings suspension and solids control for drilling fluid are adverse. The design of drilling fluid must ensure the suspension capacity while at the same time decrease density and gel strength as much as possible.

9.4.3 Analysis of Solids Control Effect

A three-stage solids control system (Fig. 9.13) with oscillating screen, desander (desilter) and centrifuge was employed for CCSD-1 Well. A complete desanding and desilting cleaning system consisting of 2E48-90F-3TA

Fig. 9.13 The flowing diagram of drilling fluid solid control system. *1, 2, 3* Oscillating screen, *4* combination of desander and desilter, *5* low speed centrifuge, *6* high speed centrifuge, *7* triangular tank, *8, 9, 10* mud tank

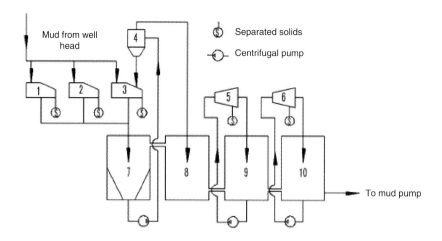

oscillating screen, ZE-10 (in.) cyclone desander and 12E-6 (in.) desilter from U.S. Derrick Company was employed. A LW500 × 1250-N low speed centrifuge with large discharge and a LW500G × 1250-N high speed centrifuge with large discharge were equipped, aiming at the fine cuttings produced by diamond drilling. Besides, LZCQ/3 deaerator was employed.

In diamond core drilling, the screen cloth of oscillating screen in solids control system was of 170 meshes (91 μm), through the outlet of the oscillating screen only a very small part of coarse cuttings could be separated out from the drilling fluid, while most of cutting chips flowed through the oscillating screen into mud tank, and small part deposited in triangular tank which could be then separated out by desander and desilter. Most of the suspended cutting chips could be separated out by perpetual working of the low speed centrifuge, and fine cutting chips could be eliminated by interrupted working of the high speed centrifuge. Almost all the cuttings and fine chips could be separated out by this solids control system and the sand content of drilling fluid could be maintained at about 0.03 %, thus the properties of drilling fluid could be guaranteed over a long period of time.

The grain distribution of the cuttings could be separated out by high and low speed centrifuge can be found in Table 9.19. The average volume grain size of the cutting chips separated out by low speed centrifuge was 50.53 μm. According to the quantity of separated cuttings it could be known that 4.83–104.32 μm was the main distribution range of the cuttings grain size produced by diamond core drilling. The average volume grain size of the cutting chips separated out by high speed centrifuge was 20.80 μm. The grain diameter of the cutting chips separated out by high speed centrifuge was mainly distributed between 3.8 and 47.21 μm. So high speed centrifuge could eliminate extremely fine cutting chips, beneficial to decreasing the shearing force of drilling fluid.

In diamond drilling, the produced cutting chips are very fine and usually smaller than 200 meshes (70 μm), so the maintenance of drilling fluid is very important. Though the quantity of the cutting chips is not much, the accumulated cutting chips disperse in drilling fluid can affect rheological property of the drilling fluid and thus causing the vicious circle of increase of cutting chips quantity, increase of solids content, increase of shear force, cuttings are hard to be separated out and then shear force increases even more, then finally lead to an abandonment of the drilling fluid. By using a combined solids control system with desander, desilter and centrifuge for CCSD-1 Well, solids content in drilling fluid was strictly controlled, tiny cutting chips was eliminated in time, so a continual stability of drilling fluid performance was guaranteed which was not only propitious to the control of drilling parameters but also benefit for stability of well wall. In the second opening (spudding-in) of the pilot hole of CCSD-1 Well, core drilling lasted from July 16th, 2001 to April 15th, 2002, for totally 274 days, an accumulated footage of 1945.54 m was completed, during which, density

Table 9.19 Cuttings grain size separated out by centrifuge

Sample	Size range (μm)			Average volume grain size (μm)	Specific surface area (m²/g)	Average surface area grain size (μm)
	D (0.1)	D (0.5)	D (0.9)			
Separated by low speed centrifuge	4.83	27.03	104.32	50.53	0.52	11.51
Separated by high speed centrifuge	3.80	13.87	47.21	20.80	0.70	8.60

Table 9.20 Drilling fluid properties for non-core drilling in the first opening (spudding-in)

Item	Parameter	Item	Parameter
Density (g/cm^3)	1.06	Funnel viscosity (s)	42
Apparent viscosity (mPas)	12.5	Plastic viscosity (mPas)	7
Initial gel strength (Pa)	4.5	10-min gel strength (Pa)	8
API filter loss (ml)	10.4	Cake thickness (mm)	0.3
Sand content (%)	≤0.3	pH value	9

of drilling fluid was maintained at about 1.05–1.07 g/cm^3, in most time at 1.05–1.06 g/cm^3 and sand content was always maintained at no more than 0.03 per cent. In the fourth opening (spudding-in), core drilling started on May 8th, 2004 and ended on March 8th, 2005, for totally 305 days, with an accumulated footage of 1533.84 m drilled, during which, density of drilling fluid was maintained at about 1.07–1.08 g/cm^3, and sand content was always maintained at no more than 0.05 %. In the whole drilling progress, no drilling fluid was replaced due to weak bearing capacity to solids (all drilling fluid replacement only happened after well cementation as drilling fluid was contaminated with cement slurry), by employing the solids control techniques, high quality circulating medium was supplied for core drilling and for down hole power engine.

9.5 Site Application of Drilling Fluid

CCSD-1 Well was started on June 25th, 2001 and field construction was all finished on March 8th, 2005, totally lasted 1353 days. Single LBM-SD drilling fluid system was used from the second opening (spudding-in) at the well depth of 101 m. LBM composite mud powder could be directly added into make-up water and agitated, then mud was prepared for use. According to the requirement of rheological property, regulating the dosage of LBM-SD could acquire satisfactory rheological property.

9.5.1 Application of Drilling Fluid in Non-core Drilling in the First Opening (Spudding-in)

A drilling tool assembly with Φ444.5 mm(17 1/2 in.) tri-cone bit and Φ228.6 (9 in.) drill collar was employed for the first opening (spudding-in) from 6.87 m deep to 101.00 m deep. Upward velocity of drilling fluid was slow because of the large clearance between drill string and well wall.

Then the problem should be solved was the cuttings carrying capacity of drilling fluid and the repeated crushing of the cuttings at hole bottom must be avoided. The formula of the drilling fluid applied was 5 % artificial sodium bentonite +0.3 % high viscosity CMC, pH value was regulated to 9 with Na_2CO_3, with the performances shown in Table 9.20.

When drilling to the hole section from 20 to 51.40 m, drilling fluid lost. 2 % GD-III plugging agent-while-drilling was added into drilling fluid and sealing was successful. 65.28 m^3 drilling fluid lost in this hole section.

9.5.2 Application of Drilling Fluid in Pilot Hole Core Drilling

PH hole section of the pilot hole (101.00–2046.54 m) started on July 16th, 2001 and finished on April 15th, 2002, a total of 274 days of core drilling was completed, with accumulated core drilling of 1945.54 m. The drilling fluid for PH hole section was produced by adding appropriate LBM-SD during drilling operation into the surplus drilling fluid from the non-core drilling in the first opening and changing it into CMC-LBM-SD composite drilling fluid system. Thus the problem of low shear force of LBM-SD drilling fluid system was solved and the requirement of low uphole velocity at upper annulus for running out of the drilling tool (the drift diameter of upper Φ244.5 mm moving casing was big) was satisfied. In PDM hydro-hammer core drilling, drilling fluid not only works as common mud, more importantly, it offers working medium and energy to down hole power engine (PDM and hydro-hammer) and swivel mechanisms of core drilling tool. Thus, the performance of drilling fluid is of vital importance to the working condition and service life of down hole power system.

Cutting chips produced in diamond core drilling were very tiny, from 5 to 100 μm, the delivery rate of drilling fluid used was from 9 to 11 L/min. Drilling fluid was required not only to be favourable to carry the cuttings during drilling but also suspend the cuttings while circulation was stopped.

Besides, drilling fluid should satisfy the driving of down hole power system and guarantee its normal working. Considering that drilling mud was prepared and managed at drill site, simple preparation technology, easy regulation of performance and convenient management should be achieved. From the considerations of saving cost and reducing discharge, single drilling fluid system was utilized at drill site and only through simple regulation of drilling fluid properties that the requirements of change of drilling

Table 9.21 Property parameters of drilling fluid in core drilling of PH hole section

Item	Parameter	Item	Parameter
Type of drilling fluid	LBM-SD	Density (g/cm^3)	1.05
Fann φ_{600}	10–12	Fann φ_{100}	3–4
Funnel viscosity (s)	28–32	Plastic viscosity (mPas)	3–5
Initial gel strength/Pa	2.0–4.0	10-min gel strength (Pa)	0.5–1.0
Yield-plastic ratio YP/PV (ks^{-1})	0.3–0.5	API filter loss (ml)	<12
Sand content (%)	≤0.1	pH value	9.0–9.5

technology could be satisfied, unnecessary to change drilling fluid system. With the increase of well depth, PDM hydro-hammer technology was employed. Circulated punching should be avoided before running out drilling tool in case of core loss. Because the cutting chips produced in drilling still dispersed in annulus, the 10-min gel strength of drilling fluid should be improved to increase its suspending capacity and to avoid drill bit bury caused by cuttings deposition.

Based upon the rock formations to be drilled, hole structure, drilling methods and down hole power system to be employed, basic parameters of drilling fluid properties in CCSD-1 Well were determined and continuously optimized in application. The controlled basic parameters of drilling fluid performance are shown in Table 9.21.

To achieve above-mentioned performances, base mud was first prepared with 3 % LBM-SD and various property parameters were detected after full agitation and hydration. Because of the influence of weather and season, the influence of temperature fluctuation to the performance of newly prepared base mud had certain of differences, but it could be appropriately regulated according to the detected results.

Because the production cycle of core drilling was comparatively long, after each coring roundtrip the drilling fluid property was comparatively less affected and the increase of well depth was rather slow, reserve base mud was just for complementing circulated drilling fluid and each time was little complemented, so maintenance was the main work for controlling drilling fluid performance.

In normal drilling production, the conventional performances of drilling fluid such as viscosity, filter loss, sand content, shear force and pH value and so on must be detected at least two times in each shift. Performances should be regulated according to the detected results to keep stable.

Density of drilling fluid was kept at 1.05–1.07 g/cm^3 and sand content was kept at 0.03 % equipped with good solids control system. In drilling process no drilling fluid was replaced because of deterioration of drilling fluid properties.

In the stage of core drilling in the pilot hole, as drilling was just beginning, each drilling method and drilling technology just under test, drilling tool assemblies which were employed changed greatly, and then the technological parameters of drilling were incessantly adjusted. Also, drilling fluid was under test for comprehensive assessment and regulation.

Besides, considering that in PDM hydro-hammer core drilling drill rod and drill collar did not rotate and there was no rotated wear on drill rod and drill collar and the problem of large rotary torque on drill rig did not exist, so lubricant was not used until well depth reached to 1837.09 m at Roundtrip PH626. 1 % GLUB was added into drilling fluid at well depth deeper than 1837.09 m to improve the working condition of drilling tools such as PDM, hydro-hammer and coring tools. As a result, drilling efficiency was improved.

9.5.3 Application of Drilling Fluid in the First Expanding Drilling in the Main Hole

The first expanding drilling in the main hole at well depth from 101.00 to 2033.00 m was started on May 7th, 2002 and completed on September 5th, 2002, with a total of 122 days lasted and accumulated footage of 1932.00 m drilled. Drilling fluid for expanding drilling was made by adding 4 % artificial sodium bentonite + 0.3 % high viscosity CMC (HV-CMC) + 2 % GLUB into the remanent drilling fluid from core drilling in the pilot hole, and the pH value was adjusted to 9 by using Na2CO3 or NaOH. Performances of the drilling fluid for expanding drilling can be found in Table 9.22.

Maintenance of drilling fluid in this hole section mainly involved appropriately adding HV-CMC, GLUB, NaOH or Na 2CO 3 and MV-PLUS (low molecular weight polyacrylate) according to performance change of drilling fluid. A little antifoam agent should be added if drilling fluid frothed.

9.5.4 Application of Drilling Fluid in the First Core Drilling in the Main Hole

The first core drilling in the MH section of the main hole from well depth of 2046.54–2982.18 m was started on October 4th, 2002 and finished on April 6th, 2003 with 185 days lasted and accumulated coring footage of 935.64 m completed. LBM-SD + GLUB lubricant was employed as drilling fluid system in this hole section. Performance

9.5 Site Application of Drilling Fluid

Table 9.22 Performance parameters of drilling fluid in the first expanding drilling of the main hole

Item	Parameter	Item	Parameter
Density (g/cm^3)	1.09	Funnel viscosity (s)	43
Apparent viscosity (mPas)	16.5	Plastic viscosity (mPas)	9
Initial gel strength (Pa)	2.5	10-min gel strength (Pa)	18.5
API filter loss (ml)	11.2	Cake thickness (mm)	0.5
Sand content (%)	≤0.3	pH value	9

Table 9.23 Performance parameters of drilling fluid in the core drilling of MH section

Item	Parameter	Item	Parameter
Type of drilling fluid	LBM-SD	Density (g/cm^3)	1.03–1.05
Fann φ_{600}	10–14	Fann φ_{100}	3.5–5.0
Funnel viscosity (s)	28–32	Plastic viscosity (mPas)	3–5
Initial gel strength/Pa	0–1	10-min gel strength (Pa)	1–2
Yield-plastic ratio YP/PV (ks^{-1})	0.3–0.5	API filter loss (ml)	9–15
pH value	9–10	Solids content (%)	3–4
Sand content (%)	≤0.1		

parameters of drilling fluid in this hole section are shown in Table 9.23.

Based upon the experience of core drilling in the later stage of the PH section, it could be seen that the lubricity of drilling fluid had an extremely significant influence on the working conditions of hydro-hammer, core drilling tool and PDM. Thus, the lubricity of drilling fluid was especially reinforced in the design of the drilling fluid for the MH section to ensure stable working of hydro-hammer and reduce the wear of the swivel type bearing of the core drilling tool.

High quality and efficient solids control played a key role in improving drilling efficiency, decreasing the wear of down hole power engine, increasing the service life of down hole power engine and keeping the stability of drilling fluid performance. Because circulated punching was forbidden before running out drilling tool, precipitable solids in drilling fluid should be reduced as much as possible in case of burying drill bit. A combined solids control system with desander and low speed centrifuge was employed for controlling sand content at approximately 0.03 % constantly.

Severe leakage happened at well depth of 2300 m with the largest leak-off rate of 83 m^3/h and total filter loss in this leakage hole section reached to 400 m^3. 11.5 t LCA and 6 t GD-III plugging agents were used one after another, both of them were of inert materials with main components of plant and mica pieces, which would not exert an obvious influence on drilling fluid rheological properties and make no disturbance to borehole log. LCA had fine particles as the main while GD-III mainly consisted of coarse particles and the total dosage of those two plugging agents was 17.5 t. Oscillating screen was stopped in the period of sealing to avoid elimination of the sealing materials by oscillating screen. After continuous sealing while drilling, the problem of leakage was effectively controlled.

9.5.5 Application of Drilling Fluid in the First Sidetrack Straightening Drilling in the Main Hole

The first sidetrack straightening drilling of the main hole between well depth of 2749.00 and 2974.59 m was started on April 11th, 2003 and finished on June 8th, 2003, with 59 days used for straightening drilling and accumulated footage of 241.30 m (not including the footage of drilling and grinding cement plug) completed.

At the initial stage of sidetrack straightening drilling, drilling fluid which had been used for core drilling in MH hole section was still employed. Drilling fluid was severely polluted in sidetrack drilling in cement plug, so part of the drilling fluid polluted with cement slurry was discharged and Na$_2$CO$_3$ was added to regulate the performance. After cement plug was drilled out, new drilling fluid was prepared with 4 % LBM-SD + 2 % GLUB + 0.5 % MV-PLUS and Na$_2$CO$_3$ or NaOH was employed for regulating the pH value to 9. Performances of drilling fluid are shown in Table 9.24.

9.5.6 Application of Drilling Fluid in the Second Core Drilling in the Main Hole

The second core drilling in the main hole (section MH-1C) from well depth of 2982.18–3 665.87 m was started on June 9th, 2003 and finished on October 2nd, 2003 with 116 days used for core drilling and accumulated footage of 710.61 m (including guide expanding drilling for small diameter straightening at hole bottom) completed.

In the process of core drilling in this hole section, straightening at hole bottom was conducted for three times

Table 9.24 Performance parameters of drilling fluid in the first sidetrack straightening drilling in the main hole

Item	Parameter	Item	Parameter
Density (g/cm^3)	1.03	Funnel viscosity (s)	34
Apparent viscosity (mPas)	9	Plastic viscosity (mPas)	7
Initial gel strength (Pa)	0.5	10-min gel strength (Pa)	6
API filter loss (ml)	22	Cake thickness (mm)	0.5
Sand content (%)	≤0.3	pH value	9

Table 9.25 Performance parameters of drilling fluid for core drilling in MH-1C hole section

Item	Parameter	Item	Parameter
Type of drilling fluid	LBM-SD	Density/(g/cm^3)	1.05
Fann φ_{600}	12–16	Fann φ_{100}	4–6
Funnel viscosity (s)	30–33	Plastic viscosity (mPas)	3–5
Initial gel strength/Pa	1–2	10-minute gel strength Pa	2–4
Yield-plastic ratio YP/PV (ks^{-1})	0.3–0.5	Relative friction coefficient	≤0.12
API filter loss (ml)	9–15	pH value	9–10
Solids content (%)	3–4	Sand content (%)	≤0.1

as a result of the increase of well deviation. Rock formation collapse and rock falling happened in hole section from 3479 to 3522 m, logging indicated that well diameter was seriously expanded and the largest well diameter reached to 444 mm at well depth of 3516.5 m.

When core drilling reached well depth of 3665.87 m, endurance crack took place on reaming shell because of rock falling from hole wall and long service time (130.62 h) of the reaming shell and finally the reaming shell fell into hole bottom together with drill bit. After 12 roundtrips of failure fishing and grinding operations, during which a severe bit jamming accident happened, well condition became extremely complicated, so it was decided at last that expanding drilling was conducted.

Based upon the satisfactory result obtained with drilling fluid in core drilling in the upper hole section, lubricity of drilling fluid was strengthened in this hole section. Moreover, considering that the well diameter was enlarged, shear force of drilling fluid was to be appropriately improved to increase the capacities of carrying and suspending cuttings in the lower hole section construction. Property parameters of drilling fluid can be found in Table 9.25.

A leakage happened when core drilling in MH-1C hole section reached to well depth of 2300 m and drilling practice proved that this leakage zone was a confined leakage zone, leakage would happen once the density of drilling fluid was higher than 1.2 g/cm^3. Therefore, it was very necessary to control the density of drilling fluid strictly to balance formation pressure. During the process of drilling construction, the density of drilling fluid was kept no more than 1.07 g/cm^3, and no more leakage happened in the latter core drilling for more than 300 m.

9.5.7 Application of Drilling Fluid in the Second Expanding Drilling in the Main Hole

The second expanding drilling in the main hole from well depth of 2028.00–3525.18 m was started on October 29th, 2003 and finished on March 14th, 2004 with 141 days used for expanding drilling and accumulated footage of 1497.18 m (not including the footage of grinding fish) drilled. Drilling fluid in expanding drilling was prepared by adding 4 % LBM-SD + 0.3 % PHP + 2 % GLUB into the remanent drilling fluid in core drilling in MH-1C hole section, and NaOH was employed for regulating pH value to 9. Performances of the drilling fluid for expanding drilling can be found in Table 9.26.

Table 9.26 Performance parameters of drilling fluid in the second expanding drilling in the main hole (2028–3400 m)

Item	Parameter	Item	Parameter
Density (g/cm^3)	1.06	Funnel viscosity (s)	41
Apparent viscosity (mPas)	20.5	Plastic viscosity (mPas)	12
Initial gel strength (Pa)	2	10-min gel strength (Pa)	8
API filter loss (ml)	9	Cake thickness (mm)	0.5
Sand content (%)	≤0.2	pH value	9

9.5 Site Application of Drilling Fluid

Table 9.27 Performance parameters of drilling fluid in the second expanding drilling in the main hole (3400–3525.18 m)

Item	Parameter	Item	Parameter
Density (g/cm^3)	1.19	Funnel viscosity (s)	40
Apparent viscosity (mPas)	22	Plastic viscosity (mPas)	15
Initial gel strength (Pa)	1.5	10-min gel strength (Pa)	7.5
API filter loss (ml)	9	Cake thickness (mm)	0.5
Sand content (%)	≤0.2	pH value	9

Table 9.28 Performance parameters of drilling fluid in the second sidetrack drilling-around in the main hole

Item	Parameter	Item	Parameter
Density (g/cm^3)	1.21	Funnel viscosity (s)	43
Apparent viscosity (mPas)	18	Plastic viscosity (mPas)	15
Initial gel strength (Pa)	1.5	10-min gel strength (Pa)	12
API filter loss (ml)	9	Cake thickness (mm)	0.5
Sand content (%)	≤0.2	pH value	9

When expanding drilling reached to well depth of 3400 m, to avoid hole wall collapse and rock falling, a small amount of LBM-SD, HV-CMC and non-dispersed high molecular polymer thinner was added into above-mentioned drilling fluid to regulate the performance. Barite powder was added to improve the density of drilling fluid. So the weighted mud was prepared with 5 % LBM-SD + 0.3 % PHP + 2 % GLUB + 0.1 % HV-CMC + 1 % non-dispersed high molecular polymer thinner, and NaOH was employed for regulating pH value to 9. Performances of drilling fluid are shown in Table 9.27.

Drilling fluid maintenance in this hole section mainly included the addition of PHP, GLUB, NaOH, thinner and barite powder into drilling fluid according to the variation of its performance. A little antifoam agent should be added if foaming happened in drilling fluid. Aiming at the long time used for each roundtrip in expanding drilling, a little anti-corrosive agent should be added into drilling fluid.

9.5.8 Application of Drilling Fluid in the Second Sidetrack Drilling-Around in the Main Hole

The second sidetrack drilling around in the main hole from well depth of 3400.00–3624.16 m was started on March 28th, 2004 and finished on April 27th, 2004 with 31 days used and accumulated footage of 287.47 m (not including the footage of drilling and grinding cement plug) completed.

At the early stage of sidetrack drilling-around, the drilling fluid for upper expanding drilling was still used in this section. In the sidetrack drilling of cement plug, drilling fluid was polluted seriously. Part of the seriously polluted drilling fluid by cement slurry was discharged and the left mild polluted part was regulated by Na$_2$CO$_3$. The drilling fluid formula of sidetrack drilling-around was 5 % LBM-SD + 0.3 % PHP + 2 % GLUB + 0.1 % HV-CMC + 1 % non-dispersed high molecular polymer thinner, NaOH was used for regulating pH value to 9. Performances of drilling fluid are shown in Table 9.28.

9.5.9 Application of Drilling Fluid in the Third Core Drilling in the Main Hole

The third coring drilling in the hole section MH-2C of the main hole from well depth of 3624.16–5118.20 m was started on May 8th, 2004 and finished on January 24th, 2005 with 262 days used for core drilling and accumulated footage of 1494.04 m completed.

PDM hydro-hammer swivel type double tube diamond core drilling technology was employed for the hole section MH-2C. Drilling fluid was still the low solids content, low density and low viscosity system prepared by adding GLUB into the LBM-SD system. Some severe hole enlargement sections happened at the well depths from 3670 to 3750 m, from 3890 to 3930 m and from 4350 to 4385 m respectively, with the largest well diameter of 330 mm. With the increase of well depth, drilling became more and more difficult, especially when well deviation was relatively large, friction drag of drilling tool would be more than 20 t, which brought about difficulty to control drilling parameters, particularly weight on bit. In line with those above-mentioned situations, strengthening the lubricity of drilling fluid and solids control was very necessary. The technical parameters of the drilling fluid used in the hole section MH2C are shown in Table 9.29.

Generally speaking, core drilling in this hole section was successful and no severe in-the-hole accident happened. However, to drilling fluid, leakage happened after drilling to 4840 m and this leakage lasted off and on to 4900 m deep,

Table 9.29 Performance parameters of drilling fluid for core drilling in the hole section MH-2C

Item	Parameter	Item	Parameter
Type of drilling fluid	LBM-SD	Density (g/cm^3)	1.05–1.07
Fann φ_{600}	14–16	Fann φ_{100}	5–7
Funnel viscosity (s)	35–40	Plastic viscosity (mPas)	7–8
Initial gel strength (Pa)	0.5–1.0	10-min gel strength (Pa)	1.0–1.5
API filter loss (ml)	0.3–0.5	Relative friction coefficient	≤0.12
Solids content (%)	10–15	pH value	9–10

with the maximum leakage of 20 m^3 per roundtrip. Leakage was relieved by adding plugging agent-while-drilling according to the situation of leakage.

In comparison with the earlier stage construction, the technical measures adopted for drilling fluid in this hole section mainly included (1) with the increase of well depth and increase of deviation, the friction drag of rotation and the friction drag of running in/out of the drilling tool became larger and larger, the rotary torque and the resistance of running in/out could be effectively reduced by improving the lubricity of drilling fluid; and (2) shearing force of drilling fluid could be properly reduced to improve its flowability.

With above-mentioned measures, upon condition that the discharge capacity ranged from 9–9.5 L/s, circulating pressure was just 12–14 MPa when drilling reached the final hole point.

In the period of drilling tool test (at well depth from 5118.20 to 5158.00 m), drilling fluid used for the hole section MH-2C was still employed. In the way of running down the PDM hydro-hammer diamond wireline core drilling tool (SK139.7 wireline drill rod) into the hole and starting pump for circulation, 101m^3 drilling fluid was lost. Leakage was relieved by changing the wireline drill rod into 89 mm drill rod and by adding GD-III plugging agent-while-drilling.

9.5.10 Application Characteristics of LBM Drilling Fluid

LBM-SD mud material and field mud technology developed aiming at the requirements of CCSD drilling technology satisfied the technological requirements of scientific drilling, especially the requirements of PDM hydro-hammer diamond core drilling, with the following characteristics:

1. Drilling fluid preparation technology was simplified and the conventional overelaborate methods of field mud preparation, regulation and management were all eliminated.
2. GLUB lubricant which had good compatibility with LBM-SD drilling fluid system could obviously improve the working condition of drilling tool, decrease the wear of drilling tool and core, and dramatically improve drilling efficiency. This drilling fluid system was the key auxiliary technology to ensure the success of PDM hydro-hammer coring technology in deep hard rock.
3. With the characteristics of low viscosity, low shearing force, low density and low filter loss, LBM-SD drilling fluid system has good rheological property. Low shearing force and circulating pressure drop were always maintained in the progress of use to improve the energy utilization ratio of down hole power engine.
4. Optimization of drilling fluid parameters not only satisfied the requirement of down hole power engine (PDM and hydro-hammer), but also guaranteed the need of carrying the cuttings at the same time. Especially when the difference between the inner diameter of the upper casing and hole diameter was comparatively large, effective carry of the cuttings could be still guaranteed.
5. Being favourable for eliminating hazardous solids, the advanced field three-staged solids control technology with oscillating screen, desander and centrifuge was the key to guarantee the stability of the properties of drilling fluid over a long period of time.
6. With good properties of temperature resistance and invasion resistance, this drilling fluid system could circulate steadily over a long period of time under down hole temperature of 140 °C, without pollution to environment and interference to borehole log.

In conclusion, the application results from drill site further proved that LBM-SD drilling fluid system could fully satisfy the requirement of drilling construction in CCSD-1 Well, thus a high performance price ratio was realized. Long term field application proved that the drilling fluid system prepared with LBM-SD as the base material and field drilling fluid technology played an important role in the implementation of PDM hydro-hammer swivel type double tube core drilling technology. It also provided valuable experience for the drilling fluid application in large diameter and deep hard rock core drilling in our country in the days to come.

Casing and Well Cementation

The casing program of CCSD-1 Well was different from that of conventional oil drilling and water well drilling in two aspects: not only were the standards of casing program different but also the moving casing techniques employed. In oil drilling, a casing program of 508.0 mm (20 in.) × 339.7 mm ($13^3/_8$ in.) × 244.5 mm ($9^5/_8$ in.) × 177.8 mm (7 in.) was generally employed, whereas in CCSD-1 Well a casing program of 339.7 mm ($13^3/_8$ in.) × 273.0 mm ($10^3/_4$ in.) × 193.7 mm ($7^5/_8$ in.) × 127.0 mm (5 in.) was adopted in design and in construction. In CCSD-1 Well, the program of "two boreholes combined into one" was realized, with four times of normal casing setting and cementation (not including wellhead conductor) and two times of setting moving casing, shown in Table 10.1.

10.1 Borehole Structure and Casing Program

10.1.1 Borehole Structure and Casing Program for the Pilot Hole

After penetrating through unstable surface layer of 100.36 m deep by non-core drilling with 444.5 mm roller cone bit, 339.7 mm intermediate casing was run and well was cemented. After that 244.5 mm moving casing was set and then diamond core drilling was conducted.

Core drilling reached to the designed depth of 2000 m, where the formation was broken, being unfavourable for setting casing. Under these circumstances, core drilling was continued, finally to the depth of 2046.54 m. The actual borehole structure and casing program of the pilot hole can be found in Table 10.2 and in Fig. 10.1.

10.1.2 Borehole Structure and Casing Program for the Main Hole

Because the quality of the pilot hole satisfied the requirement of "two holes combined into one", reaming was conducted on the basis of the pilot hole and the reamed pilot hole became a part of the main hole. 273.0 mm casing was run after reaming and cementation conducted, then 193.7 mm moving casing was set. 157 mm diamond core drilling was started from the hole depth of 2046.54 m. "Two holes combined into one" saved large funds and much time.

In the third opening, when core drilling reached to 3665.87 m, 244.5 mm reaming drilling was conducted as a result of complicated borehole conditions. When reaming drilling reached to the depth of 3525.18 m, because fish existed at the hole bottom, 244.5 mm sidetrack drilling-around was employed. 193.7 mm intermediate casing was set and hole cemented. Then 157 mm diamond core drilling was continued.

Core drilling in the fourth opening drilled to 5118.2 m and after drilling tool test reached to the final hole depth of 5158 m. Then 127 mm tail pipe was run, cementation was conducted and the borehole was completed.

The actual borehole structure and casing program of CCSD-1 Well can be found in Table 10.3 and in Fig. 10.2.

10.1.3 Casing Design

1. **Downhole temperature estimation**

The static temperature and circulating temperature (Table 10.4) at different hole depths could be estimated by using geothermal gradient of 2.5 °C/100 m, which was derived from the temperature curve of CCSD-PP2 hole and the predictive temperature curve of CCSD-1 well.

Table 10.1 Casing and cementation construction schedule

No.	Hole section	Construction	Description
1	0–100.36 m	Set 339.7 mm surface casing and cementation	Set 339.7 mm surface casing and cementation was smoothly operated
2	0–101 m	Set and withdraw 244.5 mm moving casing	Set 244.5 mm moving casing and at the lower position of the moving casing string were installed three rigid centralizers and at the upper part were three elastic centralizers
3	0–2028 m	Set 273.0 mm intermediate casing and cementation	Set 273.0 mm intermediate casing. Casing string: float shoe + float collar + one piece of short casing (5 m) + setting seat + one piece of short casing (3 m) + retaining joint assembly + casing string. Cementation was smoothly operated
4	0–2019 m	Set and withdraw 193.7 mm moving casing	Set 193.7 mm moving casing. Casing string: insert guide head + inserted sleeve + casing string + landing joint. On to the five pieces of casing near the retaining joint each was added with an elastic centralizer and on the sixth piece of casing above was added with a rigid centralizer. For other casings a centralizer was added for every two pieces, with rigid and elastic centralizer alternately added
5	0–3620 m	Set 193.7 mm intermediate casing and cementation	Set 193.7 mm intermediate casing. The position of stage collar was at 1928.11 m. Casing string: guide shoe + one piece of casing + float collar + six pieces of casing + float collar + casing string + stage collar + casing string. Staged cementation was smoothly operated
6	3523.55–4790.72 m	Set 127.0 mm tail pipe and cementation	Set 127.0 mm tail pipe. Casing string: float shoe + one piece of casing + float collar + one piece of casing + float collar + one piece of casing + ball seat + casing + joint + hanger + running- in tool + drill rod. Cementation was smoothly operated.

Note In No. 4, seventy two steel pieces of the elastic centralizer fell into the borehole after withdrawing moving casing

Table 10.2 The actual borehole structure and casing program of the pilot hole

Opening (spud-in)	Drill bit size		Drilled depth (m)	Casing size		Casing setting depth (m)	Remarks
	mm	in.		mm	in.		
1	Man digging		4.0	508.0	20	4.0	Conductor
2	444.5	17^1/$_2$	101.00	339.7	13^3/$_8$	100.36	Cementation
	Set 244.5 mm moving casing to the depth of 101.00 m						
3	157.0	6^3/$_{16}$	2046.54	The pilot hole was completed and two holes combined into one, reaming for the main hole			

2. **Basic data of casing**

The basic data of the casing employed for CCSD-1 well are shown in Table 10.5.

3. **Other data**

Elastic modulus of casing E: 207 GPa
Thermal expansion coefficient of casing k_t: 1.25×10^{-5} m/°C
Permissible tension safety factor (St): 1.8
Permissible collapsing safety factor (Sc): 1.125
Permissible internal pressure safety factor (Si): 1.1

4. **Calculation of effective internal pressure**

Effective internal pressure denotes the difference between the maximum internal pressure that inside casing may bear and the fluid column pressure outside casing. Generally speaking, when well control operation is conducted in case of well kick and well shut-in intermediate casing bear the maximum internal pressure. Therefore, intermediate casing should be designed based on the load of well shut-in and checked by using the condition of oil well kick.

Suppose well kick happens when drilling next layer and the whole borehole is full of brine, then in well shut-in, Wellhead pressure

$$P_s = P_p - P_{os} = (G_P - G_{os})H_s \quad (10.1)$$

Effective internal pressure

$$P_{be} = (P_s + G_{os}H) - G_w H \quad (10.2)$$

in which,

P_s wellhead pressure, MPa
P_p formation pore pressure, MPa
P_{os} brine column pressure, MPa
G_p formation pore pressure gradient, MPa/m
G_{os} brine column pressure gradient, MPa/m
H_s well depth where casing shoe is located in the layer, m

10.1 Borehole Structure and Casing Program

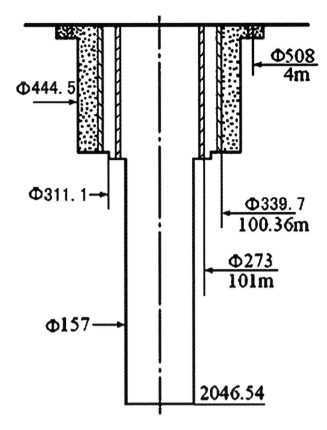

Fig. 10.1 The actual borehole structure and casing program of the pilot hole

P_{be} effective pressure at arbitrary well depth, MPa
H arbitrary well depth, m
G_w fluid column pressure gradient outside casing, MPa/m (calculated based on mud density).

5. **Calculation of effective external pressure**

Effective external pressure denotes the difference between the maximum external pressure that casing string may bear and the minimum internal pressure inside casing. The maximum external pressure that casing bears and the

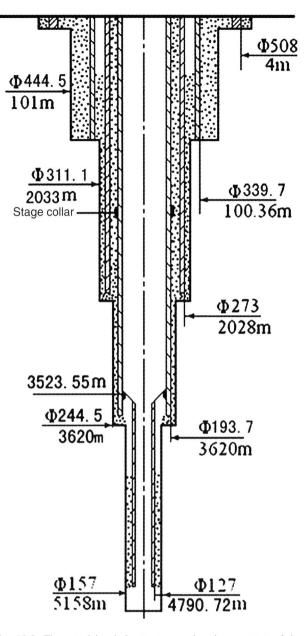

Fig. 10.2 The actual borehole structure and casing program of the main hole

Table 10.3 Borehole structure and casing program of the pilot hole of CCSD-1 Well

Opening (spud-in)	Drilling		Casing		Remarks
	Drill bit size [mm (in.)]	Drilled depth (m)	Casing (tail) size [mm (in.)]	Setting depth (m)	
First opening	444.5 (17^1/$_2$)	101.00	339.7 (13^3/$_8$)	100.36	Non-core drilling
Second opening	157 (6^3/$_{16}$)	2046.54			Core drilling
First reaming	311.1 (12^1/$_4$)	2033.00	273 (10^3/$_4$)	2028	Cementation
	193.7 mm moving casing was then run, to the depth of 2019 m				
Third opening	157 (6^3/$_{16}$)	3665.87			Core drilling
Second reaming	244.5 (9^5/$_8$)	3624.16	193.7 (7^5/$_8$)	3620	Cementation
Fourth opening	157 (6^3/$_{16}$)	5118.20			Core drilling
	157 (6^3/$_{16}$)	5158.00	127 (5)	4770.72	Drilling tool test
	127 mm tail pipe was set to the depth of 4790.72 m. The suspension position of the top of the tail pipe was at 3523.55 m. Cementation and then well completion				

Table 10.4 Downhole temperature

No.	Hole depth (m)	Static temperature (°C)	Circulating temperature (°C)
1	100	28	22
2	2000	75	60
3	3000	100	73
4	4500	138	113
5	5000	150	125

minimum internal pressure inside casing are related to casing type and formation conditions. In this design, calculation for 273 mm intermediate casing is based on 1/2 emptied while for 193.7 mm moving casing 1/3 is emptied.

$$P_{ce} = G_{wl}H - G_{ml}(H - H_{lost}) \text{ when } H > H_{lost}. \quad (10.3)$$

$$P_{ce} = G_{wl}H \text{ when } H \leq H_{lost} \quad (10.4)$$

in which,

P_{ce} effective squeeze pressure at arbitrary well depth, MPa
H arbitrary well depth, m
G_{wl} fluid column pressure gradient outside casing, MPa/m (calculated based on saturated brine gradient)
H_{lost} level depth of lost fluid, m; $H_{lost} = H_{sl}(8 - G_o/G_{ml})$
G_{ml} drilling fluid pressure gradient of the depth for setting casing when drilling to next layer, MPa/m
H_{sl} setting depth of casing shoe for next layer, m
G_o pressure gradient of formation supporting fluid column (0.0105–0.0115), MPa/m.

6. **Calculation of effective axial force**

The effective axial force of casing string is the sum of vectors of dead weight, buoyance, inertial force, impact force, frictional force, bending moment force and the additional axial force produced by the variation of in-the-hole temperature and pressure after well cementation. In the master design of CCSD-1 Well, dead weight and buoyance were the main factors to be considered. Buoyance factor method was adopted to calculate buoyance. The effective axial force T_e at arbitrary well depth is:

$$T_e = \left[\sum_{i=1}^{n} T_i + (H_x - H)Q_j\right] K_f \quad (10.5)$$

in which,

T_e effective axial force at calculation level, kN
H hole depth at calculation level, m
H_x casing running depth of calculated section, m
Q_j casing weight per meter of calculated section, kN/m
T_i casing weight of the i section below the casing of calculated section, kN
n number of casing section below the casing of calculated section
K_f buoyance factor, dimensionless, $K_f = 1 - \rho_m/\rho_s$
ρ_m drilling fluid density, g/cm^3
ρ_s casing density, g/cm^3; $\rho_s = 7.85$ g/cm^3.

7. **Calculation result in the design**

The designed strength of casing string and the calibrated results are shown in Table 10.6, from which it can be found that the casing string strength can fully satisfy the requirement of CCSD-1 Well.

10.2 Well Head Assembly

10.2.1 Well Head Assembly for the First Opening (Spud-in)

Before the first opening (spud-in) of CCSD-1 Well, a foundation pit with the area of 12.56 m × 12.56 m × 8.70 m was dug by manpower, the position of well head coordinate was fixed, and 508.0 mm conductor was set, by rubble

Table 10.5 Basic data of casing

Type of casing	339.7 mm surface casing	273.0 mm intermediate casing	193.7 mm intermediate casing	127.0 mm tail pipe
Running interval (m)	0–100.36	0–2028	0–3620	3523.55–4790.72
Steel grade	J55	J55	P110	P110
Wall thickness (mm)	10.92	10.16	9.52	7.52
Nominal weight (kg/m)	90.86	67.77	44.24	22.34
Casing O.D. (mm)	339.7	273	193.7	127
Casing I.D. (mm)	317.9	252.7	174.6	112
Run through casing size (mm)	313.9	248.8	171.5	108.8
Casing collar O.D. (mm)	365.1	298.5	215.9	141.3
Thread type	Oval-shaped	Oval-shaped	Buttress BTC	Oval-shaped

10.2 Well Head Assembly

Table 10.6 The design of casing string strength and the calibrated results

Item		339.7 mm intermediate casing	273.0 mm intermediate casing	193.7 mm intermediate casing	127.0 mm tail pipe
Casing strength	Tensile strength (kN)	4367	3180	4270	2240
	Collapsing strength (MPa)	10.6	14.4	36.8	61.1
	Internal pressure strength (MPa)	21.3	24.7	65.2	78.6
Actual safety factor/ permissible safety factor	Tensile (S_t)	55.1/1.80	3.07/1.80	2.71/1.80	25.1/1.80
	Collapsing (S_c)	17.0/1.125	1.28/1.125	1.41/1.125	1.96/1.125
	Internal pressure (S_i)	85.2/1.10	5.49/1.10	5.40/1.10	/

cemented with M20 cement slurry and on the top 200 mm reinforced concrete with 16 mm steel bar was constructed so that a well head of 3 m × 3 m × 2.2 m was formed. At the bottom of the well head was a small squared pit of 450 mm × 450 mm × 450 mm, for collecting the drilling mud in the well head.

The well head assembly for the first opening is illustrated in Fig. 10.3. The size of the conductor was 508.0 mm, with the setting depth of 4 m, acting as a casinghead for welded casing.

10.2.2 Well Head Assembly for the Second Opening (Spud-in)

After drilling to 100.36 m deep in the first opening, surface casing of 339.7 mm was set and cemented. In order to increase the upward velocity of drilling mud, 244.5 mm moving casing was run in 339.7 mm surface casing. Taking the well head of the first opening as a base, the upper conductor was cut off, a ring flange was welded, well head conductor was changed into 339.7 mm, and 244.5 mm moving casing was fixed by casing hanger (Fig. 10.4).

TF13$^3/_8$ in. × 9$^5/_8$ in −70D casing head manufactured by Jinhu Petroleum Machinery Co. Ltd. was adopted for

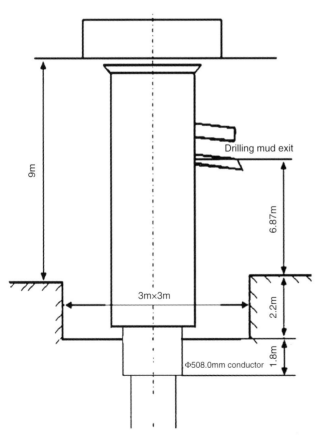

Fig. 10.3 Well head assembly for the first opening

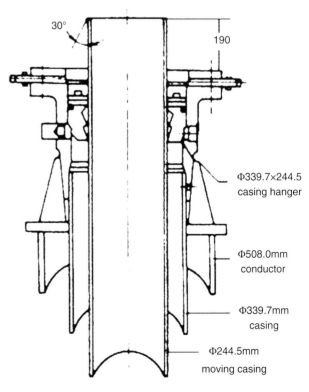

Fig. 10.4 Well head suspending system for the second opening

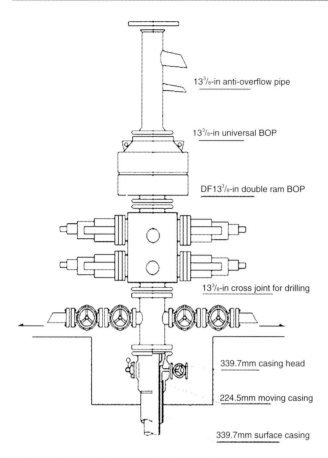

Fig. 10.5 Well head structure for the second opening

CCSD-1 Well. This well head assembly mainly consists of 339.7 mm casing head, 339.7 mm WD casing hanger, gate valve and related connection pieces. 339.7 mm casing head flange and the drift diameter of the side outlet conform to API Spec 6A standard. For 339.7 mm casing hanger, a structure of integrally opened/closed slip element was adopted. For installation, bolts are pressed by screwing in slips at the bottom of 339.7 mm casing head and the slips are excited to tightly hold casing so that a firm connection of the slips and casing is realized. The sealing between casing and casing head is realized by two BT O-rings injected with sealing oil.

Core drilling reached to 1209.61 m and a DF $13^3/_8$ double ram blowout preventer system (Fig. 10.5) was installed at well head, with the pressure testing result shown in Table 10.7. The testing pressure at its control system and choke-line manifold was 21 MPa, standing up for 10 min, with 0 MPa pressure drop. Pressure testing was qualified.

Table 10.7 Pressure testing result of the blowout preventer

BOP type	Ram 2FZ35	Annular FH35	Throttle JG-35	Well kill YG-35
Testing result (MPa)	21	21	21	21

10.2.3 Well Head Assembly for the Third and the Fourth Opening (Spud-in)

In the second opening, 244.5 mm moving casing was withdrawn after core drilling reached to the depth of 2046.54 m, then expanding drilling was conducted to 2033 m deep, 273.0 mm intermediate casing was run and cemented. To increase the upward velocity of drilling mud, 193.7 mm moving casing was run in 273.0 mm intermediate casing and the top of the moving casing was fixed by casing hanger (Fig. 10.6).

In the third opening, 193.7 mm moving casing was withdrawn after core drilling was completed and then expanding drilling was conducted. 193.7 mm moving casing was taken as intermediate casing and run into the borehole. The well head assembly for the fourth opening was just the same as that for the third opening.

10.2.4 Well Head Assembly for Well Completion

After the completion of CCSD-1 Well, the conductor, blowout preventer and cross joint for drilling at well head were removed and gate valve was fixed on 339.7 mm casing head and casing hanger, for well shut-in and open. The drift diameter of the gate valve was 177.8 mm (7 in.), satisfying the needs of in-the-hole testing and drifting after the establishment of long-term observation station. To ensure the gate valve higher than surface, a lifting sub was installed between the gate valve and the casing head and on the gate valve were fixed tight cover and rainproof cap. Moreover, locks were fixed on the tight cover to prevent foreign objects from falling into the borehole. This specially designed well head assembly is illustrated in Fig. 10.7.

10.3 Casing Running and Well Cementing Operation

10.3.1 508.0 mm Well Head Conductor

Well head conductor was installed through a method of mechanical excavation. In foundation pit excavation, measurement, location and set-out were carried out based upon the datum mark provided by local surveying information. Location and set-out were recorded, as well as the geological conditions and the variation of underground water were recorded. Peculiar situations encountered were treated in peculiar ways. Organic impurities were strictly controlled in backfilled earthwork, which were laid in layers and correspondingly compacted, with water content strictly limited.

Fig. 10.6 Structure of casing head for the third opening

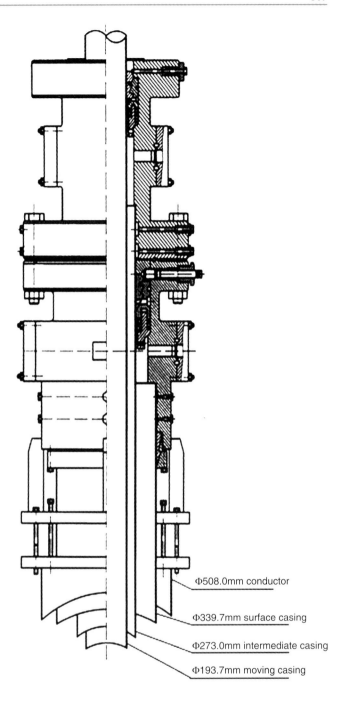

Φ508.0mm conductor
Φ339.7mm surface casing
Φ273.0mm intermediate casing
Φ193.7mm moving casing

The geographic coordinates of the well head were X = 3809.530 km and Y = 40,377.874 km. The well head foundation was in situ concreted, with the basic materials of medium grained sand, rubble, cement, reinforcing bar and water. The mixing ratio of concrete raw materials (cement:sand:stone) was: 1:2:4 for derrick foundation, to ensure that the bearing capacity of the foundation could reach 1500 tons. The construction sequence: construction preparation, location and set-out, foundation engineering and installation. In construction a principle of "underground first and afterwards surface" was abided by. The conductor was 508.0 mm (20 in.) in outside diameter, with the buried depth of 4 m (see Fig. 10.3).

10.3.2 339.7 mm Surface Casing

Non-core drilling reached to 100.36 m with hole diameter of 444.5 mm ($17^1/_2$ in.) on July 9th, 2001. 339.7 mm ($13^3/_8$ in.) surface casing was run and cemented.

Fig. 10.7 Well head assembly for well completion

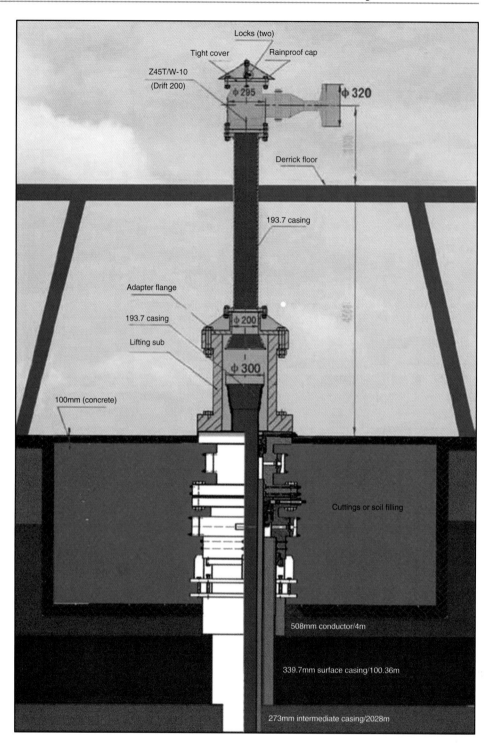

The formula of the cement slurry for well cementation was Jiahua G grade cement + 4.0 % G203 (early strength agent) + 0.6 % G301 (dispersant) + 2.0 % G201 (curing accelerator), with water-cement ratio of 0.44 and cement density of 1.86 g/cm^3.

After running 339.7 mm surface casing, cementation operation was started. Drilling mud was circulated for 3 h, injection pipeline was connected, 8 m^3 spacer fluid was injected and 22 m^3 cement slurry was injected. After injecting cement slurry, displacing slurry of 8.1 m^3 was injected under a displacement pressure of 11 MPa. The mud displacement operation was normal, cement slurry returned to well head and cementation was smoothly completed.

Fig. 10.8 The curves of caliper logging before and after 311.1 mm borehole repairing

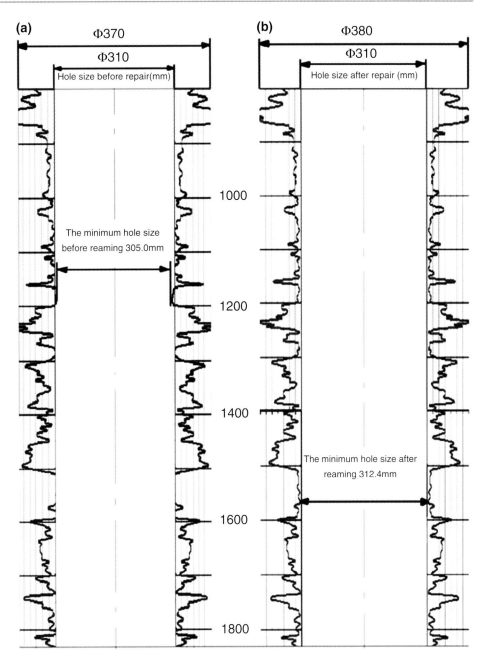

10.3.3 273.0 mm Intermediate Casing

After 311.1 mm reaming drilling to 2028 m deep, a pocket of 5 m was drilled by using 215.9 mm cone bit. Through drifting, well flushing and circulating slurry displacement, cementation operation with 273.0 mm ($10^3/_4$ in.) intermediate casing was started on September 9th, 2002. Casing running and cementation in this interval was typically an operation in small clearance in hard rock, with the following characteristics:

(1) It was difficult to run casing in small clearance

It was an exploratory construction technique for reaming drilling by using cone guide bit in ultra high pressure metamorphic crystalline rock formations with drillability from 8 to 11 grade. Generally, cone bit is suitable for drilling rock formations with drillability lower than 6 grade, whereas its gauge protection capacity is obviously inadequate in drilling ultra high pressure metamorphic crystalline rock formations with drillability from 8 to 11 grade. As a result, the actual diameter of the drilled hole is smaller than nominal size. The curves of caliper logging for CCSD-1 Well reamed by 311.1 mm cone bit are shown in Fig. 10.8a, from which it can be found that lots of hole intervals from 1000 to 1360 m and from 1512 to 1855 m had the diameters smaller than 310.0 mm, with the minimum of 305.0 mm. As the outside diameter of the collar of 273.0 mm intermediate casing was 298.5 mm, when casing was in the middle, the clearance between borehole wall and casing was 6.3 mm, with the minimum clearance of only 3.3 mm. Furthermore, borehole had dogleg to a certain extent, and this would inevitably bring about unfavourable influence on running casing and cementation.

(2) Cementing slurry leaked easily

The clearance between the collar (298.5 mm) of 273.0 mm intermediate casing and 311.1 mm hole wall and between the collar (215.9 mm) of 193.7 mm intermediate casing and 244.5 mm hole wall were both very small, thus the upward resistance of cement slurry was very large, the equivalent density of cement slurry might be larger than that of formation leakage and that of formation fracturing pressure, resulting in cement slurry leakage.

1. **Borehole preparation**

Borehole repairing, redressing and trial casing running before setting casing were the technical measures that must be adopted for running casing and cementation in small clearance and in hard rock. 311.1 mm diamond repairing tool not only had the ability of wiping, but also the ability of reaming, being favourable for wiping borehole wall and reaming borehole size. The reaming shell was composed of drill bit steel body brazed with diamond impregnated strips, which were mixed with polycrystalline diamond, natural diamond and carbonado diamond, so that the gauging property of each material could be fully utilized, for instance, abrasive resistance of polycrystalline diamond, high strength of natural diamond and high hardness of carbonado, and in this way the cutting capacity and the service life of the reaming shell were improved and satisfactory borehole repairing result obtained.

Borehole repairing tool assembly: 311.1 mm tri-cone bit + 311.5 mm diamond repairing tool (Fig. 7.11a) + 203.2 mm drill collar (one piece) + 309 mm stabilizer + 203.2 mm drill collar (one piece) + 305 mm stabilizer + 203.2 mm drill collar (four pieces) + 177.8 mm drill collar (four pieces) + 127.0 mm drill rod.

Borehole repairing tool assembly simulated the rigidity of 273 mm intermediate casing (including collar). Tri-cone bit, diamond repairing tool, the upper 309 mm stabilizer and 305 mm stabilizer all had a certain cutting capability, thus in the process of borehole repairing and redressing the flex point and its neighboring hole sections could be enlarged, dogleg degree was decreased, the resistance for running down 273 mm intermediate casing was reduced and then smooth running to the position for 273 mm intermediate casing could be ensured. The curves of caliper logging after borehole repairing can be found in Fig. 10.8b. After borehole repairing the minimum hole size was 312.4 mm, the clearance between casing collar and borehole wall was increased to 6.95 mm and therefore the resistance for running casing and the pressure for cement slurry displacement were both effectively decreased.

2. **Trial run casing**

It was decided to trial run casing in order to guarantee a successful casing setting on its first run.

The trial casing running tool assembly included guiding shoe + 273.0 mm casing (two pieces) + 177.8 mm drill collar (nine pieces) + 178 mm jar while drilling + 127 mm drill rod. As a result, trial run casing was successful, however, resistance was encountered in the hole section from 1891 to 1975 m with the maximum resistance of no more than 3 t (according to stipulation in trial running the resistance was no more than 6 t). This indicated that the condition for running casing was satisfied.

3. **Casing**

After cementing 273.0 mm intermediate casing, 193.7 mm ($7^5/_8$ in.) moving casing was to be run. The lower end of this moving casing was set on the retaining sub at the bottom of 273.0 mm outer casing. For this reason in the program of 273.0 mm intermediate casing cementation the cementing program for 193.7 mm moving casing should be considered at the same time. The process of setting 193.7 mm moving casing in 273.0 mm intermediate casing can be found in Fig. 10.9.

(1) Structure of casing string

The steel grade of 273.0 mm intermediate casing was J55, wall thickness was 10.16 mm, outside diameter was 273.0 mm and inside diameter 252.7 mm. Casing structure: float shoe + floating collar + short casing (one piece, 5 m) + setting seat + short casing (one piece, 3 m) + (moving casing) retaining sub assembly + casing string. The structure of the lower part of casing string is shown in Fig. 10.10.

(2) Casing accessories

i. Retaining sub assemblyRetaining sub assembly consisted of retaining collar, bearing seat, insert sleeve and insert guide head. The retaining collar was connected with the casing string and the bearing seat was fixed on the retaining collar, for bearing partial weight of 193.7 mm moving casing. At the upper part of the bearing seat was machined with teeth, to prevent 193.7 mm moving casing from rotating. The insert sleeve was connected in 193.7 mm moving casing string and inserted into the bearing seat. The insert sleeve was machined with teeth, which had the opposite direction to the teeth on the bearing seat, and the teeth meshing could prevent 193.7 mm lower casing from rotating and thread loosing. The structure of retaining sub assembly can be found in Sect. 10.4 of this chapter.

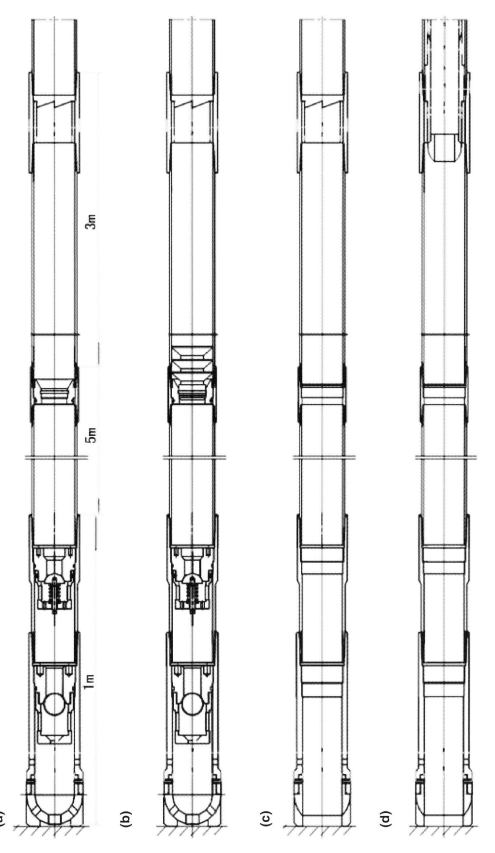

Fig. 10.9 Construction process of setting 193.7 mm moving casing in 273.0 mm intermediate casing. **a** Structure of the lower part of 273.0 mm casing. **b** Cementation setting of 273.0 mm casing. **c** After drilling off 273.0 mm casing accessaries. **d** After running down 193.4 mm moving casing

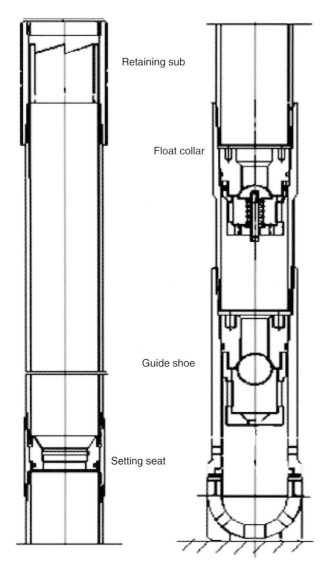

Fig. 10.10 The structure of the lower part of 273.0 mm casing

ii. Rubber plug and setting seat

Rubber plug (Fig. 10.11) was made into a structure of flexible cup, which could smoothly pass through the retaining sub. At the top of the rubber plug, spring hoop and sealing ring were designed to avoid retrace. Setting with the setting seat, spring hoop entered into the groove, to avoid rubber plug retrace. Through sealing ring to avoid back press, safety was further improved. For the core parts of the rubber plug and the setting seat (Fig. 10.12), materials with good drillability such as cast aluminium was used, for easy drill-out.

iii. Floating collar

Valve tructure was adopted and cast aluminium was used for the core parts (Fig. 10.13).

iv. Float shoe

A structure of nylon floating ball was used and cast aluminium was used for the core parts, for easy drill-out (Fig. 10.14).

The floating collar and the float shoe adopted different structures, to ensure a success of anti-back press.

(3) Casing running operation

To guarantee concentricity between 273.0 mm casing and borehole wall to obtain good cement consolidation quality, elastic centralizer was to be fixed onto the casing body of 273.0 mm casing in the original design. However, the designed outside diameter of the centralizer for 273.0 mm casing was 370 mm, whereas the borehole size was 311.1 mm, downhole accident might happen if the centralizer was run down. So, an elastic centralizer was set at 1938.44, 1947.71, 1957.04 and 2000.53 m of the casing string respectively. A technical measure of enhancing lubrication property for drilling fluid was adopted, for a smooth run-down of the casing. Special-purpose drilling fluid of 50 m^3 was prepared, with the formula of 3 % LBM + 2 % liquid lubricant + 0.5 % solid friction reducer. The solid friction reducer was a high strength 60 mesh plastic ball evenly dispersed in drilling fluid, playing the roles of supporting and rolling between casing and borehole wall, in this way drilling fluid would have extremely good lubrication property and the frictional resistance between casing and borehole wall would be effectively reduced.

The lower parts of casing string was assembled and casing was run down based upon casing cementation procedure, with run-down velocity controlled and cement

Fig. 10.11 Rubber plug for 273.0 mm casing cementation

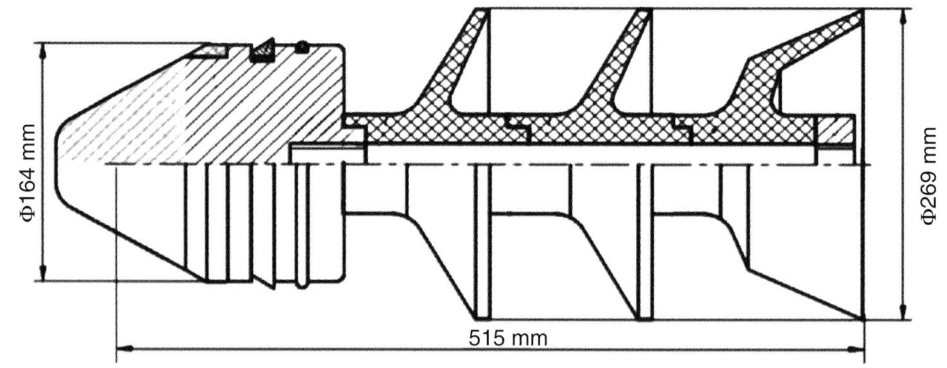

10.3 Casing Running and Well Cementing Operation

Fig. 10.12 Setting seat for 273.0 mm casing cementation

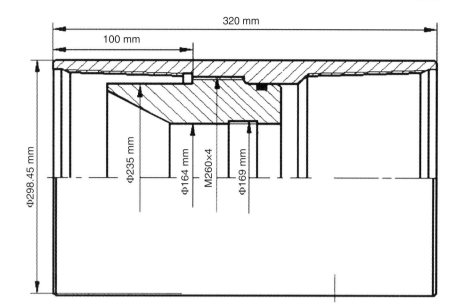

Fig. 10.13 Floating collar for 273.0 mm casing

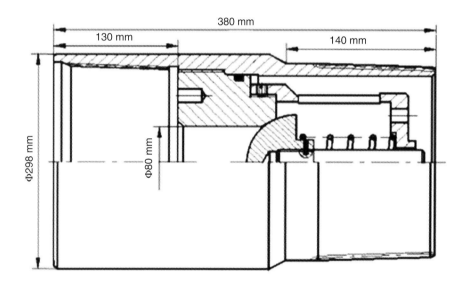

Fig. 10.14 Float shoe for 273.0 mm casing

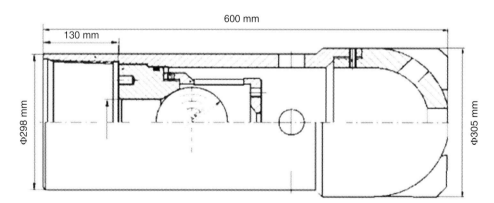

Table 10.8 Calculation of annular volume for 273.0 mm intermediate casing

Well depth (m)	Hole section length (m)	Hole diameter (cm)	Annular volume (L/m)	Volume (L)
0–100	100	31.4	18.68	1868
100–2028	1928	35	37.6	72,493

grouting was conducted once for every 20 pieces of casing. Started at 8:00 on September 9th, 2002 and ended at 18:53 on September 10th, the operation of running 273 mm casing was successfully completed, taking 36 h and 53 min.

4. **Well cementation**
(1) Design and calculation
 i. Basic data
 Drill bit size × well depth: 311.1 mm × 2028 m; casing size × running depth: 273 mm × 2028 m; returned height of cement: ground surface
 ii. Calculation of cement quantity
 Length of cementation: 2028 m; total volume: 74,361 L (Table 10.8); average annular volume: 36.7 L/m; average hole size of the opened borehole: 350 mm; borehole diameter enlargement ratio: 12.5 %.
 It was designed that the volume of leading cement slurry was 35708 L; the volume of tail cement slurry was 37605 L. Cardinal number of leading cement was 33 t and with 15 % addition, thus the total quantity of leading cement was 38 t. Cardinal number of tail cement was 50 t and with 15 % addition, thus the total quantity of tail cement was 58 t. The total quantity of cement was 96 tons.
 iii. Design of cement slurry
 In the light of well cementation in small clearance, cementation with dual density cement slurry (conventional density and low density) was utilized, to solve the technical difficulties of large pressure drop in cement injection and in displacement of cement slurry, as well as the problem of leakage.
 Leading slurry: BHST = 75 °C, BHCT = 60 °C, pressure = 35 MPa. Formula: Jiahua G grade cement + 2 % G404 (fluid loss reducer) + 1 % G106 (retardant) + 0.6 % G301 (dispersant) + 5 % G203 (early strength agent) + 0.6 % G603 (defoamer) + 15 % G605 (lightening agent) + 1 % G202 (expanding agent). Water-cement ratio: 0.58, density: 1.55 g/cm^3, bleeding ≤3.5 ml, filter loss ≤150 ml/60 °C × 7 MPa × 30 min, thickening time >200 min/60 °C × 35 MPa × 50 min, initial consistency ≤25BC, compressive strength ≥5 MPa/24 h × 25 °C × 0.1 MPa, ≥10 MPa/24 h × 75 °C × 21 MPa, rheological property n > 0.7, K < 0.3 Pa·s^n.
 Tail slurry: BHST = 75 °C, BHCT = 60 °C, pressure = 35 MPa. Formula: Jiahua G grade cement + 0.6 % G301 (dispersant) + 1.5 % G404 (fluid loss reducer) + 1.0 % G106 (retardant) + 0.6 % G603 (defoamer). Water-cement ratio: 0.44, density: 1.90 g/cm^3, bleeding ≤3.5 ml, filter loss <150 ml/60 °C × 7 MPa × 30 min,

Table 10.9 Calculation of displacing slurry volume

Well depth (m)	Hole section length (m)	Wall thickness (mm)	Annular volume (L/m)	Accumulation (L)
0–2028	2028	11.43	49.17	99,717

thickening time >200 min/60 °C × 35 MPa × 35 min, initial consistency ≤25BC, compressive strength ≥14 MPa/24 h × 75 °C × 21 MPa, rheological property n > 0.7, K < 0.3 Pa·s^n.
 iv. Design of spacer fluid and rubber plug pressing fluid
 Spacer fluid: SNC15 % + H$_2$O, 6m^3 + 2m^3, or make-up water 6–8m^3 (defined according to the test result).
 Rubber plug pressing fluid: deep well type, 2 m^3.
 v. Calculation of displacing slurry (see Table 10.9)
 vi. Calculation of pressure
 Hydrostatic fluid column pressure of annular cement slurry:
 Cement slurry: Ps1 = 0.01 × 1000 × 1.55 = 15.5 MPa
 Cement slurry: Ps2 = 0.01 × 1028 × 1.90 = 19.5 MPa
 Hydrostatic fluid column pressure of annular spacer fluid: Psp = 0.01 × 0 × 1 = 0 MPa
 Annular total fluid column pressure: Pan = 15.5 + 19.5 + 0 + 0 = 35 MPa
 Fluid column pressure in casing: Pim = 0.01 × 2028 × 1.20 = 24 MPa
 Hydrostatic column pressure difference: P = Pan − Pim = 35 − 24 = 11 MPa
 Calculation of formation equivalent density: 35 × 100/2028 = 1.73 g/cm^3.

(2) Cementation operation
Cement head (Fig. 10.15) and cementing pipeline were installed at 19:40 of September 10th, 2002 and drilling mud was circulated at 6:40 of September 11th, for 1 h and 20 min. Cement slurry injection pipeline were connected, 6 m^3 spacer fluid was injected, 37 m^3 low density (1.56 g/cm^3) cement slurry was injected, and 38 m^3 high density (1.89 g/cm^3) cement slurry was injected. Rubber plug was set, 2.4 m^3 rubber plug pressing fluid was injected. Displacing slurry was injected by large pump, 33 m^3 weighted mud (with density of 1.52 g/cm^3) was firstly displaced and then 66 m^3 mud (with density of 1.08 g/cm^3) was displaced, under the slurry displacement pressure of 13 MPa, setting pressure of 14.5 MPa, and with cement slurry returned to the well head for 3 min. Pressure releasing was normal and no mud returned to well head. Started from 19:40 of September 10th and ended at 11:00 of September 11th, the cementation operation was successfully conducted according to design, with 13 h and 40 min lasted.

10.3 Casing Running and Well Cementing Operation

Fig. 10.15 Cement head for 273.0 mm intermediate casing cementation

10.3.4 193.7 mm Intermediate Casing

In the main hole of CCSD-1 Well, when core drilling reached to the depth of 3665.87 m, the lower reaming shell in the core drilling tool was broken, and fishing failed for many times. After technical argumentation it was decided that the upper 193.7 mm moving casing was to be withdrawn for 244.5 mm reaming drilling. Reaming drilling was conducted to 3525.18 m deep, because of many fishes in the hole and hole enlargement, it was decided that sidetrack drilling around (to avoid the underground obstacles) was to be carried out. After successful sidetrack drilling around, non-core drilling reached to 3623.91 m, 193.7 mm ($7^5/_8$ in.) intermediate casing was set and cemented. In order to come up to the requirement of design (cement slurry returns to well head), two-stage dual density cement slurry cementation was adopted, with the casing stage collar (Fig. 10.16) and technical service provided from Top-Co Industries Ltd.

1. **Casing**
(1) Casing string structure

The running depth of 193.7 mm intermediate casing was 3620 m, wall thickness of the casing was 9.52 mm, and the position of stage collar was at 1901.5 m (designed at 1928 m). The structure of the casing string (Fig. 10.17) was:

Fig. 10.16 193.7 mm stage collar

193.7 mm guide shoe + 193.7 mm casing (one piece) + 193.7 mm float collar + 193.7 mm casing (six pieces) + 193.7 mm float collar + casing string + 193.7 mm stage collar + casing string.

(2) Operation of running casing

At 3:00 on April 29th, 2004 the operation of running 193.7 mm casing was started (Fig. 10.18). Based upon 193.7 mm casing program and requirement of casing running, casing was run and cement was grouted once for every five pieces of casing. From the 202nd casing (well depth below 2015.92 m), an elastic centralizer was added between every two pieces of casing. Until 4:00 of May 1st, casing was smoothly run down to 3620 m deep, with 49 h lasted.

2. **Well cementation**
(1) Design and calculation
  i. Conditions for calculation (see Table 10.10)
  ii. Calculation of cement quantity

   The first stage cementing: leading slurry 2028–2600 m, quantity of cement needed $(2600-2028) \times 29.5/(105 \times 20) = 8.0$ t, addition for technological consideration 2 t, total 10 t; Tail slurry 3620–2600 m, quantity of cement needed $(3620-2600) \times 29.5/(38 \times 20) = 39.6$ t, addition for technological consideration 5.4 t, total 45 t. Accumulative cement quantity needed was 55 t.

   The second stage cementing: 0–1928 m, in the upper casing: $1928 \times 20.7/(38 \times 20) = 52.5$ t, addition for technological consideration 5.5 t, total 58 t.

  iii. Design of cement slurry

   The first stage cementing: BHST = 2.5 °C/100 × 3620 + 15 °C = 106 °C, BHCT = BHST × 0.7 % = 74.2 °C.

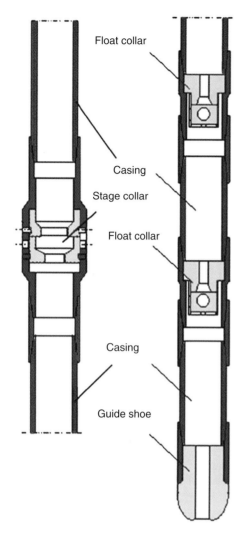

Fig. 10.17 Lower structure of the casing string for 193.7 mm two-stage cementation

Fig. 10.18 Running 193.7 mm casing

Leading slurry: 2028–2600 m, Formula: Jiahua G grade cement + 2.5 % G404 (fluid loss reducer) + 1.0 % G105 (retardant) + 1.0 % G301 (dispersant) + 0.3 % G603 (defoamer) + 3.0 % G202 (expanding agent) + 35 % G606 (high temperature thermal stabilizer) + 10.0 % G605 (lightening agent) + 20.0 % G607 (high temperature strengthening agent). Water-cement ratio: 1.15, density: 1.35 g/cm^3, bleeding ≤3.5 ml, filter loss ≤200 ml/77 °C × 7 MPa × 30 min, thickening time >300 min/77 °C × 40 MPa, initial consistency ≤25BC, compressive strength ≥5 MPa/24 h × 25 °C × 0.1 MPa, ≥ 14 MPa/24 h × 74.2 °C × 21 MPa, rheological property n > 0.7 %, K < 0.3 Pa·s^n.

Tail slurry: 3620–2600 m, Formula: Jiahua G grade cement + 2.0 % G404 (fluid loss reducer) + 1.0 % G105 (retardant) + 0.6 % G301 (dispersant) + 0.3 % G603 (defoamer) + 1.0 % G202 (expanding agent). Water-cement ratio: 0.44, density: 1.90 g/cm^3, bleeding ≤3.5 ml, filter loss ≤100 ml/74.2 °C × 7 MPa × 30 min, thickening time >200 min/74.2 °C × 48 MPa, initial consistency ≤15BC, compressive strength ≥14 MPa/24 h × 106 °C × 21 MPa, rheological property n > 0.7 %, K < 0.3 Pa·s^n.

The second stage cementing: 0–1928 m, BHST = 69 °C, BHCT = 49 °C, pressure = 36.6 MPa. Formula: Jiahua G grade cement + 2.0 % G404 (fluid loss reducer) + 0.6 % G301 (dispersant) + 0.3 % G603

Table 10.10 Calculation conditions for 193.7 mm casing cementation

Item	Parameter	Item	Parameter
Drill bit diameter (mm)	244.5	Density of cement slurry (g/cm^3)	1.35/1.90
Casing O.D. (mm)	193.7	Sack prepared grout (L)	105/38
Position of stage collar (m)	1928	Annular volume of opened hole (L/m)	29.5
Returned height, the first stage (m)	2028	Annular volume between 273.0 and 193.7 mm casing (L/m)	20.7
Returned height, the second stage (m)	Surface	Average hole diameter (cm)	27.4
Position of choke ring (m)	3550	Hole diameter enlargement ratio (%)	12

10.3 Casing Running and Well Cementing Operation

Table 10.11 Calculation of displacing fluid volume

Stage	Well depth (m)	Hole section length (m)	Casing wall thickness (mm)	Volume of casing (L/m)	Accumulation (L)
The first stage	0–3550	3550	9.52	23.95	85,022.5
The second stage	0–1928	1928	9.52	23.95	46,175.6

(defoamer) + 1.0 % G202 (expanding agent). Water-cement ratio: 0.44, density: 1.90 g/cm³, bleeding ≤3.5 ml, filter loss ≤100 ml/49 °C × 7 MPa × 30 min, thickening time >150 min/49 °C × 40 MPa, initial consistency ≤25BC, compressive strength ≥14 MPa/24 h × 69 °C × 21 MPa, rheological property n > 0.7 %, K < 0.3 Pa·s^n.

iv. Design of spacer fluid and rubber plug pressing fluid
Spacer fluid: the first stage was SNC15 % + H$_2$O, 6 m³ + 2 m³; the second stage was make-up water 4 m³. Rubber plug pressing fluid: clear water type (same for the first and second stages), 2 m³ for each stage.

v. Calculation of displacing fluid (see Table 10.11)

vi. Calculation of pressure
The first stage:
Hydrostatic fluid column pressure of annular cement slurry:
Cement slurry: Ps = 0.01 × (3620 − 2600) × 1.90 + 0.01 × (2600 − 2028) × 1.35 = 27.1 MPa
Hydrostatic fluid column pressure of spacer fluid: Psp = 0.01 × 245 × 1.0 = 2.45 MPa
Annular drilling fluid column pressure: Pm = 0.01 × 1928 × 1.18 = 19.2 MPa
Annular total fluid column pressure: Pan = 27.1 + 2.45 + 19.2 = 48.75 MPa
Fluid column pressure in casing: Pim = 0.01 × 3620 × 1.18 = 42.7 MPa
Hydrostatic column pressure difference: P1 = Pan − Pim = 48.75 − 42.7 = 6.05 MPa
Circulating pump pressure: Pz = 0.0018 × 3620 = 6.5 MPa
Maximum operating pressure: P = P1 + Pz = 12.55 MPa
Calculation of formation equivalent density: 49.15 × 100/3620 = 1.36 g/cm³
The second stage:
Hydrostatic fluid column pressure of annular cement slurry:
Cement slurry: Ps1 = 0.01 × 1928 × 1.90 = 36.6 MPa
Annular total fluid column pressure: Pan = 36.6 MPa
Fluid column pressure in casing: Pim = 0.01 × 1928 × 1.18 = 22.75 MPa
Hydrostatic column pressure difference: P1 = Pan − Pim = 36.6 − 22.75 = 13.85 MPa
Circulating pump pressure: Pz = 0.0018 × 1928 = 3.5 MPa
Maximum operating pressure: P = P1 + Pz = 17.35 MPa (displacing fluid should be weighted to decrease 3 MPa pressure)

Fig. 10.19 Operation of 193.7 mm casing cementation

Formation equivalent density: 1.90 g/cm³.

(2) Cementation operation

The operation sequence of cementation (Fig. 10.19): pressure testing → injecting spacer fluid → cement injection, first stage → pressing rubber plug → slurry displacement → setting → blowdown to check backflow → putting in gravity plug → opening circulation hole → well flushing (small displacement circulation at first, then circulating displacement was increased when pump pressure became normal, to flush out surplus cement slurry) → curing → injecting spacer fluid → cement injection, second stage → pressing rubber plug → slurry displacement → setting → pressurizing to close shut-off sleeve → pressure relief to check the closing condition → construction completed.

At 4:00 of May 1st, 2004, casing running was completed, at 4:30 drilling mud was circulated, at 9:30 6 m³ SNC 15 % and 2 m³ H$_2$O were injected, 17 m³ low density (1.35 g/cm³) cement slurry was injected, and 34 m³ high density (1.90 g/cm³) cement slurry was injected. Rubber plug was set, 2.0 m³ rubber plug pressing fluid was injected. 84.60 m³ displacing slurry (theoretical calculation 84.4 m³) was injected by large pump, and then stopped because of no setting. 0.5 m³ returned with pressure relief and this showed that the back-pressure valve was in a good condition. Cement head (Fig. 10.20) was released and gravity plug was put in and it would reach stage collar in 33 min according to calculation. At 12:00 pressure was built up to 4 MPa in cement truck and the circulation hole of the stage collar was opened for circulation. Approximate 2 m³ of cement slurry returned. At 18:00 circulation was stopped and the second

Fig. 10.20 Cement head for 193.7 mm casing cementation

stage cementation was started. 4 m³ of make-up water was injected, and 42 m³ cement slurry (1.90 g/cm³) was injected. Rubber plug was set, 2.0 m³ rubber plug pressing fluid was injected. 44.1 m³ displacing slurry was injected by large pump and the setting pressure increased from 11 to 15 MPa. In order to close the circulation hole, pressure was built up to 18 MPa with the cement truck, pressure was stabilized for 5 min for observation and then pressure was relieved. 0.4 m³ water returned, drilling mud did not flow backwards, circulation hole was closed, cement slurry returned to surface for 2 min and then the construction was completed.

From 11:00 to 20:00 of May 1st, the operation of two-stage cementation was smoothly completed according to the requirements of the design, with 9 h lasted.

10.3.5 127.0 mm Tail Pipe

On March 8th, 2005, drilling reached to the final hole depth of 5158 m. And on March 9th the job of running 127.0 mm (5 in.) tail pipe and cementation was started. The tail pipe hanger, tail pipe running-in tool, construction and technical service were provided by Shenzhen Baiqin Petroleum machinery and Technical Development Co.

1. **Casing**

(1) Structure of casing string

Casing running depth was from 3540 to 4800 m, with wall thickness of 7.52 mm. Structure of casing string (Fig. 10.21) was: float shoe + casing (one piece) + floating collar + casing (one piece) + floating collar + casing (one piece) + ball seat + casing + joint + hanger + running-in tool + drill rod.

(2) Running casing

At 10:00 of March 9th, 2005, operation of running casing was started. Tail pipe was run according to the structure of tail pipe string and the requirements of operation; cement

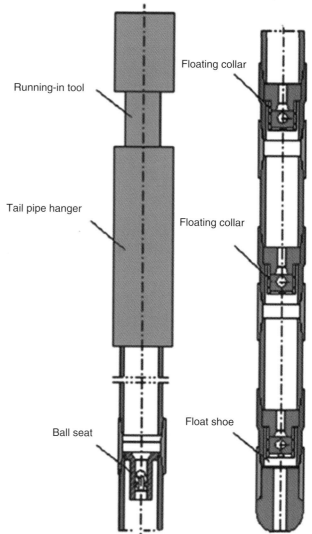

Fig. 10.21 Structure of the lower part of 127.0 mm tail pipe string

was grouted for each piece of casing and cement was fully grouted for every ten pieces of casing. An elastic centralizer was fixed at the depth of 3534, 3563.7, 3603.3, 4769.79 and 4780.06 m respectively. 127.0 mm × 193.7 mm tail pipe hanger (Fig. 10.22), tail pipe running-in tool (Fig. 10.23) and 89 mm drill rod were run into the borehole, and the suspending weight of tail pipe string was 34 t (including the

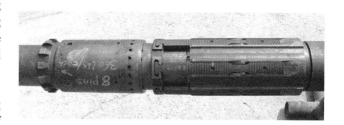

Fig. 10.22 127 mm tail pipe hanger

10.3 Casing Running and Well Cementing Operation

Fig. 10.23 Running-in tool assembly for 127 mm tail pipe

suspending weight of crane). Tail pipe was run to the position, with a length of 1271 m; and the total length of running-in tool was 3530 m, with 3.45 m kelly-up left, for connecting pipeline. 2 MPa was the maximum pressure for circulating mud, uplifting 110 t and running down 94 t (crane and swivel 14 t), dropping a ball and pumping for 15 min till the ball to the position. Building the pressure to 12.5 MPa and stabilizing the pressure for 3 min, running down the drilling tool, suspending weight decreased, and it was indicated that setting was successful. Pressing to 64 t so as to guarantee setting was reliable. The float shoe of the casing string was set to 4790.72 m and the top of the hanger to 3523.55 m. Turning clockwise for 25 rounds and back off, uplifting 1.4 m to verify drop-off, suspending weight increased to 78 t and stopped, drop-off was successful. Pressure of 24 MPa was built to reject ball seat, and circulation was established. The operation of running casing was completed at 23:45 of March 10th, with 37 h and 45 min lasted.

2. **Well cementation**

Cementing operation was started after successful setting, suspension of the tail pipe and drop-off, and establishment of circulation.

(1) Design and calculation

  i. Conditions for calculation (see Table 10.12)
  ii. Calculation of cement quantity

  Cement quantity for opened hole section: $(4800 - 4200) \times 12.57 = 7542$ L; cement quantity in casing: $(4800 - 4760) \times 9.85 = 394$ L; total cement quantity was 7936 L, 7936/50 = 159 sacks, i.e. 8 t.

  iii. Design of cement slurry

  Cement slurry: 4200–4800 m, BHST = 135 °C, BHCT = 122 °C, pressure = 54 MPa. Formula: Jiahua D grade cement + 30 % SiO_2 + 0.5 % WS + 1.2 % G106 (retardant), density: 1.85 g/cm^3, heating time: 45 min, bleeding ≤3.5 ml, filter loss <50 ml/7 MPa × 30 min, thickening time >250 min/122 °C × 54 MPa, initial consistency <25BC, compressive strength >14 MPa/24 h × 135 °C × 21 MPa, rheological property n > 0.7 %, K < 0.3 Pa·s^n.

  iv. Design of spacer fluid and rubber plug pressing fluid

  Spacer fluid: SNC15 % + H_2O, 6 m^3 + 2 m^3 or make-up water 6–8 m^3

  Rubber plug pressing fluid: make-up water, 2 m^3

  v. Calculation of displacing fluid (see Table 10.13)
  vi. Calculation of pressure

  Hydrostatic fluid column pressure of annular cement slurry:
  Cement slurry: $Ps = 0.01 \times 600 \times 1.85 = 11.1$ MPa
  Hydrostatic fluid column pressure of spacer fluid:
  $Psp = 0.01 \times 636 \times 1.0 = 6.36$ MPa

Table 10.12 Calculation conditions

Item	Parameter	Item	Parameter
Drill bit diameter (mm)	157	Density of cement slurry (g/cm^3)	1.85
Casing O.D. (mm)	127	Sack prepared grout (L)	50
Returned height of cement slurry (m)	4200	Water-cement ratio	0.58
Casing running depth (m)	4800	Annular volume of opened hole (L/m)	12.57
Cementing interval (m)	600	Volume in casing (L/m)	9.84
Position of choke ring (m)	4760	Annular volume between 193.7 and 127 mm casing/(L/m)	11.28

Table 10.13 Calculation of displacing fluid volume

Well depth (m)	Section length (m)	Wall thickness (mm)	Volume per meter (L/m)	Total volume (L)
0–3540	3540	9.35 (89 mm drill rod)	3.87	13,700
3540–4760	1220	7.52	9.84	12,004.8
Total	4760			25,704.8

Annular drilling fluid column pressure: Pm = 0.01 × 3564 × 1.08 = 38.49 MPa

Annular total fluid column pressure: Pan = 11.1 + 6.36 + 38.49 = 55.95 MPa

Hydrostatic fluid column pressure of rubber plug pressing fluid: Psp = 0.01 × 203 × 1.0 = 2.03 MPa

Drilling fluid column pressure in casing: Pm = 0.01 × 4597 × 1.08 = 49.65 MPa

Total fluid column pressure in casing: Pim = 2.03 + 49.65 = 51.68 MPa

Hydrostatic column pressure difference: P1 = Pan−Pim = 55.95−51.68 = 4.27 MPa

Circulating pressure: Pz = 0.0018 × 4800 + 3 = 11.64 MPa

Maximum operating pressure: P = P1 + Pz = 4.27 + 11.64 = 15.91 MPa

Calculation of formation equivalent density: 55.95 × 100/4800 = 1.17 g/cm³

(2) Cementation operation

The operation sequence of cementation: pressure testing → injecting spacer fluid → cement slurry injection → pressing rubber plug (to flush the cement slurry injection pipeline at the same time) → slurry displacement → setting → pressure relief to check the setting condition.

Cementation (Fig. 10.24) was started at 10:10 of March 11th, 2005. 8 m³ spacer fluid was injected, 10 m³ cement slurry was injected (2 m³ more than design), and then rubber plug was set and slurry displaced. Setting pressure increased from 11 to 15 MPa, and pressure released without water backflow. Cementation operation was completed at 15:10.

10.4 Moving Casing Techniques

Mostly used in continental scientific drilling engineering, moving casing denotes the casing run into the borehole for a certain purpose in drilling construction and can be completely retrieved from the borehole after the completion of the drilling construction. Because the complex formations would be encountered, borehole would be deep, complicated drilling technologies were to be utilized, more casing program was to be used and core drilling was required, thus in design a method of advanced open hole drilling was adopted. Generally, small sized core drilling was conducted and then reaming. In drilling construction, after the large sized casing was set in the upper section of the borehole and core drilling was conducted in the lower section, the inside diameter of the casing in the upper section of the borehole would be much larger than the diameter of core drill bit (Fig. 10.25) and under these circumstances it would be

Fig. 10.24 Cement head for 127 mm tail pipe cementation

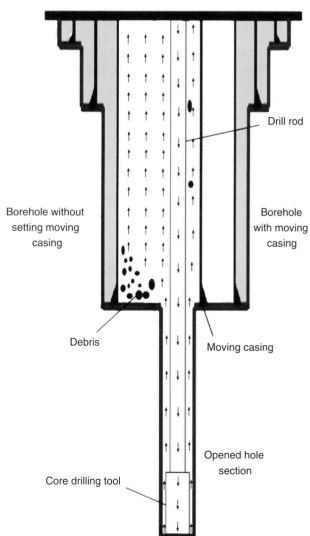

Fig. 10.25 Sketch of moving casing techniques

necessary to run moving casing, to reduce the difference between the inside diameter of the casing and the diameter of the lower opened hole section, so as to ensure a smooth return of the cuttings (Fig. 10.25).

Moving casing techniques mainly have following advantages:
 i. to reduce casing program and construction cost;
 ii. to avoid an excessive oscillation of drill string, being favourable to stable rotation;
 iii. to avoid vortex deposition of cuttings, which may cause drill bit burying (bit freezing);
 iv. borehole structure can be flexibly adjusted, being favourable for implementation of drilling technology.

10.4.1 Overall Programme

In drilling construction design of CCSD-1 Well, moving casing was to be run twice, with basic data shown in Table 10.14, in which the deepest moving casing designed was 2000 m. The overall programme design of moving casing construction was mainly based upon 193.7 mm ($7^5/_8$ in.) moving casing running to 2000 m.

In the overall programme design of moving casing, following problems should be considered:
 i. The length of moving casing will change along with the variation of the temperature of drilling fluid in borehole, and resulting in a dynamic expansion and contraction, and a stress variation during the BTC period of construction.
 ii. The fixing of moving casing, besides the fixing of top and bottom ends, the stability of the whole casing should be taken into consideration.
 iii. When rotating in high velocity in moving casing, drill string will produce dynamic frictional force in circumferential direction to the inside wall of moving casing, which may cause back-off for some casing joints.
 iv. In design, withdrawal of moving casing after usage should be taken into consideration.

The overall programme of moving casing: a method of fixing casing at both ends was adopted, i.e. some of the casing weight load was to be hanged at the top end while some of the casing weight load was to be set at the bottom end. This programme could prevent the moving casing from back-off and improve the stability of the whole casing. A casing hanger for oil drilling was adopted to hang the top end of the moving casing (Fig. 10.6) while a specially designed casing shoe was utilized to support the bottom end. The casing shoe of the moving casing was set on a specially designed fillet (sub) at the bottom of outer casing (Fig. 10.26).

The method of fixing casing at both ends required that appropriate load should be added on the both ends, mainly based on the length of moving casing and possible variation range of temperature. As the top hanging load and the bottom supporting load varied along with the temperature variation of moving casing, the expansion and contraction of the moving casing resulted from temperature variation could be adjusted by utilizing the variation of the force bearing state of the whole moving casing in the borehole (the change of top hanging and bottom supporting weights). This programme could satisfy the requirement of running long moving casing of 2000 m under hole bottom temperature no higher than 150 °C (Wenjian et al. 2003).

10.4.2 Design of Fixing Moving Casing

In the design of fixing moving casing, the length of casing hanged at top and set at bottom could be calculated according to the possible temperature variation and the weight of the moving casing, so as to guarantee that the entire state of the moving casing (moving casing bore a certain load at top and bottom) would not change when temperature varied, that is, the variation of the force borne at top and bottom of the moving casing was used to compensate for the length change of moving casing resulted from temperature variation.

In running moving casing, the designed depth was 2000 m whereas the maximum possible well depth might reach 5000 m after completion of the well. Based upon Table 10.4, it could be calculated that the maximum temperature variation during the working period of 2000 m moving casing would not exceed 30 °C.

Table 10.14 Basic data of moving casing

Type	Running in hole interval (m)	Steel grade	Nominal weight (kg/m)	O.D. (mm)	Wall thickness (mm)	I.D. (mm)	Run through size (mm)	Collar O.D. (mm)	Thread type
244.5 mm moving casing	0–101	N80	64.79	244.5	11.05	222.4	218.4	269.9	Oval-shaped
193.7 mm moving casing	0–2019	P110	44.24	193.7	9.52	174.6	171.5	215.9	Buttress BTC

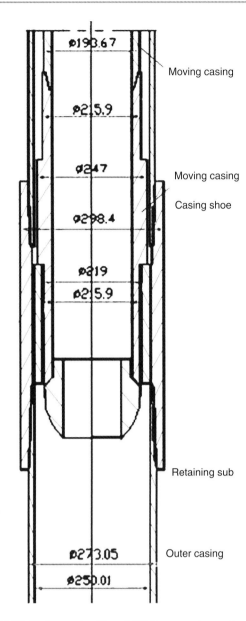

Fig. 10.26 Bottom end setting of 193.7 mm moving casing

Supposing wall thickness is 9.52 mm, length of the moving casing is 2000 m, and density of drilling fluid is 1.2 g/cm³, then 193.7 mm moving casing should be designed as follows:

Casing with the length L is pulled or pressed by its dead weight when under the vertical state, its deformation dimension should be:

$$\Delta L = (QgL^2)/(2AE) = KL^2 \quad (10.6)$$

in which,

- Q unit weight of moving casing, 44.24 kg/m
- g gravitational acceleration, 9.8 m/s²
- A cross sectional area of moving casing, 0.00551 m²
- E elastic modulus of moving casing, 207 GPa
- K calculation coefficient, in air $K = 0.1901 \times 10^{-6}$; in drill mud $K = 0.1610 \times 10^{-6}$

When moving casing has been run into borehole, supposing the casing length that hanged at top is L_1 and the casing length that set at bottom is L_2, then

$$L_1 + KL_1^2 + L_2 - KL_2^2 = 2000 \quad (10.7)$$

Supposing that 80 % of the total length of moving casing is hanged at well head and 20 % of the total length of moving casing is set at the bottom, the neutral point (where the axial force is zero) is at the position of 1600 m. Moving casing of 1600 m is hanged at well head and 400 m set at the bottom, and

$$L_1 + KL_1^2 = 1600 \quad (10.8)$$

$$L_2 - KL_2^2 = 400 \quad (10.9)$$

Based upon Formulas (10.8) and (10.9), it can be obtained that

$$L_1 = 1599.585 \text{ m}$$
$$L_2 = 400.026 \text{ m}$$

The natural total length of the moving casing to be run is 1999.611 m. The weight of the moving casing to be hanged by the hanger at well head is 60,208 kg while the weight of the moving casing to be set at the bottom is 15,052 kg.

When the average temperature of the moving casing increases to the maximum, i.e. the average temperature of the whole moving casing increases by $\Delta t = 30$ °C, suppose the casing length hanged at top is L_{11} while the casing length set at bottom is L_{21}, then

$$L_{11} + KL_{11}^2 + L_{21} - KL_{21}^2 + \Delta t \times k_t \times (L_{11} + L_{21}) = 2000 \quad (10.10)$$

$$L_{11} + L_{21} = L_1 + L_2 \quad (10.11)$$

in which, k_t denotes the thermal expansion coefficient of moving casing, 1.25×10^{-5} m/°C

From Formulas (10.10) and (10.11), it can be obtained that

$$L_{11} = 442.254 \text{ m}$$
$$L_{21} = 1557.357 \text{ m}$$

The position of the neutral point can thus be further derived

$$L_0 = L_{11} + KL_{11}^2 + \Delta t \times k_t \times L_{11} = 442.451 \text{ m}$$

That is, the weight hanged at the top is 16,642 kg and the weight set at the bottom is 58,603 kg, about 70 % of the weight is supported at the bottom.

Because the temperature variation of the whole moving casing is uneven, being a linear variation from top to bottom, with large temperature change at top and little change at bottom, hence the above calculation results should be revised. However, the temperature variation has slight influence on the neutral point, which moves downwards after the revision.

When mud loses, casing temperature decreases. Based upon an extreme condition, i.e. drilling mud completely loses at a low speed and casing is under a low temperature state, supposing the casing length hanged at top is L_{12} and the casing length set at bottom is L_{22}, then

$$L_{12} + KL_{12}^2 + L_{22} - KL_{22}^2 = 2000 \quad (10.12)$$

$$L_{12} + L_{22} = L_1 + L_2 \quad (10.13)$$

Then,

$$L_{12} = 1510.94 \text{ m}$$

$$L_{22} = 488.671 \text{ m}$$

When drilling mud loses quickly and casing is under a high temperature state, it can be obtained that

$$L_{12} = 525.491 \text{ m}$$

$$L_{22} = 1474.12 \text{ m}$$

It can be found from the calculation results that mud loss has a limited influence on the neutral point (zero stress) of moving casing. The influence of mud loss on the position of the neutral point is that the neutral point tends to the central point.

From above calculation results it could be known that the fixing programme of moving casing that 80 % hanged at top and 20 % set at bottom was feasible. In running the moving casing, 80 % was to be hanged at top and 20 % set at bottom. When mud temperature in borehole changed, casing temperature changed accordingly. When temperature reached the maximum, approximate 30 % moving casing was to be hanged at top and about 70 % set at bottom (retaining 10 % innage so as to increase the safety factor of the casing to forestall temperature change). During the entire period of scientific drilling, the fixing state of moving casing only varied quantitatively, without any qualitative variation. Therefore, the programme of fixing 2000 m moving casing was that in running casing 80 % of the casing was to be hanged by hanger at top while 20 % of the lower casing was set on the retaining sub of the outer casing.

10.4.3 Moving Casing Strength Check

The maximum load of 2000 m casing hanged at top was 88,480 kg while the maximum load of 1557.36 m casing set at bottom was 68,898 kg. The method of buoyancy factor was adopted to calculate buoyancy. The effective external collapsing force was calculated based on 50 % casing emptied; external fluid column pressure was calculated based on full borehole saturated salt water (with density of 1.15 g/cm^3) and internal pressure was calculated based on oil well kick. The checked results can be found in Table 10.15.

10.4.4 Design of Casing Shoe and Retaining Sub

1. **Casing shoe**

20–70 % of moving casing was set on the retaining sub of the outer casing and supported by casing shoe of the moving casing. Casing shoe consisted of inserted sleeve and inserting guide head. In order to guarantee a stable and reliable support by casing shoe and ensure that the normal tripping for follow-up drilling was not to be affected, casing shoe of moving casing must be designed according to in-the-hole condition when moving casing was run and the drilling tool condition for follow-up drilling.

Moving casing diameter was 193.7 mm, wall thickness 9.52 mm and length 2000 m. The weight of casing set on casing shoe was (calculated when weight was the maximum) (weight of casing centralizer can be ignored): 2000 × 70 % × (44.24 − 1.2 × 5.508) = 52,682 kg.

And compressive stress of casing was 52,682 × 9.8/5508 = 93.7 N/mm^2.

As the yield strength of P110 steel is 760 N/mm^2, so casing can satisfy the requirement for yield strength. The material similar to P110 can be selected for casing shoe, with wall thickness the same as that of casing. Based upon the bit size for drill-off after outer casing cementation, drift

Table 10.15 Checked results for strength of 193.7 mm moving casing

Item		193.7 mm moving casing
Casing strength	Tensile strength (kN)	4180
	Collapsing strength (MPa)	36.8
	Internal pressure strength (MPa)	65.2
Actual safety factor/permissible safety factor	Tensile S_t	4.7/1.8
	External collapsing S_c	1.4/1.125
	Internal pressure S_i	5.4/1.10
	Compressive S_y	7.3/1.8

Fig. 10.27 Casing shoe structure for 193.7 mm moving casing

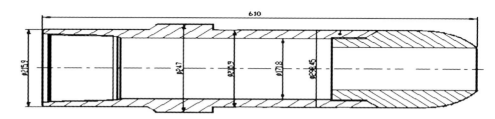

Fig. 10.28 Structure of retaining sub

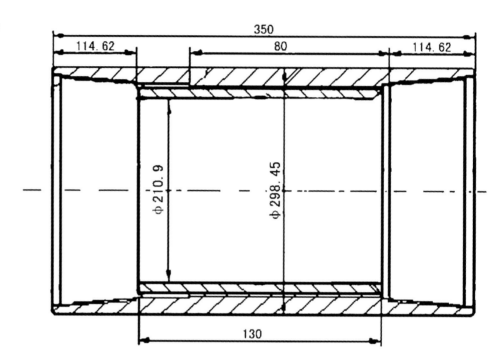

diameter requirement for follow-up drilling and other technical requirements, casing shoe for moving casing can be designed, with the structure shown in Fig. 10.27.

2. **Retaining sub**

Retaining sub consists of retaining coupling and bearing seat, fixed on the upper section of the setting seat of outer casing. According to the matching requirements for moving casing shoe, strength requirement and the structure of outer casing, the structure of retaining sub can be designed (Fig. 10.28).

The strength of bearing seat of retaining sub was checked as follows:

(1) Maximum shear strength

Shear area:

$$S = \pi dh = \pi \times 247 \times 800 = 62{,}077 \text{ mm}^2$$

in which,

d diameter of bearing seat, mm
h height of bearing seat, mm

The material selected was 45# steel, in normalized condition

Yield strength σ_s = 353 MPa
Safety factor η = 2.5
Then permissible stress $[\sigma]$ = 353/2.5 = 141 MPa
Permissible shearing stress $[\tau]$ = 0.5 $[\sigma]$ = 70.5 MPa
Thread could be calculated on the basis of approximate value, for safety:

$$Q = 0.5[\tau] \, S = 0.5 \times 70.5 \times 62{,}077 = 2{,}188{,}214 \text{ N}$$
$$= 2188 \text{ kN}$$

(2) Compressive damage

Compressive area $S = \pi(D^2 - d^2)/4 = \pi(247^2 - 219^2)/4$
$$= 10{,}247 \text{ mm}^2$$
$$[\sigma_{jy}] = [\sigma] = 141 \text{ MPa}$$
$$Q_{jy} = [\sigma_{jy}] \, S = 141 \times 10{,}247 = 1{,}444{,}827 \text{ N} = 1445 \text{ kN}$$

(3) Maximum potential bearing load

193.7 mm casing wall thickness δ = 9.52 mm
Casing unit weight q = 44.24 kg/m
2000 m casing weighs 88.5 t
70 % of the weight set on the retaining sub, 61.9 t

Even though all the weight of 193.7 mm moving casing was set on the retaining sub, its safety factor can still satisfy the requirement of strength.

10.4.5 Design of Thread Back-off Proof for Moving Casing

The whole state of moving casing is stable when hanged at top and set at bottom. To prevent moving casing from slipping off the casing head, anti-slipping block can be welded on the top for connection while to prevent moving casing from rotating at the bottom anti-slipping teeth can be machined at the connected part between the retaining sub and moving casing shoe (Fig. 10.29).

In follow-up drilling, due to the frictional force between rotating drill rod (drilling tool) and moving casing, thread loosening of some casing and large torsional deformation of long casing, thread back-off or even thread off may happen for some joints of the moving casing, and normal drilling will be affected. For this reason, measures should be adopted for the connection of the whole moving casing string, to prevent casing from thread back-off.

As the type of moving casing was determined and the casing string was rather long, mechanical and technological methods for preventing thread back-off can only obtain a limited result, with unsatisfied reliability. Based upon the experiences and measures adopted for preventing casing thread back-off in oil drilling, sticking was the only feasible method, which was realistic, well-developed and easy for use.

However, after using this method the stuck casing thread must be heated when break-out in casing removal and this heating operation was rather troublesome and would affect the reuse of the casing. Therefore, it was decided that only some casing joints where back-off easily happened should be stuck, this could not only prevent casing from back-off, but also improve the service life of casing and the

Fig. 10.29 Inserted sleeve and inserting guide head for 193.7 mm moving casing

operability of construction. Sticking glue for oil casing purpose was adopted to ensure casing safety and to avoid casing accident.

10.4.6 Design of Centralizer

Without consolidated support by cement sheath, it was a crux to keep moving casing fixed and stable in the follow-up operations. Besides the fixing at the top and the bottom, appropriate fixing in the middle was necessary, to guarantee that moving casing would not shake in the outer casing, and the best way was to install centralizer onto moving casing to improve the rigidity and stability of the moving casing string. Because moving casing would bear frequent impact and vibration from drilling tool during drilling operations, high quality centralizer, especially with good solidity would be required. For moving casing, integrally welded elastic centralizer (Fig. 10.30) should be utilized. Meanwhile, to reduce the oscillating amplitude of moving casing and decrease the force borne on the elastic centralizer, rigid centralizer (Fig. 10.31) should be added on the moving casing string.

Fig. 10.30 Structure of elastic centralizer

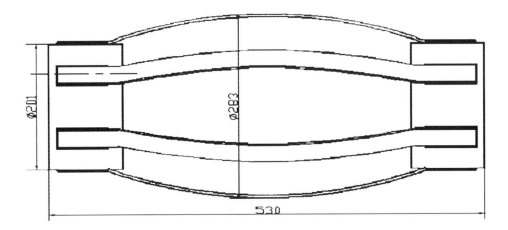

Fig. 10.31 Structure of rigid centralizer

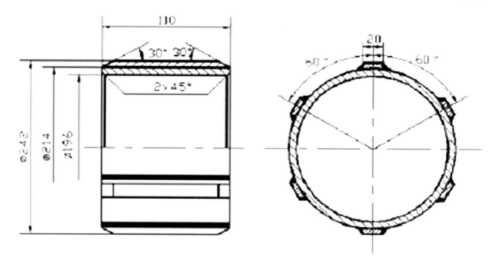

As the upper casing string was in tension, the installation interval of the elastic centralizer could be calculated by reference to the Standards of Petroleum and Natural Gas Industry SY/T5334-1996. Because of the large vibration from the upper drill rod against casing, some rigid centralizers should be installed so as to ensure the moving casing at well head stable. Because the lower casing string was under the pressure, the installation interval of the rigid centralizer could be designed and calculated in accordance with the semiwave length of casing. To reduce the vibration of the rigid centralizer on the moving casing to the outer casing, some elastic centralizers should be installed.

The structure of rigid centralizer was adopted by reference to the rigid centralizer used in oil and natural gas drilling (Fig. 10.31) while for the structure of elastic centralizer an integrally welded elastic centralizer (Fig. 10.30) for oil and natural gas drilling was utilized.

As the lower casing string was under the pressure, rigid centralizers should be installed, if not the casing would be unstable. For the installation interval of the rigid centralizer, the bending radius of the lower moving casing, i.e. the installation interval could be calculated by using the method of energy. The maximum installation interval of rigid centralizer that could keep the compressed casing string stable could be calculated with the following equation:

$$qL^3 - 2P_b L^2 + \gamma EI = 0 \quad (10.14)$$

where,

- P_b axial pressure at the bottom of casing string, N
- q   axial distributing force on casing string, N/m
- E   elastic modulus 207 GPa
- I   inertia moment of casing string cross-sectional area, m^4
- γ constraining coefficient, $\gamma = 74.63$.

A meaningful solution could be obtained:

$$L = x_1 + a_1 \quad (10.15)$$

in which,

$x_1 = 2a_1 \cos(\theta_1 + 3\pi/4)$
$\theta_1 = \arccos(-s/2t)$
$s = -2t + \gamma EI/q$
$t = a_1^3$
$a_1 = 2P_b/(3q)$

In drilling mud, $q = (44.24 - 1.2 \times 5.508) \times 9.8 = 368.774$ N/m

$$I = \pi \times (D^4 - d^4)/64 = 2.3368 \times 10^{-5} \text{ m}^4$$

In application, the axial pressure at the lowest bottom of casing string P_{b1} should be calculated first and then the installation interval of the first centralizer L_1 calculated, afterwards the installation interval of other centralizers successively calculated. The axial pressure at any lower end of casing string $P_{bi} = P_{bi} - L_{i-1} \times q$, i = 1, 2, 3… when the axial pressure at the lower end of casing was less than critical load $P_{bi} = qL_i$, calculation should be stopped.

Based upon the calculation of moving casing fixing design, under the highest temperature the maximum length of the lower casing string that under pressure was approximate 1557 m. Based on the length of the lower casing string that under pressure was 1500 m and through a trial calculation the sequence of installation interval of rigid centralizers on casing could be obtained as: 18.12; 18.23; 18.34; 18.46; 18.58; 18.71; 18.83; 18.96; 19.1; 19.23; 19.38; 19.52; 19.67; 19.83; 19.99; 20.15; 20.33; 20.5; 20.69; 20.88; 21.08; 21.28; 21.49; 21.72; 21.95; 22.19; 22.44; 22.71; 22.99; 23.28; 23.59; 23.91; 24.26; 24.62; 25.01; 25.42; 25.86; 26.34; 26.85; 27.4; 28; 28.66; 29.39; 30.19; 31.09; 32.11;

33.28; 34.63; 36.24; 38.2; 40.66; 43.92; 48.57; 56.18; and 73.68. Totally 55 rigid centralizers should be installed.

The principle of installing centralizer should be determined based upon theoretical design and calculation, and according to the actual condition of casing.

10.4.7 Operating Technology of Moving Casing

The operating technology of moving casing includes two parts: operation procedure of running moving casing and pulling moving casing.

1. **Operation procedure of running moving casing**
(1) Preparation
 i. To measure and number the casings which are to be run one by one, and to clean the thread.
 ii. To check carefully casing accessories (centralizer and casing shoe) and anyone damaged is forbidden for use.
 iii. To check the hoisting system and to confirm that the weight indicator is sensitive and error-free.
 iv. To clean the derrick floor and to fix the teeth of casing tongs.
(2) Running casing
 i. To connect casing shoe with moving casing, and to coat high temperature resistant back-off proof glue or to fasten with rivets.
 ii. An elastic centralizer is added for every piece of the five pieces of casing above casing shoe, on the casing body (fixed with collar clamp), being beneficial to the centralization of casing and then retaining sub is inserted.
 iii. To run casing in order and add elastic centralizer (fixing by collar clamp) and rigid centralizer according to the requirement. Back-up tong is used to prevent the casing string from rotating in the borehole, producing unnecessary damage to the centralizers.
 iv. After the last piece of casing connected, to run down slowly and when it approaches to the position of retaining sub the variation of weight indicator must be carefully noticed. Once bit pressure indicated, the depth and the kelly above rotary at the moment should be immediately recorded, and compared with the theoretical values.
 v. To run down continuously, to release the designed casing seating weight and to set at well head and thus the running of moving casing is completed.

2. **Operation procedure of pulling moving casing**
Following points should attract attention in case of pulling casing in drilling process:
 i. To slowly pull casing string while watch the weight indicator, and carefully analyze and find out the reason in case of abnormality happens.
 ii. In break out, back-up tong is utilized to prevent the lower casing string from rotating. When disassembling centralizer and collar clamp, well head should be well protected so as to prevent small parts from falling into the borehole.
 iii. It is recommended that the rivets near the connecting area of the accessories should be drilled off by using electric drill before breaking out casing. Check accessory (retaining sub) and analyze down hole condition.
 iv. It is recommended that all the centralizers and collar clamps should be changed when moving casing is run down again.
 v. Running procedure is just the same as above.

10.4.8 Application of Moving Casing Techniques

As advanced open hole drilling method was adopted for CCSD-1 Well drilling, the "hole size" of the upper cased hole section was much larger than that of cored open hole. Moreover, during core drilling, because the pump discharge was rather small and the upward velocity of drilling mud in the upper cased hole section was too slow, the problem of cuttings sedimentation easily happened. For this reason, moving casing must be set within the casing of the upper hole section, to decrease the annular clearance and to increase the flowing velocity of drilling fluid. When it was necessary to set another layer of casing, moving casing was to be removed, hole was to be enlarged and then casing was to be set and cemented.

In CCSD-1 Well, moving casing was used twice:
 i. 244.5 mm ($9^5/_8$ in.) moving casing, setting depth 0–101 m
 ii. 193.7 mm ($7^5/_8$ in.) moving casing, setting depth 0–2019 m

1. **Application situation**

As the setting depth was only 101 m, the operation of running 244.5 mm moving casing was rather simple. As the rock at 101 m deep was intact and hard and thus the casing was directly set on the rock at hole bottom. Three rigid centralizers and three elastic centralizers were added on the moving casing string. On July 15th, 2001, 244.5 mm moving casing was run and on May 1st, 2002 the moving casing was withdrawn, both the operations were smooth.

As the setting depth of 193.7 mm moving casing was 2019 m, both the design and running operation were complicated. Abovementioned design and running technology of moving casing were adopted. On September 10th, 2002, 193.7 mm moving casing was run and on October 22nd, 2003 the moving casing was withdrawn.

For 193.7 mm moving casing, steel grade was P110 and the wall thickness was 9.52 mm, with the moving casing string

Fig. 10.32 Elastic centralizer for 193.7 mm moving casing

Fig. 10.33 Rigid centralizer for 193.7 mm moving casing

Fig. 10.34 Running 193.7 mm moving casing

Fig. 10.35 Pulling 193.7 mm moving casing

10.4 Moving Casing Techniques

Fig. 10.36 Elastic centralizer with elastic steel strips broken off

structure of inserting guiding head (guide shoe) + inserted sleeve + 193.7 mm moving casing + landing joint.

On the basis of theoretical design and calculation and according to the actual condition of the moving casing, the principle of installing centralizer was defined: elastic centralizer (Fig. 10.32) was installed onto 193.7 mm moving casing body and limited with collar clamp. A rigid centralizer (Fig. 10.33) was added for every piece of the five pieces of casing near the retaining sub, and on the sixth piece of casing above this an elastic centralizer was installed while for the other casing an elastic centralizer and a rigid centralizer were alternately installed for every two pieces of moving casing (Fig. 10.34).

2. **The problem existed**

In general, withdrawing 193.7 mm moving casing was smooth (Fig. 10.35). However, because of no tempering treatment after the welding between elastic steel strips and centralizing ring, 90 pieces of elastic steel strip were broken at both ends due to the vibration of drill rod and the impact of tripping, for some elastic centralizers the elastic steel strips were completely broken off, and some only with one or two pieces left (Fig. 10.36). 18 pieces of steel strip were carried out with centralizing ring, rigid centralizer and specially made guiding shoe during pulling the moving casing and 72 pieces fell into the borehole, bringing about a lot of troubles for follow-up reaming drilling.

11 Drilling Data Acquisition

What CCSD-1 Well encountered were not only the deep hole to be drilled but also the complex formations to be met, and then a wide variety of drilling technologies were adopted during the difficult constructions. In order to improve drilling efficiency and reduce drilling accident, drilling process monitoring must be strengthened, so as to improve the levels of on-site monitoring, decision-making at drill site and drilling data processing. So a complete set of drilling data acquisition and processing system was indispensable for deep well drilling, which could assist to identify the bottom-hole conditions, improve the level of drilling decision-making, and reduce accidents and construction costs. At the same time, the drilling data acquisition system was also an important data source for drilling management.

As the drilling technologies for continental scientific drilling are different from others and the bottom hole conditions reflected by the drilling parameters are not the same, the drilling monitoring strategy and contents must be studied based upon the special construction characteristics of the continental scientific drilling. Drilling data acquisition system includes two parts, the surface data acquisition system and the down-hole data acquisition system. And the surface data collection consists of two systems, comprehensive geological logging system of the comprehensive geological logging company and the drilling data gauges equipped on the drill rig (with computer data acquisition system). In the actual construction, the comprehensive geological logging system had a powerful function and high specialization degree, collecting data comprehensively, therefore the comprehensive logging system was used as the primary data acquisition system, while the drilling data gauges for real-time monitoring in the drilling process.

Translated Li Haipeng.

11.1 General Situation

Since CCSD-1 Well is the first continental scientific exploration well in China, comprehensive logging instruments specifically for continental scientific drilling were unavailable in market. Although the comprehensive logging instrument for oil drilling could basically satisfy the need of continental scientific drilling, a certain improvements must be made in the software and hardware for data acquisition, and more improvements in the data processing software and application software to meet the needs of drilling data acquisition and processing at surface for continental scientific drilling.

Petroleum exploration comprehensive logging instrument is a large scale mechanical and electrical integrated apparatus for oil and gas exploration services, which can collect real-time drilling parameters, drilling fluid gas logging (CO_2, CH_4, H_2S, etc.) parameters, continuously monitor oil, gas and down-hole drilling conditions, and make evaluations. There are several manufacturers producing different brands of comprehensive logging instruments for petroleum exploration in the country. SDL9000 comprehensive logging instrument jointly developed by Petrochina Shanghai Petroleum Instrument Co. Ltd, China Technical Service Beijing Geological Logging Technology Company and US Halliburton Company, is a new generation of comprehensive logging instrument with the international advanced level. DLS-CPSIC comprehensive logging instrument jointly developed by the US International Logging Company and Petrochina Shanghai Petroleum Instrument Co. Ltd, with the international advanced level of the 21st century, can be used not only on land drilling platform, but also on offshore fixed and floating drilling platforms. Drillbyte comprehensive logging instrument produced by Shanghai Shenkai Science & Engineering Co. Ltd authorized by Beck Hughes Inteq Company is a large scale comprehensive logging instrument with the advanced Drillbyte Ver 2.3 system software and other related systems introduced from Hughes Inteq Company.

Fig. 11.1 The internal units of SK-2000FC/C comprehensive logging instrument developed by Shanghai Shenkai company

SK-2000FC/C comprehensive logging instrument (Fig. 11.1) developed by Shanghai Shenkai Science & Engineering Co. Ltd under the bases of absorbing foreign advanced technology and experience accumulated over years and depending on the scientific and technological superiority of the company and strong industrial base in Shanghai, is a new generation of important instrument for petroleum exploration, occupying a leading role in domestic market in overall design and manufacturing technology and reaching the international advanced level. Various brands of comprehensive logging instruments are similar in technical indexes and in functions, with the main differences in sensor performance, computer hardware structure and the software functions, etc. In recent years, China has introduced a large number of comprehensive logging instruments produced by foreign manufacturers, which not only have the complete comprehensive logging functions, but also have the remote transmission function of real-time data and data files. However, some of them lack the corresponding hardware equipment and software systems, and few though have a data file transmission system but can not used for real-time data transmission. Most of domestic comprehensive logging instruments do not have real-time data transmission function (Figs. 11.2 and 11.3).

Because of the high cost, the down-hole drilling parameter detection instruments are generally used only in the deep or complex well drilling constructions with greater investment. The down-hole drilling parameters detection measurements are divided into the measurement-while-drilling and non-measurement-while-drilling, and the measurement-while-drilling is divided into the wired and the wireless measurement. The wireless measurement (MWD) mainly includes the mud-pulse method, the electromagnetic method and the acoustic wave method. The main foreign MWD service companies and their products include: three MWD systems, Slim1, MWD and LWD by Anadrill Sclumberger; three MWD systems, Acuu Trak 1, Acuu Trak 2 and DMWD by Eastman Chiristensen; the DLWD MWD system by Exlog Company; two MWD systems, AGD and RGD by Halliburton; three MWD systems, slim-hole MWD, small displacement MWB and large displacement MWB by Smith Datadrill; three MWD systems, MPT, RLL, DWD/DGWD/FEED by Sperry-Sun Company; the four hole sized MWD systems by Teleco Company. The domestic MWD products include: the PMWD LWD inclinometer by Beijing Pulimen Company, and the YST-48 LWD inclinometer by Beijing Hailan Company, etc.

The down-hole data transmission technology using mud pulse (MWD) method is well developed, and many oil drilling companies in the country have the equipment and technology, which are mainly imported. The technology has certain defects in underbalanced drilling, mainly because of the gas compressibility that makes the pressure wave signals distorted, resulting in great difficulties to detect the correct signals at surface. Since the 1990s, the electromagnetic MWD technology in underbalanced drilling applications has experienced great development, and it becomes the most effective method for down-hole data real-time transmission

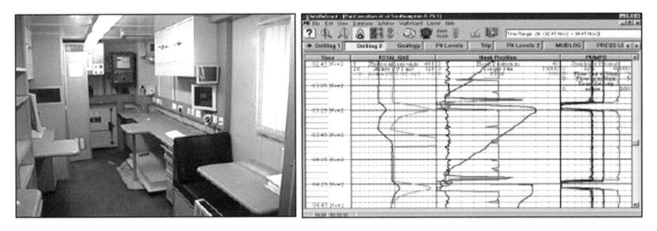

Fig. 11.2 The comprehensive logging instrument of Datalog company and the data interface

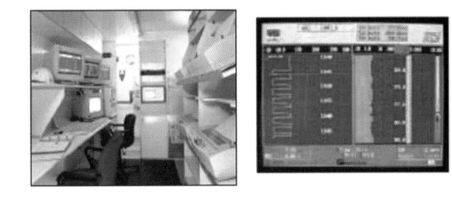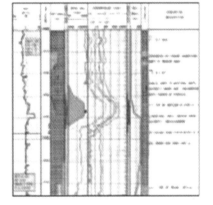

Fig. 11.3 The comprehensive logging instrument of Geoservice company and the data interface

in underbalanced drilling, which has quite a good development prospects (Fig. 11.4) because it is not restricted by deviation angle, drilling fluid (mediums) and drilling methods (rotary drilling or slide drilling).

Continental scientific drilling is a complicated system engineering, and the key to reducing the drilling costs is to drill safely and fast with high quality to decrease drilling accident rate, improve drilling efficiency and shorten drilling cycle. So the engineering and technical personnel at drill site must make and carry out the design scientifically for every aspect, and implement real-time monitoring to prevent and timely forecast the possible drilling accidents and complicated situations. Drilling process monitoring is an effective way to achieve highly efficient and safe drilling, but drilling crews in our country are still relatively weak in technical support for drilling process monitoring.

The real-time monitoring and accident diagnosis for drilling process is to identify and control the drilling process based upon the analysis of dynamic characteristics. Due to the large random disturbance of the dynamic system, the law of drilling process is difficultly described by using mathematical model. Though a single characteristic quantity can express the law of formation and development of a drilling accident, it often cannot distinguish the drilling status.

Fig. 11.4 The electromagnetic down-hole drilling parameter acquisition system from Geoservice Company

Therefore, it is required to use a variety of pattern recognition methods to make comprehensive judgments and identifications on many drilling parameters in the process of drilling. Many methods are available for pattern recognition. At present, the expert system (ES) and the artificial neural network (ANN) of the artificial intelligence (AI) for drilling real-time monitoring and accident diagnosis are most prospective, which do not need accurate mathematical models and thus are suitable for drilling process with the characteristics that experience can be relied upon, information is inadequate and imprecise.

Because most of the drill sites are in fields with inconvenient traffic, it is not easy for relevant superior department leaders or technical personnel to understand the drilling progress and conditions. Nowadays, with the continuous development of the computer network technology, the computer network and internet have been applied in all walks of life, and it is possible as well to realize remote real-time network monitoring of drilling parameters and conditions at different drill sites. At present, remote real-time monitoring system for drilling state at different drill sites in the country is still unavailable, and only DATALOG company from Canada is now conducting the research, but has not yet fully realized remote real-time monitoring based on internet.

11.2 Analysis of Data Acquisition and Processing Requirements

11.2.1 Data Acquisition System Requirements

According to the needs of full range recording and monitoring in CCSD-1 Well drilling, the main requirements for the drilling data acquisition system could be analyzed as follows:

1. **Basic data collection and drilling basic data entry**

   The basic data collection mainly includes sensor nominal data and system-setting parameters, etc.; and the drilling basic data mainly includes drilling tool structure data, borehole structure data and mud pump performance data. These data are necessary for the normal work of a comprehensive logging system.

2. **Rapid acquisition of surface drilling parameters**

   The data collected by the comprehensive logging system includes drilling parameters, hook parameters, mud properties and circulation parameters, and hole bottom fluids detection parameters, which basically contain all the data can be collected at surface (except drilling tool vibration).

3. **Rapid acquisition of bottom hole drilling parameters**

   The bottom hole drilling parameters acquisition and real-time transmission technology are complex and difficult. Based upon the available technical conditions and the drilling needs it was determined the parameters collected included vertex angle, tool face angle, temperature and pressure; and only the non real time data transmission method could be used, namely, the method of playback after lifting the drilling tool assembly.

4. **Output of real-time data**

   The output data, which include every parameter collected, are for technical personnel to monitor and analyze the drilling process. The forms of the data output mainly include data display and printing, and data curves display and printing, and the parameters of display and printing can be chosen according to needs.

5. **Drawing system of logging color map based on time and well depth**

   In oil comprehensive logging, the function of color logging map drawing is essential, while in continental scientific drilling, the logging maps, which record all technical parameters of the whole construction process in detail and basically are the records of the whole construction process, should be designed according to the practical needs.

6. **Real-time data communication of the real-time batch processing mode**

   In order to improve the decision-making and management level, the research on the real-time data transmission ways from well site to base data center has been conducted in oil drilling industry, and now a lot of oil fields have relatively well developed data transmission systems in wire and wireless ways. Real-time data transmission is an essential function of modern comprehensive logging system as well as the requirement of information management in drilling construction. The method of real-time data transmission in CCSD-1 Well can be selected based upon the conditions at drill site.

7. **Data-saving in the network database management system**

   After being transmitted to the base, the real-time data are generally saved in the network database as the basis for future analysis and decision making. To save the mass real-time data of the continental scientific drilling in the database system can provide a good environment for the follow-up data processing and utilization.

8. **Historical data processing software system**

   During the drilling construction process, it is required sometimes to make some necessary searching, processing and analysis towards the previous historical data, so various data off-line processing software should be developed according to the requirements in drilling construction.

11.2.2 Data Processing System Requirements

The requirements of the data processing system mainly include two aspects: record and monitor in the drilling process and follow-up treatment of the drilling data and comprehensive utilization of the related knowledge. The main requirements of CCSD-1 Well for the data processing system are as follows.

1. **Recording and monitoring the drilling process**

To record and monitor the drilling process is a basic requirement of the production and management in continental scientific drilling, and is a fundamental function of the comprehensive logging instruments and the requirement of the continental scientific drilling informationization as well.

2. **Abnormal data monitoring in drilling process**

Abnormal data monitoring is to monitor whether drilling parameters are abnormal, such as pump pressure, torque and drilling speed, etc. and plays the role of warning, which is mainly completed by single parameter monitoring and usually realized by computer.

3. **Data monitoring for drilling state identification**

To judge abnormalities in drilling process, basic professional knowledge combined with a number of drilling parameters are usually required to analyze and evaluate drilling condition, which provide the necessary basis for the subsequent decision. Artificial monitoring is given priority during abnormality monitoring, assisted by computer monitoring, and the job is mainly done by the technical personnel at drill site.

4. **Hole bottom fluids abnormal display monitoring**

As one of the important tasks in drilling construction, monitor of hole bottom fluids in continental scientific drilling has important scientific significance. Fluids abnormalities include hydrocarbon gases, non hydrocarbon gases and other fluids.

5. **Network monitor in drilling process to expand monitoring range and improve monitoring level**

As continental scientific drilling is a complicated project covering a wide range of advanced technologies, cybernation of drilling production management and decision-making and cybernation of drilling data processing are required as an important technical measure to improve decision-making level in drilling process and as well as an important index for modernization of continental scientific drilling.

6. **Providing software platform for subsequent software development**

In the professional computer software development, a certain software platform is usually provided for subsequent data development and utilization, and for continental scientific drilling which is rare and high-tech, data acquisition and processing system must provide a certain platform for subsequent software development to improve the system itself, and at the same time, with the development of science and technology, the subsequent software development can provide larger and better drilling data processing, management and decision-making platforms for future drilling work.

11.3 Drilling Data Acquisition System

The drilling data acquisition system for CCSD-1 Well includes two parts: surface data acquisition system and down-hole data acquisition system, the former is mainly to monitor the drilling process parameters through the surface-installed sensors, while the latter is to monitor the down-hole drilling parameters through the sensors installed in lower part of the drilling tools. The overall structure of the data acquisition system is shown in Fig. 11.5.

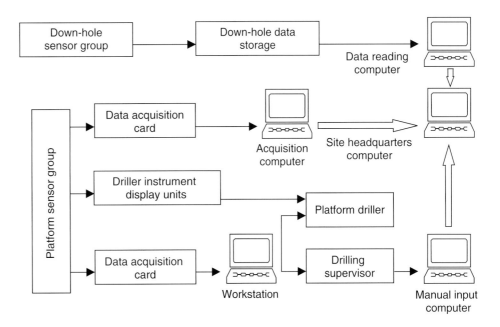

Fig. 11.5 The data acquisition system structure diagram

Fig. 11.6 The comprehensive logging instrument used in CCSD-1 well

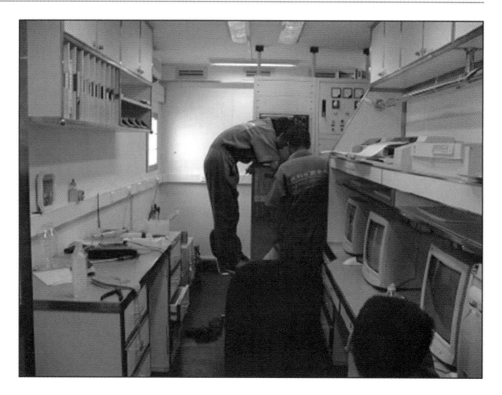

11.3.1 Surface Drilling Data Acquisition System

SK-2000FC/C comprehensive logging instrument (Fig. 11.6) developed by Shanghai Shenkai Co. Ltd. was adopted as the surface drilling data acquisition system for CCSD-1 Well, a certain improvement and supplement were made. For example: Strengthening the manual data acquisition in the drilling supervisor office to collect some data which computer cannot collect, such as the rotary speed of dual power rotary drilling and the recorded data of drilling roundtrip; Redeveloping the software to improve the data processing and preservation function, such as to calculate the torque and drilling rate of PDM and to save the real-time data into the network database.

1. **SK-2000 FC/C comprehensive logging instrument**

SK-2000 FC/C type comprehensive logging instrument can preserve and print all the data recorded in 5 s, which is the important data source to identify the bottom-hole state and diagnose accidents, with the main technical performance as follows:

(1) Power supply

Power input: Voltage 380/220 V (+10 %, −20 %), Frequency 50 Hz

Power output: Voltage (220 + 11) V, Frequency (50 ± 1) Hz, Klirr $\not> 5$ %

Relay time in case of power failure: Full load work $\not< 10$ min (applicable to the power grid supply or diesel generator power supply, and instruments with overload circuit breaking, electric leakage safety protection and lacking-phase protection)

(2) Natural gas total hydrocarbon detector and component detector

Measuring range of total hydrocarbon detector: Maximum detecting concentration ∼100 % (methane), Minimum detecting concentration 50 ppm

Measuring range of hydrocarbon component detector: Maximum detecting concentration ∼100 % (methane), Minimum detecting concentration 30 ppm (C1–C5 can be analyzed in 3 min analysis cycle)

Measuring range of non hydrocarbon component detector: CO_2 0.2–100 %, H_2 0.01–2 %; Accuracy of measurement: 2.5 % (F.S)

(3) Sensors

Measuring unit of pump stroke rate: Measuring range 30, 60, 120, 240, 480, 960, and 1920 BPM optional; Accuracy: 1 % (F.S)

Measuring unit of rotary speed of rotary table: Measuring range 30, 60, 120, 240, 480, 960, and 1920 RPM optional; Accuracy: 1 % (F.S)

Measuring unit of rotary table torque: Measuring range 0–1.6 MPa (0–50 kN m); Accuracy: 2 % (F.S)

Measuring unit of standpipe pressure and casing pressure: Measuring range 0–40 MPa; Accuracy: 2 %

Measuring unit of hook weigh parameter: Measuring range 0–6 MPa (0–4000 kN); Accuracy: 2 %

Measuring unit of drilling fluid temperature: Measuring range 0–100 °C; Accuracy: 1 °C

Measuring unit of drilling fluid density: Measuring range 0.9–2.5 g/cm^3; Accuracy: 1 %

Measuring unit of drilling fluid conductivity: Measuring range 0–50 ms/cm; Accuracy: 2 % (F.S)

Measuring unit of drilling fluid outlet flow rate: Measuring range 0–100 % (relative flow rate); Accuracy: 5 % (F.S)

Measuring unit of drilling fluid pit volume: Measuring range 0–2 m (or according to the requirements of customer); Accuracy: 1.5 % (F.S)

Measuring unit of drawworks sensor: Measuring range 0–9999 (hook position); Accuracy: 10 mm/single rope

(4) Instruments

Carbonate analyzer;

Mudstone density measuring instrument;

Fluorescence analyzer;

Recording instrument

(5) Computer system

The full set of computer software system has the functions of acquisition, display, alarming, printing and memory, etc.

i. Data real-time acquisition procedure

As the main program for real-time acquisition while drilling and the core program for the whole set of software system, the procedure has multiple functions of real-time acquisition, calculation, multi screen display, alarming and Chinese-English free switch-over, etc., which can provide eighteen function modules and display frames (such as drilling animation, main drilling parameters, initialization of the system, sensor calibration, drilling tool management, chromatographic calibration, strip chart recording, etc.).

ii. Real-time data processing procedure

The procedure is an auxiliary procedure for real-time data acquisition program, which can generate entire-meter data files from real-time data collected by the acquisition procedure, delay the entire-meter data file, establish the open database and print the LWD real-time monitoring data tables.

iii. Off-line data processing procedure

As a processing procedure for data files generated from real-time data program, the procedure, which produces the drilling tool reports, drill bit reports, gas logging diagrams, hydraulics reports, hole deviation reports and pressure forms, etc., can edit and print the morning geological report, morning engineering report, daily geological report and water horsepower report, and input, modification, print of the geologic logging data and optimization of the drilling expert system as well.

iv. Monitor data remote transmission

In order to facilitate the geological supervisor and engineering supervisor to instruct the production at drill site, monitors 100 m away from the instrument house were respectively arranged, and the monitoring pictures could be selected according to needs.

2. Improvement on the surface data acquisition function

Data collected in drilling are generally based on time and well depth, however, for continental scientific drilling, the records based on core drilling roundtrips are also important. In the operations of statistical calculation, enquiry and storage of the data for continental scientific drilling, data calculation and processing based on the roundtrips are important as well, besides based on time and well depth, including footage drilled per roundtrip, penetration rate per roundtrip and core recovery per roundtrip, etc. According to this characteristic, in order to facilitate record, enquiry and statistics, the Excel format (as shown in Fig. 11.7) was used in manual data entry of the core drilling roundtrip records.

	A	B	C	D	E	F	G	H	I
1	MH-1C-R46回次	CCSD-MH-1C取心钻进实时记录				起始井深: 3367.52			
2	钻具组合	钻头(TGS-94)+扩孔器(TGS-09)+岩心筒(8.50-7)+扩孔器(TGS-10)+单动接头+上接头+液动锤(YZX-127)+接头（311*310)+							
3	时间 (hh:mm)	井深 (m)	进尺 (m)	钻压 (kN)	转速 (r/min)	排量 (l/s)	泵压 (MPa)	[illegible]	备 注
25	2003-8-6 9:30	3370.64	0.23	20.00	176	9.65	11.80	8	泵压不稳(11.5-12.5MPa)
26	2003-8-6 9:40	3370.78	0.14	20.00	176	9.65	11.80	8	泵压不稳(11.5-12.5MPa)
27	2003-8-6 9:50	3370.91	0.13	20.00	176	9.65	11.80	8	泵压不稳(11.5-12.0MPa)
28	2003-8-6 10:00	3370.99	0.08	20.00	176	9.65	12.00	8	泵压不稳(11.5-12.0MPa)
29	2003-8-6 10:10	3371.08	0.09	20.00	176	9.65	12.00	8	泵压不稳(11.5-12.0MPa)
30	2003-8-6 10:20	3371.12	0.04	20.00	176	9.65	12.00	7	泵压不稳(11.5-12.0MPa)
31	2003-8-6 10:30	3371.24	0.12	25.00	176	9.65	12.00	6.5	泵压不稳(10.5-12.0MPa)
32	2003-8-6 10:40	3371.31	0.07	25.00	176	9.65	11.00	6.5	泵压不稳(10.5-12.0MPa)
33	2003-8-6 10:50	3371.33	0.02	25.00	176	9.65	9.00	6.5	泵压下降，活动钻具
34	2003-8-6 11:00	3371.43	0.10	25.00	176	9.65	9.00	6.5	泵压不稳(9-12.0MPa)
35	2003-8-6 11:10	3371.48	0.05	25.00	176	9.65	11.00	6.5	泵压不稳(10-11.0MPa)
36	2003-8-6 11:20	3371.57	0.09	25.00	176	9.65	9.15	6.5	泵压下降
37	2003-8-6 11:30	3371.57	0.00	25.00	176	9.65	9.15	6.5	不进尺，活动钻具
38	2003-8-6 11:40	3371.62	0.05	25.00	176	9.65	9.00	6.5	泵压下降
39	2003-8-6 11:55	3371.62	0.00	25.00	176	9.65	9.00	6.5	停钻
40	2003-8-6 12:25								上提距井底8.5m遇阻(20t)倒划眼缓慢提出钻具。
41	2003-8-7 0:05								起钻,15:36-15:58保养钻机。检查钻具用钻杆。21:
42									
43									4#岩心筒已检查（废）
44	累计钻时	回次进尺	岩心长度	平均钻压	平均转速	平均排量	平均泵压	平均扭矩	泥浆泵缸套直径
45	4.59	4.10	2.00	21.46	177.91	9.75	10.74	6.86	1#泵Φ120mm2#泵Φ120mm
46	技术指标	回次采取率	0.49	机械钻速	0.89	钻头水口	12	起钻井深	3371.62
47	回次简述:	钻进过程中泵压高、不稳、钻速慢、最终不进尺停钻。钻头（废）。下扩孔器外径157.6mm、上扩孔器外径157.7mm。岩心破碎,直径95.8mm。TGS-11#扩孔器公扣断入4#外筒							
48	现场调度:	刘玉顺（夜）李旭东（白）					驻井工程师		任海军
49									
50									

MH-1C-R45 / MH-1C-R46 / MH-1C-R47 / MH-1C-R48 / MH-1C-R49 / MH-1C-R50 / MH-1(

Fig. 11.7 The manual data entry interface

	A	B	C	D	E	F	G	H	I
1	MH-1C-R46回次	CCSD-MH-1C取心钻进实时记录				起始井深: 3367.52			
2	钻具组合	钻头(TGS-94)+扩孔器(TGS-09)+岩心筒(8.50-7)+扩孔器(TGS-10)+单动接头+上接头+液动锤(YZX-127)+接头(311*310)+螺							
3	时间 (hh:mm)	井深 (m)	进尺 (m)	钻压(kN)	转速(r/min)	排量(l/s)	泵压(MPa)	[illegible]	备注
4	2003-8-5 21:55	3367.52			转盘11+螺杆				钻具组合：普双7"+液动锤+螺杆
5	2003-8-6 2:45								下钻
6	2003-8-6 3:20								31"立柱(2430米)返浆，接方钻杆顶浆
7	2003-8-6 5:40								下钻
8	2003-8-6 6:45	3367.52			189	10.34	8.00	3	距井底7m遇阻，接方钻杆划眼
9	2003-8-6 6:50	3367.60	0.08	20.00	189	10.34	10.50	3	
10	2003-8-6 7:00	3367.76	0.16	20.00	189	10.34	10.50	5	
11	2003-8-6 7:10	3367.95	0.19	20.00	189	10.34	10.50	5	
12	2003-8-6 7:20	3368.09	0.14	20.00	189	10.34	11.00	6	
13	2003-8-6 7:30	3368.26	0.17	20.00	189	10.34	11.00	6	
14	2003-8-6 7:40	3368.41	0.15	20.00	189	10.34	11.00	6	
15	2003-8-6 7:50	3368.69	0.28	20.00	170	9.31	10.50	8	
16	2003-8-6 8:00	3368.94	0.25	20.00	170	9.31	10.50	8	

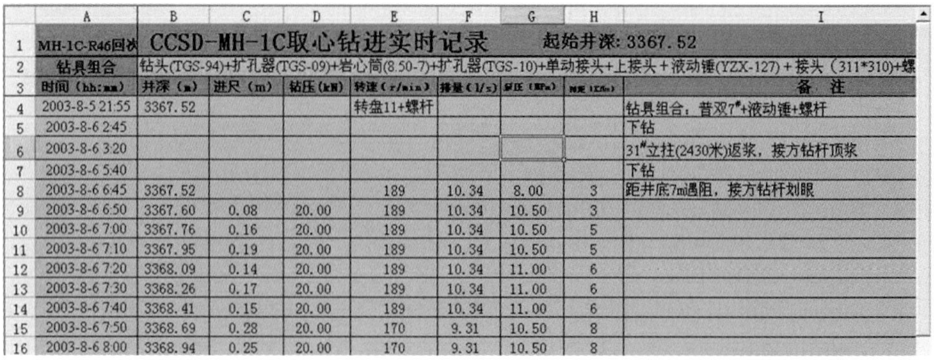

Fig. 11.8 Manual data entry and the recording mode

In the design of the data entry interface, a drilling roundtrip was recorded as one electronic working table, at the upper part of the interface was mainly recorded the basic data of a roundtrip (current roundtrip number, bottom hole assembly and initial depth), at the lower part was mainly for statistical data and important descriptions of this drilling roundtrip, and in the intermediate part was for the main drilling parameters and abnormities (recorded every 10 min). Some parameters to which comprehensive logging instruments could not collect were manually calculated and input, such as the rotary speed of dual power rotary drilling, abnormal phenomena happened during tripping (pulling out and running in), and the distance to the hole bottom where slacking-off was encountered and reaming-down was needed, etc. (Fig. 11.8).

In the redevelopment of the software system of the comprehensive logging instrument, functions of parameter calculation, data processing and storage of the downhole power drilling tool were strengthened in the aspect of data acquisition. The function of parameter calculation mainly included the calculations of the rotary speed and torque of PDM, and calculation of drilling rate, etc. The functions of data processing and storage mainly included the display of historical data and the playback of the curves, and the storage of real-time data into the network drilling database, etc. The software structure is shown in Fig. 11.9.

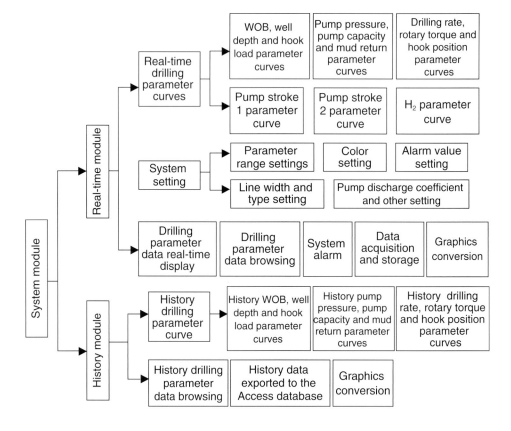

Fig. 11.9 Structure of the redeveloped software

11.3 Drilling Data Acquisition System

Fig. 11.10 The real-time data displaying interface

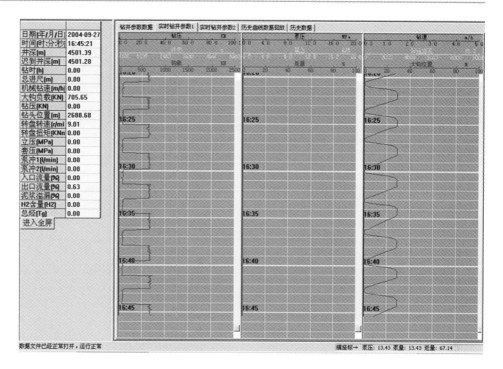

Fig. 11.11 The playback interface of historical data curves

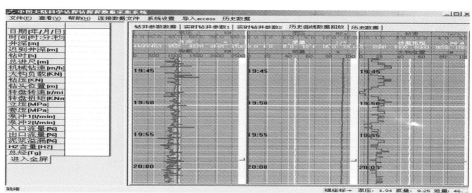

The functions of the redeveloped drilling data software include the following:

i. Function of data real-time display. After startup, the software enters the interface as shown in Fig. 11.10. On the left of the main interface, commonly used drilling parameters are real-time displayed in the drilling data list, and on the right drilling curves are real-time displayed, rolling upwards with the lapse of time.

ii. Functions of drilling data auto-saving and management. The system automatically saves the daily drilling data, with a data file formatted as the CCSD file per day, which was encrypted and opened only by the system.

iii. Function of system setting. The system settings include five parts: setting for drilling parameter curve coordinate range; setting for color; setting for line width and type; setting for alarming value; setting for pump flow coefficient and other settings;

iv. Full-screen function;

v. Function of graph saving. Graphs of drilling curves are saved in BMP format files;

vi. Function of data export. The CCSD format data in the system can be exported into path files, to be easily processed and utilized;

vii. Function of browsing historical data and curves. Historical data and curves can be browsed (Fig. 11.11).

11.3.2 Down-Hole Drilling Data Acquisition System

The information of bottom hole condition in continental scientific drilling contains the data which reflect the working state of drilling, economic indexes of drilling technology, drilling parameters and the information about the bottom hole drilling tool assembly. Therefore, monitoring these data information is of important practical significance.

On the premise of satisfying the requirements of continental scientific drilling, the design of the bottom hole drilling parameter data acquisition system should be of an economical, practical, convenient and high-precision technical scheme, and at the same time the technical feasibility and practicability should be fully considered. According to the importance of drilling parameters and bottom hole drilling tool structure, the down hole data acquisition system was researched and developed, the parameters monitored by which contained vertex angle, tool face angle, temperature and mud pressure, etc.; Detection instruments could be installed either on the overshot assembly of the wireline core drilling tool or on conventional core drilling tool; The collected data were stored in the memory of the instruments, which could be connected with a computer when pulled out of the borehole, with the data read into the computer for processing. China University of Geosciences (Wuhan), entrusted by the CCSD Centre, developed DPMA-1 downhole parameter detection playback acquisition system (Figs. 11.12 and 11.13) for detection-while-drilling of vertex angle, tool face angle, and pressure and temperature of down hole drilling fluid.

The main technical indexes:
Vertex Angle: 0–45°, with error ≤0.5°
Pressure: 0–50 MPa, with error ≤0.5 MPa
Temperature: 0 to +150 °C, with error ≤0.5 °C
Face Angle: 0–360°, with error <5°
Impact resistance: 10 g

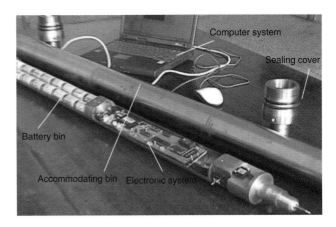

Fig. 11.13 Assembly of down-hole drilling data acquisition system

Heat resistance: 0 to +85 °C
Sealing: 50 MPa above
Data record capacity is 32 KB, in case that sampling time is set for 5 s, 8 bytes of four sets of data (2 sets of vertex angle, pressure, and temperature data) are stored each time, the memory can work for 22 h.

The main technical indexes of the hole parameters storage and fishing instrument of the auxiliary wireline fishing spear:

The instrument has the function of automatically save the data of down-hole temperature, pressure and vertex angle, with the storage capacity expanded to 64 KB;

The instrument has the function of playback the recorded data at surface, thus can make continuous measurement of multiple points.

Vertex angle test range and precision: 0–45°, with error ≤5°; temperature test range: 0–150 °C, with error ≤0.5 °C

Pressure test range: 0–50 MPa, with error ≤0.5 MPa
Impact resistance: 10 g
Sampling interval was adjustable and sampling time could be set as four groups of data could be saved each time (2 groups of vertex angle, pressure and temperature data).

Size of instrument barrel (detection tube) size: ϕ 60 × 1500 mm

The down-hole drilling data acquisition system was tested for five times at the drill site of CCSD-1 Well and the tests showed that the tested parameters of the DPMA-1 down-hole drilling parameter automatic recording and playback device were complete. The device could not only detect the borehole temperature and pressure changes (especially the dynamic changes), and provide the most direct reference basis for mud property designs, but also could judge the working conditions of the down-hole tool and the on-way resistance of mud during circulation, etc. based upon the mud dynamic and static pressure changes respectively measured by changing the installation positions of the instrument.

The tests obtained four parameters, i.e., pressure and temperature of the mud at bottom hole, vertex angle and tool

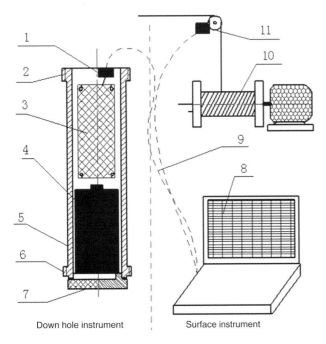

Fig. 11.12 Structure of down-hole drilling data acquisition system. *1* Playback interface. *2*, *6* Connecting thread of tension ring. *3* Double PCB. *4* Battery. *5* Instrument outer barrel. *7* End cover. *8* Computer. *9* Data playback cable. *10* Drawworks. *11* Depth synchronous recorder

Fig. 11.14 Data playback curves collected by the down-hole drilling parameter acquisition system

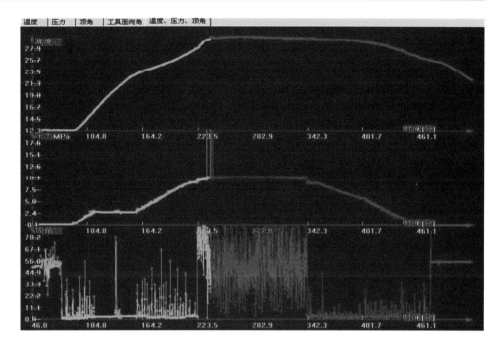

face angle during drilling in CCSD-1 Well. And the test results were accurate with repeatability and the instrument performed stably and reliably, which came up to the requirements of design and got a successful result. The playback curves are shown in Fig. 11.14.

11.4 Drilling Data Processing System

The real-time data collected by the continental scientific drilling data acquisition system contain omnibearing information in drilling process, which must be analyzed and processed to raise the monitoring level in drilling and decision making level in construction. The monitoring in drilling process is carried on according to the difficulty degrees. Firstly, the setting limited value of an important single parameter should be monitored, which can play a warning role when change takes place in drilling process, and drilling engineers should pay attention to the downhole changes; Secondly, when downhole conditions change, drilling technicians must analyze and judge the changes of the parameters based upon the previous experience and relevant theoretical knowledge or with the aid of the computer software, evaluate the downhole conditions, prevent and timely forecast the possible complicated status and drilling accidents. So, the strategy of monitor in drilling process is from easy to difficult, from single parameter monitoring to multiple parameters monitoring, as shown in Fig. 11.15.

11.4.1 Single Parameter Monitoring

Single parameter monitor includes two parts, real-time monitoring (including well-flushing and redressing) in drilling process and real-time monitoring in tripping process. Single parameter monitoring can be implemented by computer or manual work, which mainly monitor the abnormal conditions during drilling processes as an early warning.

The main contents of the monitoring in drilling process include display of pump working state (including the starting state and the pump stroke number, etc.); display of drilling fluid performance (including mud density parameters, etc.); real-time calculation of hydraulic parameters (including equivalent circulating density, etc.); display of pump pressure change (including nozzle block and nozzle piercement, judgment of drilling tool piercement, etc.); display of net variation of total pool volume (judgment of well kick, lost-circulation, etc.); display of torque change (judgment of drill bit failure or drilling tool broken, etc.); display of bit weight (bit weight overrun and drill string not well braked); display

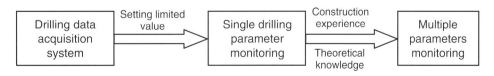

Fig. 11.15 Monitoring in drilling process

of hanging weight (main basis for judging over pull or tight pull) and display of drilling speed (rock formation changes, bit wear, down-hole drilling tool broken, etc.).

The main contents of the monitoring in tripping process include monitor of tripping speed (pressure fluctuation); monitor of the rest time of drill string in open hole section; monitor of hook load change; monitor of net variation of total pool volume, etc.

11.4.2 Comprehensive Monitoring

Comprehensive monitoring denotes the monitor of drilling process dynamics, mainly the monitor of drilling process, based on the changes of two or more parameters. The comprehensive monitoring of a number of parameters can accurately judge the down-hole states or accidents happened, etc. Comprehensive monitoring is mainly completed by engineering and technical personnel, who then make an analysis and form a judgment on the parameter changes based upon their theoretical knowledge and practical experiences.

The main contents of the comprehensive monitoring include: the changes of hook position and hook weight are used to monitor the free-fall or not-well-braked drill string; the changes of drilling rate and pump pressure are used to monitor the working status of PDM and hydro-hammer and condition of core blockage; the changes of drilling rate, bit weight, pump pressure and torque are used to monitor tool broken and bit wear; the changes of total mud pool volume and inlet and outlet flow are used to monitor well kick and lost circulation.

Computer network can be used to expand the limits of drilling process monitoring, so as to improve the monitoring level. According to the monitoring technological level, the monitor of drilling process in continental scientific drilling project can be divided into two levels: one is on-site monitoring, which makes preliminary data analysis and processing to provide necessary information for drilling supervisors and engineers for decision-making; the other one is the base monitoring, which redevelops the data and make deep analysis and processing with the aid of the related data processing software, and supplies more detailed information for the related technical personnel and leaders. Both the on-site monitoring and the base monitoring are multi-point monitoring, while the content and level of the latter are higher than those of the former. The structure of the drilling process monitoring system for CCSD-1 Well is illustrated in Fig. 11.16.

The on-site monitoring is more real-time, and is completed by drillers, drilling engineers and drilling supervisors. The base monitoring is less real-time, and is completed by senior technical personnel. During the drilling process monitoring of CCSD-1 Well, the whole drilling process monitoring system ran well.

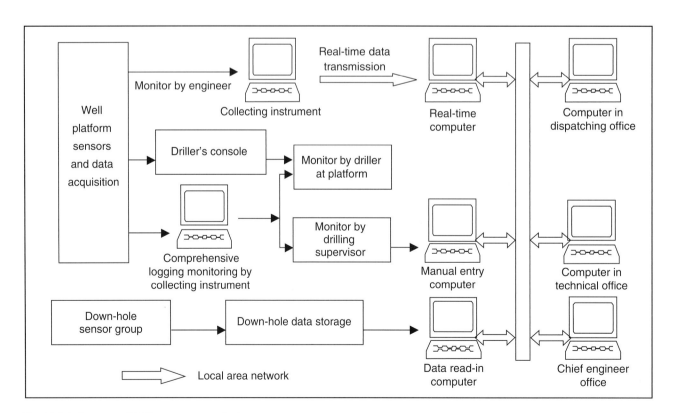

Fig. 11.16 Structure of drilling process monitoring system

11.4.3 Case History

Figure 11.17 shows the logging curves of φ 157/311.1 mm normal reaming drilling on May 9th, 2002, from which can be found the pipe connection process and time.

Figure 11.18 illustrates the drilling parameter curves before pulling out the broken lower reaming shell in core drilling on February 10th, 2002, from the curves it can be found that torque didn't change and penetration rate was very low when drilling torque was reduced and bit weight increased.

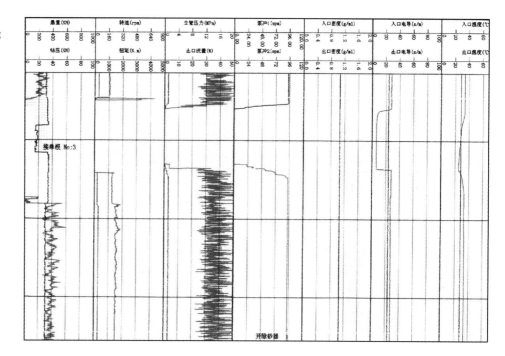

Fig. 11.17 Drilling parameter curves of 157/311.1 mm reaming drilling

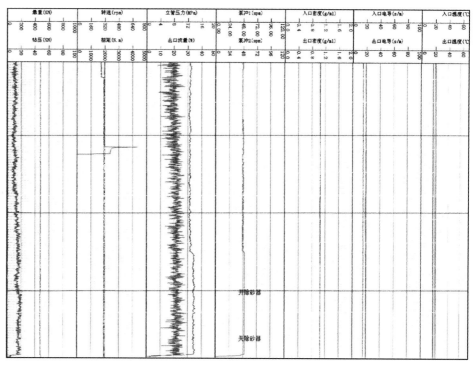

Fig. 11.18 Drilling parameter curves before pulling out the broken lower reaming shell

Fig. 11.19 Logging curves of the settlement of the drill pipe sticking accident occurred during deviation correction drilling

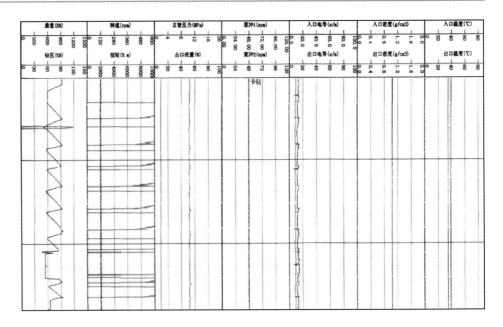

Fig. 11.20 Logging curves of redressing process

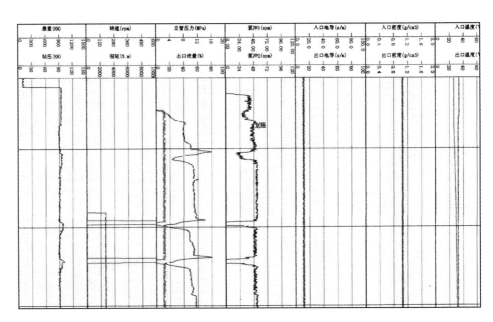

Figure 11.19 indicates the logging curves when the drill pipe sticking accident occurred during the deviation correction drilling on May 18th, 2003, in which the process of accident settlement was clearly recorded.

Figure 11.20 shows the logging curves of redressing process before core drilling when drilling tool was run down to the hole bottom on September 24th, 2003, in which was recorded the change of the drilling parameters

Fig. 11.21 Logging curves when drill bit steel body fractured during core drilling

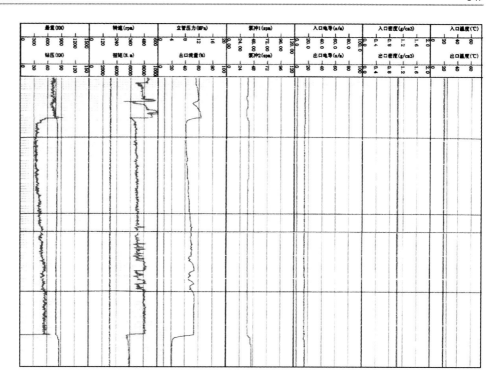

when bit bouncing occurred during PDM hydro-hammer core drilling.

Figure 11.21 indicates the last logging curves when drill bit steel body fractured during core drilling on February 6th, 2003, from the curves it can be found that the penetration rate was very low, and the torque and pump pressure increased when bit weight was increased.

12 Technical Economical Analysis

12.1 Construction Time and Cost Analysis

12.1.1 Construction Time Analysis

CCSD-1 Well was officially started on June 25th, 2001, and the designed coring drilling tasks was completed on January 23rd, 2005, when drilled to 5118.2 m. From January 25th, 2005 to March 8th, drilling tools were tested and the well was drilled to 5158 m, and the whole well drilling constructions were completed. From March 9th, 2005 to April 1st, the well completion, logging and other works were conducted. It totally took 1395 days from the spudding-in on June 25th, 2001 to the completion of CCSD-1 Well on April 19th, 2005, with a drilling depth of 5158 m and a total drilling footage of 177.71 m (Table 12.1) completed in various drilling (coring drilling, reaming, sidetrack drilling and straightening drilling, non-core drilling and grinding drilling). The curves of the construction progress of the whole drilling stages are shown in Fig. 4.3.

It took 1310 days from the spud-in to the completion of coring at the depth of 5118.2 m (effective core drilling cycle), and it took 1353 days from the spud-in to the completion of drilling to the well depth of 5158 m (effective drilling cycle). It lasted 1377 days from the spud-in to the 127 mm tail pipe completion (well completion cycle), and 1395 days to the final completion (drilling construction cycle).

The total completion time of CCSD-1 Well took 33,030.5 h, including production time 30,820 h which account for 93.31 % of the well completion cycle; non production time 2210.5 h, accounting for 6.69 %; actual drilling time 9678.12 h accounting for 29.30 %; the accidents losing time 647.65 h, accounting for 1.96 %. The production and the non-production time data can be found in Tables 12.2, 12.3 and Fig. 12.1.

12.1.2 Construction Cost Analysis

The CCSD Project totally cost 185.79 million RMB Yuan, including 136.19 million Yuan for basic construction investment from the State Development and Reform Commission, and 39.6 million Yuan from the Geological Survey Project allocated by the Ministry of Land and Resources, which constituted 175.79 million Yuan as the total state investment. In addition, the Sinopec Group sponsored 10 million Yuan in the form of deduction of drilling construction fees. The expenses of the funds above are shown in Table 12.4.

The drilling sub-project of the CCSD Project totally cost 111.59 million Yuan, and the specific expenses are shown in Table 12.5.

The project was a very complicated drilling engineering project, and a special way was adopted for the organization of the drilling construction, without any precedent in China. The main characteristic was that the Party A (the CCSD Engineering Centre) highly participated in the construction, during which, all the construction decisions were made and issued by the Party A, and the Party B (No. 3 Drilling Company, Zhongyuan Petroleum Exploration Bureau, Sinopec Group Company) constructed in strict accordance with the instructions of the Party A. According to the contract, drilling equipment (drill rig, mud pumps and supporting facilities) were provided by the Party B, while all of the other drilling equipment, tools and drilling power would be solved by the Party A. The Party A would pay the Party B the daily construction fee according to the character and progress of the daily drilling construction. Part of drill pipes, drill collars and accessory tools needed for the construction were to be leased by the Party B on behalf of the Part A, while most of the drilling tools were to be purchased by the Party A. This method played a significant role in the successful implementation of the drilling construction as well as in reducing construction time and cost, but more staffs were

Translated by Li Haipeng.

Table 12.1 The time used and footage drilled in each stage of the CCSD-1 well construction

No.	Drilling construction stage	Hole section drilled (m)		Footage drilled (m)						Time (d)	Time percentage (%)
		From	To	Coring	Reaming	Straightening	Non-core drilling	Grinding	Subtotal		
1	Non-core drilling in die 1st opening	6.87	101.00				94.13		94.13	21	1.5
2	Pilot hole coring	101.00	2046.54	1945.54					1945.54	295	21.1
3	The 1st reaming in die main hole	101.00	2033.00		1931.89			0.11	1932.00	137	9.8
4	Main hole 1st stage coring	2046.54	2982.18	934.67				0.97	935.64	198	14.2
5	The 1st sidetrack drilling in die-the main hole (straightening)	2749.00	2974.59	10.75	15.31	215.20		0.04	241.30	63	4.5
6	Main hole 2nd stage coring	2974.59	3665.87	634.85	19.33	40.20	15.39	0.84	710.61	135	9.7
7	The 2nd reaming in die main hole	2028.00	3525.18		1497.18				1497.18	145	10.4
8	The 2nd sidetrack drilling in die main hole (obstacle avoidance)	3400.00	3624.16		63.31	45.62	178.54		287.47	54	3.9
9	Main hole 3rd stage coring	3624.16	5118.20	1480.06			13.53	0.45	1494.04	262	18.8
10	Testing tools	5118.20	5158.00	3.50		5.30	31.00		39.80	43	3.1
11	Completing, logging and stop- and waiting									42	3.0
	Total			5009.37	3527.02	306.32	332.59	2.41	9177.71	1395	100.0

Note Not including the footage of drilling cement plugs and grind drilling in reaming

Table 12.2 The production time statistics

Total completion time (h)	Production time								
	Time for footage drilling								Total
	Actual total drilling (h)	Tripping (h)	Pipe connection (h)	Reaming and redressing (h)	Circulation (h)	Logging (h)	Cementation (h)	Auxiliary works (h)	
33,030.5	9678.12	13,629.96	122.35	1404.35	285.65	1139.30	494.83	4065.44	30,820.00
Percentage in the completion cycle (%)	29.30	41.26	0.37	4.25	0.86	3.45	1.50	12.31	93.31

12.1 Construction Time and Cost Analysis

Table 12.3 The non-production time statistics

Total completion time (h)	Non production time					
	Accidentsa (h)	Repair (h)	Organized stoppage (h)	Stoppage for natural reason (h)	Others (h)	Total
33,030.5	647.65	43.22	1512.46	3.17	4.00	2210.50
Percentage in the completion cycle (%)	1.96	0.13	4.58	0.01	0.01	6.69

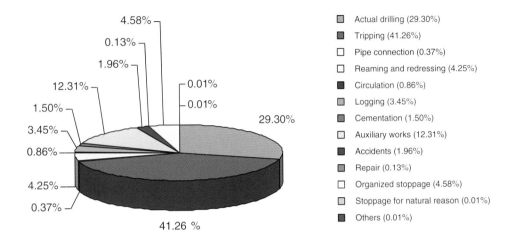

Fig. 12.1 CCSD-1 well drilling time distribution

- Actual drilling (29.30%)
- Tripping (41.26%)
- Pipe connection (0.37%)
- Reaming and redressing (4.25%)
- Circulation (0.86%)
- Logging (3.45%)
- Cementation (1.50%)
- Auxiliary works (12.31%)
- Accidents (1.96%)
- Repair (0.13%)
- Organized stoppage (4.58%)
- Stoppage for natural reason (0.01%)
- Others (0.01%)

Table 12.4 The total expenses of CCSD project

Expenditure item	Expenditure/ 10,000 Yuan	Funding percentage (%)
Drilling sub-project	11,159	60.1
Logging sub-project	1016	5.5
Geophysical sub-project	676	3.6
Analysis and testing sub-project	2803	15.1
Information sub-project	614	3.3
Civil engineering	933	5.0
Long-term observation	200	1.1
Comprehensive study	344	1.9
Engineering design	127	0.7
Management expense	607	3.3
Expense of stay-caring at drill site	100	0.5
Total expense	18,579	100.0

Table 12.5 The expenses of drilling sub-project

Expenditure item	Expenditure/ 10,000 Yuan	Funding percentage (%)
Removal and relocation	122	1.1
Preliminary work for spudding	169	1.5
Drill rig operation	3684	33.0
Drilling equipment leasing	685	6.1
Drilling equipment purchase	3423	30.7
Personnel for drilling	782	7.0
Power	637	5.7
Technology research and development	403	3.6
Transportation and environmental protection	144	1.3
Travel and office works	602	5.4
Others	508	4.6
Total	11,159	100.0

needed for the Party A in comparison with the conventional drilling project.

Because drilling lasted a long time, in order to save costs, a power-supply scheme from public power grids was adopted. In that case, the fees for power transmission and transformation facilities (included in civil engineering), together with the electricity costs (power consumption), totally cost much lower than the power consumption by diesel fuel.

To solve the drilling technical problems in construction, a number of drilling technology research and development projects were set up before drilling, and some key technical problems were tackled. During the construction, further researches were carried out based upon some technical problems encountered in drilling.

The practice of purchasing drilling materials (drilling tools and drilling materials) by the CCSD Engineering

Table 12.6 Drilling tools and materials costs for CCSD-1 well

Expenditure item	Expenditure/ 10,000 Yuan	Funding percentage (%)	Expenditure item	Expenditure/ 10,000 Yuan	Funding percentage (%)
PDM	580	16.9	Hydro-hammer technical service	258	7.5
Casing	418	12.2	Coring tool manufacture	174	5.1
Drill rod and collar	396	11.6	Jar and absorber inclination	150	4.4
Coring bit and reaming shell	542	15.8	Survey equipment and instruments	117	3.4
Reaming bit and rock bit	280	8.2	Top drive related cost	161	4.7
Mud materials	200	5.8	Total	3423	100.0
Well cementation materials	147	4.3			

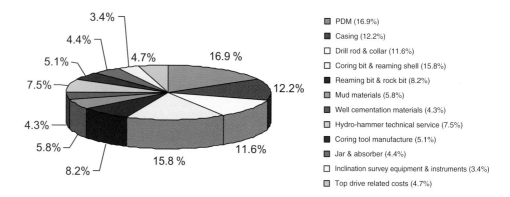

Fig. 12.2 Composition of equipment and materials cost for CCSD-1 well

Centre was a very effective way to reduce drilling cost. During the construction, the expenditure of drilling equipments and materials purchased by the Centre amounted to 34,230,000 RMB Yuan, the composition of which can be found in Table 12.6 and Fig. 12.2.

12.2 Economic Evaluation of Core Drilling Techniques

12.2.1 Evaluation Method

The construction task of CCSD-1 Well was to complete a continuously cored borehole (with depth of 5000 m and diameter of 157 mm) in hard crystalline rocks (gneiss and eclogite), and the difficulties in the construction were enormous, which were not only unprecedented in China but also rarely seen in the world. During the design phase of the project and the initial implementation, the engineering and technical personnel of the Centre faced great difficulties and pressure, and for the implementation of such an exceedingly difficult drilling project, we had neither available drilling technologies nor experiences. Thanks to the joint efforts of China's drilling circles and the unremitting research, experiment and exploration of the drilling technical personnel from the Centre, a complete set of worldwide originally created scientific deep hole drilling techniques (PDM hydro-hammer diamond core drilling by lifting drilling tool for core recovery) was eventually developed. The successful research and application of this method was a significant contribution to the world on one hand (putting out a drilling method of highly efficient, high quality, low consumption and safety for deep hole core drilling in the world), and on the other hand, created considerable economic benefits for the CCSD Project itself, which greatly shortened the construction time and saved the construction cost. Compared with conventional core drilling method, PDM hydro-hammer diamond core drilling by lifting drilling tool for core recovery (conventional coring) could save much time and cost for this 5000 m core drilling project, and had obvious advantages in reducing technology risks and ensuring construction safety as well. It is rather difficult to make direct comparisons between the core drilling results in CCSD-1 Well and the other major scientific deep drilling projects in the world, for the reason that the formation conditions, coring depth and diameters in every scientific deep drilling project are not the same. In addition, core drilling cost in every project is basically unable to obtain. The only feasible comparison way is to make evaluations through the calculation of different coring methods, which can be carried out in accordance with the following procedures:

1. Establish an index system of technical economic evaluation for core drilling, namely to determine which parameters should be used to conduct a comparative evaluation of the core drilling methods.
2. Deduce the calculation formula, and establish the relationship between the various technical and economic indexes in core drilling and the selected technical and economic indexes.
3. Through wide investigations and statistics to acquire the technical indexes of different drilling methods under different construction conditions.
4. Through market research to acquire the prices of different core drilling tools and materials.
5. Make economic evaluations, analyzing the drilling construction time and costs of different core drilling methods.
6. Make technical risk evaluations, evaluating the adverse consequences each core drilling technique applied may bring in technology.
7. Make comprehensive evaluations to compare different drilling methods, combined with the results of economy and technology risks evaluations.

This comparison can be used not only for evaluation of different core drilling methods in an executed project, but also for a selection of core drilling method in the design of a new project.

12.2.2 Index System of Technical Economic Evaluation for Core Drilling Construction

Drilling cost, either unit drilling cost C_{pm} or total drilling cost C_T is commonly used to evaluate the technical economical properties of a drilling construction.

The unit drilling cost (or cost per meter) C_{pm} is used to make an economic evaluation of a part or a certain depth of the drilling construction, while the total cost C_T is used to make an overall economic evaluation of the drilling construction. Each of them has its own characteristics and complements each other.

The unit drilling cost at any depth of the borehole can be acquired through calculation. In making calculation of the total drilling construction cost, borehole can be divided into several sections and the drilling cost in each section can be regarded as constant, the drilling cost of this section can be obtained by multiplying the length of the section with the unit drilling cost. And the total drilling construction cost of the borehole can be obtained by accumulating the drilling cost of every section.

Time for drilling construction is also an economic index often arouses general concern. However, because the factor of time will be eventually reflected in the cost, thus drilling time is considered as an auxiliary technical economic index. Same as cost calculation, drilling time can also be divided into the unit footage time (or the drilling time per meter) T_{pm} and the total drilling time T_T, the meaning and the calculation principle of which are similar to those of drilling cost.

12.2.3 Calculation of Drilling Construction Time and Cost

1. Calculation of unit footage time

The unit footage time is the time needed for drilling 1 m. The unit footage time of wireline core drilling can be calculated by using the following formula (calculated based on one run out roundtrip):

$$T_{pmw} = \frac{T_R + T_D + T_{FC}}{L_R} = \frac{\frac{2D_{ep}}{V_R} + \frac{L_R}{V_C} + \frac{L_R}{L_{FC}} t_{FC}}{L_R}$$
$$= \frac{3D_{ep}}{V_R L_{bit}} + \frac{1}{V_C} + \frac{t_{FC}}{L_{FC}} \quad (12.1)$$

where

T_{pmw} unit footage time of wireline core drilling, h/m
T_R roundtrip time, h
T_D time used for drilling a roundtrip, h
T_{FC} time used for core-fishing in a roundtrip, h
L_R drill string lifting interval, m. The drill string lifting interval in core drilling by lifting drilling tool for core recovery is equal to roundtrip length. The drill string lifting interval in wireline core drilling is closely related to the bit service life, certainly shorter than bit life, according to experiences, supposing drill string lifting interval as 2/3 of the bit service life, i.e. $L_R = 2/3\ L_{bit}$
D_{ep} borehole depth, m
V_R average velocity of tripping, m/h. Generally, the running-down speed is higher than the pulling-out speed. In addition, some auxiliary operations such as replacing tools and check-up will be made after every pulling-out and running-down, and these auxiliary operations will take half an hour at least. To simplify the calculation, the average velocity of tripping can be defined through dividing 2 times of the borehole depth by the total time used from the start of pulling-out drill string in this roundtrip to running-down drill string to the hole bottom in the next roundtrip.
V_C rate of penetration, m/h
L_{FC} core-fishing interval in wireline core drilling, m
t_{FC} time for single core-fishing in wireline core drilling, h; according to the experience of German KTB pilot

hole drilling, the relationship between the single core-fishing time t_{FC} and the borehole depth D_{ep} is $t_{FC} = 0.5 + 0.0007\ D_{ep}$

L_{bit} bit service life, m

The unit footage time of conventional core drilling (by lifting drill string for core recovery) can be calculated by the following formula:

$$T_{pmr} = \frac{T_R + T_D}{L_R} = \frac{2D_{ep}}{V_R L_R} + \frac{1}{V_C} \quad (12.2)$$

In which,

T_{pmr} unit footage time in conventional core drilling (by lifting drill string for core recovery), h/m

2. Calculation of unit drilling cost

The unit drilling cost for wireline core drilling can be calculated with the following formula (with the deduction process omitted):

$$C_{pmw} = \frac{p_{rig}}{24}\left(\frac{3D_{ep}}{V_R L_{bit}} + \frac{1}{V_C} + \frac{t_{FC}}{L_{FC}}\right) + \frac{P_{bit}}{L_{bit}} + \frac{1}{V_C}\left(\frac{P_{cb}}{H_{cb}} + \frac{P_{rem}}{H_{rem}}\right) + C_{mudm} + C_{dpm} \quad (12.3)$$

where

C_{pmw} unit cost for wireline core drilling, RMB Yuan/m
P_{rig} drill rig daily cost, including equipment depreciation cost, replacing cost for easily-worn parts, personnel cost, management cost and profit, RMB Yuan/d
P_{bit} bit unit price, RMB Yuan/p
L_{bit} bit service life, m
P_{cb} unit price of core barrel, RMB Yuan/p
H_{cb} core barrel time life, h
P_{rem} unit price of reaming shell, RMB Yuan/p
H_{rem} reaming shell time life, h
C_{mudm} mud cost per meter, RMB Yuan/m
C_{dpm} drill rod cost per meter, RMB Yuan/m

The unit cost of conventional core drilling (by lifting drill string for core recovery) can be calculated by the following formula (with the deduction process omitted):

$$C_{pmr} = \frac{p_{rig}}{24}\left(\frac{2D_{ep}}{V_R} + \frac{L_R}{V_C}\right) + \frac{P_{bit}}{L_{bit}} + \frac{1}{V_C}\left(\frac{p_{mot}}{H_{mot}} + \frac{P_{cb}}{H_{cb}} + \frac{P_{rem}}{H_{rem}}\right) + C_{hhm} + C_{mudm} + C_{dpm} \quad (12.4)$$

where

C_{pmr} unit cost of conventional core drilling (by lifting drill string for core recovery), RMB Yuan/m
P_{mot} unit price of down-hole motor, RMB Yuan/p
H_{mot} down-hole motor time life, h
C_{hhm} hydro-hammer drilling tools cost per meter, RMB Yuan/m, the use of hydro-hammer belonged to technical service, price calculated by the footage drilled

3. Calculation of total drilling cost and total time

By using the formula (12.3) and (12.4), unit drilling cost for wireline core drilling and conventional core drilling (by lifting drill string for core recovery) can be separately calculated. Drilling construction is carried out in sections and every pulling-out trip is a drilling construction section. To multiply each section length (footage drilled in a roundtrip) by unit drilling cost in this section, the drilling cost in this section (roundtrip) can be obtained, and by accumulating the drilling cost of all sections (roundtrips) the total drilling cost of the borehole can be acquired.

The total drilling cost can be calculated by the following formula:

$$C_T = \sum_{i=1}^{n} C_{Ti} = \sum_{i=1}^{n} C_{pmi} \times L_{Ri} \quad (12.5)$$

in which,

C_T total drilling cost, RMB Yuan
C_{Ti} drilling cost of section i (borehole has n sections), RMB Yuan
C_{pmi} unit drilling cost of section i, RMB Yuan/m
L_{Ri} footage drilled in the roundtrip in section i, m
C_{pmi} in the formula (12.5) is C_{pmw} in the formula (12.3) or C_{pmr} in the formula (12.4). Substituting the formula (12.3) or (12.4) into the (12.5), the total drilling cost for wireline core drilling or for conventional core drilling (by lifting drill string for core recovery) can be obtained. When different drilling methods are used in one borehole, the formula (12.5) can be adopted to calculate total drilling cost as well

Likewise, calculation of drilling time can be expressed as:

$$T_T = \sum_{i=1}^{n} T_{Ti} = \sum_{i=1}^{n} T_{pmi} \times L_{Ri} \quad (12.6)$$

in which,

T_T total drilling time, h
T_{Ti} time used for drilling section i (borehole has n sections), h
T_{pmi} drilling time for 1 m in section i, h/m
L_{Ri} footage drilled in the roundtrip of section i, m

Substituting the formula (12.1) or the (12.2) into the (12.6), the total drilling time of wireline core drilling or conventional core drilling (by lifting drill string for core recovery) can be calculated.

12.2.4 Technical Economical Indexes of Different Core Drilling Methods

The basic conditions of core drilling design for CCSD-1 Well included:
(1) Design well depth: 5000 m
(2) Coring requirement: continuous coring
(3) Core drilling diameter: 157 mm
(4) Formation condition: Hard crystalline rock, mainly gneiss and eclogite, rock drillability grade:
(5) generally 8–9 (minority reaching grade 10).

As there was no precedent in China to implement a highly difficult drilling project like CCSD-1 Well, and no well-developed coring method available. Referring to the foreign experiences, combining with China's own situations and the previous drilling technology research and development conditions, the CCSD Engineering Center established a set of core drilling scheme including a variety of methods. During the construction of CCSD-1 Well, ten kinds of core drilling methods were tested, including PDM single tube conventional core drilling method (tested in order to reduce the rock fragmentation area of the drill bits) and the rest nine kinds of coring methods (as follows) adopting swivel type double tube coring tools.
(1) rotary table conventional core drilling
(2) rotary table hydro-hammer conventional core drilling
(3) top drive conventional core drilling
(4) PDM conventional core drilling
(5) PDM hydro-hammer core drilling by lifting drill string for core recovery (conventional coring)
(6) top drive wireline core drilling
(7) top drive hydro-hammer wireline core drilling
(8) PDM wireline core drilling
(9) PDM hydro-hammer wireline core drilling

Among above mentioned methods, rotary table conventional core drilling and top drive conventional core drilling were the traditional coring methods for petroleum drilling. PDM conventional core drilling and top drive wireline core drilling were learned from foreign experiences. PDM hydro-hammer core drilling by lifting drill string for core recovery, top drive hydro-hammer wireline core drilling, PDM wireline core drilling and PDM hydro-hammer wireline core drilling were developed by the CCSD Engineering Center.

The technical and economic indexes obtained in the CCSD-1 Well construction by using the abovementioned latter six methods are shown in Table 12.7. When confirming and selecting the figures of the parameters in the table, some data were referred, including the data from some domestic wireline core drilling projects (Zhang and Liu 1996) and hydro-hammer drilling projects (Wang et al. 1988) and some foreign scientific deep drilling projects, such as Cola super deep drilling project (Kozlovsky 1989), German continental scientific deep drilling project (Engeser 1996), Hawaii scientific drilling project (Zhang 1999) and the CCSD-1 Well construction data (Wang and Zhang 2005).

12.2.5 Economic Evaluation

In economic evaluation the formulas (12.3)–(12.6) are adopted for calculating the unit footage drilled, the total drilling cost and the total time cost, because in economic evaluation of CCSD-1 Well three indexes, the unit drilling cost, the total cost of 5000 m core drilling and the total time cost are mainly put into consideration. The calculation results are shown in Tables 12.8, 12.9 and Fig. 12.3.

From Fig. 12.3 and Table 12.9, it can be found out that the unit drilling cost and the total drilling cost and the total time cost of PDM conventional core drilling were obviously higher than other methods, and the next was PDM wireline coring method. The most economical method was PDM hydro-hammer core drilling by lifting drill string for core recovery.

12.2.6 Technical Risk Evaluation

Indexes adopted for technical risk evaluation include drill string safety coefficient, influence on hole wall stability, influence on coring effect, possibility of drilling string break-off, possibility of drill rod sticking, reliability of core drilling tools, possibility of formation leakage and capability to handle drill rod sticking. Six kinds of core drilling methods were evaluated one by one according to the indexes above.

1. **Drill string safety coefficient**

The six methods could be divided into two categories according to the drill strings used:

In two conventional core drilling methods (lifting drill string for core recovery), conventional oil drill string meeting API (American Petroleum Institute) standard was used, with steel grade of S135, outside diameter 89 mm, wall thickness 9.35 mm, unit weight 19.8 kg/m, tensile strength 222 t and safety coefficient at 5000 m deep was 2.1.

Four wireline coring methods utilized specially developed drill string by, with outside diameter of 139.7 mm, wall thickness of 8.1 mm, unit weight 25.4 kg/m, tensile strength of 224 t and safety coefficient at 5000 m deep was 1.59.

Obviously, the drill string adopted by the conventional coring methods was safer, not only because of the higher safety coefficient, but also because the drill string was well developed and reliable. The drill string for wireline coring

Table 12.7 Technical and economic indexes of different core drilling methods

Core drilling method	PDM conventional coring	PDM hydro-hammer coring by lifting drill string	PDM wireline coring	PDM hydro-hammer wireline coring	Top drive hydro-hammer wireline coring	Top drive wireline coring
Rig daily cost (¥/d)	50,000	50,000	50,000	50,000	58,000	58,000
Tripping rate (m/h)	400	400	350	350	350	350
String lifting interval (m)	3	7	20	20	20	20
Penetration rate (m/h)	0.7	1.2	0.6	0.9	0.9	0.7
Core-fishing interval (m)			3	4	4	3
Single wireline core-fishing time (h)			$1 + (7/10{,}000) \times D_{ep}$		$0.5 + (7/10{,}000) \times D_{ep}$	
Bit unit price (¥/p)	15,000	15,000	18,000	18,000	16,000	16,000
Bit life (m)	35	30	30	30	30	30
Down-hole motor unit price (¥/p)	70,000	70,000	70,000	70,000		
Down-hole motor life (h)	80	80	70	70		
Core barrel unit price (¥/p)	25,000	30,000	40,000	40,000	30,000	25,000
Core barrel life (h)	100	100	100	100	100	100
Reaming shell unit price (¥/p)	10,000	10,000	10,000	10,000	10,000	10,000
Reaming shell life (h)	100	100	100	100	100	100
Hydro-hammer cost per meter (¥/m)		700		500	500	
Mud cost per meter (¥/m)	300	300	300	300	300	300
Drill rod cost per meter (¥/m)	400	400	1200	1200	1800	1800

Notes 1. D_{ep}—borehole depth
2. According to the results of wireline core drilling in German KTB pilot hole, if common wireline core drilling method was used, the single wireline core fishing time = $0.5 + (7/10{,}000) \times D_{ep}$. For PDM hydro-hammer wireline coring and PDM wireline coring methods, due to the complex operations on the rig floor, more time was to be needed. So the single wireline core fishing time of the two methods = $1 + (7/10{,}000) \times D_{ep}$

Table 12.8 Unit drilling cost of different coring methods at different depth (*unit* RMB yuan)

Hole depth (m)	PDM conventional coring	PDM hydro-hammer coring by lifting drill string	PDM wireline coring	PDM hydro-hammer wireline coring	Top drive hydro-hammer wireline coring	Top drive wireline coring
200	6549	4551	8983	7294	6788	7239
500	7591	4997	9307	7582	7122	7616
1000	9327	5741	9848	8062	7678	8243
2000	12,799	7229	10,929	9022	8792	9497
3000	16,271	8717	12,011	9982	9905	10,752
4000	19,744	10,205	13,092	10,942	11,019	12,006
5000	23,216	11,694	14,173	11,901	12,132	13,260

was newly developed, and thus more risky even though the safety coefficient was equal to API drill string.

2. **Influence on hole wall stability**

PDM drive method has very small negative influence on hole wall stability because the drill string does not rotate or rotate very slowly during drilling construction.

Top drive method, with rotary speed up to 200–300 r/min, produces significant striking effect on borehole wall, with obviously negative influence on hole wall stability and easily causing drill pipe sticking accident.

In addition, compared with the conventional core drilling methods, wireline core drilling method has large negative

Table 12.9 Total drilling cost and total time for 5000 m core drilling by using different core drilling methods

Core drilling method	Total drilling cost/ 10,000 RMB Yuan	Total drilling time/day
PDM conventional coring	7272	1167
PDM hydro-hammer coring by lifting drill string	4226	547
PDM hydro-hammer wireline coring	4756	525
PDM wireline coring	5740	688
Top drive hydro-hammer wireline coring	4680	501
Top drive wireline coring	5069	607

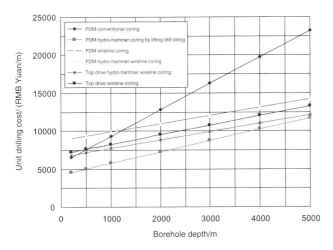

Fig. 12.3 Relationship between unit drilling cost and hole depth by using different methods

influence on hole wall because of the large pressure surge resulted from tripping, as its larger diameter drill pipe and smaller annulus between drill string and borehole.

Therefore, PDM hydro-hammer coring by lifting drill string (conventional coring) and PDM conventional coring (by lifting drill string for core recovery) are the methods which have the smallest negative influence on hole wall stability.

3. **Influence on coring effect**

Compared with top drive methods, drill string and drill bit work more stably in down hole drive core drilling, with higher core recovery. This effect becomes more obvious in broken and oversized hole sections where drill string and drill bit shake and swing dramatically when top drive method is adopted, resulting in low core recovery, because core becomes slim and easily falls off. Also, drill bit service life reduced.

After 2046.54 m, top drive diamond wireline core drilling method was adopted in CCSD-1 Well. Because hole enlargement section was encountered, the application result of drill bits was very poor. Drill bits were all polished only in one trip for eight continuous roundtrips, and four new drill bits only obtained average service life of 0.52 m (normal bit life was 30 m or so). Basically no core was recovered. It even happened that the female thread root of SK146 drill collar broke off (in MH18 roundtrip). After adopting PDM downhole drive method, this situation was immediately improved.

4. **Possibility of drilling string break-off**

In the hole enlargement sections, drilling tool is easily broken in case of top drive method is adopted.

5. **Possibility of drill rod sticking**

Drill rod sticking is easily happened if adopting top drive methods, because the high speed rotating drill rod knocks at borehole wall, making the wall rocks easily to fall, which in turn leads to drill pipe sticking.

6. **Reliability of core drilling methods and tools**

Has been widely applied in foreign countries, top drive wireline coring method is the most reliable and simple method among the abovementioned wireline core drilling techniques. Even so, the method still has its own reliability problems. After repeated improvements to the wireline coring tools applied in the KTB pilot hole, the failure rate caused by inner tube not reaching to the position or being stuck at halfway was limited to 10.2 % (running in for 90 times). In the operations of lifting the inner tubes which had not been placed in position or stuck at halfway, 27 additional trippings of wireline drill string were carried out due to wire rope break, which totally took 16.8 days (Engeser 1996).

Application experiences of top drive hydro-hammer wireline core drilling for deep drilling was still unavailable. The principle of this method is similar to that of top drive wireline coring technique. Added with hydro-hammer, ROP and trip length were improved, however, with slight decrease of reliability.

The reliability problem of PDM wireline core drilling system was not yet solved. It often occurred that the pump pressure abruptly increased during drilling, which led to drilling stoppage and short trip length, and thus the advantages of wireline core drilling could not be reflected.

The structure of the inner tube assembly of PDM hydro-hammer wireline core drilling system is rather complex, and any problem in core barrel, hydro-hammer or PDM will adversely affect drilling and coring. Meanwhile, inner tube assembly is very long and heavy, and it is rather difficult for fishing. Therefore, the reliability of the method is a relatively prominent problem.

The results of drilling practices indicated that PDM conventional core drilling and PDM hydro-hammer core drilling by lifting drill string for core recovery (conventional coring) were two methods with high reliability, the latter had slightly less reliability than the former due to the more complicated drilling tool structure. However, the latter method could be applied as a normal drilling system, thanks to the continuous improvements and perfection of the method in CCSD-1 Well, during which the problem of reliability was solved.

7. **Possibility of formation leakage**

Wireline core drilling has an obvious weakness, the annulus between drill rod and borehole wall is small and results in high pump pressure in drilling, especially in deep hole section, which, on one hand, requests for higher pressure resistance for mud pump and drill string components, and on the other hand causes a risk of formation leakage, for it had happened during the construction of CCSD-1 Well. If formation leakage happens, normal mud circulation cannot be established, cuttings cannot be carried out of borehole, and the rock pieces falling off from borehole wall cannot be suspended, which easily lead to drill rod sticking and bit burying.

Besides the adverse factor of adopting wireline drill rod, PDM hydro-hammer wireline coring method needs a large pump discharge to drive PDM, which is the way most likely causing formations leakage.

8. **Capability to handle accidents**

In six methods, top drive methods have much better capability of dealing with accidents than other methods, with the following reasons:

(1) Except lifting up and running down drill string, top drive can rotate at any time. As for PDM, if the sticking point is at the large diameter portion of the drilling tool under PDM, due to the relative movement of PDM stator and rotor, the torque from rotary table cannot be transmitted to or effected on the sticking point, thus the rotating function will lose its efficiency.

(2) Drill rod can be connected at any position in case top drive method is used, while PDM drive method is adopted drill rod can only be connected when drill rod sub is seated at rotary table.

12.2.7 Comprehensive Evaluation

After respectively evaluating the economic and technical risks of above six core drilling methods, we can comprehensively consider the two evaluation results, and make a comprehensive evaluation on the drilling methods of this 5000 m core drilling project. The results and conclusions are collected in Table 12.10.

It should be emphasized that reliability is an important index in technical and economic evaluation. When low reliable methods are used, the economic evaluation results should be discounted. For economic evaluation on drilling tool, the various technical and economic indexes given are all ideal indexes, which can only be realized when drilling tool is reliable and well developed. These ideal technical and economic indexes cannot be reached if something often goes wrong with the drilling tool in operation. For instance, although PDM wireline core drilling method was well developed in principle, because of the poor reliability, the technical economic indexes obtained were also very poor. And during the test of this method in CCSD-1 Well, in eight roundtrips the average ROP was only 0.33 m/h, and the average trip length was only 0.84 m, which were greatly different from the data listed in Table 12.7 (the ideal indexes). According to present data, PDM wireline core drilling method is the most uneconomic one. Drilling cost and time data in Table 12.9 show that even if all the technical problems of this method are solved and the method becomes perfect and mature, the technical and economic indexes acquired are still inferior to those of most other methods. Nevertheless, it is very difficult to achieve a perfect and mature result and reach the technical and economic indexes in Table 12.7.

PDM hydro-hammer wireline core drilling system is the most unreliable method, for its structure is too complicated, with too many affecting factors; hence, the results of its economic evaluation should be largely discounted. It is very difficult to solve the reliability problem of this method and make it be applied as a well developed core drilling method.

Top drive hydro-hammer wireline core drilling method is similar to top drive wireline core drilling method, with reliability slightly decreased after added with hydro-hammer. As the problems of hydro-hammer used for deep hole have been thoroughly solved, the reliability of this system has been basically solved. Moreover, as adding hydro-hammer can significantly improve ROP and roundtrip length, and thus be advantageous to reducing hole deviation as well, top drive hydro-hammer wireline core drilling method has better prospects than top drive wireline coring method.

Top drive wireline core drilling techniques and PDM conventional core drilling are the main methods adopted in foreign scientific deep hole drilling projects, however, with economic and technical risk evaluation results inferior to those of PDM hydro-hammer core drilling by lifting drill string for core recovery (conventional coring).

Among the six methods, PDM hydro-hammer core drilling by lifting drill string for core recovery was selected to be the most suitable method for the 5000 m hard rock core drilling project because of its best economy and least technical risks. During the construction of CCSD-1 Well, PDM hydrohammer core drilling by lifting drill string for core recovery (conventional coring) was used to drill 4043.25 m, with very good indexes and stable effects obtained in penetration rate, footage drilled per roundtrip and core recovery.

The results of the technical and economic analysis came to the following conclusion:

Table 12.10 Comprehensive evaluation of core drilling methods

	Core drilling method	PDM conventional coring	PDM hydrohammer coring by lifting drill string	PDM wireline coring	PDM hydrohammer wireline coring	Top drive hydro-hammer wireline coring	Top drive wireline coring
Technical risk evaluation	Drill string safety factor	High	High	Low	Low	Low	Low
	Negative influence on hole wall stability	Little	Little	Relatively little	Relatively little	Large	Large
	Negative influence on coring effect	Little	Little	Little	Little	Relatively large	Relatively large
	Possibility of drilling tool break	Small	Small	Relatively small	Relatively small	Large	Large
	Possibility of drill rod sticking	Small	Small	Small	Small	Large	Large
	Working reliability of drilling tool	Relatively high	Relatively high	Low	Low	Relatively high	High
	Possibility of formation leakage	Small	Small	Relatively large	Large	Relatively large	Relatively large
	Capability to handle drill rod sticking	Weak	Weak	Weak	Weak	Strong	Strong
Technical risk evaluation result		Good	Good	Poor	Poor	Relatively poor	Relatively poor
Economic evaluation	Total core drilling cost/ 10,000 Yuan	7272	4221	5740	4756	4680	5069
	Total core drilling time/day	1167	547	688	525	501	607
Economic evaluation result		Poor	Good	Poor	Relatively good	Relatively good	Relatively poor
Comprehensive evaluation result		Relatively poor	Good	Poor	Relatively poor	Relatively good	Relatively poor

PDM hydro-hammer core drilling by lifting drill string for core recovery (conventional coring) successfully invented and developed by the CCSD Engineering Centre is an efficient and safe core drilling method with high quality and low cost. The application of this method indicated that in the 5000 m hard rock core drilling project this method is superior to any other existing core drilling methods in the world. The success of research and development of this method is an important contribution of China to the world in the field of drilling techniques.

References

Cheng, H. (2001). *Management of engineering project*. Beijing: China Construction Industry Publishing House.

Chien, S. F. (1971). *Annular velocity for rotary drilling operations*. Austin, Texas (USA): Proceeding of rock mechanics conference.

Chilingarian, G. V. (1987). *Drilling and completion fluids*. Beijing: Petroleum Industry Publishing House.

Chunbo, Z., & Feng, L. (1996). Current status of wireline core drilling techniques in China. *Drilling and Excavation, 4*

Da, W., Wei, Z., et al. (1995). Technical situation and characteristics of scientific deep drilling in Russia—technical investigation report, No. 1. *Drilling and Excavation, 1*

Da, W., Wei, Z., et al. (1995). Technical situation and characteristics of scientific deep drilling in Russia—technical investigation report, No. 4. *Drilling and Excavation, 4*

Da, W.,& Wei, Z. (2005). An introduction to the drilling techniques for CCSD-1 well. *China Geology, 32*(2)

Deli, G., et al. (2004). *Drilling techniques of deep well and ultra-deep well under complicated geological conditions*. Beijing: Petroleum Industry Publishing House.

Engser, B. (1990). Die Kernbohrstrategie fuer die KTB Hauptbohrung. *Erdoel Erdgas Kohle, 12*, 496–500.

Engeser B (1996) Das Kontinentale Tiefbohrprogram der Bundesrepublik Deutschland—KTB Bohrtechnische Dokumentation (KTB REPORT 95-3). Niedersaechsischen Landeamt fuer Bodenforschung

Guangzhi, L. (1991). *Diamond drilling handbook*. Beijing: Geological Publishing House.

Guangzhi, L. (1998). *Scientific and technological history of drilling in China*. Beijing: Geological Publishing House.

Guangzhi, L. (2005). *On scientific drilling*. Beijing: Geological Publishing House.

Guolong, Z. (2002). Developing course of China continental scientific drilling project (part 1). *Drilling and Excavation, 4*

Guolong, Z. (2002). Developing course of China continental scientific drilling project (part 2). *Drilling and Excavation, 5*

Haishi, L., & Guoqiang, F. (1993). *Coring techniques in drilling*. Beijing: Petroleum Industry Publishing House.

Honghai, F. (2003). Real-time monitoring in drilling engineering and the development of well site information system. *Petroleum Drilling Techniques, 10*

Hongxiang, Q. (2000). Application of comprehensive logging techniques in drilling engineering. *Exploration Engineering in West China, 6*

Huaiwen, Z., & Zhixi, C. (1995). *Drilling Machinery (for speciality of drilling engineering)*. Beijing: Petroleum Industry Publishing House.

Jialang, C., et al. (1997). *Flow principle of drilling fluids*. Beijing: Petroleum Industry Publishing House.

Jinzhou, Z., & Guilin, Z. (2005). *A Handbook of Drilling Engineering and Techniques*. Beijing: China Petroleum and Chemical Industry Publishing House.

Jiren, X., & Zhixin, Z. (2005) Crustal velocity structure of Dabie-Sulu ultra-high pressure metamorphic belt and its subduction and exhumation mechanisms. *China Geology, 32*(2)

Jun, J., & Kaihua, Y. (2005). The influence of solid control and lubrication of drilling fluid on core drilling in CCSD pilot hole. *Earth science, 30* (supplementary issue).

Junwen, C., et al. (2004). Stress field of brittle deformational structure above 2,000 m of CCSD main hole. *Rock Journal, 20, 1*.

Keqin, Z., & Leliang, C. *A handbook of drilling techniques (II) (drilling fluid)*. Beijing: Petroleum Industry Publishing House

Kexiang, L., & Junchang, X. (1990). *Drilling handbook (Party A)*. Beijing: Petroleum Industry Publishing House.

Kozlovsky, E. A. (1989). *Super deep hole in Cola peninsula (I)*. Translated by Zhang Qiusheng. Beijing: Geological Publishing House.

Kozlovsky, E. A. (1989). *Super deep hole in Cola peninsula (II)*. Translated by Zhang Qiusheng. Beijing: Geological Publishing House.

Lauterjung, J. (2000). The International Continental Scientific Drilling Program (ICDP). *ICDP Newsletter, 2*, 3–6.

Lianjie, W., et al. (2005). Determination of current earth stress status by acoustic emission method in CCSD main hole. *China Geology, 32*(2)

Miaorong, L. (1997). Systematic analysis on turbodrill. *Journal of Jianghan College of Petroleum, 19*(3), 56–60.

Moore, P. L. (1974). *Drilling practice manual*. Tulsa (USA): The Petroleum Publishing Co.

Ning, Z. (2004). Application of new type turbodrill in deep well. *Oil Drilling and production Technologies, 3*

Pingquan, W., & Shiliang, Z. (2003). *Drilling fluid conditioners and their action principle*. Beijing: Petroleum Industry Publishing House.

Renjie, W., Rongqing, J., & Junzhi, H. (1988). *Hydro-percussive rotary drilling*. Beijing: Geological Publishing House.

Research group of "Developing Strategy of Earth Sciences in China" of the Department of Earth of the Chinese Academy of Sciences. (2002). *Earth sciences: review of the 20th century and prospects for the 21st century*. Shandong: Shandong Education Publishing House.

Scholz, C. A. (2006). The 2005 Lake Malawi scientific drilling report. *Scientific Drilling, 2*, 17–19.

Science Planning Commission of IODP (2003). *Earth, ocean and lives (IODP Initial Scientific Plan 2003–2013)*. Translated by the Department of Earth Sciences, the National Natural Sciences Foundation. Shanghai: The Publishing House of Tongji University.

Vozdvizhenski, Б. И. (1985). *Core drilling*. Beijing: Geological Publishing House.

Wei, Z. (1999). Drilling techniques and construction of Hawaii scientific drilling project. *Drilling and Excavation, 4*

Wei, Z., & Wenwei, X. (2005). Hydro-hammer non-core drilling for large diameter borehole in hard rock. *Earth science, 30* (supplementary issue)

Wenjian, Z., Jianyuan, Z., & Peifeng, Z. (2003). Research on the construction of moving casing for China continental scientific drilling activities. *Drilling and Excavation,* (supplementary issue), 468.

Wenwei, X., Changshou, S., & Yiquan, M. Application of YZX127 hydro-hammer in the pilot hole of CCSD-1 well. *Earth science, 30* (supplementary issue)

Wuhan College of Geology. (1980). *Drilling technologies.* Beijing: Geological Publishing House.

Xiaoxi, Z., Yongyi, Z., et al. (2005). Drilling techniques in CCSD-1 pilot hole. *Earth science, 30* (supplementary issue)

Xiwen, J. (2002). *Accidents and complicated problems in drilling.* Beijing: Petroleum Industry Publishing House.

Yarui, Z. (2005). Initial study on application of detailed construction quantity list for entering a public bidding. *Journal of Zhongshan University (natural sciences edition),* 6

Yijin, Z., & Jianli, L. (2005). Current status and developing trend of deep and ultra-deep well drilling techniques. *Petroleum Drilling Techniques, 33*(5), 2.

Yinao, S. (2001). *Research and application of PDM drilling tool.* Beijing: Petroleum Industry Publishing House.

Yinao, S. (1998). Working properties of PDM drilling tool. *Oil Drilling and production Technologies, 20*(6)

Yinao, S. (2003). *Anti-deviation and quick drilling techniques for oil and gas straight well—theory and practice.* Beijing: Petroleum Industry Publishing House.

Yixiong, N., Heping, P., Wenxian, W., et al. *Geophysical logging of 0–2 000 m in CCSD main hole.* Wuhan: Publishing House of China University of Geosciences, Wuhan.

Zhemin, T. (2005). Eclogite and gneiss foliation and microfault occurrence characteristics of 100–2,950 m in CCSD main hole. *Rock Mineralogy Journal, 24*(5)

Zhiqin, X., Ruilun, G., Qinghui, X., & Wei, Z. (1996). *Early study of China continental scientific drilling.* Beijing: Metallurgical Industry Publishing House.

Zhiqin, X., Jingsui, Y., Zemin, Z. et al. (2005). Completion of China continental scientific drilling and research progress. *China Geology, 32*(2)

Zhisheng, A., Li, A., et al. (2006). Lake Qinghai scientific drilling project. *Scientific Drilling, 2,* 20–22.

Zhongfeng, Y., & Yuanming, L. (2003). Study on improving the benefit of a drilling project by using flexible incentive mechanism. *Journal of Jianghan College of Petroleum (social sciences edition),* 6